ARENA2036

Series Editor
ARENA2036 e.V., *ARENA2036 e.V., Stuttgart, Germany*

Die Buchreihe dokumentiert die Ergebnisse eines ambitionierten Forschungsprojektes im Automobilbau. Ziel des Projekts ist die Entwicklung einer nachhaltigen Industrie 4.0 und die Realisierung eines Technologiewandels, der individuelle Mobilität mit niedrigem Energieverbrauch basierend auf neuartigen Produktionskonzepten realisiert. Den Schlüssel liefern wandlungsfähige Produktionsformen für den intelligenten, funktionsintegrierten, multimaterialen Leichtbau. Nachhaltigkeit, Sicherheit, Komfort, Individualität und Innovation werden als Einheit gedacht. Wissenschaftler verschiedener Disziplinen arbeiten mit Experten und Entscheidungsträgern aus der Wirtschaft auf Augenhöhe zusammen. Gemeinsam arbeiten sie unter einem Dach und entwickeln das Automobil der Zukunft in der Industrie 4.0.

The book series presents the results of an ambitious research project in automotive production. The goal of the project is the development of a sustainable Industry 4.0 and the realization of a technology shift that will realize the mobility of the future with low energy consumption based on innovative production concepts. The key is provided by intelligent and flexible forms of production. Sustainability, safety, comfort, individuality and innovation are conceived as a unity. Scientists from various disciplines work together with experts and decision-makers from industry on an equal footing. Together, they work under one roof and develop the automobile of the future in Industry 4.0.

Daniel Holder · Frederik Wulle · Jannik Lind
Editors

Advances in Automotive Production Technology – Digital Product Development and Manufacturing

Stuttgart Conference on Automotive Production (SCAP2024)

Editors
Daniel Holder
ARENA2036 e.V.
Stuttgart, Germany

Frederik Wulle
ARENA2036 e.V.
Stuttgart, Germany

Jannik Lind
ARENA2036 e.V.
Stuttgart, Germany

ISSN 2524-7247   ISSN 2524-7255 (electronic)
ARENA2036
ISBN 978-3-031-88830-4   ISBN 978-3-031-88831-1 (eBook)
https://doi.org/10.1007/978-3-031-88831-1

© The Editor(s) (if applicable) and The Author(s) 2025. This book is an open access publication.

Open Access This book is licensed under the terms of the Creative Commons Attribution 4.0 International License (http://creativecommons.org/licenses/by/4.0/), which permits use, sharing, adaptation, distribution and reproduction in any medium or format, as long as you give appropriate credit to the original author(s) and the source, provide a link to the Creative Commons license and indicate if changes were made.

The images or other third party material in this book are included in the book's Creative Commons license, unless indicated otherwise in a credit line to the material. If material is not included in the book's Creative Commons license and your intended use is not permitted by statutory regulation or exceeds the permitted use, you will need to obtain permission directly from the copyright holder.

The use of general descriptive names, registered names, trademarks, service marks, etc. in this publication does not imply, even in the absence of a specific statement, that such names are exempt from the relevant protective laws and regulations and therefore free for general use.

The publisher, the authors and the editors are safe to assume that the advice and information in this book are believed to be true and accurate at the date of publication. Neither the publisher nor the authors or the editors give a warranty, expressed or implied, with respect to the material contained herein or for any errors or omissions that may have been made. The publisher remains neutral with regard to jurisdictional claims in published maps and institutional affiliations.

This Springer imprint is published by the registered company Springer Nature Switzerland AG
The registered company address is: Gewerbestrasse 11, 6330 Cham, Switzerland

If disposing of this product, please recycle the paper.

Preface

The automotive industry is undergoing a profound transformation driven by the convergence of digital innovation, data-centric approaches, and sustainable practices. The shift toward advanced automotive production is not merely a response to technological advancements but a necessity dictated by evolving market demands, environmental concerns, and the pursuit of operational excellence. This volume, Advances in Automotive Production - Digital Product Development and Manufacturing, captures the essence of this transformation by delving into cutting-edge developments in digitalization, data-driven technologies, and sustainable manufacturing, all while exploring their impact on the industrial and automotive value chain.

The 3rd Stuttgart Conference on Automotive Production (SCAP) in 2024 was organized by ARENA2036 in close cooperation with the Institute for Control Engineering of Machine Tools and Manufacturing Units (ISW) at the University of Stuttgart. ARENA2036 is the innovation platform for future technologies, where science and industry work together to develop new and potentially disruptive future technologies. The conference's focus on digital product development and manufacturing is based on the thematic focus of the research project DigiTain (**Digi**talization for Sus**Tain**ability). DigiTain aims to develop processes, methods, and models for fully digital product development and certification of electric drive architectures, taking into account both competitive performance and environmental and economic sustainability criteria from the outset. The research project is supported by the German government and the European Union as part of the economic stimulus package "New, innovative products as the key to vehicles and mobility of the future".

This book is structured along the four main topics of the conference SCAP2024:

1. Digital Methods and Models.
2. Digitalization of the Industrial and Automotive Value Chain.
3. Data-driven Technologies.
4. Sustainability and Circular Economy.

The first section, "**Digital Methods and Models**," examines the transformative role of digital tools in product development and manufacturing. Industry 4.0 technologies such as digital twins and simulation-driven design enable faster prototyping, improved accuracy, and cost savings. These approaches are no longer theoretical but are driving real-world advancements in areas like electric vehicle design and production line optimization. The section begins with foundational modeling methods, progresses to application-driven simulations, and concludes with real-time and advanced modeling innovations.

The second section, "**Digitalization of the Industrial and Automotive Value Chain**," focuses on how digital integration is reshaping supply chain management, production, and customer interactions. Smart factories, powered by IoT and robotics, are enhancing efficiency and resilience, while digital platforms improve collaboration and transparency through real-time data sharing. This section covers the evolution from digitalization frameworks to advanced applications in industrial systems and automotive manufacturing.

The third section, "**Data-driven Technologies**," explores how data-centric methods are driving innovation and optimization. Tools such as artificial intelligence, machine learning, and big data analytics are used to predict trends, enhance quality, and personalize customer experiences. Predictive maintenance and advanced algorithms also support autonomous systems, a key area for future mobility. The section addresses foundational data handling strategies and advanced applications of machine learning in industrial contexts.

The fourth section, "**Sustainability and Circular Economy**," highlights the growing importance of environmentally responsible practices. Circular economy principles, such as material recycling and remanufacturing, are becoming integral to automotive production. As sustainability shifts from optional to essential, this section explores fundamental concepts, sustainable technologies, and their applications in sectors like automotive and urban transport.

This book presents a comprehensive view of these interconnected themes, offering insights from leading researchers, industry experts, and practitioners. By showcasing recent advancements and fostering dialogue, it aims to inspire innovative approaches that will shape the future of automotive production.

Contents

Investigating the Potential of Higher-Order 3D-Shell Finite Elements
in Stress Analysis of Laminated Structures 1
   Maximilian Schilling, Tolga Usta, Tobias Willmann, Malte von Scheven, and Manfred Bischoff

Validation of the *CONSTRAINED_SPR3 Joint Formulation
for Isogeometric Shell Models 13
   Philipp Bähr, Lukas Leidinger, Silke Sommer, and Stefan Hartmann

Convergence Studies to Compare the Induced Forming Defects in FE
Based Simulations, Point Clouds and in Actual Formed Parts 31
   Muhammad Saeed, Sheharyar Faisal, Eiman Nadeem, Markus Wagner, Boris Eisenbart, and Matthias Kreimeyer

Novel Slicing Algorithm for Hybrid Manufacturing on Non-planar
Surfaces with Robotic SEAM 39
   Nicolas Unger, Pradnil Kamble, and Timo Huse

A Two-Level Architecture for Mobile Grasping 58
   Troy McMahon, Alfred Thoft Christiansen, Elias Thomassen Dam, Casper Schou, and Ole Madsen

Towards Automotive Manufacturing Efficiency: Enhanced Virtual
Commissioning Simulation for Dynamic Sheet Metal Handling
Optimization 68
   Stefan Klare, Volodymyr Shramenko, Lars Klingel, Bernd Lüdemann-Ravit, and Alexander Verl

Online Real-Time Simulation for Collision Avoidance in Robotic Wire
Arc Additive Manufacturing 76
   Lars Klingel, Maximilian Nistler, Daniel Mantz, Martin Werz, and Alexander Verl

Real-Time Online Simulation at Field Level in Industrial Automation 88
   Darius Deubert, Andreas Selig, and Alexander Verl

Comparative Analysis of Machine Learning Models in Production
Environments Through Residual Distributions 104
   Jan A. Zak and Christian Weißenfels

A Baseline Model for Nugget Diameter Prediction Based on Process
Parameters for Aluminum Resistance Spot Welding 112
    Jan A. Zak, Jose M. Araya-Martinez, and Christian Weißenfels

Modeling the Aging Behavior of the Catalyst Layer in PEM Fuel Cells 121
    Theresa Uhlemayr, Joachim Scholta, and Markus Hölzle

Dynamic Process Reconfiguration Through Digital Product Passports:
A Framework for Adaptive Production Control 137
    Samed Ajdinović, Moritz Walker, Rebekka Neumann, Nicolai Maisch,
    Michael Neubauer, Armin Lechler, and Oliver Riedel

EtherCAT Tunneling Through Time-Sensitive Networks: An Experimental
Evaluation 150
    Marc Fischer, Moritz Walker, Philipp A. Neher, Michael Neubauer,
    Armin Lechler, and Alexander Verl

Dependable Cyber-Physical Matrix Production Systems Utilizing Holonic
Multi-agent Systems 160
    Jonathan Bartels, Simon Komesker, William Motsch,
    Katharina Hengel, Achim Wagner, and Martin Ruskowski

Improving Automated Manufacturing Processes by Applying Agent-Based
Planning and Control 175
    Simon Komesker, Jonathan Bartels, Bastian Lang, Achim Wagner,
    and Martin Ruskowski

Measuring Resilience: A New Perspective on Assessing Production
Facilities Through an OEE Based Resilience Metric 183
    Jonas Knüpper, Alexander Haas, Bernd Lüdemann-Ravit,
    Fabian Schmitt, Pascal Grieser, Andreas Hanzelmann,
    Miriam Schleipen, and Dimitrios Genikomsidis

From Market Research to Manufacturing: A Conjoint Analysis
and Reconfigurable Manufacturing System Framework for Product-Line
Optimization 191
    Sascha Voekler and Ulrich Berger

Exploring Interactions in Autonomous Vehicles: A Comprehensive
Evaluation of Various Interaction Methods for 2D and 3D Content 206
    Zack Walker, Ansgar Gerlicher, Axel Braun, Lea Pinnow,
    Daniel Heinemann, Simon Janik, and Elias Merzhäuser

Enhancing an Autonomous Vehicle Simulation Through Holoride
Technology Integration to Reduce Motion Sickness and Increase
Immersion: A Proof of Concept and Empirical Evaluation 223
    Zack Walker, Korbinian Kuhn, Ansgar Gerlicher, and Axel Braun

Qualitative Comparison of Tools for Handling Unstructured IIoT Data 243
    Makki Ben Salem, Philipp Niklas Rosenthal, and Abdelmajid Khelil

An Approach for a Human-Assisted Data Loop in Connected
Manufacturing Systems 257
    Matthias Weiß, Alexander Schön, Matthias Lück, Maximilian Schnierle,
    Stefan Carosella, Nasser Jazdi, Carmen Constantinescu,
    Peter Middendorf, and Michael Weyrich

Cycle Time Measurement Using AI-Based Object Detection and Tracking
in Industrial Processes 268
    Tim Staudenrausch and Bernd Lüdemann-Ravit

A Data-centric Evaluation of Leading Multi-class Object Detection
Algorithms Using Synthetic Industrial Data 283
    J. Moises Araya-Martinez, Sarvenaz Sardari, Mats Lambert,
    J. Alexander Zak, Florian Töper, Jörg Krüger, and Jens Lambrecht

Comparison of Active Learning and Self-Training as Adaptation Strategies
for Robust Classification in a Dynamic Production Environment 303
    Andreas Seitz, Florian Liebgott, Dominik Track, Daniel Kessler,
    and Hans-Peter Beise

Robotic Wiring Harness Bin Picking Solution Using
a Deep-Learning-Based Spline Prediction and a Multi-stereo
Camera Setup 319
    Manuel Zürn, Carsten Schmerbeck, Andreas Kernbach, Mara I. Kläb,
    Alper Yaman, Daniel Bragmann, Michael Heizmann, Marco Huber,
    Werner Kraus, Armin Lechler, and Alexander Verl

Reinforcement Learning to Improve Finite Element Simulations for Shaft
and Hub Connections 335
    Muhammad Saeed, Hassaan Muhammad, Narmeen Sabah, Jan Falter,
    Markus Wagner, Boris Eisenbart, and Matthias Kreimeyer

Design Automation of Fibre Composite Parts via Graph-Based Design
Languages 344
    Jonas Braiger, Johannes Baur, Jakob Gugliuzza, Stephan Rudolph,
    Stefan Carosella, and Peter Middendorf

Modeling and Optimization of Sustainability Criteria Along the Product
Engineering Process of Handling Systems 353
   Johannes Scholz, Florian Koessler, and Jürgen Fleischer

The Application of LCA Data Uncertainty Analysis in the Sustainable
Development Process 368
   Ruiyang Deng and Sebastian Kilchert

The Role of Metal Additive Manufacturing in a Circular Economy 383
   Matthias Duve, David Petasch, Bernd Lüdemann-Ravit,
   and Frieder Heieck

Overview of the Challenges in High-Pressure Type V Hydrogen Tanks
for Automotive Applications 402
   Santwana Pati, Akshay Deshmane, Maximillian Korff, and Tobias Dickhut

Sustainable and Affordable Strategies to Reduce Traffic Emissions
in Urban Areas 415
   Ali Khan Muhammad and Ali Khan Majid

Software-Defined Value Networks: Industrial Requirements and Research
Gap 426
   David Dietrich, Manuel Zürn, Ann-Kathrin Briem, David Koch,
   Werner Lober, Jannik Lind, Armin Lechler, and Alexander Verl

Author Index 443

Investigating the Potential of Higher-Order 3D-Shell Finite Elements in Stress Analysis of Laminated Structures

Maximilian Schilling[✉], Tolga Usta, Tobias Willmann, Malte von Scheven, and Manfred Bischoff

University of Stuttgart, Institute for Structural Mechanics,
Pfaffenwaldring 7, 70550 Stuttgart, Germany
schilling@ibb.uni-stuttgart.de

Abstract. The accurate prediction of stress in fiber-reinforced laminates through finite element analysis is of critical importance for the design of lightweight automotive components, as it allows for better prediction of damage such as delamination. However, standard (Reissner–Mindlin) shell finite elements, while efficient, consider only a reduced stress state and neglect the transverse normal stress. This contribution investigates the potential of higher-order 3D-shell elements to improve stress prediction in laminates. In order to assess the efficacy of higher-order 3D-shell elements, their results are compared with those obtained from standard shell finite element simulations and fully 3D solid element simulations. The findings demonstrate that the use of cubic 3D-shell elements can be beneficial for simulating laminated materials, as they lead to a more accurate stress prediction in comparison to standard shell finite elements. In particular, the shear stress and normal stress in thickness direction of the laminate can be predicted with a higher degree of accuracy.

Keywords: shell finite elements · stress analysis · fiber reinforced polymers · laminates

1 Introduction

Fiber-reinforced laminate structures are used in lightweight design due to their tailorable properties. In the automotive context, they are utilized for structural members, car body components, and pressure tanks in hydrogen-powered vehicles. In order to accurately anticipate damage such as delamination in these components, it is essential to achieve an accurate stress prediction in finite element analyses. Research by Czichos et al. [1] emphasizes the necessity of considering a fully 3D stress state, particularly in impact scenarios.

In the context of larger explicit finite element simulations in the automotive industry, such as full vehicle crash or multi-component analyses, standard

(Reissner–Mindlin) shell elements are typically employed due to their numerical efficiency. However, standard shell elements consider only a reduced stress state, neglecting transverse normal stress. This and the underlying kinematic assumptions result in a low accuracy of the distribution of the transverse stress components over the thickness coordinate of the laminate [2]. While the use of multiple solid elements to discretize the laminate in the transverse normal direction ensures that the fully 3D stress state is considered, this approach becomes computationally too expensive for larger simulations.

To address this challenge, this study investigates the potential of using 3D-shell elements with higher-order through-the-thickness kinematics. The element formulation was developed by Willmann et al. [3] for sheet metal forming simulations to obtain an improved resolution of the stress distribution in comparison to standard shell finite elements.

The objective of our research is to investigate the effectiveness of higher-order 3D-shell elements in improving the stress prediction for fiber-reinforced laminate structures while maintaining computational efficiency. This is accomplished by comparing the stress results from higher-order 3D-shell element simulations with those of standard shell element simulations and with those of fully 3D solid element simulations.

2 Shell Models and Finite Elements for Laminates

In order to introduce the higher-order 3D shell element, we first provide an overview of the key shell models and their associated shell finite elements, which are commonly available in commercial finite element software. For a comprehensive review of shell models and shell elements, readers are directed to Bischoff et al. [4].

The Reissner–Mindlin shell model forms the basis for most widely used shell elements. This model accounts for the independent rotation of cross-sectional fibers relative to the midsurface, with the underlying kinematic assumption that the fibers remain straight during the deformation. Consequently, the model incorporates constant transverse shear strain. Due to the number of independent variables, three displacements and two rotations, it is also referred to as the 5-parameter model. However, this model considers only a reduced stress state, assuming zero transverse normal stress. Shell elements derived from this model can be found in the works of Hughes et al. [5] and Belytschko et al. [6].

The so-called 7-parameter shell models introduce two additional independent parameters, accounting for both a constant and a linear transverse normal strain. Shell elements based on this model are often referred to as shell elements with thickness stretch or linear/ordinary 3D-shell elements. The term "3D-shell" highlights that these elements capture a fully 3D stress state. This shell model was independently proposed by several researchers, including Kühhorn and Schoop [7] and Parisch [8], and its implementation as a shell element was for example carried out by Cardoso et al. [9]. However, because shell elements with thickness stretch are less commonly used than Reissner–Mindlin shell elements, they will not be included in the subsequent comparison of element types.

In contrast to these shell models, the higher-order 3D-shell model developed by Willmann et al. [3] allows for quadratic and cubic cross-sectional warping and a higher-order strain field with respect to the (transverse) thickness coordinate. It therefore also considers a fully 3D stress state. In the following, the shell element derived from this model will be referred to as a cubic 3D-shell element (3DSH-cub). Figure 1 illustrates selected deformation modes of the element. For more fundamental information on the cubic 3D-shell element, we refer to Willmann et al. [3]. A number of techniques have been implemented to prevent a variety of locking effects. This includes using reduced integration techniques and hourglass stabilization, the assumed natural strain method, and an additional incompatible quartic strain field in transverse direction. For a complete overview of the differences in the element formulation between this work and that of Willmann et al. [3], see Schilling et al. [10].

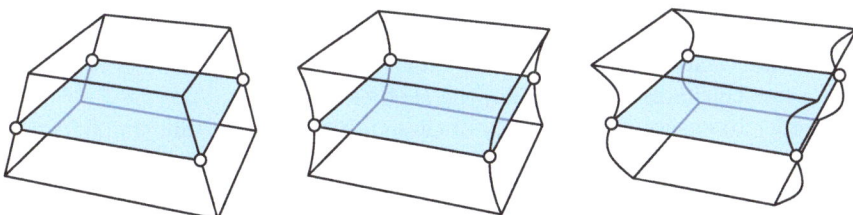

Fig. 1. Visualization of selected deformation modes of the cubic 3D-shell element, showing (from left) linear, quadratic, and cubic deformations of cross-sectional fibers

Lastly, we briefly discuss how laminated structures can be represented using shell elements. For a thorough overview of modeling laminates with shell elements, readers are referred to Reddy [11]. There are two primary approaches: laminate layers can be modeled either through the kinematics of the shell element or through the material model used. The first approach extends the presented shell elements by introducing additional degrees of freedom that capture the unique kinematics of laminates. This includes, for example, degrees of freedom that allow for the independent rotation of cross-sectional fibers in each layer. In the second approach, the shell element retains the standard kinematics of its respective model, but each integration point through the thickness of the laminate is assigned different material properties, depending on the material properties and fiber orientation of each layer. An example for this can be found in [12]. In this work, we adopt the second approach for both the standard shell elements and the higher-order 3D-shell elements, as the first approach is often prohibitively expensive.

3 Stress Analysis of Laminated Structures

In Sect. 3, we investigate the applicability of cubic 3D-shell elements for the stress analysis of laminated structures through a series of benchmark tests. We compare the simulation results obtained using standard (Reissner–Mindlin) shell elements (LS-DYNA Shell 16 [12]) and those from higher-order 3D-shell elements (3DSH-cub) with a reference solution obtained from a fully 3D simulation employing solid elements (LS-DYNA Solid −2 [12]). For the simulation results with solid elements, the mesh is not fully converged, but the results serve as a manageable, fully 3D reference solution that adheres to computational constraints and complies with standards in industrial applications.

For the simulations, we assign different material properties through the thickness of the laminate depending on the fiber angle of each layer. In the case of solid elements, this is achieved by assigning different material properties for each element through the thickness of the laminate. For the simulations with shell elements, their integration points through the thickness of the laminate are assigned the corresponding material properties. Notably, the number of integration points in the shell elements matches the number of solid elements used through the thickness in the reference solution. The thickness of the solid element layers also varies in the thickness direction. This ensures that the average stress within each solid element corresponds to the stress at the integration point at the same thickness coordinate in the shell elements.

In order to ensure the feasibility of these initial tests, certain simplifications have been made. We employ a layerwise anisotropic material model (LS-DYNA *MAT_022 [13]) and exclude in-plane damage (fiber or matrix rupture) and out-of-plane damage (delamination), although both can be incorporated in future analyses. The chosen material properties for the carbon-fiber composite material are provided in Table 1 in the Appendix. The objective of our tests is to cover standard load cases, with an initial focus on a tensile test, a bending test, a plate under uniform pressure, and an impact scenario.

3.1 Tensile Test

Initially, the mechanical response of a carbon fiber laminate of length l, width b, and thickness $t = 1.5$ mm under tensile loading is investigated. The simulation setup is illustrated in Fig. 2. During the simulation, the specimen is clamped at the left end, and a displacement in x-direction is applied to all nodes at the right edge. The laminate is composed of two outer layers and one inner layer, with respective thicknesses of $t_o \approx 0.40$ mm and $t_i \approx 0.70$ mm. The fibers in the two outer layers are oriented in x-direction (fiber angle $\beta = 0°$), while the fibers of the middle layer are oriented in y-direction ($\beta = 90°$). This is visualized in the side view of the lay-up in Fig. 2.

Three different simulation approaches are compared. Firstly, a simulation employing nine solid elements through the thickness of the laminate, with three elements per layer, is conducted. Subsequently, simulations utilizing standard (Reissner–Mindlin) shell elements (Shell 16) or cubic 3D-shell elements (3DSH-cub) with nine integration points in the z-direction are performed. For all simulations, the laminate is discretized with square elements with an edge length $h = 5$ mm. It is noted that the solid elements and standard shell elements are fully integrated. For better comparison, the stress is averaged across the eight integration points per solid element. The standard shell element utilizes four in-plane integration points, for which the stress is also averaged. This yields nine stress values over the thickness for each element formulation. The evaluation was performed near the center of the specimen; see Fig. 2.

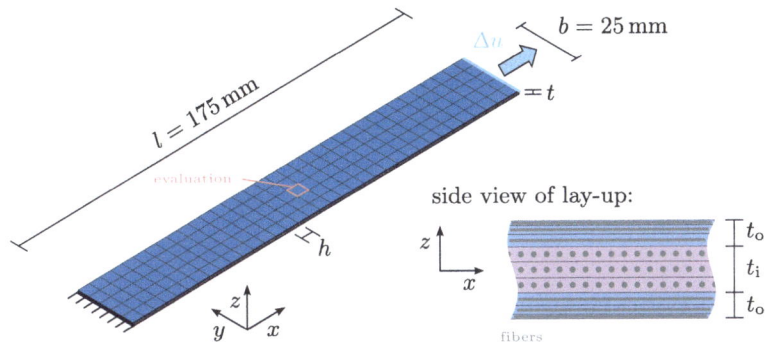

Fig. 2. Set up and solid finite element mesh of the tensile test

Fig. 3 shows the normal stress σ_{xx} and normal stress σ_{yy} for the highlighted element over the dimensionless thickness coordinate $\tilde{z}$ of the laminate. The gray background indicates the two outer layers with a fiber angle of $\beta = 0°$.

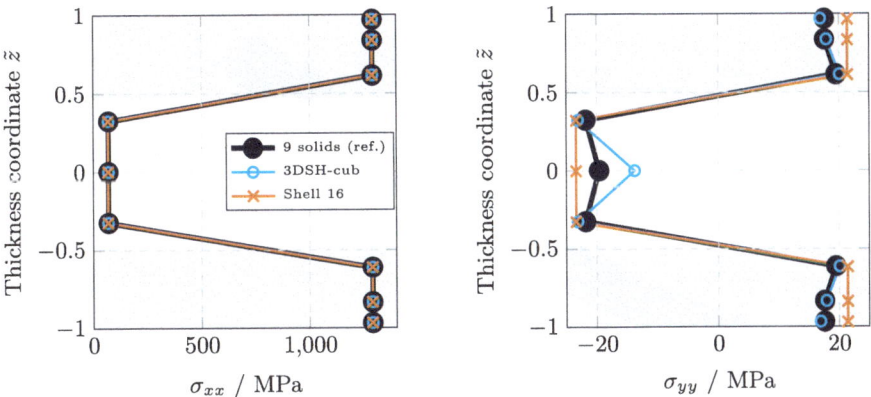

Fig. 3. Plot of normal stress σ_{xx} and normal stress σ_{yy} against dimensionless thickness coordinate $\tilde{z}$ of the laminate for different element formulations evaluated near the center of the tensile specimen

The results in Fig. 3 indicate that both shell element formulations predict a normal stress σ_{xx} that is nearly identical to the reference solution in the case of uniaxial tension. This is due to the simple loading scenario where the deformation can already be captured by a standard shell element. In y-direction, the orthotropic laminate induces a normal stress σ_{yy} due to the larger Poisson effect for the top and bottom plies. This results in a compressive stress σ_{yy} in the middle layer. For the stress component σ_{yy} the cubic 3D-shell elements give a prediction closer to the reference solution than the standard shell elements. However, it should be noted that all solutions for this stress component indicate a discretization that is still too coarse in the thickness direction since the stress-free condition is violated at the top and bottom surfaces for all element formulations.

In summary, the cubic 3D-shell element has been validated for tensile tests with laminated materials. Due to the self-induced stress in y-direction, cubic 3D-shell elements can give a prediction of normal stress σ_{yy} closer to the reference solution than standard shell elements, even in the case of symmetrical laminates under tensile loading.

3.2 Three-Point Bending Test

In the second benchmark, the mechanical response of a carbon fiber laminate of length l, width b, and thickness $t = 1.5\,\text{mm}$ subjected to a three-point bending test is analyzed. The simulation setup is depicted in Fig. 4. For loading, a cylindrical punch with a diameter of $d = 6\,\text{mm}$ is displaced by $\Delta u = 5\,\text{mm}$ in the negative z-direction. The laminate is identical to the laminate of the tensile specimen in Subsect. 3.1. Consequently, the fiber orientations are once again $\beta = \{0°, 90°, 0°\}$ relative to the x-direction.

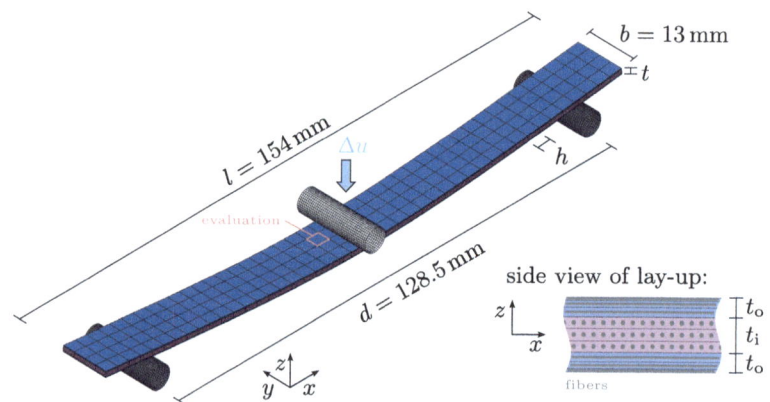

Fig. 4. Set-up and solid finite element mesh of the three-point bending test

Again, the three different simulation approaches are compared. A simulation with nine solid elements through the thickness of the laminate is used as the reference solution. For both shell elements, nine integration points are utilized in the z-direction. For all simulations, the specimen is discretized with rectangular elements measuring 3.5 mm in the x-direction and 3.25 mm in the y-direction.

Figure 5 shows the transverse shear stress σ_{xz} against the dimensionless thickness coordinate of the laminate for the three element formulations. The shear stress is evaluated near the punch, as illustrated in Fig. 4. For each stress component σ_{ij}, an error relative to the maximum stress is calculated at each integration point IP using the following equation:

$$e_{ij,IP}^{\text{shell}} = \frac{\left|\sigma_{ij,IP}^{\text{solid}} - \sigma_{ij,IP}^{\text{shell}}\right|}{\max\left|\sigma_{ij}^{\text{solid}}\right|}. \tag{1}$$

The results in Fig. 5 indicate that the standard shell element overestimates the transverse shear stress in the outer layers of the laminate (max $e_{xz,IP}^{\text{Shell 16}} = 39.8\%$, avg. error $\bar{e}_{xz}^{\text{Shell 16}} = 18.4\%$). In contrast to this, the cubic 3D-shell element exhibits a much closer alignment of the shear stress σ_{xz} with the reference solution (max $e_{xz,IP}^{\text{3DSH-cub}} = 14.9\%$, avg. error $\bar{e}_{xz}^{\text{3DSH-cub}} = 6.1\%$).

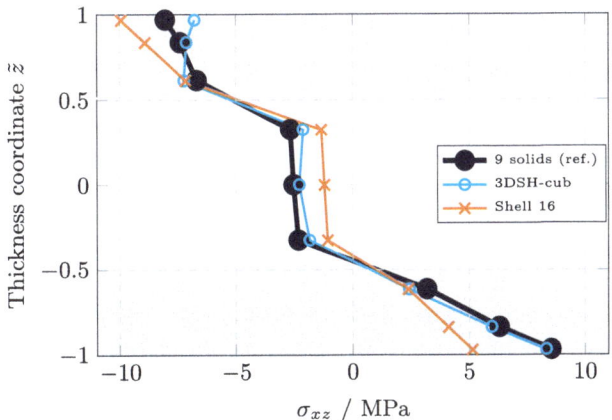

Fig. 5. Plot of the transverse shear stress σ_{xz} against the dimensionless thickness coordinate $\tilde{z}$ of the laminate for different element formulations evaluated near the punch

3.3 Clamped Plate Under Pressure

Next, the mechanical response of a laminated plate of length l, width b, and thickness $t = 2.0\,\text{mm}$ subjected to a uniform pressure load p is analyzed. The simulation setup is depicted in Fig. 6. The plate is clamped at all four edges, thus fixing all degrees of freedom at the edges. For the shell models, the pressure is applied to the midsurface. For the reference solution with solids, the pressure is distributed across the thickness coordinate, resulting in a constant load over the thickness. The laminate consists of four layers: two outer layers with thicknesses $t_o \approx 0.34\,\text{mm}$ and two inner layers with thicknesses $t_i \approx 0.66\,\text{mm}$. Going through the layers in positive z-direction, the fibers are oriented with

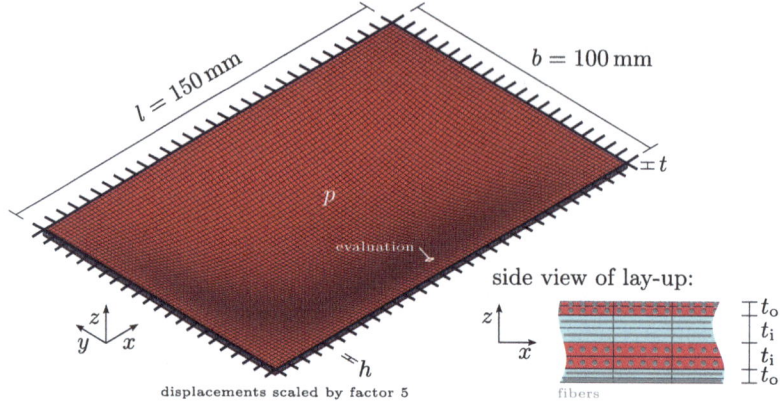

Fig. 6. Set-up and finite element mesh of the clamped plate under pressure

$\beta = \{0°, 90°, 0°, 90°\}$ relative to the x-direction, resulting in an asymmetrical laminate.

Again, the same three different simulation approaches are compared. A simulation with eight solid elements through the thickness of the laminate is used as the reference solution. For comparability, the shell elements use eight integration points in z-direction. For all simulations, a square in-plane discretization with edge length $h = 1.5$ mm is used.

Figure 7 shows the transverse shear stress σ_{yz} and the normal stress in thickness direction σ_{zz} against the dimensionless thickness coordinate of the laminate for the three element formulations. Both stress components are evaluated in the vicinity of the long edge of the plate, as illustrated in Fig. 6. The results

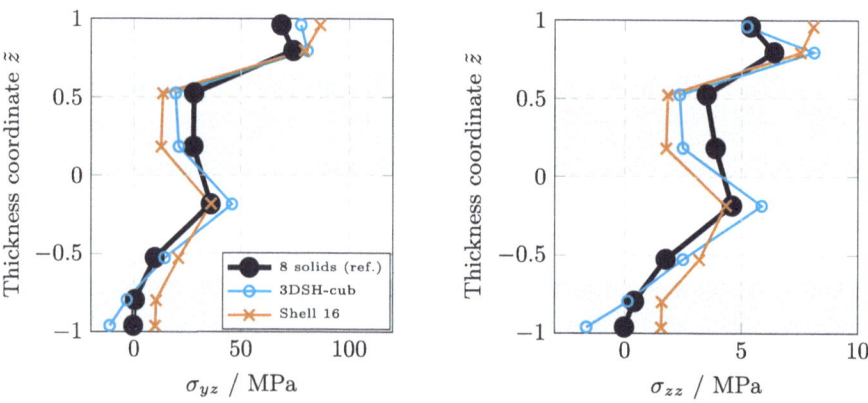

Fig. 7. Plot of the shear stress σ_{yz} and the normal stress σ_{zz} against the dimensionless thickness coordinate $\tilde{z}$ of the laminate for different element formulations evaluated near the long edge of the plate

in Fig. 7 indicate that the cubic 3D-shell element exhibits a closer alignment of the shear stress σ_{yz} with the reference solution (max $e_{yz,IP}^{\text{3DSH-cub}} = 15.1\%$, avg. error $\bar{e}_{yz}^{\text{3DSH-cub}} = 10.2\%$) than the standard shell element. In contrast, the standard shell element exhibits a larger maximum error and average error (max $e_{yz,IP}^{\text{Shell 16}} = 24.2\%$, avg. error $\bar{e}_{yz}^{\text{Shell 16}} = 14.2\%$).

The results for the normal stress σ_{zz} are comparable. The cubic 3D-shell element continues to align better with the reference solution, but some discrepancies, particularly in the third layer from the bottom, are apparent. This results in the following errors in comparison to the reference solution: max $e_{zz,IP}^{\text{3DSH-cub}} = 26.7\%$, $\bar{e}_{zz}^{\text{3DSH-cub}} = 16.3\%$ / max $e_{zz,IP}^{\text{Shell 16}} = 42.2\%$, $\bar{e}_{zz}^{\text{Shell 16}} = 23.7\%$.

3.4 Plate and Impactor

In the last benchmark, the mechanical response of a laminate of length l, width b, and thickness $t = 2.0\,\text{mm}$ subjected to an impact is investigated. The impactor with a diameter d has a mass of $m = 5.998\,\text{kg}$ and an initial velocity of $v_0 = 0.25\,\text{m/s}$ in negative z-direction. Figure 8 depicts the simulation setup. The plate is clamped at all four edges. The laminate consists of four layers: two outer layers with thicknesses $t_o \approx 0.34\,\text{mm}$ and two inner layers with thicknesses $t_i \approx 0.66\,\text{mm}$. Going through the layers in positive z-direction, the fibers are oriented with $\beta = \{0°, 45°, -45°, 90°\}$ relative to the x-direction, resulting in an asymmetrical laminate.

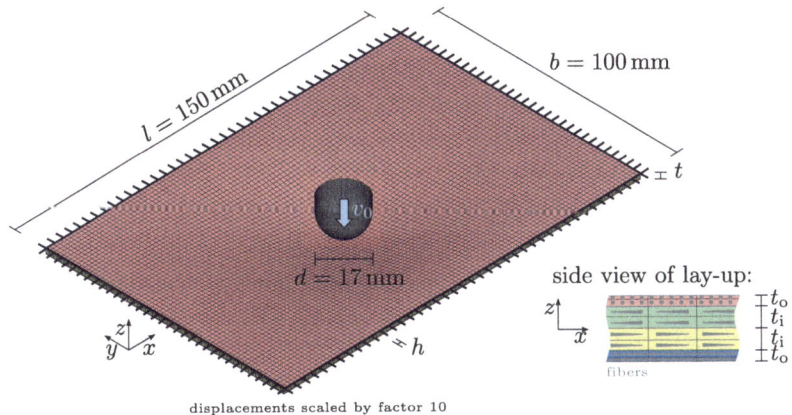

Fig. 8. Set-up and solid finite element mesh of the plate impact test

A simulation with eight solid elements across the thickness of the laminate is used as the reference solution. The shell elements use eight integration points in z-direction. For all simulations, a square in-plane discretization with edge length $h = 1.5\,\text{mm}$ is used.

Figure 9 shows a contour plot of the normal stress in thickness direction σ_{zz} in the 45°-layer of the laminate for all three element formulations. The results show that the stress distribution from the cubic 3D-shell element exhibits a significantly closer alignment with the reference solution than the stress distribution from the standard shell element. This is observed for the region near the impactor, where the highest normal stress in thickness direction arises. The same holds true for the clamped edges of the plate.

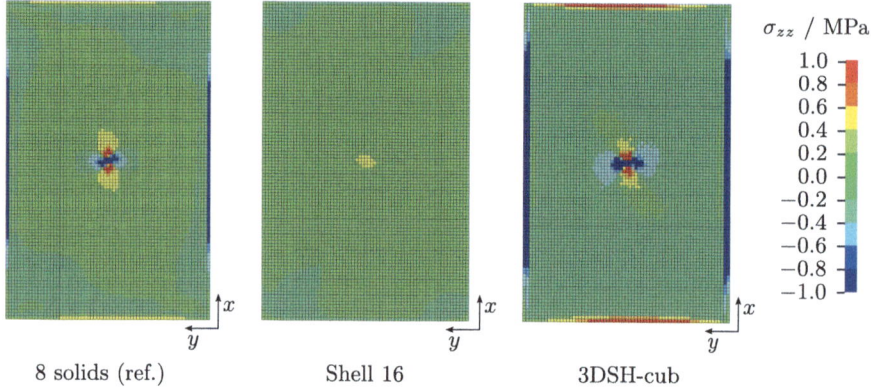

Fig. 9. Contour plot of the normal stress σ_{zz} in the 45°-layer of the laminate for different element formulations with an impactor penetration of $i \approx -0.93$ mm

Therefore, the cubic 3D-shell element is able to qualitatively capture the stress distribution. The element is capable of correctly differentiating between tensile stress and compressive stress in the normal direction in all relevant areas of the plate. In contrast to this, the standard shell element significantly underestimates the normal stress σ_{zz} in comparison to the reference solution. The same observations can be made for all three other layers of the laminate.

4 Summary and Outlook

The enhanced predictive capability of a higher-order 3D-shell element for stress in laminated composites is demonstrated for a range of standard load cases. Results from numerical tests indicate that the cubic 3D-shell element is more effective than Reissner–Mindlin shell elements in representing the stress in laminated composites. The simulation results from the cubic 3D-shell elements are more closely aligned with those of fully 3D simulations with solid elements while requiring significantly less computing time.

This contribution establishes a foundation for further investigation into the potential of higher-order 3D-shell elements to accurately capture the mechanical behavior of laminated materials under a wider range of loading conditions and more complex laminates. On the other hand, this contribution also shows that further improvements in the stress prediction might be possible if the higher-order 3D shell element is enhanced with the capacity to model discontinuities in the thickness direction of the element.

Acknowledgement. This research has been funded by the project DigiTain 19S22006K by the Federal Ministry of Economic Affairs and Climate Action based on a resolution of the German Bundestag. This support is gratefully acknowledged.

5 Appendix

Table 1. LS-DYNA Material Card for IM7-8552, adapted from [14]

Param.	Value	Param.	Value	Param.	Value	Param.	Value
ro	$1.58 \cdot 10^{-9}$	ea	165000.0	eb	9000.0	ec	9000.0
prba	0.0185	prca	0.0185	prcb	0.45	gab	5600.0
gbc	2800.0	gca	5600.0	kfail	0.0	aopt	2.0
macf	1	atrack	0	xp	0.0	yp	0.0
zp	0.0	a1	1.0	a2	0.0	a3	0.0
v1	0.0	v2	0.0	v3	0.0	d1	0.0
d2	1.0	d3	0.0	beta	0.0	sc	90.0
xt	2560.0	yt	73.0	yc	185.0	alph	0.0
sn	73.0	syz	90.0	szx	90.0		

Table 1 provides the material card for the carbon-fiber composite used in all of the finite element simulations in this contribution. A layerwise anisotropic material model (LS-DYNA *MAT_022) is employed. For a detailed explanation of the parameters, see [13].

References

1. Czichos, R., Bergmann, T., Moldering, F., Middendorf, P.: Comparison of numerical modelling approaches for the residual burst pressure of thick type IV composite overwrapped pressure vessels related to low-velocity impact. Int. J. Press. Vessels Pip. **199**, 104770 (2022)
2. Främby, J., Brouzoulis, J., Fagerström, M., Larsson, R.: Prediction of through-thickness stress distribution in laminated shell structures. In: Proceedings of the NSCM-27: the 27th Nordic Seminar on Computational Mechanics (2014)
3. Willmann, T., Wessel, A., Beier, T., Butz, A., Bischoff, M.: Cross-sectional warping in sheet metal forming simulations. In: Proceedings of the 13th European LS-DYNA Conference, Ulm, Germany (2021)

4. Bischoff, M., Ramm, E., Irslinger, J.: Models and finite elements for thin-walled structures. In: Encyclopedia of Computational Mechanics Second Edition, pp. 1–86 (2018)
5. Hughes, T. J., Liu, W. K.: Nonlinear finite element analysis of shells: part I. Three-dimensional shells. Comput. Meth. Appl. Mech. Eng. **26**(3), 331–362 (1981)
6. Belytschko, T., Lin, J.I., Tsay C.-S.: Explicit algorithms for the nonlinear dynamics of shells. Computer Meth. Appl. Mech. Eng. **42**(2), 225-251 (1984). https://doi.org/10.1016/0045-7825(84)90026-4
7. Kühhorn, A., Schoop, H.: A nonlinear theory for sandwich shells including the wrinkling phenomenon. Arch. Appl. Mech. **62**(6), 413–427 (1992)
8. Parisch, H.: A continuum-based shell theory for non-linear applications. Int. J. Numer. Meth. Eng. **38**(11), 1855–1883 (1995)
9. Cardoso, R.P., Yoon, J.W.: One point quadrature shell element with through-thickness stretch. Comput. Meth. Appl. Mech. Eng. **194**(9–11), 1161–1199 (2005)
10. Schilling, M., Willmann, T., Wessel, A., Butz, A., Bischoff, M.: Higher-order 3D-shell elements and anisotropic 3D yield functions for improved sheet metal forming simulations: part I. In: Proceedings of the 14th European LS-DYNA Conference, Baden-Baden, Germany (2023)
11. Reddy, J. N.: Mechanics of Laminated Composite Plates and Shells: Theory and Analysis. CRC Press (2003)
12. ANSYS, Inc.: LS-DYNA Theory Manual – LS-DYNA R15 (2024). https://lsdyna.ansys.com/manuals/
13. ANSYS, Inc.: LS-DYNA Keyword User's Manual – Volume II Material Models – LS-DYNA R15 (2024). https://lsdyna.ansys.com/manuals/
14. Cherniaev, A., Montesano, J., Butcher, C.: Modeling the axial crush response of CFRP tubes using MAT054, MAT058 and MAT262 in LS-DYNA. In: Proceedings of the 15th International LS-DYNA Users Conference, Detroit, United States of America (2018)

Open Access This chapter is licensed under the terms of the Creative Commons Attribution 4.0 International License (http://creativecommons.org/licenses/by/4.0/), which permits use, sharing, adaptation, distribution and reproduction in any medium or format, as long as you give appropriate credit to the original author(s) and the source, provide a link to the Creative Commons license and indicate if changes were made.

The images or other third party material in this chapter are included in the chapter's Creative Commons license, unless indicated otherwise in a credit line to the material. If material is not included in the chapter's Creative Commons license and your intended use is not permitted by statutory regulation or exceeds the permitted use, you will need to obtain permission directly from the copyright holder.

Validation of the *CONSTRAINED_SPR3 Joint Formulation for Isogeometric Shell Models

Philipp Bähr[1][✉], Lukas Leidinger[2], Silke Sommer[1], and Stefan Hartmann[2]

[1] Fraunhofer - Institut für Werkstoffmechanik IWM, Wöhlerstr. 11, 79108 Freiburg, Germany
philipp.baehr@iwm.fraunhofer.de
[2] DYNAmore GmbH, an Ansys Company, Industriestr. 2, 70565 Stuttgart, Germany

Abstract. Isogeometric Analysis (IGA) uses higher-order and higher-continuity spline basis functions known from Computer Aided Design (CAD) to describe the geometry and the solution field of the simulation model (mainly Non-Uniform Rational B-splines). This leads to a more accurate geometry description, a smooth solution field and therefore superior simulation properties compared to traditional Finite Element Analysis (FEA). Using the same geometry description for CAD and IGA also speeds up the modeling process for the simulation. Real components consist of multiple sheet metal parts connected by point-like joints (spot-welds, rivets, screws). These joints significantly influence the component behaviour under crash load. Therefore, their properties must be accurately described within the component simulation. Due to minimum time step requirements a detailed modelling of every single joint is impossible. Thus, substitute models are used in component simulations, which describe the joints behaviour based on constrained conditions between the joining partners. However, these substitute models are developed for traditional FEA. To enable the application of IGA to vehicle simulations, this paper investigates whether existing constrained-based substitute models (e.g. *CONSTRAINED_SPR3 spotweld elements) can be combined with isogeometric shell models without further modifications. Therefore, specimen and component tests are simulated with IGA and FEA. The simulation results are compared to each other as well as experimental test results. It can be shown that IGA achieves a very good agreement with the experimental results, with a prediction quality comparable to the traditional FEA. This allows a straightforward replacement of existing FEA shell components with their IGA counterparts in vehicle simulations.

Keywords: Isogeometric Analysis (IGA) · crash simulation · modelling of joints

1 Introduction

Isogeometric Analysis (IGA) is a relatively new Finite Element Analysis (FEA) technology based on splines, for example B-Splines or Non-Uniform Rational B-Splines (NURBS), which are the standard in Computer Aided Design (CAD). In fact, IGA was developed by Hughes et al. [1] in 2005 with the goal to alleviate the time- and work-intensive model conversion between CAD and analysis. While having a consistent geometry description and data structure for design and analysis is still a major advantage, IGA also offers attractive analysis properties like a smooth solution field or a larger explicit time step size [2]. These properties originate from the fact that the higher-order and higher-continuity spline basis functions are not only used to describe the geometry, but to also describe the solution field (isoparametric concept).

Beginning from 2010, a wide range of IGA capabilities has been implemented in LS-DYNA and successively extended. At present, LS-DYNA supports various shell and solid elements, explicit, implicit and modal analysis, MPP and SMP, damage and element erosion, time step estimation and mass scaling, contact, various connection technologies, and the possibility to combine IGA models with conventional FEA models within one simulation [3–5]. The latter strongly facilitates the application of IGA in industry as it allows replacing a conventional FEA component with its IGA counterpart, for example in a vehicle crash simulation. Furthermore, IGA in LS-DYNA is implemented in a way that enables a simple one-to-one component exchange without changing the existing joint connections or tied contacts, see [3, 4]. In a recent study, Bauer et al. [6] took an existing body-in-white (BIW) model, replaced hundreds of FEA shell components with their IGA counterparts and ran it successfully in a full vehicle crash simulation. The IGA shell models were generated from CAD models using ANSA [7]. In this vehicle model, thousands of spotwelds were modeled with *CONSTRAINED_SPR3 in LS-DYNA, see Sect. 2.2 for more details on SPR3. Since the SPR3 definition in LS-DYNA is based on the PIDs of the involved sheets, no modification was necessary for the IGA shell model, as long as the correct PIDs were kept [3]. The mentioned study [6] assessed the crash behavior on a global level and reported good overall agreement between the pure FEA vehicle model and the hybrid IGA/FEA vehicle model in a front and side crash scenario. However, no detailed validations of the *CONSTRAINED_SPR3 connection for IGA shell models was done.

The present contribution aims at providing such detailed investigations of the *CONSTRAINED_SPR3 connection for IGA shell models by comparing with experimental and FEA results for LWF-KS-2-specimens and a component test.

2 Preliminaries

2.1 Isogeometric Analysis (IGA) in LS-DYNA

IGA uses spline basis functions of higher order and higher continuity, mainly B-Splines and NURBS, to describe the geometry and the solution field, see [1] for more details. This contrasts with conventional FEA, where mainly linear C0 Lagrange polynomials are used as shown by the comparison in Fig. 1. As can be seen, for the same number of elements, the geometry is captured much more accurately in the IGA model on the right. In IGA, the degrees of freedom (DOFs) are defined on the so-called control points, which are the equivalent to nodes in FEA. Note that the control points are in general not interpolatory, that is, not located on the model geometry.

In LS-DYNA, various isogeometric shell element formulations are available, see *SECTION_IGA_SHELL in [8]. In this paper, a six DOF Reissner-Mindlin shell element formulation with fibers at the integration points (ELFORM $=$ 3) is used [9]. For the

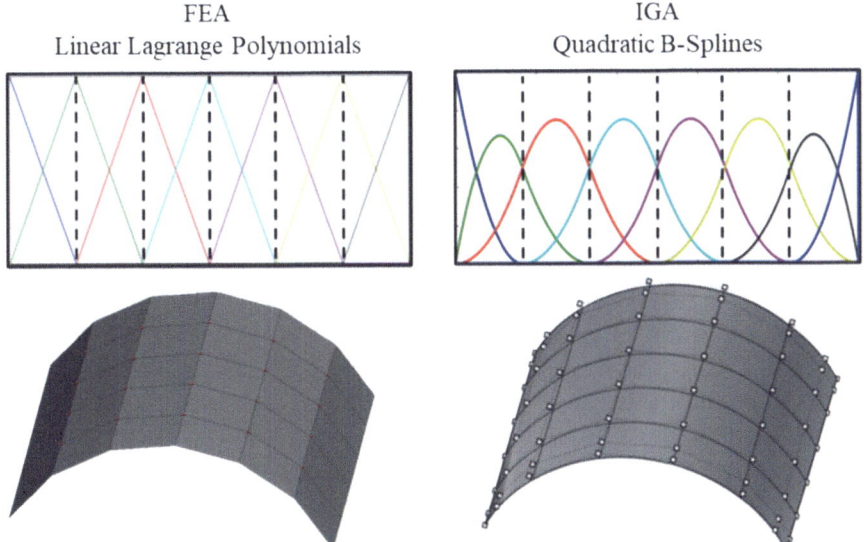

Fig. 1. Comparison of element basis functions and geometry between FEA with linear Lagrange polynomials (left) and IGA with quadratic B-splines (right)

IGA model definition, the *IGA_ keyword structure containing keywords for geometry, topology and analysis information is used, see [3, 8] for more information.

When running an IGA model, LS-DYNA automatically generates a so-called interpolation mesh consisting of linear finite elements on top of the spline geometry, see Fig. 2. This is done for two purposes: (i) Visualization of the deformed IGA model including stresses, strains, etc. in standard FE postprocessors, and (ii) handling contact effectively and efficiently through the well-established FE contact algorithms. Referring to the latter, LS-DYNA first evaluates the contact on the interpolation mesh and then maps the forces on the interpolation nodes to the control points of the corresponding IGA elements. Please note that the interpolation elements are not evaluated since their nodes are fully constrained to the IGA model, do not possess DOFs and do also not affect the stable time step size. The number of interpolation elements per IGA element in r- and s-direction (see Fig. 2) can be controlled by the parameters NISR and NISS, respectively (NISR = NISS = 1 equals 1×1 interpolation element, NISR = NISS = 2 equals 2×2 interpolation elements, etc.). In the remainder of this paper, NISR and NISS are chosen identically and simply denoted as NIS.

Similar to contact, also tied contacts or connections can be handled via the interpolation mesh. The *CONSTRAINED_SPR3 connection as described in detail in the next section is also implemented such that it acts on the interpolation mesh of the IGA shell model. In that way, the existing FEA connection technology can also be directly applied to IGA without further adaptations.

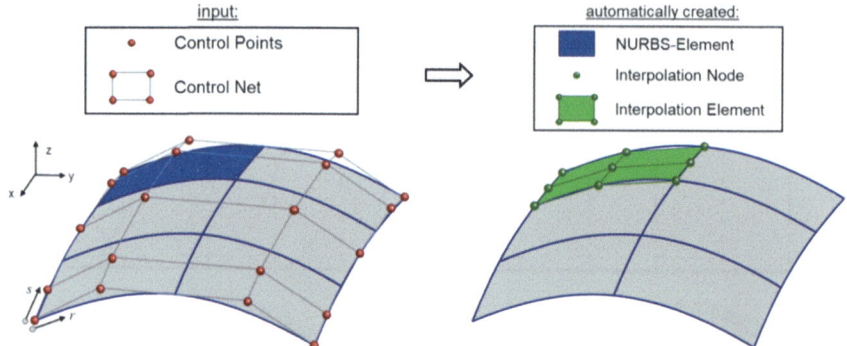

Fig. 2. IGA model input with NURBS surface (left) and automatically generated interpolation mesh on top of the NURBS surface (right)

2.2 *Constrained_SPR3

The *CONSTRAINED_SPR3 is a constraint-based model which can be used to simulate the properties of point-type joints in component simulation [8]. Due to its constraint-based formulation, no extra finite elements need to be modeled to represent the joint. To define the joint, one node and a domain of influence must be specified (Fig. 3 left). This information is used to automatically find nodes on the two joining partners and link them via constraints. The *CONSTRAINED_SPR3 is based on the flow and failure model shown on the right-hand side in Fig. 3.

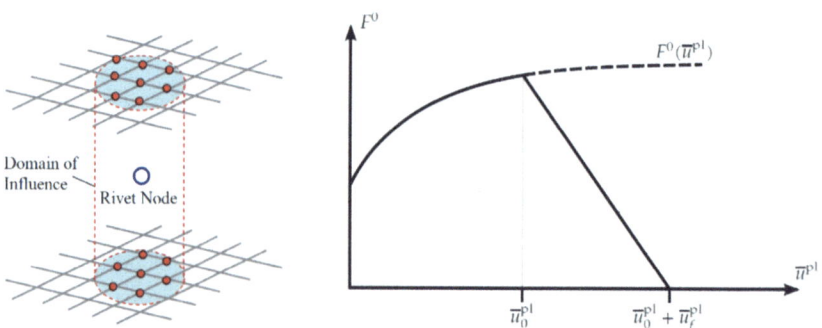

Fig. 3. Left: Schematic representation of the *CONSTRAINED_SPR3 model [10], right: flow and failure behavior of the *CONSTRAINED_SPR3 model [8]

In this work a modified definition of the *CONSTRAINED_SPR3 presented in [11] is used (model 2). This modified variant includes an adapted flow behavior to consider the influence of an asymmetrical load on the joining point as well as the possibility to consider a strain rate dependency of the load-bearing capacity. To describe the deformation and failure behavior of the joint the relative displacement vector $\vec{u}$ is calculated. This vector comprises the relative displacement in normal δ_n and tangential δ_t direction between the two connected sheets (Eq. 1).

$$\vec{u} = (\delta_n, \delta_t) \tag{1}$$

Additionally, the relative rotation angle ω_b is calculated from the normal vectors $\vec{n}_s$ and $\vec{n}_m$ of the connected shell elements.

$$\omega_b = \cos^{-1}\left(\frac{\vec{n}_s \vec{n}_m}{|\vec{n}_s||\vec{n}_m|}\right) \tag{2}$$

The elastic force vector $\vec{f}$ is calculated from the relative displacement vector $\vec{u}$ using the elastic stiffness C (Eq. 3). This vector also comprises a component in normal f_n and tangential f_t direction.

$$\vec{f} = C\vec{u} = (f_n, f_t) \tag{3}$$

To describe plastic deformation the yield function Φ is defined as a function of the force vector $\vec{f}$ and the relative plastic displacement $\bar{u}^{pl}$ (Eq. 4). In this equation P is the plastic potential and F^0 is the isotropic hardening curve which can be defined by the user.

$$\Phi\left(\vec{f}, \bar{u}^{pl}\right) = P\left(\vec{f}\right) - F^0\left(\bar{u}^{pl}\right) \leq 0 \tag{4}$$

The plastic potential P is calculated from the force vector $\vec{f}$ according to Eq. 5. The parameters R_n and R_s describe the load-bearing capacity of the joint in normal and tangential direction. The exponent β_1 defines the load-bearing capacity under mixed loading and the parameter α_1 scales the influence of the relative rotation angle ω_b on the load-bearing capacity under normal loading.

$$P(\vec{f}) = \left[\left(\frac{f_n}{R_n(1 - \alpha_1 \omega_b)}\right)^{\beta_1} + \left(\frac{f_s}{R_s}\right)^{\beta_1}\right]^{\frac{1}{\beta_1}} \tag{5}$$

Failure of the connection is considered by using the damage variable d, which is calculated from the relative plastic displacement $\bar{u}^{pl}$, the relative plastic displacement at damage initiation $\bar{u}_0^{pl}$ and the relative plastic displacement at failure $\bar{u}_f^{pl}$ (Eq. 6).

$$d = \frac{\bar{u}^{pl} - \bar{u}_0^{pl}}{\bar{u}_f^{pl}} \tag{6}$$

A linear scaling of the force vector $\vec{f}$ is done with the damage variable d until the connection finally fails (Eq. 7).

$$\vec{f}^* = (1 - d)\vec{f} \tag{7}$$

In order to consider the influence of a mixed load on the failure behavior of the joint, the load angle φ is calculated from the force components f_n and f_s in normal and tangential direction according to Eq. 8.

$$\varphi = \tan^{-1}\left(\frac{f_n}{f_s}\right) \tag{8}$$

The relative plastic displacement at damage initiation $\bar{u}_0^{pl}$ and the relative plastic displacement at failure $\bar{u}_f^{pl}$ can be calculated with the load angle φ and the user defined parameters $\bar{u}_{0,ref}^{pl,n}$, $\bar{u}_{0,ref}^{pl,s}$, $\bar{u}_{f,ref}^{pl,n}$ and $\bar{u}_{f,ref}^{pl,s}$ (Eq. 9 and 10). The parameters $\bar{u}_{0,ref}^{pl,n}$ and $\bar{u}_{0,ref}^{pl,s}$ describe the relative plastic displacement at damage initiation under pure normal ($\bar{u}_{0,ref}^{pl,n}$) and pure shear load ($\bar{u}_{0,ref}^{pl,s}$). The parameters $\bar{u}_{f,ref}^{pl,n}$ and $\bar{u}_{f,ref}^{pl,s}$ do the same for the relative plastic displacement at failure. The exponents β_2 and β_3 define the values of $\bar{u}_0^{pl}$ and $\bar{u}_f^{pl}$ under mixed load. The parameters α_2 and α_3 scale the influence of the relative rotation angle ω_b on the relative plastic displacement at damage initiation and failure under normal load.

$$\bar{u}_0^{pl} = \left[\left(\frac{\sin\varphi}{\bar{u}_{0,ref}^{pl,n}(1-\alpha_2\omega_b)}\right)^{\beta_2} + \left(\frac{\cos\varphi}{\bar{u}_{0,ref}^{pl,s}}\right)^{\beta_2}\right]^{-\frac{1}{\beta_2}} \tag{9}$$

$$\bar{u}_f^{pl} = \left[\left(\frac{\sin\varphi}{\bar{u}_{f,ref}^{pl,n}(1-3\omega_b)}\right)^{\beta_3} + \left(\frac{\cos\varphi}{\bar{u}_{f,ref}^{pl,s}}\right)^{\beta_3}\right]^{-\frac{1}{\beta_3}} \tag{10}$$

3 Experimental Database

The experimental database on which this project is based originates from the AiF/FOSTA research project IGF-No. 20116 N/P 1262 [12]. All experimental tests were carried out at the Laboratory for material and joining technology (LWF) of the University of Paderborn. The following investigations are carried out on an exemplary self-piercing riveting (SPR) joint between HCT780X and a HC340LA. Both sheet metals have a thickness of 1.5 mm. Figure 4 shows a microsection of the investigated SPR joint.

To determine the load-bearing capacity of the joint KS-2-tests have been carried out at the LWF. A schematic representation of the LWF-KS-2 test can be seen on the lefthand side in Fig. 5. The LWF-KS-2-specimen consists of two u-profiles joined in the middle. These specimens are tested under varying load angles, whereby a load angle of 0° corresponds to a pure shear loading of the joint and a load angle of 90° corresponds to a pure tensile load. The load angles 30° and 60° result in a combined tensile and shear loading of the joint. Additionally, a peel loading of the joint is investigated with a modified version of the LWF-KS-2-specimen, where one of the flanges of the specimen is removed at the radius. The measured force-displacement curves of the investigated joint can be seen on the righthand side in Fig. 5. The LWF-KS-2-0°-specimens have the highest load bearing capacity with an average value of 10.9 kN. With increasing load

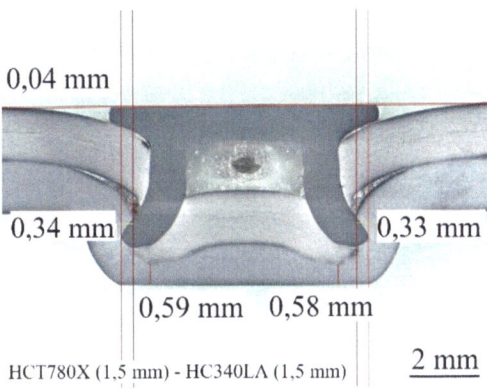

Fig. 4. Microsection of the investigated SPR joint between HCT780X and HC340LA [12]

angle, a decrease of the load bearing capacity can be observed, whereby the 90° specimen has a load-bearing capacity of 4.5 kN on average. The LWF-KS-2-peel-specimen has the overall lowest load-bearing capacity with an average value of 1.9 kN. The fracture displacement behaves in the opposite direction, with the lowest fracture displacement for the 0° specimen and the highest for the peel specimen. Additionally, to the LWF-KS-2 tests lap-shear tests have been performed. As shown in Fig. 5, the lap-shear tests have a lower load-bearing capacity compared to the KS-2-0° tests.

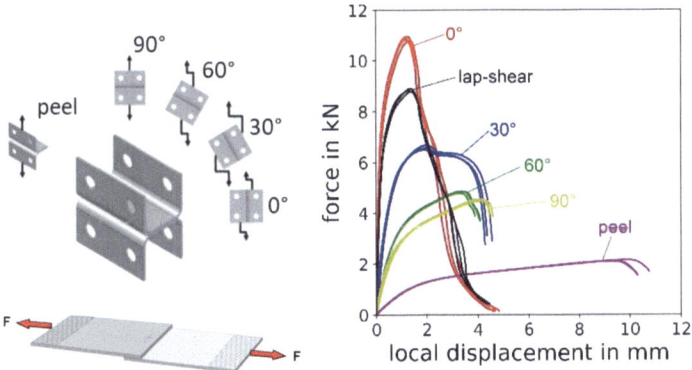

Fig. 5. Left: schematic representation of the LWF-KS-2 specimen [12], right: experimental results of LWF-KS-2 tests for the investigated SPR joint [12]

For validation of the LWF-KS-2-tests, component tests have been performed. The component used for these tests can be seen on the lefthand side in Fig. 6 and Fig. 7. The component is a T-joint made of sheet metal profiles, which is loaded longitudinally and laterally by a punch. The resulting force-displacement behavior together with the failure order of the critical joints for the two investigated load-cases can be seen on the right in Fig. 6 and Fig. 7. Under lateral load the joints 2 and 4 fail first, followed by failure of joints 1 and 3. The critical joints are mainly loaded by tensile forces with a superimposed bending moment. Under longitudinal load, joint 5 fails first, followed by the failure of points 1, 2 and 6. Joints 5 and 6 are mainly loaded under shear, whereas joints 1 and 2 are loaded by tensile forces with a superimposed bending moment.

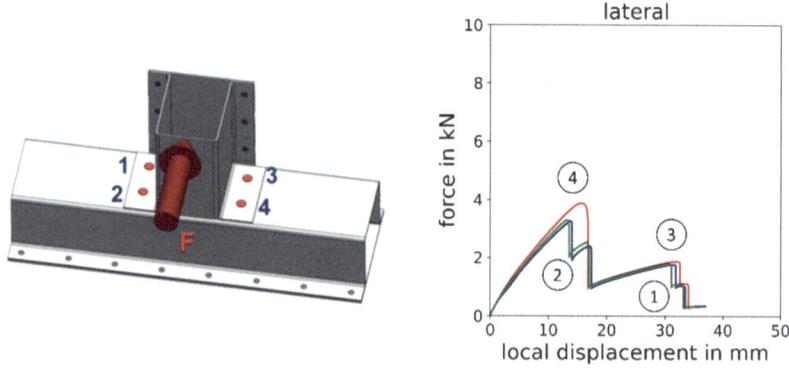

Fig. 6. Left: schematic representation of the component test under lateral load [12], right: experimental results of component tests under lateral load [12].

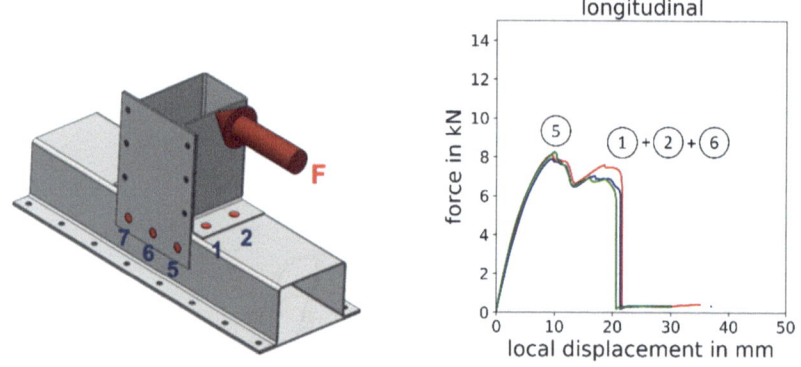

Fig. 7. Left: schematic representation of the component test under longitudinal load [12], right: experimental results of component tests under longitudinal load [12].

4 Simulation

In the following section simulations of the specimen and component tests described in Sect. 3 are performed. The sheet metal profiles of the LWF-KS-2 and component specimens are modeled using traditional FEA and IGA shell elements. The deformation and failure behavior of the SPR-joints is modeled using the *CONSTRAINED_SPR3 model described in Sect. 2.2. The parameters of the *CONSTRAINED_SPR3 are taken from the before mentioned AiF/FOSTA research project IGF-No. 20116 N/P 1262 [12], where they have been calibrated for FEA shell elements. These parameters are directly used for IGA shell elements to assess whether they can be transferred or need to be adjusted for the changed simulation approach. A comparison of the simulation results with FEA and IGA as well as the experimental results is performed.

4.1 Setup of the Simulation Models

To simulate the LWF-KS-2 and component tests different simulation models have been created, using traditional FEA shell elements and IGA shell elements. The FEA models are meshed with fully integrated shell elements with an element size of 2.5 mm (lefthand side of Fig. 8 and Fig. 9). The IGA models consist of multiple untrimmed IGA patches with matching discretization (patch coupling in a strong sense), quadratic elements, and an element size of also 2.5 mm (righthand side of Fig. 8 and Fig. 9). For the base model, two interpolation elements are used per IGA element. The IGA discretization is generated in LS-PrePost [13], parametric points and boundary conditions are applied in ANSA [7]. A comparison between the FEA and the IGA shell models is shown in Fig. 8 and Fig. 9. The SPR-joints are modeled with the *CONSTRAINED_SPR3 model described in Sect. 2.2. A radius of R = 6 mm is used for the *CONSTRAINED_SPR3 element, which is approximately twice as large as the head diameter of the used rivet.

The clamping of the KS-2-specimens is modeled with rigid bodies for both FEA and IGA. The load on the specimens is displacement-controlled with a pre-defined displacement of one of the rigid bodies. The transferred force is measured using a cross-sectional plane through the specimen in which the transmitted force is determined. The displacement is measured locally on the specimen using the relative displacement of nodes on the specimen's surface. In the FEA model the displacement of mesh nodes can be used, in the IGA model additional nodes are attached to the IGA shell elements via the *IGA_POINT_UVW keyword. With the described models of the LWF-KS-2-tests the different loading angles shown on the left in Fig. 5 are simulated as well as the LWF-KS-2-peel specimen.

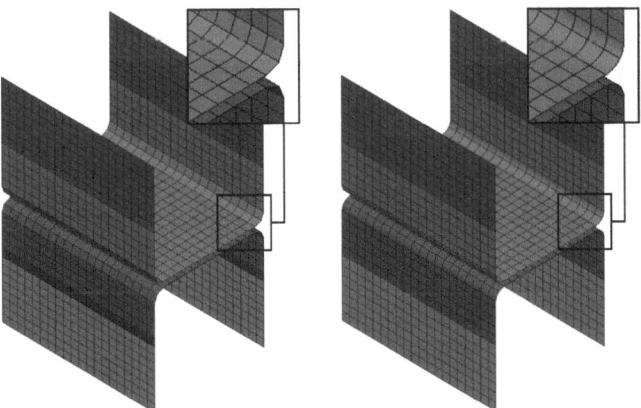

Fig. 8. Comparison of an FEA (left) and an IGA shell model (right) for the LWF-KS-2-tests.

In the simulation of the component tests, the parts representing the test device (punch and wedge) are modeled as rigid bodies using solid elements. The load on the component is applied by a defined punch displacement. Two different load scenarios are considered, one in which the load is applied longitudinal to the crossbeam and one where the load is applied lateral to the crossbeam (Fig. 6 and Fig. 7). The punch force and displacement are measured to evaluate the simulation. The critical joints connecting the crossmember and the column are modeled with *CONSTRAINED_SPR3. The parameters of the *CONSTRAINED_SPR3 model are the same as in the simulations of the LWF-KS-2 tests, whereby the maximum force of the lap-shear specimen is assumed for the shear tensile load-bearing capacity (Fig. 11). This allows the component behavior to be better reproduced as the load situation in the critical joints corresponds to the lap-shear specimen.

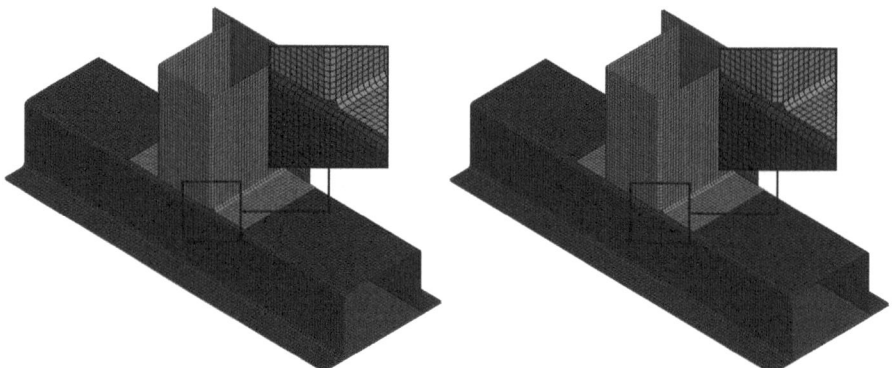

Fig. 9. Comparison of an FEA (left) and an IGA shell model (right) for the component test

4.2 Simulation Results of the LWF-KS-2-Tests

A comparison of the calculated force-displacement behavior using FEA and IGA with the experimental results of the LWF-KS-2 tests can be found in Fig. 10. It can be shown that the *CONSTRAINED_SPR3 model is able to reproduce the experimental behavior of the LWF-KS-2-tests with high accuracy. The maximum forces and the energy absorption are very well reproduced by the model. Between the FEA and the IGA model only minor differences can be observed for the LWF-KS-2-30°, −60°, −90° and peel specimens. The differences are mainly seen in the calculated fracture displacement, which is slightly higher for the simulation models using IGA. In the simulation of the LWF-KS-2-30° specimen deviations in the plastic behavior of the *CONSTRAINED_SPR3 model can be observed, which lead to an underestimation of the experimental results. For the lap-shear specimen an overestimation of the experimental load-bearing capacity can be observed for the FEA and IGA model. The calibration process of the *CONSTRAINED_SPR3 model is directly responsible for this overestimation. During calibration the shear strength of the model has been fitted to the LWF-KS-2-0° specimen. In the experimental characterization the LWF-KS-2-0° specimen shows a higher load-bearing capacity compared to the lap-shear specimen. The reason for this deviation can be found in the different stiffness of the two specimen types. The lap-shear specimen has an overall lower stiffness compared to the LWF-KS-2-0° specimen which leads to a different load situation locally at the joint. The *CONSTRAINED_SPR3 model is unable to capture these differences in the local load situation and therefore predicts a load-bearing capacity comparable to the LWF-KS-2-0° specimen for the lap shear specimen.

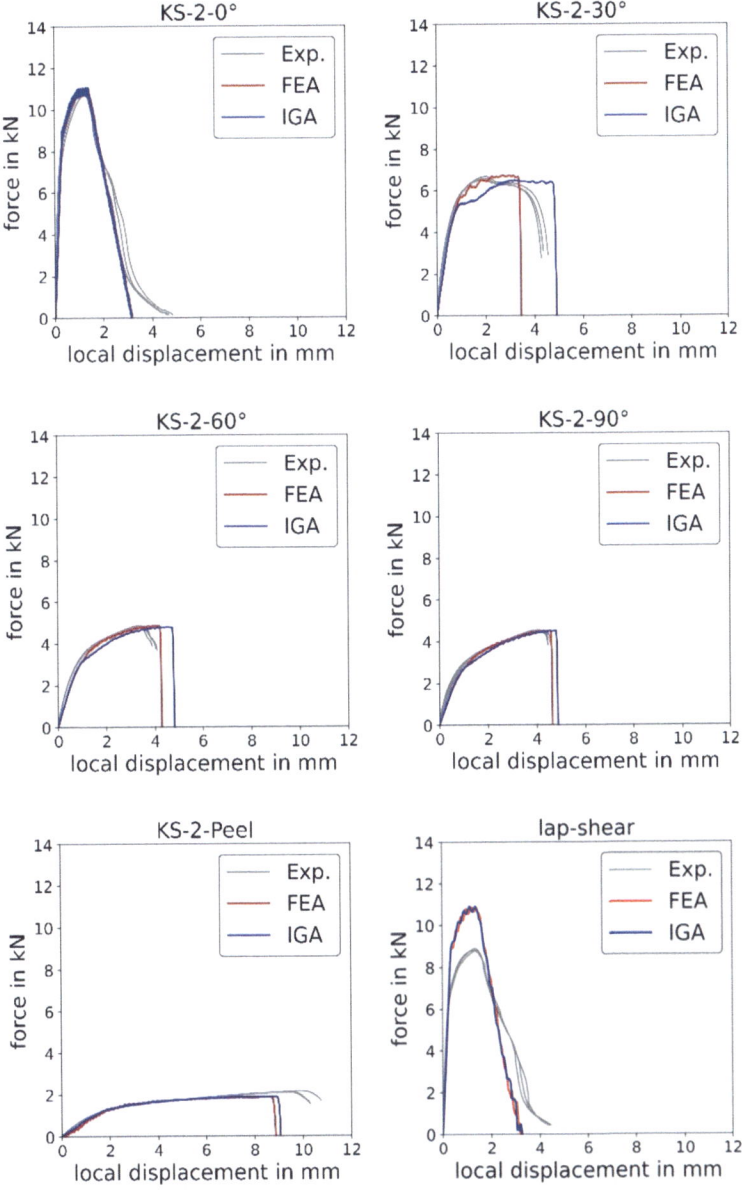

Fig. 10. Comparison of the simulated force-displacement behavior with the experimental results for the LWF-KS-2-tests with traditional FEA and IGA

To better reproduce the behavior of the lap-shear specimen the load bearing capacity under shear of the *CONSTRAINED_SPR3 model can be calibrated based on the experimental results of the lap-shear specimen. As Fig. 11 shows on the right this approach leads to a good agreement of the simulation results with the experimental behavior for the lap shear-specimen. It should be noted that this method leads to an underestimation of the load bearing capacity of the LWF-KS-2-0° specimen, as the maximum force of the lap-shear specimen is now being calculated here.

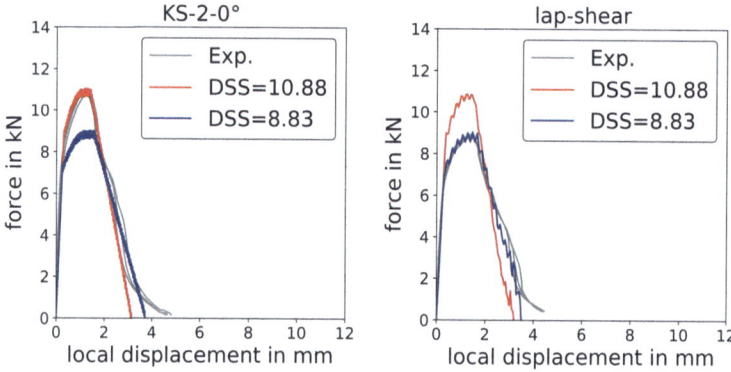

Fig. 11. Comparison of FEA simulation results for the KS-2-0° (left) and lap-shear specimen (right) using different shear strength values and the experimental results

To further investigate the behavior of the IGA model, several variations of the model are considered. In Fig. 12 a comparison of IGA simulations with a varying number of interpolation elements per IGA element can be found. A comparison of the simulation results shows that the value of NIS has only a small influence on the behavior of the *CONSTRAINED_SPR3 model. The only differences can be seen in the fracture displacement of the LWF-KS-2-30° specimen which increases with higher values of NIS. In summary, it can be noted that the influence of the NIS parameter is negligible. If the IGA mesh is fine enough, the joint behavior can be described with a reduced number of interpolation elements, since the IGA elements describe the local deformation behavior with a higher accuracy which will be shown in the following.

A comparison of simulation results using IGA and FEA for a reduced radius of R = 3 mm of the *CONSTRAINED_SPR3 model can be found in Fig. 13. A radius of 3 mm corresponds to the diameter of the rivet used in reality. The reduced radius of 3 mm results in less nodes connected by the *CONSTRAINED_SPR3 model for both FEA and IGA, although more nodes are connected in the IGA model compared to the FEA model due to the finer interpolation network. As a result, the stiffness of the coupling of the *CONSTRAINED_SPR3 is reduced which is most evident in the simulation result of the KS-2-0° with FEA. The reduced stiffness of the coupling results in an underestimation of the load bearing capacity of the LWF-KS-2-0° specimen. With IGA elements the local deformation is better captured (Fig. 14), which is why the IGA simulation matches the experimental behavior better. Figure 14 on the right shows the interpolation mesh of the IGA shell model, the actual element size corresponds to 2.5 mm and is therefore equal to FEA shell elements. For the remaining specimen types the differences between IGA and FEA are less pronounced. In summary, the radius of the *CONSTRAINED_SPR3 mainly affects the stiffness of the coupling. Since the IGA elements capture the local deformation better the stiffness of the coupling is less important, and the experimental force-displacement behavior can be better captured. In addition to the investigations described above, the influence of the element size of the FEA and IGA shell elements

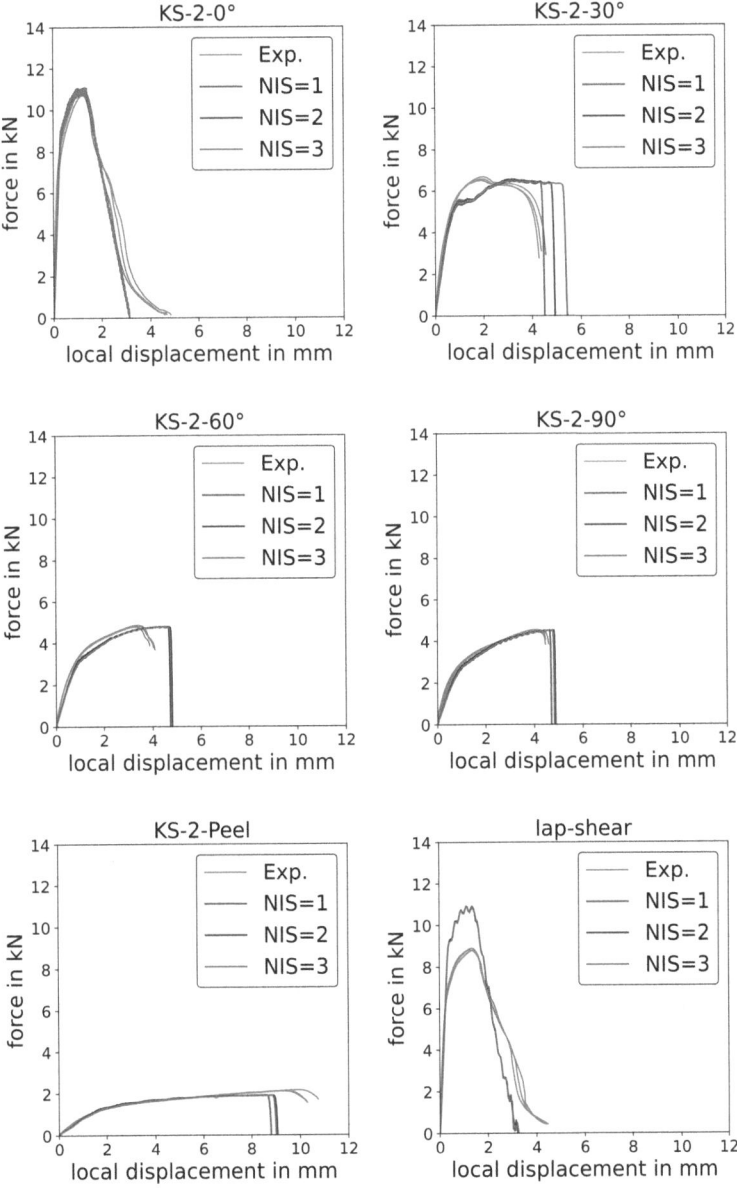

Fig. 12. Comparison of different IGA models with a varying number of interpolation elements per IGA element

was also examined. However, this did not reveal any new findings, which is why the results are not presented here.

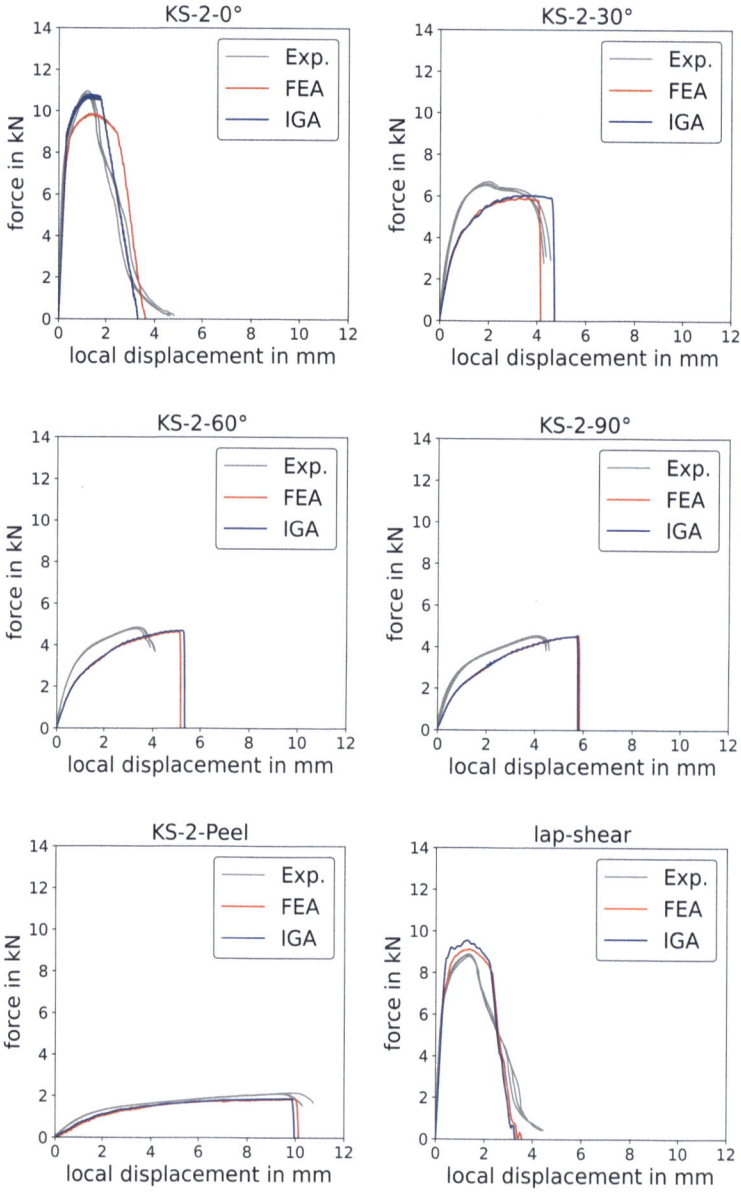

Fig. 13. Comparison of the simulated force-displacement behavior with the experimental results for the LWF-KS-2-tests with traditional FEA and IGA with a reduced radius of 3 mm for the *CONSTRAINED_SPR3 model

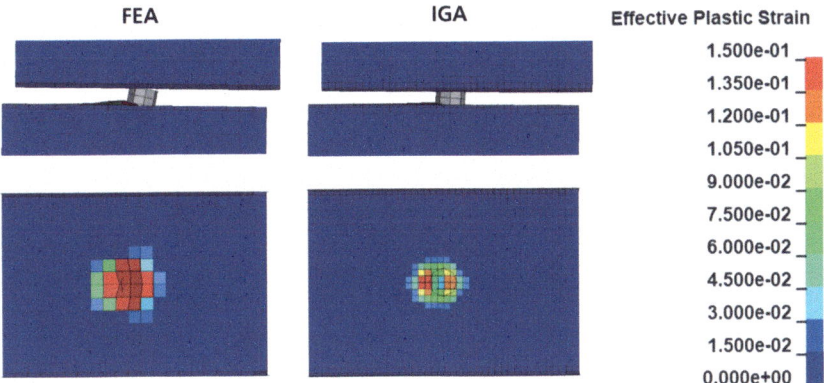

Fig. 14. Comparison of the calculated plastic strain for the LWF-KS-2-0° specimen with traditional FEA (left) and IGA (right) with a radius of 3 mm for the *CONSTRAINED_SPR3

4.3 Simulation Results of the Component Tests

To validate the *CONSTRAINED_SPR3 models under complex loading conditions simulation of the component tests are performed with FEA and IGA. Figure 15 shows a comparison for the two investigated load-cases of the component simulated with FEA and IGA. Overall, the simulations show good agreement with the experimental results, both in terms of the force-displacement curve and the failure sequence of the individual joints. Under lateral load the experimental results show two levels of load bearing capacity which is due to the failure sequence of the joints. In the test with the higher load-bearing capacity, joint 2 and 4 fail simultaneously, whereas in the tests with the lower load-bearing capacity they fail successively. In the simulations joint 2 and 4 always fail simultaneously, which is why the calculated load-bearing capacity corresponds to the upper level of the experimental results. Under longitudinal load, the load-bearing capacity, which is determined by the failure of joint 5, is reproduced very well. Subsequently, joint 7 fails

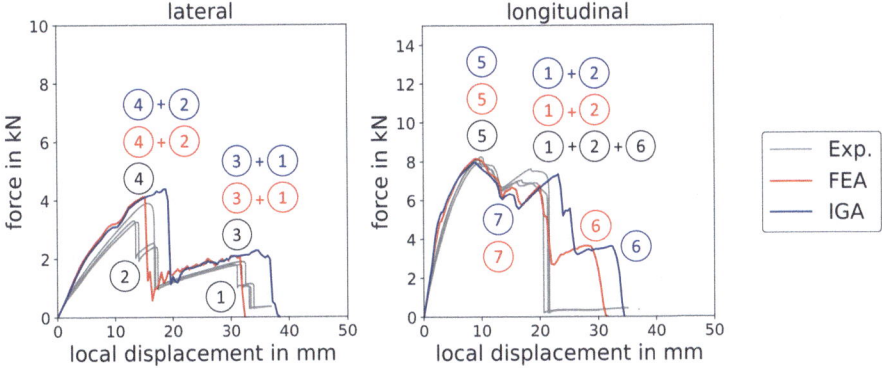

Fig. 15. Comparison of the experimentally measured force-displacement behavior for the component tests under longitudinal (left) and lateral load (right) and simulation results with traditional FEA and IGA

in the simulations, which does not occur in the experiment. This applies to both the FEA and the IGA model. The successive failure of joints 1 and 2 is captured by the models. In comparison to the experimental results, joint 6 fails later in the simulations. A comparison between the IGA and FEA results shows that the joints under tensile load (1, 2, 3 and 4) fail later in the IGA model compared to the FEA model. This results in a higher energy consumption of the IGA model compared to the FEA model.

To further investigate the behavior of the IGA model, a model with a reduced radius (R = 3 mm) of the *CONSTRAINED_SPR3 model is simulated. The results of this simulation can be seen in Fig. 16. The model with the smaller *CONSTRAINED_- SPR3 radius of 3 mm shows a similar behavior to Fig. 15. The force-displacement behavior and the failure sequence of the joints agrees with the original model. However, the difference in the failure behavior of the joints under tensile load is less pronounced than in the original model. Furthermore, the stiffness of the components is also better reproduced with the reduced *CONTRAINED_SPR3 radius.

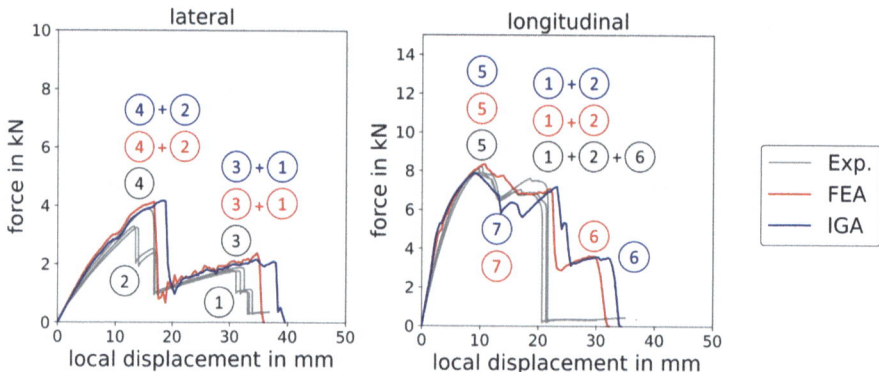

Fig. 16. Comparison of the experimentally measured force-displacement behavior for the component tests under longitudinal (left) and lateral load (right) and simulation results with traditional FEA and IGA with a reduced radius of 3 mm for the *CONSTRAINED_SPR3 model

5 Conclusion

This paper investigated the modeling of joints in component simulations utilizing IGA shell elements. The aim here was to check whether common constrained-based substitute models (*CONSTRAINED_SPR3) developed for traditional FEA elements can be used together with IGA. For this purpose, FEA and IGA models of joint KS-2- and component specimens were created and simulated using the *CONSTRAINED_SPR3 element. The simulation results were compared to each other as well as to experimental results. It was shown that the *CONSTRAINED_SPR3 elements calibrated for traditional FEA show good results when used with IGA shell elements. It was also shown that the number of linked nodes in particular influences the behavior of the model. Depending on the setup of the simulation model, the number of these nodes differs between IGA and FEA. Several parameters influence the number of linked nodes, for example the number of interpolation elements per IGA element, the radius of the *CONSTRAINED_SPR3 model, and the element size of the IGA and FEA elements. The simulation accuracy can be further optimized by adjusting these parameters. The parameters of the SPR model itself were not varied in this work, although the simulation results may be further improved as a result. Corresponding investigations are planned and will be carried out in the future. In summary, it can be shown that *CONSTRAINED_SPR3 models parameterized based on FEA simulations can be used with IGA elements and show good results. However, by using IGA shell elements, the local deformation behavior around the joint can be better reproduced, which can result in advantages compared to traditional FEA models. Due to the transferability FEA parts can be replaced by IGA equivalents in component models, making it easy to implement the use of IGA in component simulations.

Acknowledgement. The basis for this publication was developed as part of the project "DigiTain – Digitalization for Sustainability". The project "DigiTain – Digitalization for Sustainability" is funded by the European Union and the German Federal Ministry for Economy and Climate Protection within the framework of the economic stimulus package no. 35c in module b on the basis of a decision by the German Bundestag under the funding number 19S22006.

References

1. Hughes, T.J.R., Cottrell, J.A., Bazilevs, Y.: Isogeometric analysis: CAD, finite elements, NURBS, exact geometry, and mesh refinement. Comput. Methods Appl. Mech. Eng. **194**, 4135–4195 (2005)
2. Leidinger, L.: Explicit isogeometric B-rep analysis for nonlinear dynamic crash simulations. Ph.D thesis, Technical University of Munich, Germany (2020)
3. Leidinger, L., Hartmann, S., Benson, D., Nagy, A., Rorris, L., Chalkidis, I., Bauer, F.: Hybrid IGA/FEA vehicle crash simulations with trimmed NURBS-based shells in LS-DYNA. In: 13th European LS-DYNA Conference 2021, Ulm, Germany (2021)
4. Hartmann, S., et al.: Isogeometric analysis in LS-DYNA R13 - key steps towards industrial applications. In: 13th European LS-DYNA Conference 2021, Ulm, Germany (2021)
5. Leidinger, L., et al.: Enabling productive use of isogeometric shells in LS-DYNA. In: 14th European LS-DYNA Conference 2023, Baden-Baden, Germany (2023)
6. Bauer, F., Yugeng, T., Leidinger, L., Hartmann, S.: Experience with crash simulations using an IGA body in white. In: 14th European LS-DYNA Conference 2023, Baden-Baden, Germany (2023)
7. Rorris, L., Chalkidis, I.: Latest ANSA developments for IGA modeling. In: 14th European LS-DYNA Conference 2023, Baden-Baden, Germany (2023)
8. ANSYS: LS-DYNA Keyword User's Manual R15.0: Volume I (2024). https://lsdyna.ansys.com/manuals/
9. Benson, D.J., Bazilevs, Y., Hsu, M.C., Hughes, T.J.R.: Isogeometric shell analysis: the Reissner – Mindlin shell. Comput. Methods Appl. Mech. Eng. **199**, 276–289 (2010)
10. Hanssen, A.G., Olovsson, L., Porcaro, R., Langseth, M.: A large-scale finite element point-connector model for self-piercing rivet connections. Eur. J. Mech. A Solids **29**(4), pp. 484–495 (2010)
11. Bier, M., Sommer, S.: Simplified modeling of self-piercing riveted joints for crash simulation with a modified version of *CONSTRAINED_INTERPOLATION_SPOTWELD. In: 9th European LS-DYNA Conference (2013)

12. Rochel, P., Sommer, S., Olfert, V., Meschut, G.: Impact of production-related tolerances of the failure and deformation behaviour of mechanical joints under crash loading. Final report IGF-No. 20116 N. In: Forschung für die Praxis P 1262. Forschungs-vereinigung Stahlanwendung e. V., Düsseldorf (2021)
13. LS-PrePost. https://lsdyna.ansys.com/knowledge-base/ls-prepost/. Accessed 24 June 13

Open Access This chapter is licensed under the terms of the Creative Commons Attribution 4.0 International License (http://creativecommons.org/licenses/by/4.0/), which permits use, sharing, adaptation, distribution and reproduction in any medium or format, as long as you give appropriate credit to the original author(s) and the source, provide a link to the Creative Commons license and indicate if changes were made.

The images or other third party material in this chapter are included in the chapter's Creative Commons license, unless indicated otherwise in a credit line to the material. If material is not included in the chapter's Creative Commons license and your intended use is not permitted by statutory regulation or exceeds the permitted use, you will need to obtain permission directly from the copyright holder.

Convergence Studies to Compare the Induced Forming Defects in FE Based Simulations, Point Clouds and in Actual Formed Parts

Muhammad Saeed[1,2], Sheharyar Faisal[3(✉)], Eiman Nadeem[3(✉)], Markus Wagner[1(✉)], Boris Eisenbart[2(✉)], and Matthias Kreimeyer[1(✉)]

[1] University of Stuttgart, Pfaffenwaldring 9, 70569 Stuttgart, Germany
msaeed@swin.edu.au
[2] Swinburne University of Science and Technology, Melbourne, VIC 3122, Australia
[3] National University of Science and Technology, Islamabad 12 44000, Pakistan

Abstract. The automotive and aerospace industries have increasingly relied on Finite Element (FE) simulations to optimize their forming processes and reduce the costs and time associated with physical prototyping. However, ensuring the accuracy of these simulations in predicting actual forming defects remains a critical concern. To address this issue, a comprehensive convergence analysis, meticulously conducted, systematically compares the induced forming defects in FE-based simulations, point cloud representations, and actual formed parts. The study involves utilizing FE simulations to model various forming processes, capturing intricate details of formability, material-tool interactions, and varying key simulation parameters, such as material and process parameters. The study aimed to assess the sensitivity of the FE simulations to these parameters and to identify convergence criteria that lead to results closely resembling the actual formed parts and point cloud data. To achieve this, convergence studies were conducted to systematically vary these parameters and compare the simulated results against point cloud data obtained through advanced scanning technologies, providing a high-fidelity representation of the formed components. The study's comparative analysis included a detailed examination of common forming defects such as wrinkles, bridging, and gaps, employing quantitative metrics to measure the deviation between the simulated, scanned, and actual formed parts.

Keywords: Finite Element Analysis · Simulation · Point Cloud · Convergence · Forming Defects

1 Introduction

Inspecting and validating structural components, such as composite parts, is not just a crucial element of the manufacturing process, but an urgent necessity. If overlooked, it can result in a reduction in product quality and potential component rejection. Ensuring the structural integrity of manufactured components is particularly vital for large quantities of applications. This is a critical issue in modern manufacturing, demanding an effective solution. This paper delves into convergence studies of a defect detection framework proposed by Faisal, Saeed [1] and presents its validation through image-based analysis. With its direct and practical implications for your work, the framework is centred around image comparison techniques and accentuates structure analysis. It proposes a method for validating structural integrity using comparative images of ideal and defective surfaces, a practical approach that can significantly enhance your manufacturing process. The crux of the framework lies in techniques employing structural similarity index measures to generate a heat map that highlights critical areas within the image. Additionally, it advocates utilizing Structural Similarity Index Measures (SSIM) and robust image preprocessing to verify the structural similarity between two images and demonstrate the outcome.

2 Background

The Unidirectional (UD) composite laminate can be formed in numerous ways despite the challenges of forming and producing composite parts. These challenges, including manufacturing defects such as wrinkling, misalignment, kinks, and thickness variations, result in high-induced residual stresses [2] as shown in Fig. 1. An accurate forming model for composite-based tape laminates is essential in determining the material's formability. The draping order, dependent on the techniques of stacking tapes, significantly impacts the quality and defects in the final part. Furthermore, the draping strategy also determines the layup time and the formation of defects during the forming process [3].

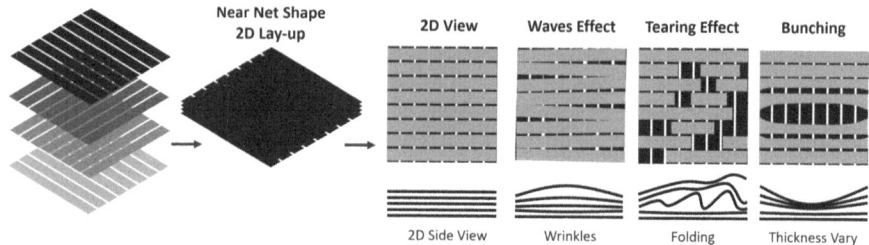

Fig. 1. Major defects occurring during the forming process

Dörr [4, 5] provides a thorough explanation of the intricate nature of stamp-forming processes, illustrating the various common defects that occur. These include void formation, crack propagation, instability during plastic deformation, material defects, and tape slippage in the composite. Wrinkling, tearing, necking, edge cracking, and bunching are just a few examples of the defects observed during deep drawing. The manufacturing processes often require numerous adjustments, achieved by varying parameters and design specifications, such as the tool's geometry and material selection [6].

In double diaphragm formation, void growth is closely linked to the hydrostatic tension created, leading to microscopic internal damage that eventually results in internal cracks [7]. Pre-existing void inclusions in the material before plastic deformation during forming and the formation of new voids even after plastic deformation are critical to note.

Understanding void growth and crack formation is vital for identifying, understanding, and analyzing the origins of failure. The sensitivity analysis for these defects, known as fractography, and formability analysis using scanning electron microscopy (SEM) are imperative [8].

2.1 Visual Inspections

In this research, all the parts are visually inspected as well to better understand defect patterns. The non-defected part does not contain wrinkles and buckling. Similarly, all the images are analyzed before executing them in the framework adopted in semi-automation and automation approaches. All the defective and non-defected parts are inspected, and their images are saved for comparative analysis. This research incorporates Image Quality Assessment (IQA), the transition from traditional metrics such as Mean Squared Error (MSE) and Peak Signal-to-Noise Ratio (PSNR) to perceptually oriented metrics like the Structural Similarity Index (SSIM) and Feature Similarity Index (FSIM) that represents a significant advancement [9, 10]. Although computationally straightforward, traditional metrics often need to capture perceptual quality more effectively than more advanced techniques. This discrepancy is well-documented in the comparative study by Sara, Akter [11] highlighting the superior performance of SSIM and FSIM in reflecting human visual perception. This research incorporates SSIM to evaluate and compare ideal and defective images, capitalizing on its capability to assess structural similarity, which is crucial for accurate defect detection. The methodology is further enhanced by integrating gradient magnitude maps, which refine the detection of subtle structural differences. This dual approach improves perceptual accuracy and provides a robust framework for visualizing defects. Moreover, the research includes pre-processing steps such as resizing and optional blurring, essential for normalizing image inputs and mitigating noise. These steps ensure that the comparison is not influenced by extraneous factors, thereby increasing the reliability of the results. This methodology proves particularly effective in industrial applications where precise and reliable quality control is critical. For instance, in manufacturing settings where defect detection is paramount, accurately identifying structural discrepancies can significantly enhance quality assurance processes. By utilizing SSIM and gradient-based techniques, this research offers a comprehensive and perceptually aligned assessment that surpasses traditional methods. This capability is precious in detecting subtle defects that might be overlooked with more straightforward metrics. The study by Wang [12] supports the effectiveness of SSIM over traditional metrics like MSE and PSNR, underscoring its alignment with human visual perception. The approach presented in this research builds on these findings, demonstrating practical applications in defect detection where accuracy and visual interpretability are essential. The dual approach of using SSIM and gradient magnitude maps ensures a more detailed and perceptually accurate comparison, enhancing the detection of crucial structural differences in quality assurance processes. Different methods of image processing have been used in the context for defect or critical region identification. However, many of these approaches assume [13, 14] images to be aligned. This poses an issue with industrial manufacturing where parts are often misaligned and have different scales.

3 Initial Evaluations

The proposed framework, which serves as a versatile validation tool and offers an alternative approach to our previous research [1], involves the comparison of two images: an ideal image of the structure and the actual defective image of the structure, as illustrated in Fig. 2. Initially, the framework preprocesses the image by ensuring equal dimensions and normalizing its intensity distribution. Additionally, it allows for vertical cropping based on specific requirements. To ensure adaptability across various scenarios, the framework utilizes a state-of-the-art image alignment algorithm that accommodates rotations, translations, and scaling transformations. This ensures image alignment by minimizing the mean squared error difference between both images, increasing their similarity.

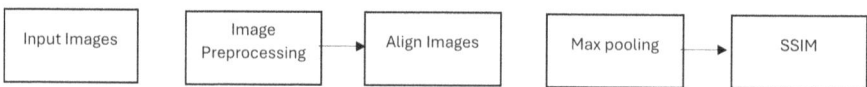

Fig. 2. Flowchart of the framework convergence for Comparative Analysis

The framework significantly enhances both images' pixel intensity and contrast by incorporating a max pooling layer. Equation 1 represents the standard formula used for max pooling in images. The max pooling layer takes advantage of bright pixels having a lower grey scale value. By downscaling the image while retaining the bright pixels, this layer ensures a substantial increase in contrast between different parts of the structure. This enhancement makes the defective part of the image more prominent, thereby boosting the framework's effectiveness in enhancing image quality.

$$P(i,j) = \max\{I(x,y) | x \in [i, i+k-1], y \in [j, j+k-1]\} - 1 \quad (1)$$

Here:
- $P(i,j)$ is the output value at each point (i,j) in the pooled image.
- $\max\{I(x,y)\}$ denotes the maximum Intensity (I) value at coordinates (x, y) within the window.
- For each point (i,j) in the pooled image, the expression is defined by local $k \times k$ neighborhood.

The selection of a comparison metric poses an important challenge in comparing different structures. Metrics such as mean square error, absolute error, or peak signal-to-noise ratio compare each pixel with the other without considering the structure or geometry of the image. To address this, the core comparison metric the approach relies on is the feature of SSIM, a full reference metric that checks the luminance, contrast, and structure in consideration of an image. This role of SSIM in image analysis is crucial, as it allows a metric to be better suited for cases involving the analysis of different structures and components. Equation 2 shows the formula to compute the SSIM of each pixel in an image.

$$SSIM(x,y) = \frac{(2u_x u_y + c_1)(2\sigma_{xy} + c_2)}{(u_x^2 + u_y^2 + c_1)(\sigma_x^2 + \sigma_y^2 + c_2)} - 2 \quad (2)$$

Here:

- u_x and u_y are the sample mean of the images (x, y) within the window
- σ_x^2 and σ_y^2 is the variance of (x, y) within the window
- σ_{xy} is the covariance of (x, y)
- C_1 and C_2 are the small constants added to avoid 0 division.

To validate the approach stated in [1], we introduced an additional image preprocessing function before the SSIM function. This function performs another layer of preprocessing, masking the defective image of the ideal and extracting a new image representing the region of the ideal image overlaid by the mask. The key step is the comparison of these two resulting images using the SSIM function. The obtained percentages, along with their corresponding pixel locations, are then analyzed to determine the percentage accuracy of the defect detection framework [1].

4 Results

First, the framework compares the ideal, predicted and original defective images. This involves applying Gaussian Blur to the results obtained from both images. Subsequently, the processed images are further evaluated by comparing their similarity, explicitly focusing on the defective pixel locations. The entirety of this process is visually depicted in Fig. 3. The results using the approach mentioned in [1] are compared with the ideal point cloud using the SSIM approach. Simultaneously the ideal point cloud is directly compared with the defected point cloud using the same approach. Comparing these two results indicate the accuracy of [1]. This approach has proven to be an effective method to identify potential critical regions in manufactured parts using SSIM.

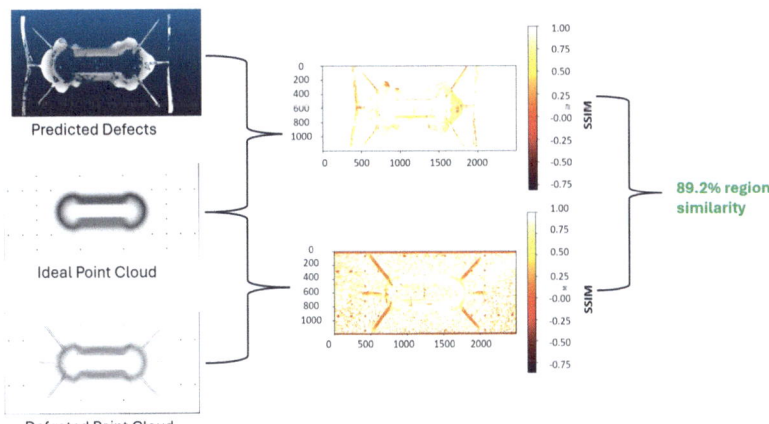

Fig. 3. Overall Framework Process for the Convergence

The output consists of two main components: a Structural Similarity Index (SSIM) metric and a heatmap representation of the SSIM map, visually depicting the similarities between images. The result for a specific case can be observed in Fig. 4. The output indicates a high regional similarity of 88.92%, highlighting the effectiveness of the proposed framework, which achieves an similarity of 88.92%. A higher similarity metric indicates a high accuracy of the framework [1] in predicting defects. This also validates the SSIM approach's accuracy in identifying the critical region.

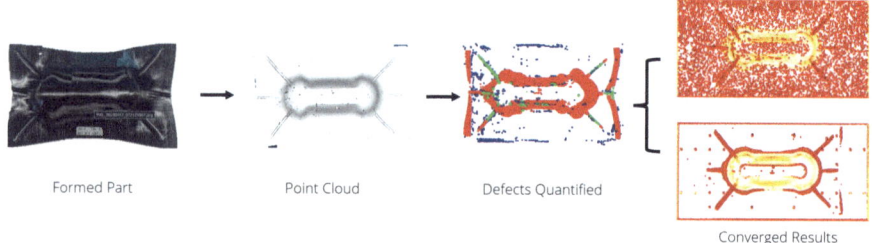

Fig. 4. Comparative Results for Predicted Defects, Actual Defects and Converged Defects

The results for Case 2 are shown in Fig. 5. The output depicts 78.6% similarity between the predicted results obtained using [1] and the output of the proposed SSIM based approach. The accuracy of the results is highly dependent on the preprocessing and post processing techniques applied to each image to identify potential defects at a pixel level. The results show that image comparison, identifies pixel level differences whereas [14] relies on clusters of pixels and hence overlook scattered pixels.

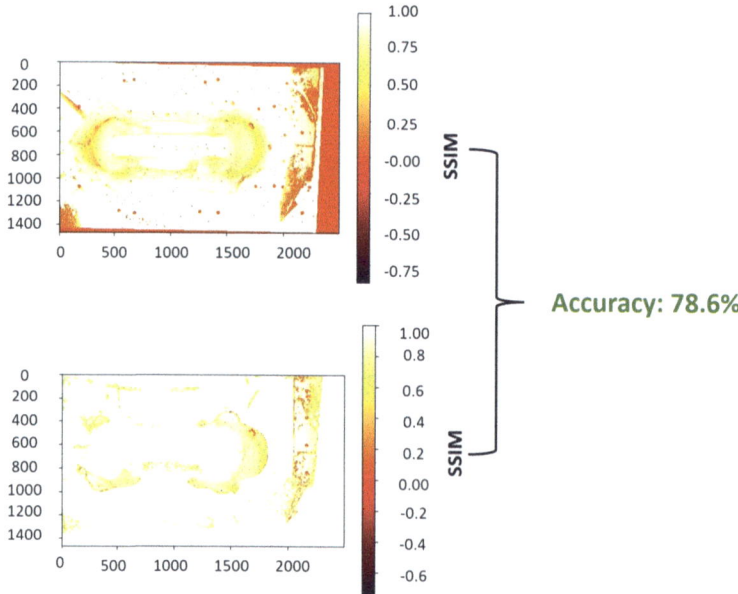

Fig. 5. Structural Similarity Index for quantified Defects

5 Conclusion and Future Outlook

This framework is valuable for conducting quality inspections of composite parts using images. It was verified in a study [1], by demonstrating an accuracy of over 70% in various scenarios. In Fig. 4, the integration of both frameworks is illustrated, illustrating how the point cloud of the actual geometry is passed through the defect detection framework, with the results then corroborated using the convergence framework. The convergence framework produces an SSIM percentage and a heat map, offering a visual tool for identifying critical regions in the defects. Future research will explore different variations of SSIM, such as CWSSIM, to achieve more robust results for image transformation. Additionally, efforts will be made to enhance image alignment between different structures and optimize the preprocessing sub-routine, ensuring improved accuracy and efficiency.

Acknowledgement. The authors acknowledge the support of the Global Innovation Linkage (GIL) grant awarded by the Australian Federal Government and the Institute for Engineering design and industrial design (IKTD) at Stuttgart University, the Swinburne University of Technology and Students from National University of Science and Technology.

References

1. Saeed, M.S., et al.: A novel heuristic approach to detect induced forming defects using point cloud scans. Proc. Des. Soc. **4**, 723–734 (2024). https://doi.org/10.1017/pds.2024.75
2. Saeed, M.S., et al.: Parameter study and optimization of forming simulations for tape-based fiber layups. In: Kiefl, N., Wulle, F., Ackermann, C., Holder, D. (eds.) SCAP 2022. ARENA2036, pp. 266–281. Springer, Cham (2023). https://doi.org/10.1007/978-3-031-27933-1_25
3. Belnoue, J.P.-H., et al.: Understanding and predicting defect formation in automated fibre placement pre-preg laminates **102**, 196–206 (2017). 1359-835X. https://doi.org/10.1016/j.compositesa.2017.08.008
4. Dörr, D.: Simulation of the thermoforming process of UD fiber-reinforced thermoplastic tape laminates. Karlsruhe Institute of Technology, Karlsruhe (2019)
5. Bártová, B., Bína, V.: Early defect detection using clustering algorithms. AOP **27**(1), 3–20 (2019). https://doi.org/10.18267/j.aop.613
6. Margossian, A., Dumont, F., Beier, U.: Validation of macroscopic forming simulations of a unidirectional pre-impregnated material through optical measurements. KEM **554–557**, 465–471 (2013). https://doi.org/10.4028/www.scientific.net/KEM.554-557.465
7. Yu, F., Chen, S., Harper, L.T., Warrior, N.A.: Investigation into the effects of inter-ply sliding during double diaphragm forming for multi-layered biaxial non-crimp fabrics. Compos. A Appl. Sci. Manuf. **150**, 106611 (2021). https://doi.org/10.1016/j.compositesa.2021.106611
8. Hussain, G., Al-Ghamdi, K.A., Khalatbari, H., Iqbal, A., Hashemipour, M.: Forming parameters and forming defects in incremental forming process: Part B. Mater. Manuf. Processes **29**(4), 454–460 (2014). https://doi.org/10.1080/10426914.2014.880457
9. Jovančević, I., Pham, H.-H., Orteu, J.-J., Gilblas, R., Harvent, J., Maurice, X., Brèthes, L.: 3D point cloud analysis for detection and characterization of defects on airplane exterior surface. J. Nondestruct. Eval. **36**(4), 1–17 (2017). https://doi.org/10.1007/s10921-017-0453-1

10. Yang, Z., et al.: Image classification for automobile pipe joints surface defect detection using wavelet decomposition and convolutional neural network. IEEE Access **10**, 77191–77204 (2022). https://doi.org/10.1109/ACCESS.2022.3178380
11. Sara, U., Akter, M., Uddin, M.S.: Image quality assessment through FSIM, SSIM, MSE and PSNR—a comparative study. JCC **07**(03), 8–18 (2019). https://doi.org/10.4236/jcc.2019.73002
12. Wang, Z., Bovik, A.C., Sheikh, H.R., Simoncelli, E.P.: Image quality assessment: from error visibility to structural similarity. IEEE Trans. Image Process. Publ. IEEE Signal Process. Soc. **13**(4), 600–612 (2004). https://doi.org/10.1109/tip.2003.819861
13. Bahadure, N.B., Ray, A.K., Thethi, H.P.: Image analysis for MRI based brain tumor detection and feature extraction using biologically inspired BWT and SVM. Int. J. Biomed. Imaging **2017**, 9749108 (2017). https://doi.org/10.1155/2017/9749108
14. Wang, G., Warren Liao, T.: Automatic identification of different types of welding defects in radiographic images. NDT E Int. **35**(8), 519–528 (2002). https://doi.org/10.1016/S0963-8695(02)00025-7

Open Access This chapter is licensed under the terms of the Creative Commons Attribution 4.0 International License (http://creativecommons.org/licenses/by/4.0/), which permits use, sharing, adaptation, distribution and reproduction in any medium or format, as long as you give appropriate credit to the original author(s) and the source, provide a link to the Creative Commons license and indicate if changes were made.

The images or other third party material in this chapter are included in the chapter's Creative Commons license, unless indicated otherwise in a credit line to the material. If material is not included in the chapter's Creative Commons license and your intended use is not permitted by statutory regulation or exceeds the permitted use, you will need to obtain permission directly from the copyright holder.

Novel Slicing Algorithm for Hybrid Manufacturing on Non-planar Surfaces with Robotic SEAM

Nicolas Unger[✉], Pradnil Kamble, and Timo Huse

German Aerospace Center (DLR e.V.) Institute of Vehicle Concepts, Pfaffenwaldring 38-40, 70569 Stuttgart, Germany
nicolas.unger@dlr.de

Abstract. Combining conventional with additive manufacturing processes opens up new possibilities for the flexible, economical production of complex automotive components and structures with varying geometries. This hybrid manufacturing approach can be realized by 3D-Printing functional structures onto pre-existing parts and surfaces with Robotic Screw Extrusion Additive Manufacturing (RSEAM). RSEAM combines the dexterity of industrial robotic arms with the productivity and extensive material range of a screw extruder to provide a highly flexible additive manufacturing process. This paper addresses a critical challenge in hybrid manufacturing: creating suitable manufacturing programs for printing on irregular, non-planar surfaces, that deviate significantly from ideal CAD input. Deviations are classified as macro or micro based on their magnitude relative to the desired printing layer height threshold. We propose a slicing and path-planning algorithm that adapts ideal paths to scanned surfaces and generates optimized trajectories to handle both macro and micro deviations via distinct methods. Macro deviations are handled by offsetting the scanned surface to create layers, smoothing to remove micro deviations and morphing each layer from the ideal CAD onto them. Micro deviations are then accounted for in the first layer by calculating the local non-uniform layer height and locally adjusting the robot velocity. The resulting smooth robot trajectories ensure accurate local bead geometry and correct over-all part geometry adaptation. The algorithm's output is validated and optimized within a digital twin of the manufacturing system to avoid collisions and singularities. The implemented algorithms are validated and demonstrated on a real-world automotive application.

Keywords: Hybrid Part Manufacturing · Non-Planar Additive Manufacturing · Robotic Screw Extrusion Additive Manufacturing · Adaptive Slicing Algorithm · Non-Planar Slicing Algorithm · Robotic Trajectory Planning · Digital Twin Simulation · Automotive · Rhino Grasshopper

1 Introduction

Additive Manufacturing (AM) has revolutionized the way complex geometries are fabricated by enabling layer-by-layer material deposition to produce parts directly from CAD data [10]. In traditional AM, the process typically starts with finding a fixed orientation for an input geometry which facilitates the slicing of geometry to obtain successive curves as deposition paths. The most common slicing algorithm is planar uniform slicing, where the geometry is sliced with successive equidistant planes along the machine Z+ axis. The distance between these planes is the desired layer height. Multiple algorithms to further optimize planar uniform slicing have been researched [4,19]. However, a significant limitation of this way of slicing is revealed when dealing with tilted or non planar surfaces. These limitations cause a staircase effect, which degrades the surface finish of the product [15] and introduces overhangs, which require the addition of supporting structures [8].

The convergence of robotics and AM in recent years have transformed the landscape of production with AM [26,27]. Robots provide additional degree-of-freedom (DOF) which enables flexibility to vary the orientation of the applicator. This allows the robot to fabricate hybrid parts by printing on pre-existing components. Moreover, this can also help in eliminating the need of supporting structures [5]. One approach to robot assisted additive manufacturing is Robotic Screw Extrusion Additive Manufacturing (RSEAM).

RSEAM utilizes the flexibility of industrial robots to position and operate an extrusion-based additive manufacturing module. This module, a small single-screw extruder, is used for plastification of injection-molding granules and direct extrusion of technical and high-temperature thermoplastics, with the possibility of high reinforcement contents. In addition to diverse and cost-effective material availability, RSEAM also addresses several constraints inherent to conventional AM including limitations with part size, material properties, and production speed [13,22,24]. Due to these features, RSEAM can be utilized to manufacture complex structures onto pre-existing surfaces to produce hybrid parts, e.g. in Automotive [13,14,25].

Despite these advantages, the success of hybrid part manufacturing depends strongly on slicing algorithms that direct the robots effectively over non-planar surfaces. Moreover, the surfaces and parts designated to printing are often non-ideal, deviating from nominal CAD data or having surface defects. This poses significant challenges in accurately translating paths generated by slicing the nominal CAD to the actual components. While strategies to obtain non-planar slices and manufacturing on non-planar surfaces has been discussed in the literature, strategies for handling deviations or surface irregularities have been marginally studied. This paper introduces a novel slicing algorithm specifically developed to enhance robot assisted additive manufacturing capabilities, particularly for handling non-planar surfaces that deviate significantly from CAD.

The developed adaptive slicing algorithm adapts ideal paths to scanned surfaces and generates optimal trajectories handling macro and micro deviations. Further, the algorithm is demonstrated with an automotive use case. The algorithm is developed within Grasshopper, a parametric programming plugin within Rhinoceros 3D [2]. The generated trajectories are then checked and optimized for collisions within a digital twin of the production cell in RoboDK [1].

2 Related Work

Additive manufacturing (AM) has undergone significant advancements, facilitated by the advancements of pre-processing steps aimed at enhancing geometrical accuracy and manufacturing efficiency. Pre-processing in additive manufacturing majorly include slicing algorithms and trajectory planning. While planar slicing algorithms are commonly used for conventional AM, non-planar algorithms can be adopted with robot assisted AM [3].

Challenges of Non-planar Slicing.

Hardware constraints The application of non-planar slicing requires more than 3DOF for the tool head to perform. This results in limitations or adaptability requirements for conventional 3D printers which do not have the DOF required to execute non-planar effectively [17,21]

Algorithmic complexity The complexity of developing slicing algorithms that efficiently manage non-ideal, free-form surfaces is high, requiring sophisticated computational models and significant processing power [5–7]

Collisions During tool path generation for non-planar slicing, since the high DOF paths and additively produced profiles vary, it becomes essential to generate paths that consider collisions between applicator and previously deposited material [7]

Non-Ideal surface and geometries Achieving consistent quality and mechanical properties across various setups and materials using non-planar slicing is challenging due to the increased likelihood of defects if not managed correctly.

Previous Work to Overcome the Challenges. The hardware limitations can be resolved by integrating robotic systems in AM processes increasing the flexibility of movement, essential for effective non-planar printing. In fact, utilization of a robot or external axis setup can also eliminate the need for support structures. Dai et al. [5] suggest a multi-facet convex front based algorithm for robot assisted multi-axis printing of support free structures. Their algorithm simplifies the process in two steps: extracting curved surface layers from the volume and then applying curved tool-paths to these layers. This strategy reveals the effectiveness of robotic AM to produce complex parts without support structures.

The algorithmic complexity of non-planar slicing is often high, which is why most commercial slicing software do not support non-planar and non-uniform slicing. Lettori et al. [15] present a review of multiple works related to representation of geometries and algorithms with complexity for multi-axial deposition in robot assisted AM.

Fortunato et al. [9] present a slicing algorithm that combines both planar and non-planar layers, which has been evaluated by printing on a complex substrate using a 5DOF robot-based extrusion AM system. A floating 2D plane, composed of printing curves, is projected onto the non-planar surface. This projection effectively captures the 3D characteristics of the surface, creating a point cloud. The point cloud can then be decomposed to extract point data with the correct Z component for further processing.

Another comparable method is presented by Li et al. [16]. They introduce a study where RSEAM is employed to manufacture support-less twisted hollow tubes, arc-like thin shells and honey comb infill patterns on a curved surface. Their strategy is to use the DOF of the robot to constantly pivot the nozzle orthogonal to the deposition surface to produce non-planar layers. Doing so results in better structural integrity of the produced parts with RSEAM.

An alternative method to create non-planar layers in a CAD model can be by offsetting the bottom surface facets of an STL model along their normal directions to form a slicing surface. This offsetting process is sequentially repeated from one offset surface to the next, allowing the entire model to be sliced into curved layers [12].

Furthermore a volume decomposition based curved surface non-planar slicing algorithm is proposed by Zhao et al. [27]. Their algorithm divides the part into volumes using concave loops and then offsetting the original surface to slice them. This generates curved slices for each volume. Tool-paths are then generated on the trimmed curved surfaces with paths aligned to the normal vector of the curved surfaces.

There are multiple non-planar algorithms [15] designed for free-form surfaces that do not specifically consider manufacturing on a non-planar substrate. However, they are notable for their novel strategies in searching for non-planar deposition layers. Some noteworthy approaches used are volume decomposition and multi-planar slicing [20], as well as convex-fronts advancing[5].

Additionally, path planning on non-planar substrates extends beyond AM to include robotic applications like welding, subtractive manufacturing etc. In robotic welding, vision sensors [18, 23] are employed to extract joint features and volumes. Similarly, Guo et al. [11] present a vision-guided method for subtractive processing on curved substrate, utilizing NURBS curves to define tool paths.

3 Problem Statement

The aforementioned approaches have crucial limitations when it comes to printing on pre-existing parts or substrates. These substrates, especially ones with large dimensions, often have significant deviations from their nominal CAD geometry. Significant means, that if printing paths are sliced based on the nominal geometry, manufacturing errors, such as collisions, will occur and the print will fail.

The presented research addresses this challenge: slicing and path generation for manufacturing on pre-existing surfaces that deviate from the nominal CAD geometry. The algorithm and method should handle large surface deviations as well as small surface defects. The geometry to be printed should automatically be adapted to the scanned pre-existing surface, without the need of designing a customized conformal geometry. In addition, the capabilities of the industrial robot positioning system should be exploited and it's limitations considered. It should be noted that the deposition rate should be maintained consistent, and the trajectories are adjusted to achieve the desired bead geometry.

4 Methodology

This section describes how the printing paths are created by the novel adaptive slicing algorithm based on a scan of the real printing substrate and subsequently checked and optimized in the digital twin of the manufacturing system.

4.1 Adaptive Slicing Algorithm for Non-Planar Robotic Hybrid Manufacturing

In accordance with the problem statement, the slicing method needs to fulfill the following requirements:

- Scanning of pre-existing surface as a substrate for printing
- Non-planar slicing of nominal CAD geometry
- Automatic adaption from the nominal CAD to the actual substrate
- Handling of small and large deviations between scan and CAD
- Creation of printing paths in accordance with the capabilities and limitations of the manufacturing system
- Output path positions, tool orientations and velocities

First, it is necessary to understand the deviations, that occur between a real part and the nominal CAD geometry. Figure 1 shows an example of a nominal CAD surface and a real surface (Sub), each with 2 layers that are offset uniformly by one layer height (L1 and L2). In the upper half, the deviations of the real surface have a magnitude below one layer height. This is what we consider "Micro" deviations. To print on a surface with only micro deviations it is possible to use the layers, which are planned based on the nominal CAD without causing a collision between the tool and the substrate.

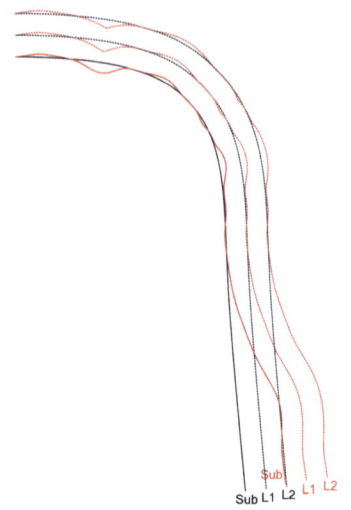

Fig. 1. Comparison of ideal CAD (black) and scanned part (red) with substrate (Sub), Layer1 (L1), Layer2 (L2)

In contrast to that are deviations with a magnitude larger than the desired layer height, which we consider "Macro". These can be identified in Fig. 1 where the scanned substrate line (red Sub) crosses the Layer 1 line of the CAD (black L1). At this point, a collision would occur between the tool and substrate, therefor a adaption of the tool path is mandatory.

Figure 2 shows a flow diagram of the proposed slicing method. The necessary steps are explained below.

Scan, Mesh Creation, Deviation Calculation To begin, the actual substrate surface needs to be scanned and the point cloud is cleaned, tessellated and optimized to acquire the substrate mesh. This mesh is oriented with a best-fit algorithm to best match the nominal CAD substrate. Subsequently the surface deviations can be calculated. These steps are taken in a commercial 3D-scanning software. Based on these deviations and their magnitude with respect to the desired layer height the following path is decided.

The subsequent slicing algorithm is implemented in the CAD software Rhinoceros and it's parametric programming tool Grasshopper.

Slicing Based on CAD. The slicing principle used in this method is similar to Huang et al. [12], however, it is performed on NURBS instead of a STL mesh. The substrate surface is selected and offset by the desired layer height, which is set by the user, to obtain the next layer surface. The layer surface is then intersected with the geometry to be printed to obtain the non-planar section curves. This process is repeated until the intersection between layer surface and geometry returns no result, signalling that the whole part was sliced. The whole

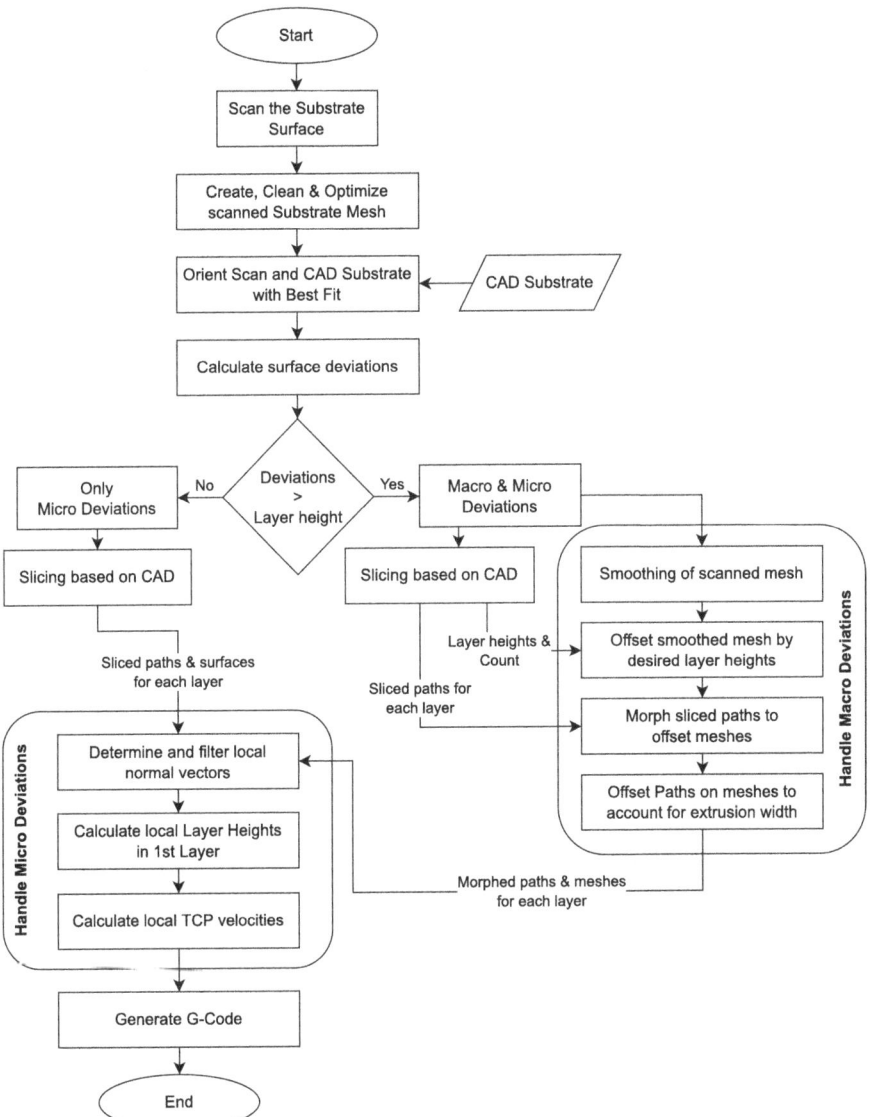

Fig. 2. Slicing Method

algorithm is fully parametric and also allows a variation of layer heights between layers.

Handle Macro Deviations. If the calculated deviations between CAD and scanned mesh exceed the desired layer height, the existing macro deviations need to be handled, before micro deviations can be addressed.

Smoothing and Cleanup of scanned Mesh. First, the scanned mesh from the 3D-scanner is re-meshed, smoothed and subsequently rebuilt to remove errors such as non-uniform normals or welding errors. The mesh smoothing step is crucial for the manufacturability, since it removes micro deviations and prevents them from propagating through subsequent layers and, more importantly, smooths the surface normals. This reduces the necessary tool reorientations and allows for higher quality deposition with the desired velocity and without sudden jerks. A comparison of the surface normals of the original scanned surface and after smoothing are displayed in Fig. 4(a).

Offset Mesh by Layer Heights. In the next step, the mesh surface is offset by the desired layer height and rebuilt again to ensure mesh integrity. The number of iterations of this step, as well as the layer height for each offset are received from the CAD slicing. This way each CAD slice has one matching offset mesh surface.

Morph sliced Paths to offset Meshes. The section curves from CAD slicing are now pulled towards their respective mesh layers with the "Pull to Mesh" Command. This command gets a polyline approximation of the input curve and moves its control points to the closest point on the mesh. It then connects the control points over the mesh edges, creating a polyline [2]. Then, the start points of each polyline are aligned iteratively to minimize travel moves between layers. This reduces manufacturing time and the chance of collisions. In addition, the polyline directions are aligned to avoid for the robot to perform 180° turns on layer changes, which would lead to velocity drops and thereby poor print quality.

Offset Curve on Mesh Now, the section curves need to be offset on their respective mesh layer towards the inside of the part by half of the desired bead width to ensure dimensional accuracy in the printed layer. Since no function for this exists in Grasshopper, an algorithm was developed to perform the offset of a curve on a mesh. The algorithm takes an incremental approach to offset the curve in multiple steps in a loop to ensure high accuracy and avoid errors. The offset distance, number of increments used and the curve approximation tolerance are parameters of the function. Figure 3 shows the Grasshopper implementation of a single Offset increment.

In each increment, first the section curves are converted to polylines with the specified tolerance and split into individual segments. The endpoints of each segment are projected to the closest point on the respective layer mesh, where the normal vector is evaluated. By calculating the cross product of the segment tangent vector and the mesh normal vector, the offset vector for each endpoint is calculated. Now, each endpoint is offset by the desired distance of the increment in the direction of its specific offset vector. Next, the line segments are rebuilt by their endpoints, connected and joined to a continuous curve and cleaned up. Finally, the curve is pulled back onto the mesh to complete the increment.

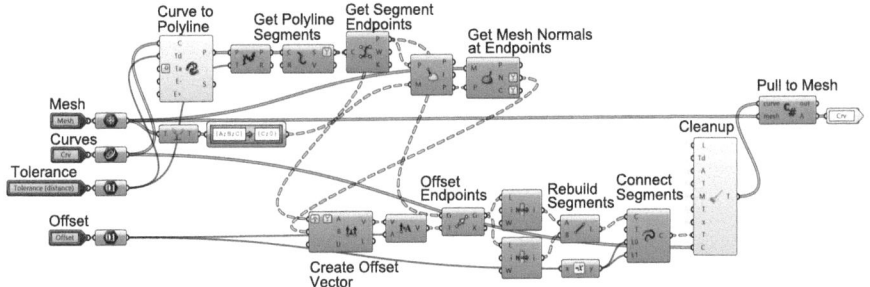

Fig. 3. Grasshopper Implementation of Algorithm to Offset Curve on Mesh

Handle Micro Deviations. The micro deviations are handled in the first layer printed on the substrate by adjusting the amount of extruded material at each point to it's respective local layer height to achieve the desired uniform bead width. The following steps take the sliced paths and surfaces or the morphed paths and meshes as input. The procedure is similar for both cases.

Determine Local Tool Normal Vectors. The provided curves are converted into polylines with a distance and angle tolerance as well as a minimum edge length. The last parameter is especially important because the spacing of the points has a direct impact on the manufacturability. Having the points too close together results in jerks and velocity drops by the robot. If they are too far apart, information about orientations or necessary adjustments in velocity is too sparse. The vertices of the polylines are projected to the closest point on their respective layer mesh/surface, where the normal vector is evaluated. To comply with restrictions of the manufacturing system the normal vectors are filtered to not exceed a user specified angle between the tool normal and the vertical direction.

Local Layer Heights and Velocities. To determine the local layer height, each vertex is projected along its inverted normal vector onto the previous layer surface/mesh. The distance between the projected point and its original vertex is calculated, resulting in the local layer height LH_{local}. From this and the nominal layer height LH_{nominal} and velocity v_{nominal}, which are set in the slicer, the local Tool Center Point (TCP) Velocity v_{local} is calculated via Eq. 1. The deviations of LH_{local} with respect to LH_{nominal} are illustrated in Fig. 4. This shows, that micro surface deviations are handled in the first layer, as in the second layer LH_{local} closely matches the LH_{nominal}.

$$v_{\text{local}} = \frac{LH_{\text{nominal}}}{LH_{\text{local}}} * v_{\text{nominal}} \qquad (1)$$

G-Code Generation. In the last step, the previously created paths, normals and velocities are combined to a G-Code File. All printing moves are labeled G1 and all travel moves with G0 as per G-Code standard. The comments, Layer

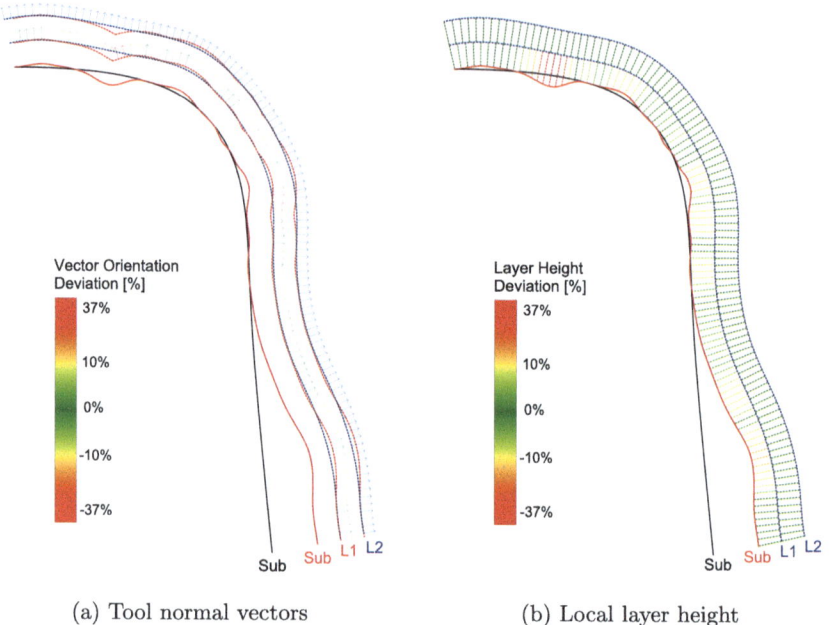

Fig. 4. Deviation of normal vectors and layer heights: (a) Vectors of original surface on L1 (color graded by deviation from smoothed) and smoothed vectors on L2 (blue), (b) layer height deviation with respect to nominal layer height

and Curve, are automatically generated within G-Code to track the respective layer and curve progressions. Such annotations facilitate targeted simulations and analyses of individual layers. One additional safety feature was implemented, which is to mark long travel moves in the G-Code with a M300 command to be considered during program testing in the digital twin. These often require the manual addition of auxiliary points to ensure a safe transition. The example code given in Listing 1 shows two different layer changes with travel moves. The change to Layer 18 is considered to be safe since the distance between the point in Line 2 and the next point in Line 5 is below the threshold set by the user. The Layer change to Layer 19 however requires a long distance travel move with significant reorientation, so it is marked for inspection with the M300 command.

```
1  ...
2  G1 X280.4 Y75.2 Z197.8 I−0.002 J−0.125 K0.992 F1200
3  ;Layer 18
4  ;Curve 0 in Layer 18
5  G0 X280.4 Y75.7 Z198.4 I−0.003 J−0.123 K0.992 F6000
6  G1 X280.4 Y76.4 Z198.5 I−0.004 J−0.123 K0.992 F1200
7  ...
8  G1 X280.4 Y75.7 Z198.4 I−0.003 J−0.123 K0.992 F1200
9  ;Layer 19
10 ;Curve 0 in Layer 19
11 M300
12 G0 X187.1 Y76.9 Z55.6  I−0.994 J0.022  K0.105 F6000
13 G1 X186.9 Y77.2 Z54.3  I−0.995 J0.021  K0.101 F1200
14 ...
```

Listing 1. G-Code Example

Once the data of normals and vectors are embedded in G-Code, the next critical step is to simulate the robotic trajectories that will lead to the manufacturing program.

4.2 Digital Twin Simulation

To confidently generate the manufacturing program, a digital twin simulation is required to validate the robotic trajectories.

Digital Twin Setup. Given the G-Code data, a digital twin of the production line is used to assess the feasibility of reaching the target locations. The purpose of the digital twin is to mirror the behavior of the physical twin, ensuring an accurate representation of the production environment. For non-planar operations with RSEAM the goals of the digital twin simulation include:

- Collision detection
- Further optimization of paths for travel moves
- Orientation and configuration optimization
- Component placement as a function of robots envelope and safe workspace
- Generation of manufacturing programs

The SpaceA Digital Twin (DT), as depicted in Fig. 5, is a calibrated digital replica of the SpaceA system developed within RoboDK. SpaceA DT includes each cell component, scenes, workpiece and cell confinement. It features two high-accuracy KUKA robots, each with 6 DOF, named Bender and Flexo. Additionally, the setup includes a 2 DOF build platform. Each KUKA robot is equipped with a screw extruder, serving as the tool for the SEAM process. The workpiece and the clamping setup, if necessary, are imported from CAD. Prior to scanning the actual part surface, a proper placement of the workpiece is validated within the digital twin with the nominal CAD file for robots reachability. The inclusion of all mentioned elements in the digital twin allows for a comprehensive simulation of the production process to meet the listed goals.

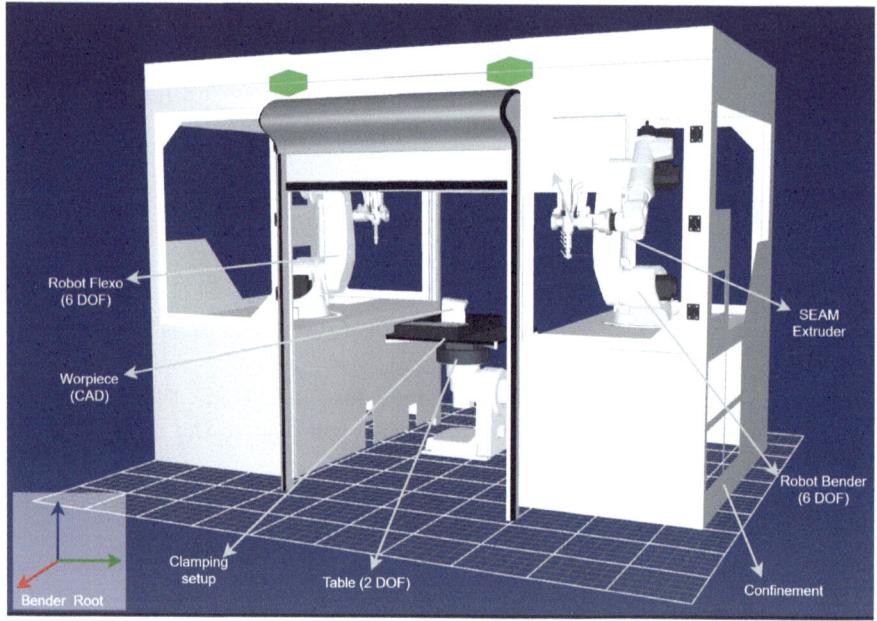

Fig. 5. SpaceA Digital Twin

Simulation Process. The simulation process begins with importing the CAD file of the scanned surface onto which the manufacturing will take place, along with the clamping setup. For the following example, the robot selected is Bender. The active base for manufacturing is the reference base set during the scanning phase, which is duplicated within the digital twin with respect to the robots fixed coordinate system i.e. Bender Root. Once the part is positioned according to the actual setup, the feasibility of the targets derived from the G-Code is assessed.

This assessment is conducted by importing the G-Code into a RoboDK Robot Machining Project. The reference base of the imported surface is selected as the active base for the machining project, while the active TCP is set to the calibrated center point of the extruder tool. This ensures that the robot's movements are accurately aligned with the real production setup.

After successfully importing and compilation of the simulation program, the path feasibility is checked, and the simulation is initiated. The simulation visualizes the active tool tracing the paths on the scanned surface. Figure 6 and Fig. 7 illustrate two potential collisions that may occur in this scenario.

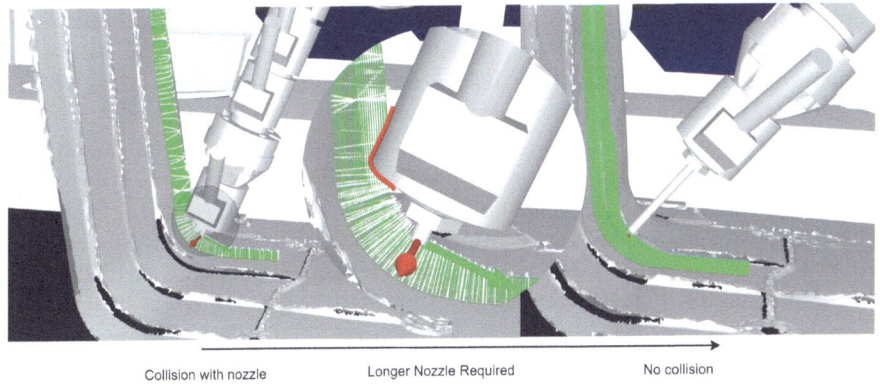

Fig. 6. Nozzle Size Selection

The collision displayed in Fig. 6 is not detected by the RoboDK collision checker module, as this collision occurs between the nozzle and the deposited material. RoboDK does not simulate material deposition and hence cannot compute collisions involving it, however, it can display the paths. The robot paths display the target points and the interpolation between them as embedded within the G-Code. If there is a possible collision between the tool and the previous robot paths, it will likely indicate a collision between the nozzle and the deposited material. It is evident from Fig. 6 that one solution to such a collision can be choosing a different applicator, such as a longer nozzle. However, the collision shown in Fig. 7 requires further strategizing.

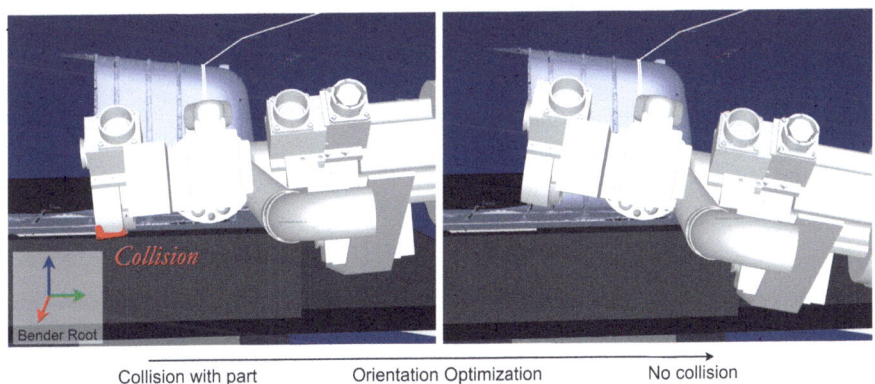

Fig. 7. Part and tool collision

The solution for collision displayed in Fig. 7 is investigated by developing an iterative collision check and orientation optimization algorithm. This algorithm iteratively checks for collisions within the generated simulation program and tries to tune the tool orientation. We choose tuning of the orientation around the Z-axis of the robot, which is the value A from the KUKA XYZABC pose parameters. Since the Z-axis typically represents the rotational axis of the tool's end effector. By adjusting the rotation around this axis, the tool can be reoriented to avoid collision without altering its position relative to the workpiece. This is with consideration to preserve position and orientation for orthogonally depositing the material on the substrate while being adjusted to fit within tight spaces. Adjustments in any other degrees of freedom can be more complex to handle especially in the need of high reorientations tasks with RSEAM. The implementation developed within RoboDK Python API is displayed in Algorithm 1.

Algorithm 1. Iterative Collision Check and Orientation Optimization Algorithm

1: Fetch program, active bases, and tool using `Robolink.Connect`
2: Trigger collision check `Robolink.Collisions()`
3: **for** each instruction in the program **do**
4:     Move the robot linearly `robot.MoveL`
5:     **if** collision detected **then**
6:         Fetch current pose and convert to KUKA XYZABC `Pose_2_KUKA`
7:         **if** previous dA (optimized rotation around Z) exists **then**
8:             Set new A to dA and check collision
9:             **if** collision avoided **then**
10:                 Collision flag off, jump to line 14
11:             **end if**
12:         **end if**
13:         Tune Z rotation (A) in small steps to find new dA
14:         **if** collision flag is off for a certain dA **then**
15:             Update instruction and dA `prog.setInstruction(ins_id, args)`
16:             Print note for later cross-validation
17:         **else**
18:             Raise exception with details if collision persists
19:         **end if**
20:     **end if**
21: **end for**

Additive manufacturing on existing surfaces often requires moving around the workpiece with non-printing travel moves without collisions. Tagging these travel moves beforehand with M300 in the G-Code helps in locating them within the simulation program. Within the RoboDK user interface, the simulation program can be overwritten or updated with additional joint and linear moves to produce auxiliary points for safe transitions. Figure 8 shows the overall simulated path to produce the desired layers on the surface, including other target coordinate systems for continuous travel moves.

Once the complete simulation is validated, a deployable manufacturing program is generated with the post processor code that translates the robot motions into Spline motion commands in KUKA Robot Language (KRL). Overall, simulation within SpaceA DT not only ensures issues are identified and addressed before physical implementation, but also helps to generate ready to deploy programs, thereby optimizing the manufacturing process and reducing the risk of errors.

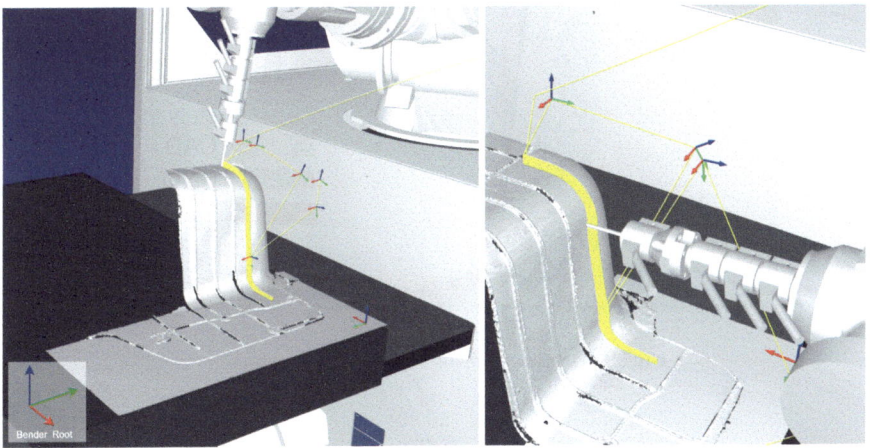

Fig. 8. Manufacturing program simulation

5 Validation and Demonstration

To validate the adaptive non-planar slicing method proposed in the methodology, an experiment was conducted on the SpaceA RSEAM system of the DLR Institute of Vehicle Concepts located in the ARENA2036 research campus. The objective was to assess the practical viability of the slicing method by manufacturing a rib on a physical non-ideal substrate with Micro and Macro deviations. The rib was printed with polypropylene granules with 40 Vol.-% glass short fibre content on a 3D organo-sheet using the manufacturing program developed through the outlined digital process chain. The substrate was scanned with the optical 3D-scanner ATOS Core 300 by Carl Zeiss GOM Metrology GmbH and the data was cleaned and tessellated into an STL mesh in the ZEISS INSPECT Optical 3D software. Figure 9 shows the nominal CAD and the scanned surface with the adapted paths, demonstrating the functionality of the novel slicing algorithm.

The manufacturing program generated within the digital twin of the SpaceA production line was tested for physical implementation. Since the nozzle length was increased compared to the setup earlier shown in Fig. 7, the algorithm (Algorithm 1) for iterative collision check and orientation optimization was bypassed due to no collision triggers. Without any changes or further modifications, the

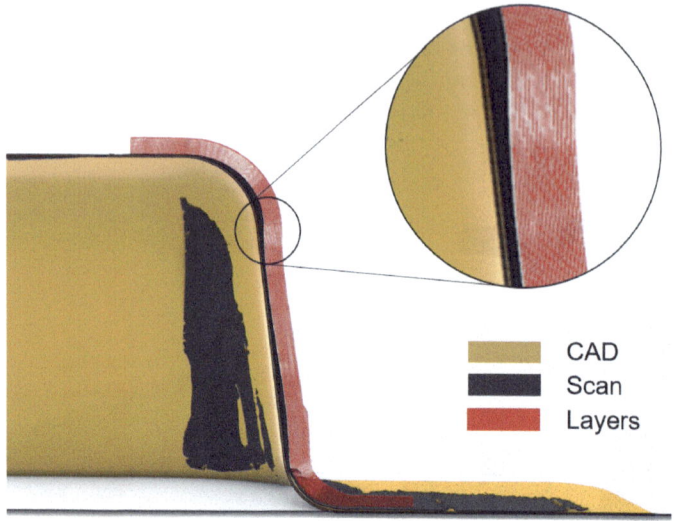

Fig. 9. Adapted paths on scanned surface

paths in the physical twin matched those observed within the digital twin, as displayed in the Fig. 10. This validates the G-Code generated with the proposed slicing algorithm and the digital twin's ability to accurately visualize paths. The digital process chain from the slicing algorithm to the digital twin proves to overall reduce the risk of errors with RSEAM for non-planar manufacturing.

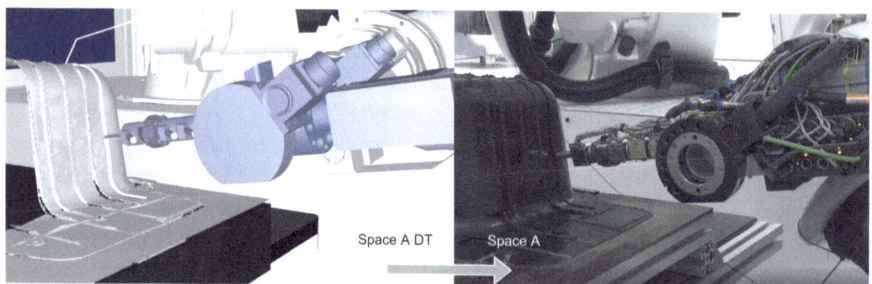

Fig. 10. Implementation within SpaceA DT and SpaceA

The results of produced part are illustrated in Fig. 11, which shows the printed rib from its virtual production to its physical production. The visual comparison between the digital simulation and the produced rib reveals a high degree of similarity, demonstrating that the physical rib closely resembles its digital twin. This observation confirms that the proposed adaptive non-planar slicing strategy effectively translates the virtual designs into accurate physical components, ensuring the intended geometries are preserved during the manufacturing process.

The experiments highlight the slicing algorithm's capability to handle non-planar geometry in the presence of micro and macro deviations. This ensures reliable adaptive material deposition that adapts the slicing strategy from nominal CAD to actual substrate. It especially demonstrates the algorithms capability to handle single- and double-curved geometries alike, since the nominal substrate was locally single-curved whereas the actual substrate was deformed to be double-curved.

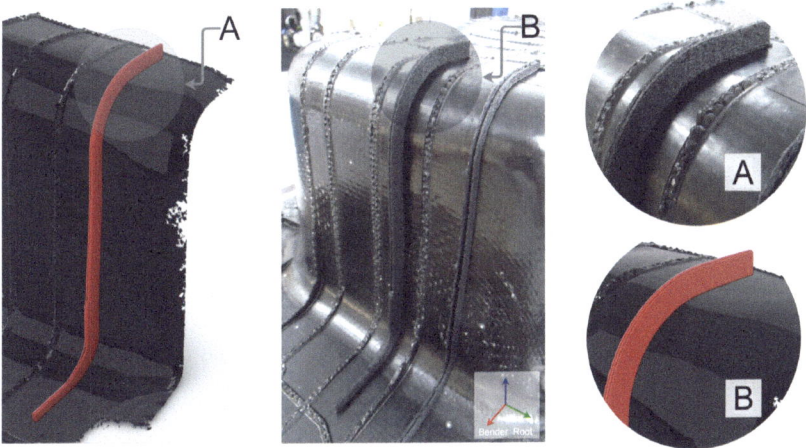

Fig. 11. Printed rib from virtual design to physical production

6   Conclusion and Future Work

In this paper, the development and demonstration of a novel slicing algorithm to manufacture geometries on non-planar non-ideal substrates with RSEAM was presented. The offered algorithm, implemented in Rhinoceros Grasshopper, automatically adapts paths created from nominal CAD to an actual scanned substrate, handling both micro and macro deviations. The whole algorithm is fully parametric and adaptive to accommodate to the non-ideal characteristics of the substrate. While the deposition rate is kept constant the degrees of freedom of a 6DOF robot is employed to position the extruder with varying velocity for desired bead geometry. The generated trajectories are optimized and validated inside a digital twin of the RSEAM production system.

Overall the complete process chain of 3D scanning, slicing, path planning, optimization and manufacturing on non-planar pre-existing substrates to create hybrid parts is demonstrated. The developed algorithm is capable of handling single- and double-curved nominal geometries, as well as single-curved nominal geometries that are deformed to actual double-curved geometries.

In future work, improvements on the existing process chain can be made in several ways. The employment of variable extrusion rates in addition to the variation of the positioning velocity offers potential for further increase in part quality. Utilizing the DOF of the build platform widens the scope of manufacturability for certain geometries. And finally, the optimization algorithm can be extended with more collision avoidance and optimization strategies, such as cone limitation or Z-axis adaption. Furthermore, future validation methods may involve 3D scanning of produced items to compare with desired specifications, tracking comprehensive production data, and creating digital replicas for G-Code comparison or training mathematical models.

References

1. Simulator for industrial robots and offline programming - RoboDK — robodk.com. https://robodk.com/. Accessed 10 May 2024
2. R. M. . Associates. Rhinoceros 3D — rhino3d.com. https://www.rhino3d.com/. Accessed 10 May 2024
3. Chen, L., Chung, M.-F., Tian, Y., Joneja, A., Tang, K.: Variable-depth curved layer fused deposition modeling of thin-shells. Robot. Comput. Integr. Manuf. **57**, 422–434 (2019)
4. Choi, S., Kwok, K.: A tolerant slicing algorithm for layered manufacturing. Rapid Prototyping J. **8**(3), 161–179 (2002)
5. Dai, C., Wang, C.C., Wu, C., Lefebvre, S., Fang, G., Liu, Y.J.: Support-free volume printing by multi-axis motion. ACM Trans. Graph. **37**(4) (2018)
6. Darweesh, B., Gutierrez, M.P., Schleicher, S.: Non-planar granular 3D printing. Constr. Robot. **7**(3), 291–306 (2023)
7. Ding, D., Pan, Z.S., Cuiuri, D., Li, H.: A tool-path generation strategy for wire and arc additive manufacturing. Int. J. Adv. Manuf. Technol. **73**, 173–183 (2014). https://doi.org/10.1007/s00170-014-5808-5
8. Dumas, J., Hergel, J., Lefebvre, S.: Bridging the gap: automated steady scaffoldings for 3D printing. ACM Trans. Graph. (TOG) **33**(4), 1–10 (2014)
9. Fortunato, G.M., Nicoletta, M., Batoni, E., Vozzi, G., De Maria, C.: A fully automatic non-planar slicing algorithm for the additive manufacturing of complex geometries. Addit. Manuf. **69**, 103541 (2023)
10. Gibson, I., Rosen, D.W., Stucker, B., Khorasani, M.: Additive manufacturing technologies, vol. 17. Springer, (2021)
11. Guo, W., Huang, X., Qi, B., Ren, X., Chen, H., Chen, X.: Vision-guided path planning and joint configuration optimization for robot grinding of spatial surface weld beads via point cloud. Adv. Eng. Inform. **61**, 102465 (2024)
12. Huang, B., Singamneni, S.: Curved layer adaptive slicing (CLAS) for fused deposition modelling. Rapid Prototyping J. **21**, 354–367 (2015)
13. Huse, T., Unger, N., Kamble, P., Kittler, H.: Accelerating the product development process of overmolded structures via robotic screw extrusion additive manufacturing. In: Automotive Circle Car Body Parts 2023 (2023)
14. Kamble, P., Unger, N.: Development of a digital process chain for hybrid manufacturing using robotic screw extrusion additive manufacturing. Master's thesis, Technische Universität Hamburg (2022)

15. Lettori, J., Raffaeli, R., Bilancia, P., Peruzzini, M., Pellicciari, M.: A review of geometry representation and processing methods for cartesian and multiaxial robot-based additive manufacturing. Int. J. Adv. Manuf. Technol. **123**(11), 3767–3794 (2022)
16. Li, X., et al.: Supportless 3D-printing of non-planar thin-walled structures with the multi-axis screw-extrusion additive manufacturing system. Mater. Des. **240**, 112860 (2024)
17. López-Arrabal, A., Guzmán-Bautista, Á., Solórzano-Requejo, W., Franco-Martínez, F., Villaverde, M.: Axisymmetric non-planar slicing and path planning strategy for robot-based additive manufacturing. Mat. Des. **241**, 112915 (2024)
18. Manorathna, R. P., et al.: Feature extraction and tracking of a weld joint for adaptive robotic welding. In: 2014 13th International Conference on Control Automation Robotics and Vision (ICARCV), pp. 1368–1372 (2014)
19. Minetto, R., Volpato, N., Stolfi, J., Gregori, R.M., da Silva, M.V.: An optimal algorithm for 3D triangle mesh slicing. Comput. Aided Des. **92**, 1–10 (2017)
20. Murtezaoglu, Y., Plakhotnik, D., Stautner, M., Vaneker, T., van Houten, F.J.: Geometry-based process planning for multi-axis support-free additive manufacturing. Procedia CIRP **78**, 73–78 (2018)
21. Schaechtl, P., Schleich, B., Wartzack, S.: On the potential of slicing algorithms in additive manufacturing for the optimization of geometrical part accuracy. Proc. CIRP **114**, 215–220 (2022)
22. Stier, T., Heinz, C., Lammert, N., Unger, N.: Additive manufacturing based on screw extrusion: machines, material and applications in the automotive industry. In: VDI-Kongress PIAE 2020 (2020)
23. Tran, C.-C., Lin, C.-Y.: An intelligent path planning of welding robot based on multisensor interaction. IEEE Sens. J. **23**(8), 8591–8604 (2023)
24. Tseng, J.-W.: Screw extrusion-based additive manufacturing of peek. Mater. Des. **140**, 209–221 (2018)
25. Unger, N., Huse, T.: Hybrid manufacturing of functionally integrated car body parts via robotic screw extrusion additive manufacturing. In: Automotive Circle Car Body Parts 2022 (2022)
26. Urhal, P., Weightman, A., Diver, C., Bartolo, P.: Robot assisted additive manufacturing: a review Rob. Comput. Integr. Manuf. **59**, 335–345 (2019)
27. Zhao, G., Guocai, M., Jiangwei, F., Wenlei, X.: Nonplanar slicing and path generation methods for robotic additive manufacturing. Int. J. Adv. Manuf. Technol. **96**(9–12), 3149–3159 (2018)

Open Access This chapter is licensed under the terms of the Creative Commons Attribution 4.0 International License (http://creativecommons.org/licenses/by/4.0/), which permits use, sharing, adaptation, distribution and reproduction in any medium or format, as long as you give appropriate credit to the original author(s) and the source, provide a link to the Creative Commons license and indicate if changes were made.

The images or other third party material in this chapter are included in the chapter's Creative Commons license, unless indicated otherwise in a credit line to the material. If material is not included in the chapter's Creative Commons license and your intended use is not permitted by statutory regulation or exceeds the permitted use, you will need to obtain permission directly from the copyright holder.

A Two-Level Architecture for Mobile Grasping

Troy McMahon[✉], Alfred Thoft Christiansen, Elias Thomassen Dam, Casper Schou, and Ole Madsen

Department of Materials and Production, Aalborg University, Aalborg, Denmark
`troymcMahon1@gmail.com`

Abstract. We propose an architecture for grasping on the move in order to facilitate pick and place operations while a mobile manipulator's base is in motion. Our architecture eliminates the need for the robot to stop and localize in order to perform a pick and place operation which improves efficiency in setting such as industrial automation.

Our framework applies a two-level planner where a base planner is used generate to trajectories for the base, and an arm planner is used to obtain a grasping trajectory given the planned motion of the base. We propose methods for combining and synchronizing these trajectories along with an online method for adjusting the trajectory of the arm to account for any deviation or error in the motion.

We show application to a mobile manipulator robot, which consists of a UR5 arm mounted on a MiR mobile base. We show application to kitting and packing tasks in an industrial setting. Our results demonstrate that the robot is able to reach specified end-effector positions required to perform grasping while the base of the robot is in motion.

1 Introduction

Mobile manipulation consists of a robot in an environment where it must navigate while performing required tasks, such as grasping or manipulation. It has been applied to a wide variety of areas, including manufacturing and industrial robotics, home care and assisted living, space exploration, and search and rescue [SA22]. Within industrial robotics, it can be applied to tasks such as kitting, where the robot must run around collecting components required for later assembly in, for example, a production cell.

Mobile manipulators generally consist of a manipulator, or arm that is mounted to a mobile base (Fig. 2). For industrial automation tasks the environment consists of a factory floor that may include dynamic obstacles such as people or other robots. Manipulation tasks generally consist of picking and collecting objects, with more complex tasks being performed by dedicated assembly cells.

Separately, mobile robotics and manipulation are mature fields with many proposed solutions, however combining these solutions to do mobile manipulation has proven difficult. Integrating planning of the base and the arm increases the complexity of the problem, which makes planning far more difficult. Moreover, the context in which these the arm and base operate is very different and requires different planning tools. Application to real world robots also introduces uncertainty due to inconsistency in the motion of the base, which translates into considerable error in the position of the arm and makes precise positioning required for grasping very difficult. It also introduces physical factors including the inertia of the arm which impacts the motion of the base. In practice, most real-world mobile manipulation is accomplished by moving the base of the robot to an appropriated docking position then performing the manipulation task while the base is stationary, thus transforming the problem into a fixed based manipulation task. While such approaches are effective, they waste time by requiring the robot to dock and by serializing the motion of the base and arm which could be done in parallel. Those methods that do allow for motion while grasping, such as [BLLLC22, HSC22], make use of reactive methods that do not consider obstacle avoidance and do not provide any completeness guarantees. Other approaches including [SA22] use traditional motion planners, however such methods can only be used reliability in a controlled environment and have a significant failure rate when applied to real robots.

We propose a two-level planner which consists of a planner for the base that finds base paths over which the grasping/manipulation task can be performed, and a manipulation planner that performs grasping/manipulation given the path of the base. This approach is advantageous because it allows us to use existing mobile robot planners to plan for the base, but comes with the complication that the manipulation planner used by the arm must take into account the motion of the base. To this end, we propose an arm planner that generates motions which incorporate the motion of the base. We also propose methods for combining and synchronizing the motion plans for the base and arm, along with an online method for adapting to any deviations in the motion of the arm or the base. We apply our approach to a *little helper* mobile manipulator that consists of a *UR5* arm mounted on a *MiR* base in a manufacturing setting doing kitting and packing tasks.

2   Preliminaries

We divide the problem into a planning phase and an execution phase. We formulate the planning phase as a kinodynamic planning problem in a dynamic environment $E(t)$. The robot is located at an initial configuration s_0 and must grasp an object, such as a part of a conveyor belt. The plan for the base is defined by a trajectory ρ_{base} and a corresponding control for the robot's wheels, τ_{base} which may be given or computed by any planner appropriate for mobile robots. A solution is a plan for the arm, $\{\rho_{arm}, \tau_{arm}\}$, which runs in parallel with $\{\rho_{base}, \tau_{base}\}$ to grasp the object. During run time we execute $\{\rho_{arm}, \tau_{arm}\}$ and $\{\rho_{base}, \tau_{base}\}$, however there may be noise in the motion of the base which results in deviation. We must therefore adaptively update the motion of the arm in order to compensate for this deviation. This process is shown in Fig. 1

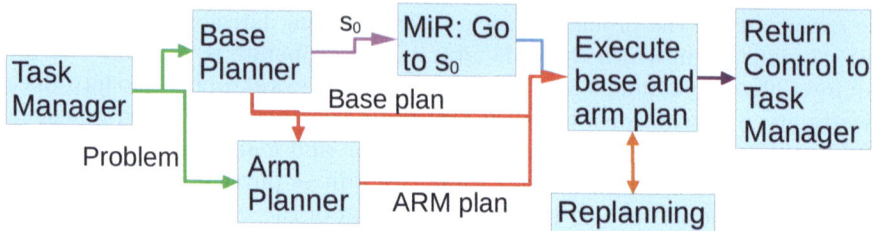

Fig. 1. Control Flow: The task manager maintains a model of the environment and initiates grasping operations by sending a grasping problem to the base planner and the arm planner. The base planner sends computes a base plan and sends it to the arm planner, which computes an arm plan. While this is happening the base planner sends the start state, s_0 to the robot. Upon completion, control passes to the controllers which executes the arm and base plan. As it executes it does replanning based on sensor input. Once the motion has been executed it returns control to the task manager.

3 Proposed Method

Our architecture (Fig. 3) consists of a planning component that is run in preprocessing and an execution component that performs the grasp. A task manager maintains a model of the environment and during the preprocessing time it generates a grasping task in for form of a motion planning problem which consists of an Environment E, a start state s_O, the position of the object being grasped, γ, and a goal positron, s_G. The *Base Planner* generates a path $\rho_{base}(t)$ and corresponding control $\tau_{base}(t)$, which starts at S_O, passes within reachable distance of γ and terminates in s_G. The *arm planner* takes as input E, $g(t)$, $\rho_{base}(t)$ and $\tau_{base}(t)$, and generates a path $\rho_{arm}(t)$ and corresponding control $\tau_{arm}(t)$ for the arm which intersects with $g(t)$ resulting in a grasp. We define the state where this intersection occurs as s_γ.

During execution time, the base and arm controllers simultaneously execute the the base and arm path to perform the grasp. This is done by transmitting controls to the base, $\tau_{base,t}$ and to the arm, $\tau_{arm,t}$. As it executes the path, it uses the robot's sensors, $\sigma(t)$ to estimate the robot's state, $s_\sigma(t)$. The base controller adjusts the path of the base based on this observation. The base controller also uses $s_\sigma(t)$ to estimate the time required for the base to reach s_γ, denoted as t_γ, which it sends to arm controller. The arm controller uses t_γ to control the timing of the motion of the arm in order to ensure that both the arm and the base reach s_γ at the same time. As the robot approaches s_γ the arm controller sends an activation signal, Γ to the gripper.

3.1 Base Planner

As the robot is a concatenation of a MiR robot as the base and a UR5 as the manipulator, the navigation of the base is done separately from the UR manipu-

Fig. 2. Mobile Manipulator

lator. This is achieved with the use of ROS2 and the Navigation 2 [MMWGC20] ROS2 packages.

The Nav2 package handles both the navigation and the control of the robot. For the robot to navigate, it uses two LiDAR sensors, one at each corner, and a pre-generated map of the workspace. From the map and the sensor input, the robot estimates its pose in the map frame using Adaptive Monte Carlo Localization (AMCL). The robot needs an estimated initial pose, and as it moves, it obtains a more accurate pose.

To move the robot, a trajectory is generated from the start point, through two points determined by the location of the item to grasp, and to the end point using a Nav2 planner server plugin. The trajectory is then passed to a controller plugin. The controller plugin uses the local cost map to fine-tune the trajectory to the local environment and send out motor velocities.

To generate the path, as mentioned, the initial pose, the item position, and the end pose are used together with the mobile robot's configuration space, to specify the top-level plan for the motion of the mobile base. Before computing the complete path, an intermediate step is used to find the pre-grasp pose and the post-grasp pose of the mobile robot based on the item position, the configuration space of the mobile robot, and the initial position of the robot. This is done by sampling trial paths around the item to grasp to find a viable path where the robot manipulator can do a grasp while the base is in motion. From these four points, a "continuous" path is generated with the use of the Nav2 Hybrid Smac Planner due to its ability to generate smoother paths where the turning

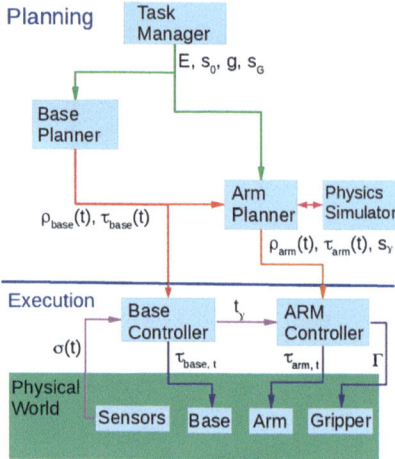

Fig. 3. Planner Architecture

rate can be limited for a more predictable traversal of the path, improving the predictability of the mobile robot's path following (Fig. 4) .

To follow the path, the Nav2 Regulated Pure Pursuit Controller [MSMG23] is used as it can be configured to move the mobile robot at a consistent velocity.

3.2 Physic Simulation

Motion planning algorithms rely on the ability to accurate predict the effects of applying a control to the robot. These predictions must take into account the dynamics of the robot as well as the effect of inertia and changing weight distribution inherent to the motion of the arm (Fig. 5) . To this end we model robot motion using a physics simulator (e.g. Bullet [CB19]). This simulator also provides collision detection required for motion planning.

3.3 Arm Planner

Planning for the arm is done using a tree-based kinodynamic planner (e.g. RRT [LaV98], RRT* [KF11]) which has been modified to accommodate the motion of the base. This planner models the entire robot however it only plans in the space of the arm, while base controls are set according to the plan given by the base planner. We define the goal for the planner to be an end effector position/orientation denoting the grasping position.

In order to ensure path quality we use an asymptotically optimal planner (i.e. DIRT [LB18]) which is guaranteed to converge to an optimal path. Such a planner will produce shorter motions that require less time to execute, meaning that we can begin executing the motion closer to the grasping point.

As an optimization, we specify a two-step planning problem consisting of a pre-grasping position followed by a grasp position. We specify a pre-grasping

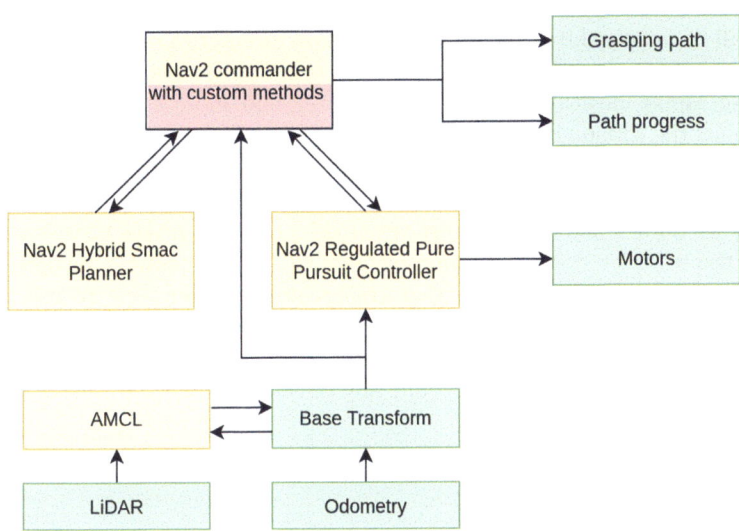

Fig. 4. System diagram of the base planner and controller. The yellow boxes are Nav2 plugins, the green boxes represent ros2 topics, and the red part of the Nav2 commander represents custom code added to make it generate the pre-grasp and post-grasp poses for the path generation

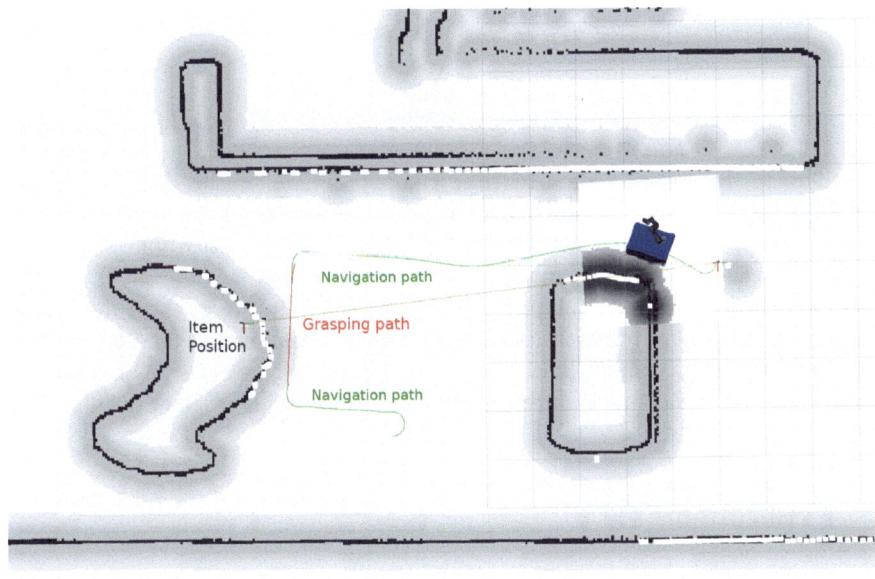

Fig. 5. A screenshot from Rviz2 where the item position and the different path sections have been annotated

pose that is above and ahead of the object being grasped, which ensures that the robot will be approaching the object from above in order to minimize the chance of collision. This also allows us to specify a rapid movement in order to get to the pre-grasping position followed by a slower and more stable motion while performing the grasp. From a planning standpoint, this is also advantageous because it allows us to specify different parameters for planning the pre-grasp and grasp. The pre-grasp problem is longer and more likely to include collisions, however we do not need to precisely reach the pre-grasp position, which means we can use a larger resolution and step size in our simulator. Conversely, the grasping problem is shorter but requires greater precision, so we use a finer resolution and step size. As a further optimization we can do the planning for the grasp motion while the pre-grasp motion is being executed and we can adapt the grasp planing to take into account any deviations detected during the pre-grasp motion.

3.4 Controller

To synchronize the motion of the arm and the base we set up a *keypoint*, which is the point on the base's path where we expect the grasp to occur. The base planner uses AMCL the time to the keypoint. This computation is described by the following equations:

$$t_p = \frac{d_p}{v_s} \qquad (1)$$

$$v_s = \frac{\sum_{i=0}^{n} v_i}{n} \qquad (2)$$

where t_p denotes the time to the pickup point, d_p is the distance to the pickup point, v_s is the smoothed velocity, v_i is the velocity between the ith and the previous point, and n denotes the number of measurements done so far.

In the arm controller, we specify a set of waiting points where the controller waits for the to the keypoint to reach a specified value (e.g. the time it will take for the arm plan to reach its gasping point). Our architecture allows us to set multiple waiting points along the path of the arm, however in practice we found that setting a single waiting point at the beginning of the arm plan was sufficient.

4 Experiments

We run experiments using a *Little Helper* robot (Fig. 2) that is applied to a table grasping application.

4.1 Experimental Setup

Little Helper Robot:

The *Little Helper* mobile manipulator consists of a UR5 robot arm [UR522] mounted on a MiR Autonomous Mobile Robot (AMR) [MiR24]. The MiR is a mass-produced industrial robot designed to perform product movement tasks in an industrial setting. Its sensor suite consists of a pair of LiDAR sensors for navigation.

The UR5 is a 6°C of freedom robotic arm that is capable of performing most grasping and manipulation tasks. It provides a maximum payload capacity of 5 kg with a maximum reach of 850 mm and a maximum joint speed of 180° per second. For grasping, we used a magnetic gripper.

Experimental Environment:

The **Table Grasp** application (Fig. 6) requires the robot to drive by a table and grab a stationary object (e.g. a 5 cm by 10 cm phone case) from a table. The robot must follow a path by the table and time the movement of the arm so it intercepts and grasps the object. This task demonstrates that our method can accommodate the noise and variation introduced by the motion of the base as well as the effects of inertia from the motion of the arm.

Experimental Architecture:

All experiments were run using an Intel(R) Core(TM) i7-4712HQ CPU @ 2.30GHz processor with 16 GB of RAM. Communication with the robot is facilitated using ROS2 and control of the UR5 are was done using the Universal Robots ROS2 Driver [UR524]. Control of the base is done using the NAV2 stack from ROS2.

Results:

Successful application of our method is shown in our video attachment. These results demonstrate that the robot is able to grasp and pick up the phone case while the robot's base is traveling at speed. Grasps are achieved without any need to stop or even slow down the base. We also demonstrate that our setup can accommodate high speed arm motions which allows us to initiate the arm motion very close to the grasping position. Moreover, the base controller is able to adapt to any disturbances caused by the motion of the arm in order to successfully reach the final goal position.

We show that the controller can successfully perform the base plan while using input given by the robot's sensors to adapt to deviations in the motion. Consequently, the robot is able to reach both the grasping configuration and the final configuration. We also show that the arm controller is able to synchronize its motion with the motion of the base in order to reach the grasping configuration at the same time, resulting in a successful grasp.

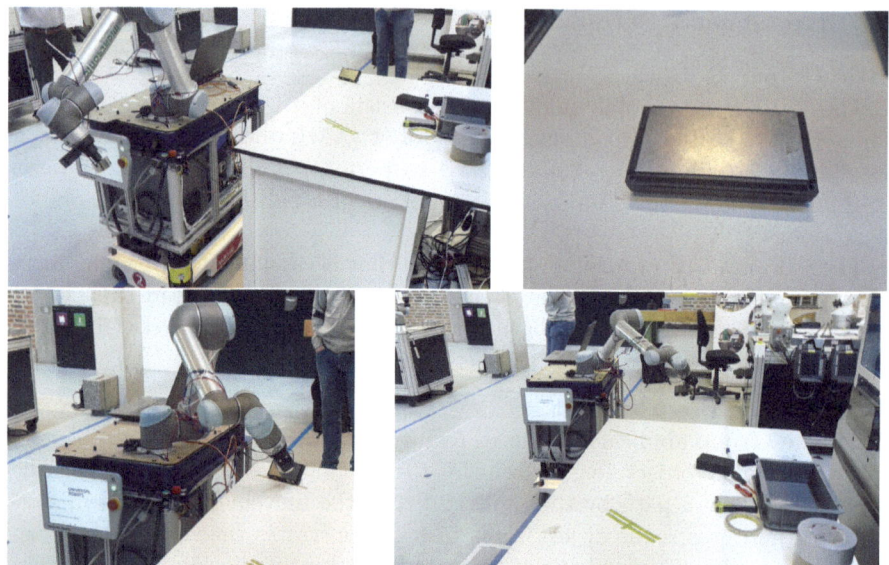

Fig. 6. Experimental Setup: ul) Initial pose, ur) object to be grasped, ll) successful grasp and lr) final pose

5 Conclusion

We presented a method for motion planning for a robotic arm that is mounted on a mobile base. We then presented an architecture for controlling and coordinating the motion of the arm and the base. We applied this architecture to a mobile grasping task in order to demonstrate that we are able to successfully coordinate the motion of the arm and the base. We showed that our method is able to successfully grasp an object while the base is in motion.

Acknowledgement. This work was supported by Innovations Fond Denmark through the project "5G-ROBOT - 5G Enabled Autonomous Mobile Robotic Systems"

References

BLLLC22. Burgess-Limerick, B., Lehnert, C., Leitner, J., Corke, P.: An architecture for reactive mobile manipulation on-the-move. In: ICRA (2022)

CB19. Coumans, E., Bai, Y.: PyBullet, a python module for physics simulation for games, robotics and machine learning. http://pybullet.org (2016–2019)

HSC22. Haviland, J., Sunderhauf, N., Corke, P.: A holistic approach to reactive mobile manipulation. IEEE Rob. Autom. Lett. **7**(2), 3122–3129 (2022)

KF11. Karaman, S., Frazzoli, E.: Sampling-based algorithms for optimal motion planning. Int. J . Rob. Res. **30**(7), 846–894 (2011)

LaV98. LaValle, S.: Rapidly-exploring random trees : a new tool for path planning. Annu. Res. Rep. (1998)

LB18. Littlefield, Z., Bekris, K. E.: Efficient and asymptotically optimal kinodynamic motion planning via dominance-informed regions. In: IROS (2018)

MiR24. Mobile industrial robots. www.rarukautomation.com/mobile-robots/mobile-industrial-robots (2024)

MMWGC20. Macenski, S., Martín, F., White, R., Clavero, J.G.: The marathon 2: a navigation system. In: 2020 IEEE/RSJ International Conference on Intelligent Robots and Systems (IROS) (2020)

MSMG23. Macenski, S., Singh, S., Martín, F., Ginés, J.: Regulated pure pursuit for robot path tracking **47**(6), 685–694 (2023)

SA22. Thushara, S., Marcelo H. A.: Motion planning for mobile manipulators: a systematic review. Machines **10**(2) (2022)

UR522. e-series cobots — collaborative robots by universal robots. www.universal-robots.com (2022)

UR524. Universal robots ROS2 driver. https://github.com/UniversalRobots/Universal_Robots_ROS2_Driver (2024)

Open Access This chapter is licensed under the terms of the Creative Commons Attribution 4.0 International License (http://creativecommons.org/licenses/by/4.0/), which permits use, sharing, adaptation, distribution and reproduction in any medium or format, as long as you give appropriate credit to the original author(s) and the source, provide a link to the Creative Commons license and indicate if changes were made.

The images or other third party material in this chapter are included in the chapter's Creative Commons license, unless indicated otherwise in a credit line to the material. If material is not included in the chapter's Creative Commons license and your intended use is not permitted by statutory regulation or exceeds the permitted use, you will need to obtain permission directly from the copyright holder.

Towards Automotive Manufacturing Efficiency: Enhanced Virtual Commissioning Simulation for Dynamic Sheet Metal Handling Optimization

Stefan Klare[1(✉)], Volodymyr Shramenko[1], Lars Klingel[2], Bernd Lüdemann-Ravit[1], and Alexander Verl[2]

[1] Hochschule für angewandte Wissenschaften Kempten, Institut für Produktion und Informatik, Bahnhofstraße 61, 87435 Kempten, Germany
stefan.klare@hs-kempten.de

[2] Universität Stuttgart, Institut für Steuerungstechnik der Werkzeugmaschinen und Fertigungseinrichtungen, Seidenstraße 36, 70174 Stuttgart, Germany

Abstract. Automated sheet metal handling in the automotive industry using robot manipulators is a standard in modern production. However, the desire of automotive companies to speed up the production process on the assembly line and at the same time to reduce expensive hardware components poses new challenges for robotics. Excessively rapid movement of flexible parts or sheet metal can either lead to its plastic deformation or increase the decay time of the vibrations to such an extent that it is necessary to wait before the part can be further processed, for example by welding. The traditional approach to solving these problems is to add fixing points for the part and/or to allow for waiting times at the end of the robot movement to ensure that the sheet metal vibration subsides. This means that more effort than necessary has to be put into the hardware setup or a poor cycle time has to be accepted. In our work, we propose to improve virtual commissioning systems by conducting a deeper analysis of the dynamics of sheet metal parts during its movement by the robot. To achieve this, a three-step optimization strategy is proposed. The first step is a structural transient analysis of the thin metal part under disturbances that arise during its movements. The second step is the creation of a substitute model, which is trained on the basis of the data obtained in the first phase and considerably reduces the computing time required compared to the time needed for a finite element method (FEM) simulation. Subsequently, this is used to optimize robotic handling efficiency by optimizing the robot trajectory. By addressing the challenges posed by sheet metal dynamics, enhanced process control, reduced cycle times, and ultimately, improved manufacturing outcomes are anticipated.

Keywords: Virtual Commissioning · Surrogate Modeling · Dynamic Sheet Metal Handling · Machine Learning

1 Introduction

Automobile manufacturing processes have evolved significantly over the years, with automation playing a pivotal role in enhancing efficiency and productivity. One critical aspect of automotive manufacturing is the handling of sheet metal, which forms the basis of vehicle components such as body panels and chassis. However, the dynamic behavior of elastic metal sheets poses a challenge to automated handling systems, often leading to inefficiencies and increased cycle times. The vibrations of the sheets are generated by the movements of the robots that move the sheets. As it is difficult to predict how long it will take for the vibrations to subside, waiting times are usually included at the end of the robot movement. These waiting times are chosen conservatively and are therefore not optimal, which results in a deterioration of the cycle time. Traditional simulation methods like they are used in virtual commissioning struggle to accurately account for these dynamics, highlighting the need for innovative solutions to optimize sheet metal handling processes. In the context of this problem, we want to address two questions. On the one hand, we want to investigate methods to predict the vibration behaviour under certain excitations by the robot. This should be possible in real time or at least near real time. Secondly, we want to use this knowledge to find optimal robot trajectories so that vibrations on the sheet metal can be avoided as far as possible. Once these problems have been addressed, the knowledge generated will be used to create new tools for production planning and virtual commissioning.

A general overview about path planning and trajectory planning methods can be found in [1], while a survey about model-based manipulation planning of deformable objects is shown in [2]. In [3,4] a methodology for optimal trajectory planning of compliant sheet metal parts is presented. With the methodology the path geometry and the motion parameters are optimized simultaneously. Maximum deflections and stress is considered as constraints in formulation of the optimization problem. These parameters are gained by an offline FEM. However the methodology is only applied to a simple geometry (flat rectangular plate), and it only deals with translational movements and no rotation. In [5] a variation analysis of assemblies containing compliant parts is performed. Since this analysis is very time-consuming due to the many FEM simulations involved, the authors propose a method that reduces the calculation times of the FEM simulations by avoiding remeshing. However, remeshing is not necessary for our problem, since in our case the type of excitation is varied, whereas in [5] the parts to be excited are varied. In [6] Kaczor et al. proposed a method for optimizing a trajectory for the handling of elastically coupled objects via reinforcement learning and flatness-based control. Their approach using reinforcement learning is interesting as the procedure is applicable to complicated systems. In the publication their methods were only applied to a simple mass-spring-damper system. Other examples where reinforcement learning is used for trajectory optimization or path planning are [7–9]. If the vibration behaviour of elastic sheet metal is to be optimized using a suitable robot trajectory, finding a model of the sheet metal represents a major challenge. In the previously mentioned publications only simple systems were considered. Thus, when aiming for controlling the vibration of a sheet metal, one needs to find a way to describe the transient behaviour in a form which is as simple as possible and accurate enough to solve the desired task. For complex systems, it is often difficult or impossible to find an analytical description. So-called surrogate models offer a possibility to find a system description nevertheless. In [10] Kudela et al. give a good overview about recent ongoing work in this field. "The principle purpose of using surrogate models (or metamodels) is to describe expensive-to-evaluate high-fidelity models, such as a FEM-based model, employing computationally less costly statistical models" [10]. In [11], a comparative analysis is conducted on six different surrogate models that have been the most popular and promising alternatives to FEM in recent years. These models are tested on the structural analysis problem of a thin metal plate with a hole, and their results and computation times are compared with those of classical FEM. The best results in terms of accuracy and time were achieved by the Multi-layer Perceptron (MLP) and Physics-Informed Neural Network (PINN). In the test example, these models provided a time advantage of more than a hundred times compared to FEM.

Work [12] is dedicated to the application of PINN to the structural analysis of shells. Training such a network requires a detailed understanding of the theory of elastic shells and a careful selection of hyperparameters. The authors concluded that this approach allows to avoid several known issues of FEM simulations for very thin shells, such as the phenomenon of "locking", and its further development could provide a significant computational time advantage compared to FEM.

The results of the aforementioned works coincide with the announced performance of the new ANSYS AI software, which promises a speed increase in structural analysis by 10 to 50 times depending on the task.

2 Challenges in Sheet Metal Handling

When manipulating flexible parts, it is crucial to consider the deformations that occur during accelerated movement and rotation. The desire to reduce the production cycle time can lead to plastic deformations of the material and an increase in the decay time of vibrations at the end of the trajectory. This necessitates a deep understanding of the processes occurring with the plate or shell under loads. Given that the part can have quite complex geometry and that securing the part with vacuum suction cups is far from classical boundary condition setups, structural static and dynamic analysis during manipulation is only feasible through FEM simulations. Each of these FEM simulations can require from several minutes to several hours of computing time.

However, the main problem arises when attempting to optimize the manipulator's trajectory. The goal is to minimize the total cycle time for moving the flexible part and damping vibrations at the end of the trajectory, without causing plastic deformations or collisions. Imagine one possible trajectory, which is a broken line with constant accelerations in each sequence. An object in 3-dimensional space can generally be manipulated with 6 degrees of freedom (3 translational and 3 rotational degrees of freedom). Conventional industrial robots also have 6 degrees of freedom, which is why we assume that the sheet metal under consideration can experience excitations in all 6 degrees of freedom. If we discretize each degree of freedom by dividing the ranges of accelerations into n parts, then even for a single manipulator action on the plate (specifying the translatory and rotatory acceleration), there can be n^6 variations. For k sequential manipulator movements, we get n^{6k} possible variations. Naturally, not all scenarios are meaningful, but even a significantly reduced number of them cannot be simulated using classical FEM within any reasonable time frame.

There are several approaches to accelerating FEM simulations. The new tools like physics-informed neural networks (PINN) or graph neural networks (GNN) can reduce computation time by tens of times and provide a complete picture of elastic body's behavior. Another approach is generally known as RS (Response Surface) and is applied in situations where a complex system has several input parameters, and it is necessary to build a dependence of some output parameter on them. This approach is reasonable since we do not need the exact shape of the part under loads; we only need two indicators: maximum von Mises stress and maximum amplitude. It is worth noting that finding an appropriate optimization method to minimize the overall cycle time (movement and vibration damping time) under a set of nonlinear constraints, each of which requires structural analysis, is also a complex task.

3 Proposed Methodology

To address the issue of reducing cycle time when manipulating a flexible part with a robot, we propose the following three-step approach. The overall concept of this approach is illustrated in Fig. 1.

In the first step, we gather information about the part: the CAD model describes the geometry; density, Young's modulus, Poisson's ratio, and damping coefficient describe the material properties. Points of Interest (POI) are also determined. POIs can be e.g. mounting points. Evaluating the damping of vibrations at these points is of greatest interest. From the documentation of the manipulator robot, we obtain ranges for each degree of freedom, based on which we generate sets of movement scenarios using the Latin Hypercube Sampling method: a vector of translatory and rotatory accelerations. For each scenario, we conduct FEM simulations for the structural static analysis of the plate. In paper [3], when modeling the plate's response to disturbances, the load was always applied to a flat plate, although after the first disturbance it no longer has its original shape. We perform a two-step simulation, where the initial shape is modified after the first step. In reality, the shape of a flexible body is influenced by all sequential manipulations, but the two-step structure is a compromise between computational complexity and model accuracy.

In the second step, we train a surrogate model. In our case, we can use a response surface model (RSM) for finding the maximum displacement and

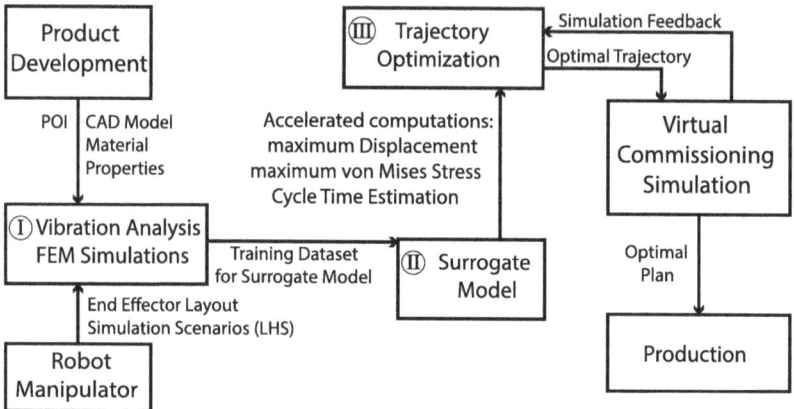

Fig. 1. Methodology for simulating elastic parts/sheet metals for virtual commissioning

maximum von Mises stress, similar to the technique applied in work [3]. Selecting, training, and validating the model that best fits the specific task is quite a delicate process. Therefore, we will choose the one from the models described in the review work [10], that offers the highest accuracy and requires a smaller training dataset. The candidates include Kriging, Support Vector Regression (SVR), and Multi-layer Perceptron (MLP). Whether one or the other method works successfully depends heavily on the model to be described. Furthermore, each method contains specific hyperparameters that must be tuned manually.

In the third step, we formulate and solve a nonlinear optimization problem, where the objective function is the total time for the manipulator to move the flexible part combined with the decay time of the part at the end of the movement. Nonlinear constraints include: limits on acceleration changes (no jerks), constraints on linear and angular accelerations (robot's limitations), and the maximum allowable von Mises stress (threshold for plastic deformation). The solution to the nonlinear problem can be determined using reinforcement learning, for example, and thus be solved in close interaction with the virtual commissioning simulation which can introduce additional conditions like avoiding collisions during the production cycle. Since we are dealing with a non-linear optimization problem, it is difficult at this stage to predict which optimization method is best suited to solving the optimization problem. For example, we have no knowledge about the gradient of the object function and its convexity. Furthermore, we also do not know whether the constraints are convex. This requires a more detailed analysis. However, reinforcement learning can be applied to many similar optimization problems, which is why this optimization method must be considered in any case. The combined time required to run the trajectory and the subsequent decay of the sheet metal vibration excited by the movement should be taken into account in the reward function of the reinforcement optimization. The maximum sheet deflection is indirectly included in the decay time. Decay can be defined in such a way that the maximum sheet deflection is below a threshold value. Material stress should also be included in the reward function. In particular, stresses that would lead to plastic deformation of the sheet metal must be taken into account. Furthermore, collisions and the violation of the robot's workspace can be handled in the reward function. It should be noted here that the maximum sheet deflection is also included in the evaluation of collisions. Other optimization methods that must be considered in addition to reinforcement learning are the trust region method [13] and differential evolution method [14] or its further development [15], as these methods can be applied to nonlinear optimization problems and do not require knowledge of the gradient. These methods can also be used for optimization problems where no knowledge about the convexity is given.

4 Expected Outcomes

The proposed methodology is expected to deliver several key outcomes. By optimizing handling trajectories, we anticipate a reduction in both robot movement time and oscillation decay, leading to shorter cycle times in manufacturing. Integrating nonlinear constraints such as material stress limits and collision avoidance will enhance process safety and reliability, reducing the risk of part deformation.

Reinforcement learning (RL) will dynamically optimize the robot's trajectory, managing the complex behavior of flexible parts. Alternative optimization methods like trust region and differential evolution will offer additional approaches for handling nonlinearities, providing a robust framework for trajectory optimization.

5 Conclusion

This concept paper proposes a three-step approach to optimize robotic handling of flexible parts, aiming to reduce cycle times while improving safety and precision. By combining FEM simulations, surrogate modeling, and reinforcement learning, the methodology addresses the challenges posed by the elastic behavior of sheet metal in automation.

Future work will focus on refining these techniques and applying them to real-world scenarios. This research provides a foundation for more efficient and adaptable manufacturing processes, enhancing productivity and part integrity in industries like automotive manufacturing.

References

1. Gasparetto, A., Boscariol, P., Lanzutti, A., Vidoni, R.: Path planning and trajectory planning algorithms: a general overview. In: Carbone, G., Gomez-Bravo, F. (eds.) Motion and Operation Planning of Robotic Systems. MMS, vol. 29, pp. 3–27. Springer, Cham (2015). https://doi.org/10.1007/978-3-319-14705-5_1
2. Jiménez, P.: Survey on model-based manipulation planning of deformable objects. Rob. comput. Integr. Manuf. **28**(2), 154–163 (2012)
3. Li, H., Ceglarek, D.: Optimal trajectory planning for material handling of compliant sheet metal parts. J. Mech. Des. **124**(2), 213–222 (2002)
4. Glorieux, E., Franciosa, P., Ceglarek, D.: End-effector design optimisation and multi-robot motion planning for handling compliant parts. Struct. Multidiscip. Optim. **57**, 1377–1390 (2018)
5. Gerbino, S., Franciosa, P., Patalano, S.: Parametric variational analysis of compliant sheet metal assemblies with shell elements. Proc. CIRP **33**, 339–344 (2015)
6. Kaczor, D., Bensch, M., Schappler, M., Ortmaier, T.: Trajectory optimization for the handling of elastically coupled objects via reinforcement learning and flatness-based control. Springer, Berlin Heidelberg (2020)
7. Xie, J., Shao, Z., Li, Y., Guan, Y., Tan, J.: Deep reinforcement learning with optimized reward functions for robotic trajectory planning. IEEE Access **7**, 105669–105679 (2019)
8. Ullrich, M., Saad, C., Jacob, D.: Robot system as a testbed for AI optimizations. In: International Conference on Computer Aided Systems Theory, pp. 562–568. Springer (2022)
9. Jaensch, F., Csiszar, A., Sarbandi, J., Verl, A.: Reinforcement learning of a robot cell control logic using a software-in-the-loop simulation as environment. In: 2019 Second International Conference on Artificial Intelligence for Industries (AI4I), pp. 79–84. IEEE (2019)
10. Kudela, J., Matousek, R.: Recent advances and applications of surrogate models for finite element method computations: a review. Soft. Comput. **26**(24), 13709–13733 (2022)
11. Hoffer, J.G., Geiger, B.C., Ofner, P., Kern, R.: Mesh-free surrogate models for structural mechanic fem simulation: a comparative study of approaches. Appl. Sci. **11** (20), 9411 (2021)
12. Bastek, J.H., Kochmann, D.M.: Physics-informed neural networks for shell structures. Eur. J. Mech. A Solids **97**, 104849 (2023)
13. Conn, A.R., Gould, N.I., Philippe, L.: Trust region methods. SIAM (2000)
14. Storn, R., Price, K.: Differential evolution-a simple and efficient heuristic for global optimization over continuous spaces. J. Global Optim. **11**, 341–359 (1997)
15. Tanabe, R., Fukunaga, A.S.: Improving the search performance of shade using linear population size reduction. In: 2014 IEEE Congress on Evolutionary Computation (CEC), pp. 1658–1665. IEEE (2014)

Open Access This chapter is licensed under the terms of the Creative Commons Attribution 4.0 International License (http://creativecommons.org/licenses/by/4.0/), which permits use, sharing, adaptation, distribution and reproduction in any medium or format, as long as you give appropriate credit to the original author(s) and the source, provide a link to the Creative Commons license and indicate if changes were made.

The images or other third party material in this chapter are included in the chapter's Creative Commons license, unless indicated otherwise in a credit line to the material. If material is not included in the chapter's Creative Commons license and your intended use is not permitted by statutory regulation or exceeds the permitted use, you will need to obtain permission directly from the copyright holder.

Online Real-Time Simulation for Collision Avoidance in Robotic Wire Arc Additive Manufacturing

Lars Klingel[1(✉)], Maximilian Nistler[1], Daniel Mantz[1], Martin Werz[2], and Alexander Verl[1]

[1] Institute for Control Engineering of Machine Tools and Manufacturing Units (ISW), University of Stuttgart, Seidenstr. 36, 70174 Stuttgart, Germany
lars.klingel@isw.uni-stuttgart.de

[2] Materials Testing Institute (MPA), University of Stuttgart, Pfaffenwaldring 32, 70569 Stuttgart, Germany

Abstract. In manufacturing processes involving industrial robots, collisions are a significant risk due to the high number of degrees of freedom. This is particularly the case in complex processes such as wire arc additive manufacturing, where metal is printed using a welding unit mounted on the robot. Simulations can be used before the operating phase to identify potential collisions between the industrial robot and itself, the workpiece clamping, or the workpiece. Even after intensive testing, a collision can still occur during the operating phase if any environmental condition differs from the simulation. This could be due to a change in motion caused by actual measured values or the modification of parameters by humans or an adaptive system. This work presents a concept for online real-time collision avoidance in robotic wire arc additive manufacturing. The concept is based on adopting simulations from virtual commissioning for the use in the operating phase. The concept is then realized with the help of a real industrial robot. The resulting solution's novelty is the virtual commissioning-based online collision avoidance of wire arc additive manufacturing with industrial robots.

Keywords: Virtual Commissioning · Digital Twin · Robotics · Additive Manufacturing

1 Introduction

Production engineering must address a number of developments such as individualization and resulting small lot sizes, uncertain supply chains and sustainable manufacturing. These developments impose requirements on today's production systems [1,2]. Consequently, plants are becoming increasingly complex, with software assuming a significant role in modern production systems. The use of software enables adaptive and flexible production systems that can help address the above developments. However, as software becomes more prevalent, the potential for errors increases. The elevated number of kinematic degrees of freedom and the prevalence of highly dynamic msotion sequences in production systems with robots and other motion controlled machines contribute to an increased risk of downtime or even damage due to collisions. One method to mitigate this risk is virtual commissioning (VC), which enables the detection of errors in the control software of a production system through the use of simulation [3].

Once the commissioning phase of a production system has been completed, it is ready to enter operation with the control software that has previously been validated in VC. However, if changes occur during the operating phase, there is no guarantee that the production system will continue to operate without collisions [4]. Such changes or interventions into the control system, which cannot be tested beforehand through VC, are caused by humans, adaptive systems or complex processes [5]. The latter is particularly common in the manufacturing industry and is difficult to handle as processes are often dependent on various parameters and environmental influences. Such alterations to the process, and the deviations in motion that result from them, can give rise to collisions.

In this work, the process of wire arc additive manufacturing in combination with an industrial robot for motion generation is considered. The contribution of this work is the modeling of the robot-based additive manufacturing process in real-time and the transfer of the simulation to the operating phase. This paper is structured as follows: The process and the experimental setup are presented in the following section. Then the state of the art is considered before the concept and realization of this work are presented. The paper concludes with a summary and an outlook for further work.

2 Robotic Wire Arc Additive Manufacturing

Wire Arc Additive Manufacturing (WAAM) is an advanced additive manufacturing technique that utilizes an electric arc as a heat source to melt wire feedstock, layer by layer, to build parts. WAAM is particularly advantageous for producing thick-walled components due to its high deposition rate, which significantly reduces production time compared to traditional manufacturing methods. The process also benefits from the use of relatively inexpensive wire feedstock, making it cost-effective for large-scale parts [6].

One of the key advantages of WAAM is its high flexibility when integrated with robotic systems and rotary swivel units. This combination allows for the fabrication of complex geometries that are challenging to achieve with conventional subtractive methods [7]. Robots provide multi-axis movement capabilities, and the addition of rotary swivel units enhances this flexibility, enabling the deposition head to reach difficult angles and orientations. This capability is essential for producing parts with intricate features and overhangs, which are common in aerospace, automotive, and structural applications.

However, the complexity of fabricating such parts brings forth significant challenges in path planning. Effective path planning is crucial to ensure that the deposition head follows an optimal trajectory, which maximizes build quality while minimizing defects and manufacturing time. Incorrect path planning can lead to collisions with the rotary swivel units or the partially printed part, which can cause catastrophic failures in the manufacturing process.

Figure 1 shows on the left side the layout of the test facility used to analyse the results in this publication. The test facility is being operated as part of the project 'Safety assessment of additively manufactured thick-walled components' in cooperation with the Materials Testing Institute (MPA) of the University of Stuttgart. The project is investigating the use of WAAM as a production process for the manufacture of spare parts for nuclear facilities. The use of traditional manufacturing processes is currently costly, so WAAM, which is particularly viable for low-volume production, is seen as a promising alternative. However, high safety requirements for nuclear facilities remain a challenge. The cell consists of a Kuka robot KR120 R2700-2 robot with a nominal payload of 120 kg and a Kuka KP2-HV500 rotary swivel unit with a nominal payload of 500 kg. A Fronius MIG/MAG welding unit is mounted on the robot flange in combination with the Fronius TPS 400i welding unit, which is used for additive manufacturing of metal components. As part of the project, a digital image of the system has been created, which will be used to simulate the robot movement and map the process. Figure 1 shows the simulation model on the right.

Collision prevention in WAAM typically involves two key strategies. For static geometries collision avoidance is well-documented and relatively straightforward. These fixed parts are accounted for in the initial path planning stage, and their positions do not change throughout the manufacturing process. The more challenging aspect of collision prevention arises from the need to dynamically update the collision geometries of the moving parts and the incrementally manufactured part in particular. As the part grows layer by layer, its geometry changes, necessitating real-time adjustments to the path planning algorithm. This dynamic environment requires sophisticated sensing and computational techniques to continuously monitor the evolving geometry and predict potential collisions. Both aspects still pose major challenges at present and will be analysed in more detail below.

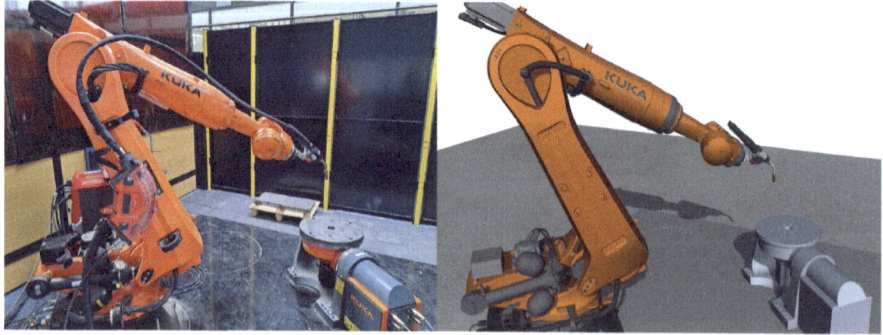

Fig. 1. A picture (left) and simulation model (right) of the experimental setup

3 State of the Art in Online Collision Avoidance

Like WAAM, many processes in industrial manufacturing are prone to collisions due to multi-axis kinematics and variable motion sequences. This section outlines the current state of the art in collision avoidance for production systems, using the example of robotics.

One possible solution is optical monitoring. For example, cameras are used to detect whether a person or another obstacle is in the working area of a robot arm [8]. One problem with these procedures is that it is never possible to monitor everything in the working area and there are always hidden areas or process-related restrictions such as cooling lubricants or undercuts. Another problem is that the configuration and costs of these additional sensors are often too high.

Cobots on the other hand are robots that are specially developed to work in interaction with humans. They use a collision detection method based on the measurement of torque acting on the robot and thus detect contact [9]. They are safe by design. The problem of such robots is the limitation regarding speed and load capacity. Another approach is the online use of geometrical simulations with virtual collision hulls. These hulls surround areas around physical components of the system. If several hulls overlap, a collision is detected [4]. Due to the use of a large industrial robot in this work and the requirement to detect all collisions even with undercuts and strong optical disturbances due to welding, the approach with virtual hulls is considered further here.

The utilization of virtual collision hulls for the purpose of safeguarding during the operating phase is based on an online simulation. The simulation is running synchronously with the real system in real-time. With collision hulls that are precisely adapted to the component geometry of the system, a stop would only be initiated as soon as a collision with the real system occurs. One approach to detect a collision before it occurs is to define the collision geometry to be larger than the physical object. In order to determine exactly how much larger the collision hull must be, the minimum braking distance required can be calculated using the maximum braking speed of the system. One disadvantage of this method is that a stop is initiated when the machine moves in the vicinity of other objects without there necessarily being a risk of collision. This makes it impractical for additive manufacturing with a layer height of a few millimeters and thus a distance to previously printed geometries. To simultaneously minimize the stopping area around an object and safely prevent collisions, a motion prediction is necessary. In industrial applications this can be realized by mathematical extrapolation, a virtual control system or by using internal data of the motion control [5]. An industrial solution from ModuleWorks [10] is available for the simulation-based collision monitoring of single-channel milling machines. In robotics, there are proprietary industrial products for online collision avoidance which use simplified geometries for the hulls and are therefore inflexible.

An additional process simulation is necessary to avoid collisions in the processes regarding the workpiece. ModuleWorks also provides simulation libraries for material removal and additive simulation. A solution for the collision avoidance of milling machines is also described by Schumann [11]. The difficulty with these process simulations in combination with collision detection is that virtual hulls are added or removed dynamically during runtime. So far, there are no solutions for online real-time collision avoidance with robotic WAAM. This is realized in this work using the application example.

4 Concept of the Online Real-Time Simulation

The idea of reusing simulations from the engineering phase in the operating phase goes back to Kain et al. [12] and was taken up by Klingel et al. [5] and applied to online collision avoidance for production systems. In VC, the control system is first connected to the simulation and then later to the real system. In principle, it is also possible to switch back and forth between these two configurations. However, the combined operation of both approaches is not part of VC. Figure 2 illustrates the concept of a combined architecture for the operating phase using the application example of this work for the illustration.

In this concept, the control system is connected to the drives and other field devices via the fieldbus as in the normal case without the online real-time simulation. In the case of the application example, this means that the robot, the rotary swivel unit and the welding device are part of this fieldbus communication with the control system. In addition, the control system is also connected to a simulation PC via a real-time capable fieldbus. The simulation PC and the model from the VC, which is real-time capable and has a fieldbus interface, are used for this. In addition to the process image (which is the communication between the control system and the real components) further internal control signals are transmitted to the simulation.

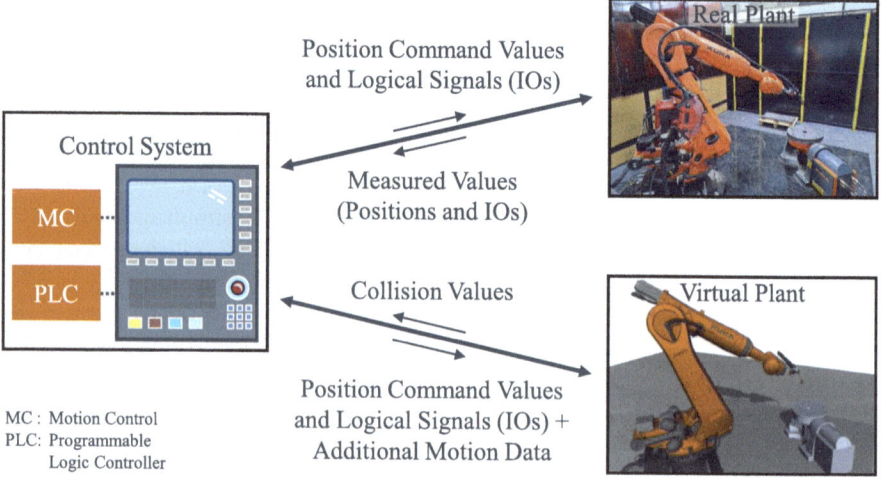

Fig. 2. Schematic illustration of the concept of the online real-time simulation

The openness and interfaces of the motion control system are the limiting factors here. Ideally, future position values can be read from a lookahead buffer of the motion control shortly before execution. A prediction horizon for this should range from a few hundred milliseconds to several seconds.

With the future axis positions a predictive real-time collision detection can be calculated in the simulation. In VC simulation systems collision detection algorithms, which are based on virtual hulls, are usually available and can be used for this purpose. Such virtual hulls are statically generated before the simulation is executed. What is not necessarily part of a VC simulation are process simulations such as simulating the additive manufacturing process. This process simulations may therefore need to be integrated into a VC simulation. A major challenge regarding this process simulations is the deterministic calculation in real-time because virtual hulls must be added dynamically during runtime.

If a collision is detected in the online simulation, this information is returned to the control system and processed there by the programmable logic controller (PLC). The PLC then stops the motion sequence safely before the real collision would occur or initiates another reaction such as slowing down the movement.

As described, this concept addresses the field level of a production system. This real-time approach increases the complexity of the system and requires an open motion control. Another option would be to use the path information from the control program as an input of the simulation. Figure 3 shows why this approach has considerable disadvantages compared to the selected concept in Fig. 2. Although the Tool Center Point (TCP) of the robot system moves to the same position, the robot is in a different joint configuration, so called pose. This means that no collision check can be carried out, as the exact position of the virtual bodies is unknown. To define the robot position reliably, the current angles of the robot joints are required, which are provided every millisecond in the approach presented.

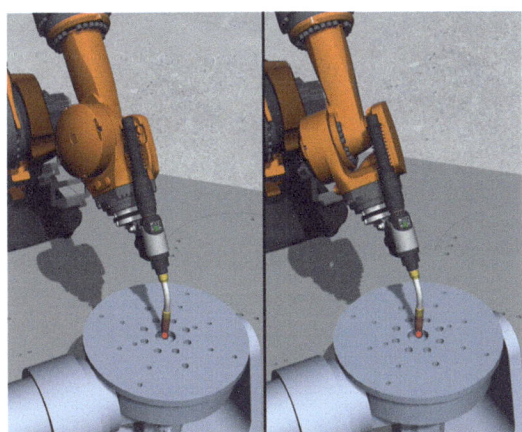

Fig. 3. Different axis positions for the same TCP

5 Implementation

As descriped in the previous section TCP-based robot controls are not optimal for real-time collision simulation in additive manufacturing because of the following reasons: Robot controllers are less suitable for additive manufacturing, where, like milling processes, high path accuracy is required. Robot tasks are mainly limited to pick-and-place applications, where the aim is to get from point A to point B as quickly as possible. The intermediate path is optimised for speed, not for path accuracy. Second, TCP-based control instead of joint-based control allows for ambiguous position which leaves it vulnerable to collision. And third, robot controllers work with a low lookahead of around 3 blocks, which makes collision simulation based on future axis positions difficult, as it is not possible to look far enough into the future to avoid collisions.

As a solution to these challenges, a computerized numerical control (CNC) was connected upstream of the robot controller. The resulting architecture is shown in Fig. 4. CNCs meet the requirements in terms of path accuracy and sufficient lookahead, and with G-Code a standardised interface for specifying print paths, as it is common in additive manufacturing. In this case, the Beckhoff TwinCAT3 control system with the ISG kernel was used as the CNC. Within the CNC, the kinematic transformation of the TCP positions supplied by the G-Code into robot joint angles takes place. In addition, a PLC runs within TwinCAT3, which performs the welding control within a state machine. The TwinCAT3 controller is then connected to the robot controller via the EtherCAT fieldbus, and the target and actual joint angles and the welding parameters are exchanged cyclically.

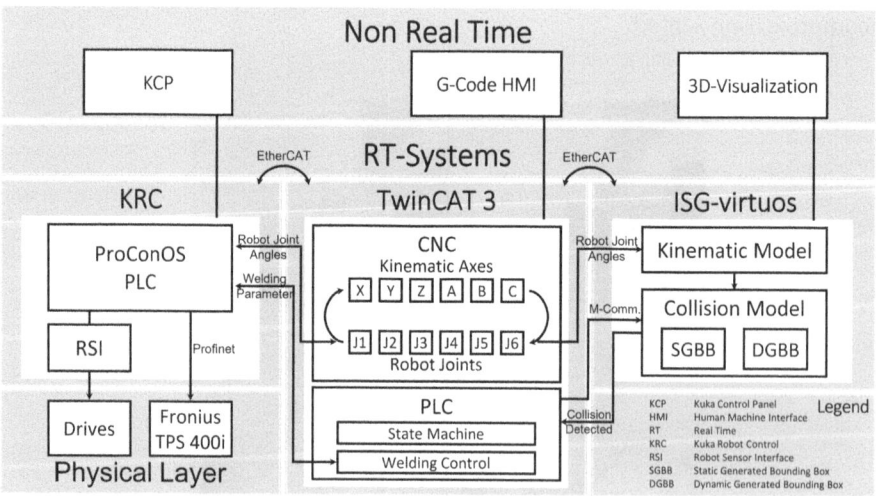

Fig. 4. Industrial control architecture for communication between CNC, robot and simulation

On the robot side, the ProConOS PLC integrated in the Kuka controller acts as an interface, which in turn transmits the joint angles to the motors via the Robot Sensor Interface and controls the Fronius welding unit via the Profinet fieldbus. In addition to the connection to the real robot, the TwinCAT3 controller is also connected to the simulation model.

The simulation consists of a real-time capable behaviour model and a non real-time visualization. The behavior model in particular consists of a kinematic model and a collision model. During the simulation, the collision detection checks cyclically whether the modeled collision objects overlap, which would be equivalent to a collision. In the case of the simulation tool ISG-virtuos which is used in this implementation, collision objects are modeled with so-called bounding boxes. In Fig. 5 the bounding boxes of the robot and the rotary swivel unit are shown. In order to model the real geometries accurate as possible with bounding boxes, the voxelization approach can be used, which means that a body is modeled with several smaller boxes. What is used here, for example, at critical points such as the welding unit.

Fig. 5. Static generated bounding boxes in ISG-virtuos

All bounding boxes in ISG-virtuos are statically generated during the modeling phase before the simulation is started. In Fig. 4 those are refereed to as static generated bounding boxes (SGBB). In addition to this SGBBs dynamic generated bounding boxes (DGBB) are necessary for the collision calculation with the workpiece, as its geometry and thus also the collision hulls changes during the manufacturing process. For this purpose an extension was implemented in form of a C++ function block in ISG-virtuos. In this function DGBBs are added when the plant is currently printing a workpiece. In order to maintain the real-time capability, the DGBBs are added to the simulation in a discrete form. A new DGBB is added as soon as the direction vector of the TCP changes significantly.

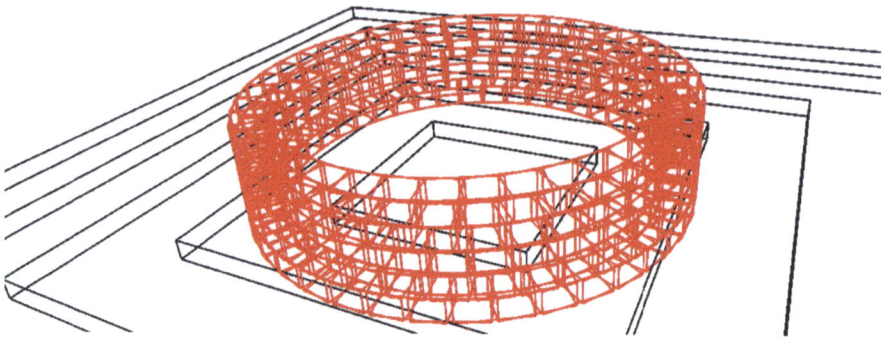

Fig. 6. Visualization of the dynamic generated bounding boxes while printing a cylinder

In the event of a constant change in direction, for example in the case of a circle, a threshold is set after which new DGBBs are created again at the earliest. A visualization of such DGBBs is shown at Fig. 6 using the example of a cylinder.

The future axis positions are read from the CNC via the ESA-function of TwinCAT3 as described above in order to recognize impending collisions in advance before they actually occur. The simulation therefore contains the current actual positions and the future positions of the robot axes. In this implementation, this is used to simulate both the current behavior and the predictive behavior. This is shown in Fig. 7. The current position of the robot is shown in full color and the prediction model is visualized transparent. If the transparent predictive robot collides in the simulation, the real system can be stopped or slowed down in advance to the collision. Should the simulation with the actual robot positions collide in a worst case scenario the whole plant will be stopped and has to be reset to the initial state by a human.

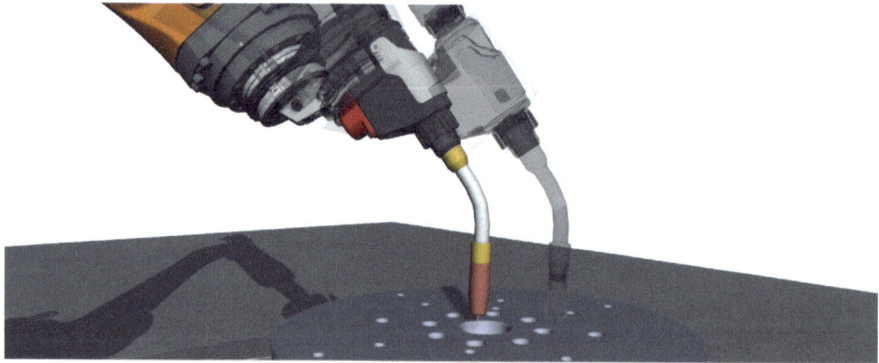

Fig. 7. Online visualization of the actual robot position (full colored) and the prediction (transparent)

6 Results and Discussion

The functionality of the implementation is demonstrated using an example component and a forced movement which would lead to a collision. A cylinder is used as a workpiece in this example, as it is more difficult to approximate due to the bounding boxes. In the current implementation, ISG-virtuos cannot visualize the DGBBs dynamically. The visualization in Fig. 6 was therefore added after a simulation run. In order to display the result better, a visualization of the current boxes was inserted in ISG-virtuos after the bounding boxes were generated and before the collision scenario was evaluated.

Individual time excerpts from a video of the evaluation scenario are shown in Fig. 8. In this scenario, a injected fault in the CNC program would lead to a collision between the workpiece and the welding unit. In the bottom part of the picture the real robot is shown and in the middle part the simulation with the current position and the prediction is presented. Additionally the axis enable value is shown. In this scenario the axis release is withdrawn in case of a virtually detected collision which leads to a stop of the robot.

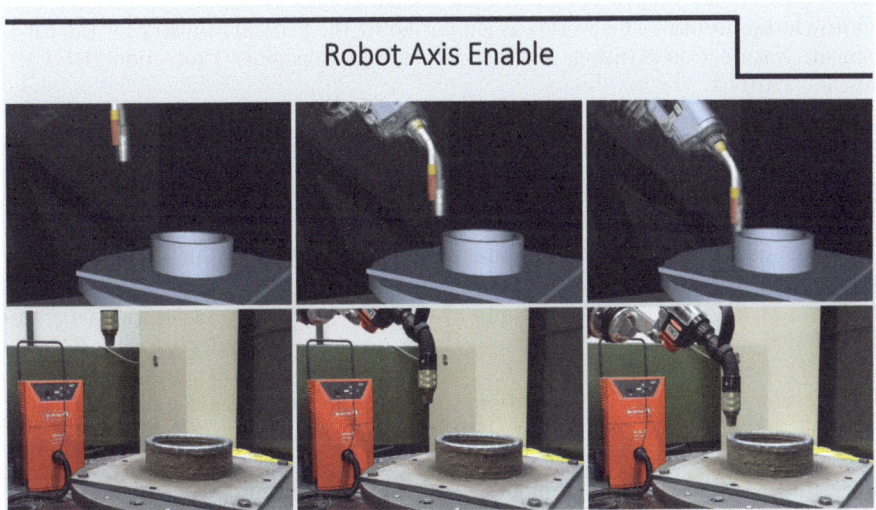

Fig. 8. Image excerpts from a video of the evaluation scenario with the real system on the bottom and the simulation above

With the current solution, errors in the control program and external interventions in the control system by humans or other systems which would lead to a collision can be well protected. This was proven by the evaluation on one example workpiece so far. In the future, however, additional measured process variables that may have an influence on the component geometry must be measured and included in the DGBB generation. In addition, an online visualization of the DGBBs should be developed.

For the collision calculation of DGBBs, it must also be investigated how physics engines or libraries such as the one from ModuleWorks address this problem and to what extent they can be integrated to further improve this approach.

7 Conclusion

In this work, a simulation-based online collision avoidance is designed using the example of WAAM and realized with the help of industrial components. The transfer of models from virtual commissioning is used in this approach and an extension for dynamically generated collision hulls is implemented. Based on an experimental evaluation on the real test setup, it was shown that the solution works. For this purpose, a movement of the robot was executed that would lead to a collision between the robot tool and the workpiece and could be prevented by the system developed in this work. In the future, further components could be used for validation. Furthermore, it must be investigated how geometrically more accurate collision hulls can be used in real-time, especially in the simulation of the additive manufacturing process.

Acknowledgements. This work was supported by the Federal Ministry for the Environment, Nature Conservation, Nuclear Safety and Consumer Protection (BMUV), grant no. 1501654.

References

1. Koren, Y.: The Global Manufacturing Revolution. Wiley, 2010. ISBN 9780470583777. https://doi.org/10.1002/9780470618813
2. Jones, M. D., Hutcheson, S., Camba, J. D.: Past, present, and future barriers to digital transformation in manufacturing: a review. J. Manuf. Syst. **60**, 936–948 (2021). ISSN 02786125. https://doi.org/10.1016/j.jmsy.2021.03.006
3. Pritschow, G., Röck, S.: Hardware in the loop simulation of machine tools. CIRP Anna. **53** (1), 295–298 (2004). ISSN 00078506. https://doi.org/10.1016/S0007-8506(07)60701-X
4. Schumann, M., Witt, M., Klimant, P.: A real-time collision prevention system for machine tools. Proc. CIRP **7**, 329–334 (2013). ISSN 22128271. https://doi.org/10.1016/j.procir.2013.05.056
5. Klingel, L., Heine, A., Acher, S., Dausend, N., Verl, A.: Simulation-based predictive real-time collision avoidance for automated production systems. In: 2023 IEEE 19th International Conference on Automation Science and Engineering (CASE), pp. 1–6 (2023). https://doi.org/10.1109/CASE56687.2023.10260637
6. Srivastava, M., Rathee, S., Tiwari, A., Dongre, A.: Wire arc additive manufacturing of metals: a review on processes, materials and their behaviour. Mater. Chem. Phy. **294**, 126988 (2023). ISSN 02540584. https://doi.org/10.1016/j.matchemphys.2022.126988
7. Attaran, M.: The rise of 3-D printing: the advantages of additive manufacturing over traditional manufacturing. Bus. Horiz. **60** (5), 677–688 (2017). ISSN 00076813. https://doi.org/10.1016/j.bushor.2017.05.011

8. Schmucker, U., Haase, T., Schumann, M.: Digital engineering and operation, pp. 283–375 (2015). https://doi.org/10.1007/978-3-662-48266-7_5
9. Haddadin, S., de Luca, A., Albu-Schaffer, A.: Robot collisions: a survey on detection, isolation, and identification. IEEE Trans. Rob. **33** (6), 1292–1312 (2017). ISSN 1552-3098. https://doi.org/10.1109/TRO.2017.2723903
10. ModuleWorks. Collision avoidance system (2024). https://www.moduleworks.com/moduleworks-collision-avoidance-system/
11. Schumann, M.: Methodik einer Echtzeit-Kollisionserkennung und -vermeidung für Werkzeugmaschinen: Berichte aus dem IWU. Verlag Wissenschaftliche Scripten (2021)
12. Kain, S., Dominka, S., Merz, M., Schiller, F.: Reuse of HiL simulation models in the operation phase of production plants. In: 2009 IEEE International Conference on Industrial Technology, pp. 1–6 (2009). https://doi.org/10.1109/ICIT.2009.4939562

Open Access This chapter is licensed under the terms of the Creative Commons Attribution 4.0 International License (http://creativecommons.org/licenses/by/4.0/), which permits use, sharing, adaptation, distribution and reproduction in any medium or format, as long as you give appropriate credit to the original author(s) and the source, provide a link to the Creative Commons license and indicate if changes were made.

The images or other third party material in this chapter are included in the chapter's Creative Commons license, unless indicated otherwise in a credit line to the material. If material is not included in the chapter's Creative Commons license and your intended use is not permitted by statutory regulation or exceeds the permitted use, you will need to obtain permission directly from the copyright holder.

Real-Time Online Simulation at Field Level in Industrial Automation

Darius Deubert[1,2](✉), Andreas Selig[1], and Alexander Verl[2]

[1] Bosch Rexroth AG, Bgm.-Dr.-Nebel-Straße 2, 97816 Lohr am Main, Germany
darius.deubert@boschrexroth.de
[2] Institute for Control Engineering of Machine Tools and Manufacturing Units, University of Stuttgart, Seidenstraße 36, 70174 Stuttgart, Germany

Abstract. In industrial automation, simulation is extensively used during the engineering phases, but only rarely during the operational phase. The simulation of a manufacturing system in parallel to its operation, with the simulation system integrated into the mechatronic system, is known as online simulation and is beneficial for system monitoring, diagnosis, predictive analyses, decision support or online optimization. This paper focuses on the implementation of real-time online simulation at field level, reusing virtual commissioning simulation models. Main architectural aspects are the integration of the simulation system towards a control system and the management of the data distribution within a multi-functional edge device or control system. An exemplary realization in the form of a demonstrator is presented, showcasing the successful implementation of an online simulation using a comprehensive virtual commissioning model of a servo drive. The findings of this research contribute to the aim of using extended virtual commissioning simulation models in the operation phase of manufacturing systems and can be seen as indication for further realizations of online simulation systems.

Keywords: simulation · digital twins · manufacturing · robotics and automation

1 Introduction

Considering the shortening of product life cycles and the increasing complexity of machines and product diversity, simulation plays a vital role in optimizing manufacturing systems concerning time, energy, and costs. However, simulation of industrial components and machines is mainly used for the design and commissioning, but there is still a huge potential of using simulation in the operation phase [4]. While the idea of online simulation has been used for production planning for a long time, it is an object of research at the machine level [3]. Many theoretical approaches exist for online simulation, but its realization for mechatronic systems in discrete manufacturing is still an unsolved topic. This paper focuses on the realization of real-time online simulation at field level, which profits from the reusability of Hardware-in-the Loop (HiL) virtual commissioning models and tools and is directly connected to a real industrial control system.

1.1 Hardware-in-the-Loop Simulation

The original aim of HiL simulation was to test hardware within the development of mechatronic systems [8]. The term HiL is always referring to a device under test. In industrial automation, from the perspective of machine builders, the device under test usually is a control system, like a programmable logic controller (PLC) or a computer numerical control (CNC). The goal of a HiL simulation in this context is that it can be coupled to the control system without any changes in software or hardware, so that the functionality of the control system could be tested without the necessity of coupling it to the real machine [11].

1.2 Virtual Commissioning

Using a simulation model to develop and validate automation systems is also known as virtual commissioning. It increases quality and efficiency, as the automation system can be validated and optimized before real commissioning [1,6]. Typically, an industrial control system is tested with simulated components, where simulation can be realized with an emulated control system (Software-in-the-Loop simulation) or a real hardware controller (HiL simulation). By virtual commissioning, basic kinematics of the manufacturing system can be tested and the automation solutions could be validated with respect to different functionalities, special scenarios, or performance. However, virtual commissioning usually does not include dynamic process simulation including energy analysis [6].

1.3 Online Simulation

The simulation of a manufacturing system in parallel to its operation, with the simulation system coupled to the mechatronic system, is known as online simulation. Through an integration at field level, the online simulation system becomes a part of the machine. Ideally, the simulation is seamlessly incorporated following the plug-and-simulate principle.

To save costs and efforts it is reasonable to reuse existing models from the design and commissioning. The idea of using virtual commissioning models in the operation phase goes back to Kain et al. [4] and was already taken up by Klingel et al. [5] in the context of online collision avoidance. However, there is still a lack of concepts regarding online simulation reusing virtual commissioning models. While Klingel et al. [5] use data from the control system and kinematic as well as collision models, this work focuses on the reflection of dynamic and logic behavior of field level automation components.

1.4 Paper Outline

Within this paper, firstly, architectural considerations, especially on the integration of a simulation system at field level and the data distribution within multi-functional edge devices, are described (see Sect. 2). Based on this, the implementation of a demonstrator for online simulation at field level is presented in Sect. 3. The results are validated by considering a positioning procedure and discussed also regarding implementation drawbacks (see Sect. 4).

2 Architectural Considerations

In contrast to simulation technologies at the state of the art, online simulation has to be coupled to a mechatronic system to reflect its current state and behavior. Hence, the main architectural challenge for the realization of an online simulation system is to integrate the simulation into the manufacturing system.

2.1 Integration at Field Level

To integrate a simulation into a mechatronic system, real-time simulation environments are necessary, as simulation results based on online data have to be available for the control system in real-time. Moreover, to reflect the real interfaces in simulation, it is reasonable to use the prevalent fieldbuses and IoT protocols also for the integration of online simulation into mechatronic systems. Virtual commissioning models are suitable to be connected to a control, as they are mainly used in this configuration for HiL simulation to test the control functionality. They already support connections via fieldbus or IoT protocols and meet real-time requirements [6]. For the integration of online simulation into a mechatronic system at field level, there are mainly two possibilities (see Fig. 1), which already were proposed by [5] and are analyzed further in the following. We assume an exemplary industrial control system, which provides at least a PLC application to control the manufacturing process, a fieldbus driver for the real-time communication with field devices, like industrial servo drives, and a data layer or middleware as interface for the applications at the control system.

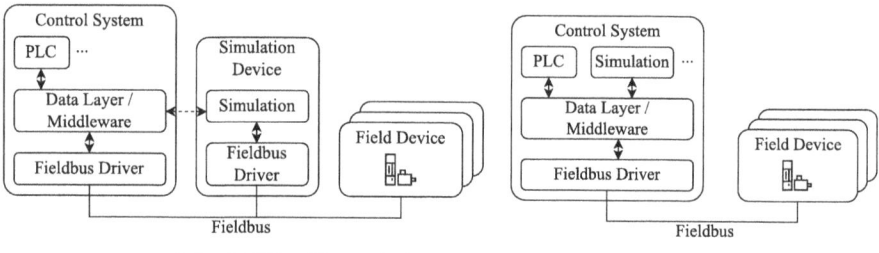

Fig. 1. Integration of online simulation at field level (Based on [5])

Dedicated Simulation Device. One possibility is to use a dedicated simulation device, which is integrated into the existing fieldbus (see Fig. 1a). In this case, the simulation device can either spy on the telegrams that are addressed to the regular field devices or receive its own telegrams. The specific implementation is depending on the used fieldbus.

Depending on the scope of the online simulation, there may be additional properties, which are not included in the regular fieldbus process data. These properties could concern the manufacturing systems' environment or process specific information, which the regular field devices do not need to know. The additional data may be distributed by the PLC application itself but may also come from an additional instance on the control system, which manages data and information on the manufacturing system and process. To include data, which is not present in the existing fieldbus configuration, this data could either be distributed by additional telegrams on the fieldbus or by establishing an additional interface between the control system and the simulation. Using the fieldbus for additional data exchange benefits from the robustness and real-time capability of industrial fieldbuses. However, they increase the fieldbus load and effect unwanted reconfiguration in brownfield applications. Using additional interfaces is more flexible but may also increase the overall system complexity.

Another option is not using the existing fieldbus at all. In this case, the necessary fieldbus data has to be distributed separately, which is inconvenient if field devices are simulated including their logic behavior.

An advantage of using a dedicated simulation device is that it could also be integrated into a brownfield application. Furthermore, the simulation device provides computational resources dedicated for the simulation and does not impede the operation of the control system. The main disadvantages are that an additional device is necessary and communication is getting more complicated, which both are related to increased costs and effort.

Control Integrated Simulation. The approach of using a simulation within a control system goes back to Sekler [14], where it was used to simulate oscillation behavior to reduce vibration in lightweight machines. If the simulation is directly integrated into the control system (see Fig. 1b), a direct communication with low latency is possible [5]. Furthermore, the data could be exchanged without the need for cyclic or acyclic fieldbus communication. An integration of the simulation into the control system has the advantage that no additional simulation device is necessary. Moreover, the exchange of data only via a data layer or middleware is more flexible and easier to implement compared to external communication. The disadvantage of this method is that it is often not possible to be applied on brownfield applications. The integration has to be supported by the control systems, which is not popular yet but likely to gain relevance in the future. Additionally, the computational load of the control device is significantly increased by the simulation, while resources often are strongly limited.

2.2 Data Distribution Within Multi-Functional Edge Devices

In industrial automation, control functionalities like PLC applications are increasingly handled by software oriented multi-functional edge devices (MFEDs) recently, which are hosting diverse tasks besides traditional control and are accessible via information technology (IT) and operational technology (OT) infrastructure [2]. Comparable systems are also known, for example, as IoT-PLC [7] or edge controllers [15]. Considering such MFEDs, the distribution of process data is not trivial, as there are diverse peripheral components and multiple applications on the MFED, which are exchanging data (see Fig. 2).

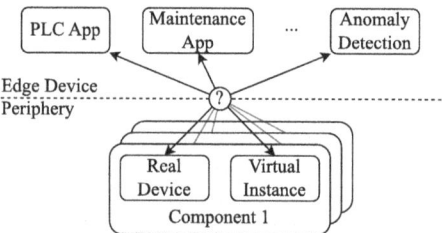

Fig. 2. Data distribution in an exemplary MFED.

The problem is intensified due to the representation of components by multiple instances. Each device could have multiple digital twins based on different information models or behavioral simulations, which run on different devices. Each of these representations contains different information about the same properties of the represented system, so they have to be distinguished carefully by some applications, like for anomaly detection or maintenance applications. Conversely, other applications, like the PLC or CNC, do not consider multiple instances of a component. These applications should be able to address the corresponding instances either directly or on a logical level.

As a solution for this problem, we propose using a middleware or data layer on the MFED, where the applications could access a data space regarding the particular component, which they want to address. Data distribution services take care of the communication with the peripheral devices. For this, the following possibilities are suitable.

Hierarchical Data Layer. In a hierarchical data layer the components' data spaces are divided into logical, virtual, or real subspaces (see Fig. 3a). The applications as well as the data distribution services can directly access these spaces. The real and virtual subspaces are directly related to a real device, which could be a field device, a simulation device, or any other implementation of a digital twin, and managed by the data distribution services, while the logical subspaces reflect the data objects from real or virtual subspaces and therefore rely on a mapping.

Multi-dimensional Data Layer. A multi-dimensional data layer is handling the real peripheral devices separate from the components' instances. Therefore, two layers are used, where the application data layer manages only logical instances to be accessed by the applications and the communication data layer manages the real peripheral devices for the data distribution services (see Fig. 3b). The logical instances of components at the application data layer reflect real or virtual representations of the components and rely on a mapping to the communication data layer. If a device should be addressed directly by an application, a corresponding logical instance of this device has to be established. By this, logical instances of components are strictly separated from the real devices.

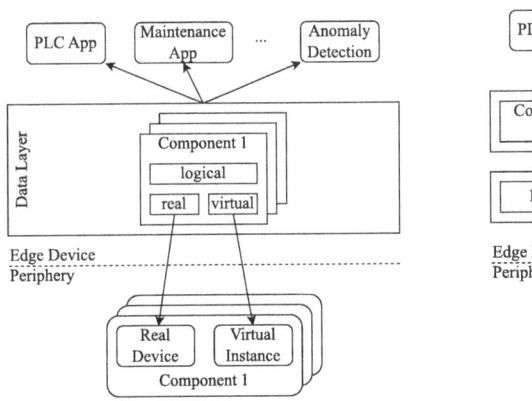
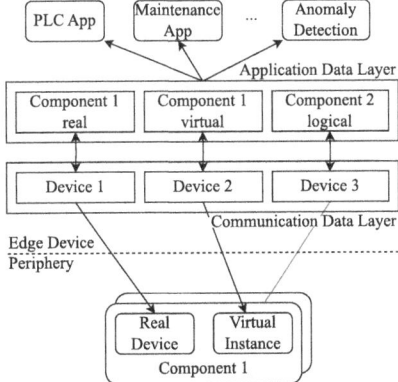

(a) Exemplary solution with hierarchical data layer.

(b) Exemplary solution with multi-dimensional data layer.

Fig. 3. Data distribution approaches for multi-functional edge devices

3 Implementation

For the demonstration of real-time online simulation at field level, a setup is chosen, where an industrial control system controls an electrical servo axis (positioning axis) of an imitated conveyor belt, whose behavior is replicated by a load motor, which applies the load to the positioning axis that the axis would face if it was connected to the real mechanics. By this, no complex machine is necessary, but the setup – including the considered positioning axis – is consisting of real automation components and therefore sufficiently realistic to demonstrate an implementation of real-time online simulation at field level.

This system is enhanced by an online simulation, which, in contrast to [5], is running on a dedicated simulation device, as this enables the setup to be applied to a variety of different control systems by different suppliers. The existing fieldbus is used as exclusive interface between the simulation and the control system, as this ensures reliable real-time data exchange.

3.1 System Overview and Communication

The considered system consists of a control system, a servo drive with a servo motor as a field device, and a simulation device that is also connected via the fieldbus (see Fig. 4). The control system is used for the management of the servo axis as field device and the execution of positioning tasks, utilizing an EtherCAT fieldbus with a cycle time of 2 ms. EtherCAT is used in the demonstrator for:

– configuration of the real and virtual servo axes via their engineering interface by using the control to forward Ethernet telegrams via Ethernet over EtherCAT (EoE)
– cyclic and acyclic communication between the control system and the servo axes via the servo drive profile over EtherCAT (SoE)

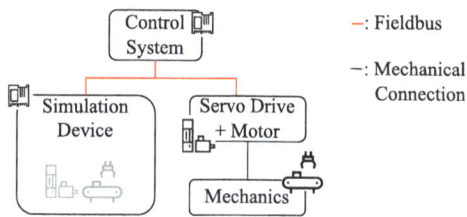

Fig. 4. Overview of online simulation demonstrator

The cyclic communication is used for command values and control words, which specify the movement and operation modes of the servo axes, as well as for the status words and actual values that inform the control system about the state of the drives. To distinguish between the real and the virtual positioning axes, the real and the virtual drive have individual EtherCAT addresses. If the real and the virtual drive should be treated exactly the same, this leads to the necessity of copying the master data for both telegrams, but also enables to consider the responses separately.

Individual data objects, like the load mass, the conveyor feed rate, and the conveyors gradient angle, are transmitted from the control to the simulation by the configuration of digital and analog I/O modules, which are managed by the simulation device and only present in the virtual space. Using virtual I/O modules as containers for additional data objects has the advantage, that the process data does not underlie any given data profile and that the data is transmitted simultaneous with the existing process data. However, for more complex additional data objects, an individual profile should be preferred.

As the used simulation device effects jitter on the fieldbus, the EtherCAT distributed clock synchronization mechanisms are not working reliably. This means that precise synchronization across the field devices is not possible reliably, so the demonstrator uses operation modes, where no accurate synchronization is necessary. Operation modes, where the interpolation of a trajectory is done at the control system require the fieldbus to reliably transport new command values in every NC cycle. Otherwise, the drive shows unwanted behavior, such as using the same command value twice or errors due to excessive deviation between actual and command values.

3.2 Control System

As control system, a ctrlX COREplus X3 by Bosch Rexroth is used, which incorporates a PLC application and an EtherCAT master instance. The PLC application comprises the control of the positioning, the control of the load imitation, handling of the EtherCAT process data, visualization of the process and collection of data, and user interface via visualization elements. All tasks are performed in the EtherCAT cycle of 2 ms.

To realize the online simulation, the virtual positioning axis has to receive the same inputs as the real positioning axis. For this demonstrator, this is implemented by copying the fieldbus master data for the virtual drive from the real drive in the PLC but sending separate EtherCAT telegrams. The drives' responses are handled separately anyway. Assuming consistent initial conditions and sufficiently accurate simulation models, the virtual axis will behave correspondingly to the real positioning axis.

3.3 Positioning Axis

The positioning axis consists of a ctrlX DRIVE XMD1-W1616 inverter and a MSK050B-600 synchronous servo motor by Bosch Rexroth. The drive is addressed by the control system as an EtherCAT slave and runs in the drive-controlled positioning operation mode, where the interpolation of a trajectory is done at the drive and the positioning profile is specified by parameters given by the control system. The profile is defined by a positioning velocity, as well as a maximum acceleration and jerk. The drive receives a position command value and a positioning control word, which triggers the positioning. The current status is fed back to the control via a positioning status word and the actual position and velocity values.

3.4 Mechanics

As mentioned before, for the demonstration of online simulation with a focus on the implementation and integration of a simulation system into a real mechatronic system, it is important that real automation components from industrial practice are used and that there is a dynamic behavior with real electronic components and movement so that the components are applied in the way they would be in industrial practice. The particular dynamic behavior is substitutable, as long as the resulting torques underlie a realistic behavior.

This is why a conveyor belt is chosen as a basic example within this work. Instead of using a real conveyor belt, the physical behavior is imitated by applying a corresponding load torque to the positioning axis. This is done by coupling the positioning servo motor to a load motor that is controlled by another servo drive and applies a torque corresponding to the load the positioning motor would experience with a real conveyor belt. The load axis is realized with a ctrlX DRIVE XMS1-W0036 and a VEM K21R 71 G 2 three phase AC motor.

The load torque is determined by a model, which runs on the control system. As the complexity of the mechanic behavior is irrelevant within the scope of this work, only a certain inertial load and a gradient effecting a static influence of the gravitation are modeled (see Fig. 5). The resulting load torque is determined by adding the transformed gravitational force $F_G = \sin(\alpha) \cdot m \cdot g$, where α is the gradient angle, m is the load mass, and g is the gravitational constant, to the resulting force, which can be determined by multiplying the moved mass m with its acceleration, and transforming this sum via the radius r to the torque T, which has to be applied at the load motor to imitate these forces. This model does not consider friction and inertial mass that is not affected by the gravitation, like rotating symmetric parts, as representing a real application accurate is not necessary to demonstrate the implementation of an online simulation system. To avoid oscillation caused by the delay between acceleration measurement and reaching the torque command value, the model does not consider the measured acceleration but the command value of the positioning axis.

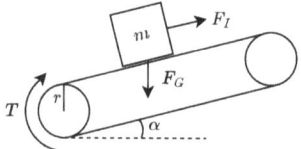

Fig. 5. Simple conveyor belt model

3.5 Simulation System

The simulation system must guarantee a deterministic response time, as the communication is realized by a real-time fieldbus. Hence, the simulation on standard operating systems is not suitable [11]. Within this research, ISG virtuos has been chosen as a simulation platform, which is known as a tool for virtual commissioning and therefore provides a lot of functionalities that are also beneficial for online simulation [13]. ISG-virtuos provides the possibility to distribute the simulation on a Windows target, as well as on a real-time target (RTT) with a Beckhoff TwinCAT runtime environment via multiple solvers. The RTT operates on hard real-time conditions and also provides the fieldbus interface.

In the case of EtherCAT, the simulated slave devices have to be the first slaves in the fieldbus strand. As a complete image of the EtherCAT master configuration is arranged at the simulation device, the real slaves remaining on the EtherCAT strand have to be deactivated in the configuration at the simulation device and a second network adapter has to be configured to forward the telegrams to the real EtherCAT slaves.

Figure 6 shows how the models are arranged. The interface to the control system is the fieldbus. The process data are handled by the fieldbus driver, which is implemented with a TwinCAT solution including an EtherCAT simulation device. These data are accessed by the simulation solvers so that they could be processed by the simulation models. As the solvers are not synchronized to the fieldbus cycle, they have to run with twice the frequency of the fieldbus cycle. In this case, the fieldbus data is processed by a comprehensive model of the ctrlX DRIVE inverter, which includes the logic behavior by incorporating large parts of the real drive firmware and uses the fieldbus data as interface towards the control system. Corresponding to the real drive firmware, the drive model bases on parameter sets, which define the state and configuration of the drive. To reflect the real positioning axis in the simulation, the original parameter sets are taken to initialize the simulated drive. By this, the configuration of the real drive is transferred into the simulation except minor adjustments, like optimized control parameters, which are necessary because of the divergent control cycles.

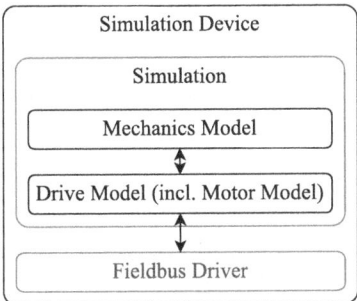

Fig. 6. Overview of simulation composition

The used drive model also incorporates a simple motor model, which is necessary to model the delay of the torque controller until a demanded actual torque is reached. Within this work, this is done via a PT1 element. The actual torque, which is the output of this PT1 element, is fed to the mechanics model, which models the current load and determines a resulting acceleration by considering the motor torque, gravitation, and inertia. This acceleration is then integrated to determine the resulting velocity and position. The position is fed back to the drive model, where it is processed by a virtual encoder. By this, the actual position and velocity are also fed back to the control system as fieldbus data. The models are run in a TwinCAT solver on the RTT.

Summing this up, the simulation reflects the behavior of the positioning axis' drive, motor, and mechanics. Moreover, it uses the same interfaces as the real positioning axis: the drives fieldbus data. This enables the control system to provide the virtual axis with the same inputs as the real axis and therefore is a suitable configuration for an online simulation system.

4 Results and Discussion

4.1 Evaluation of Positioning Application

To show that the implemented online simulation system reflects the behavior of a real application, a typical positioning task is considered in this section. The conveyor belt is configured with a load mass of 30 kg, a feed rate of 0.36 m/r, and gradient angle of 4°. Note that within this work, the results are interpreted qualitatively, as the demonstrator focuses on implementing the principles of online simulation with components from the industrial practice and not on highly accurate system modeling. Reflecting the real behavior quantitatively as accurate as possible in the online simulation would be the next step when it comes to the application in a real machine.

Figure 7 shows the positioning behavior of the previously described demonstrator. The PLC defines a positioning command value, which is included in the cyclic master data telegram for the real positioning drive and therefore also copied to the master data telegram for the simulated positioning drive. The positioning is triggered by the positioning control word, which is also identically sent to the real and simulated positioning drives. Subsequently, the axes accelerate to the positioning velocity and decelerate to stop at the desired position. The corresponding trajectory is interpolated by the drives following the velocity and acceleration profile given by the master data.

The positioning was recorded via the PLC application with a sampling time of 2 ms. While the position diagram shows the original position values of the drives, the velocity and the torque of the real drives are adjusted by a moving average filter over ten adjacent values corresponding to a 20 ms time window to filter the noise of the physically measured signals for the visualization.

As a comprehensive drive model is used, the calculation of the trajectory corresponds well between the real and simulated drive. The cascaded position, velocity, and current control loops reliably ensure that the drives follow the determined trajectory. Hence, also the actual position values correspond well between the real and simulated drive (see Fig. 7a). Comparing the velocity (see Fig. 7b), different overshoot behavior occurs when the target velocity and the target position are reached, respectively. This is due to the divergent control loops and can also be seen in the torque diagram. Nevertheless, also the velocity curve is sufficiently reflected by the online simulation.

To compare the dynamic behavior of the simulated axis to reality, the torque curves (see Fig. 7c) are of special interest. During standstill, the real and virtual axes are affected by a static torque respectively due to the conveyor's gradient, which correspond well. When the acceleration and deceleration start and end, torque overshoot can be seen at the simulated positioning axis due to the divergent control loops in the simulated drive. During the movement, the torque of the simulated axis corresponds well to the load, which is determined for and applied at the load axis. However, the torque measured at the real positioning axis slightly diverges from this load, even though they are mechanically coupled. The reasons for this are discussed in the following section.

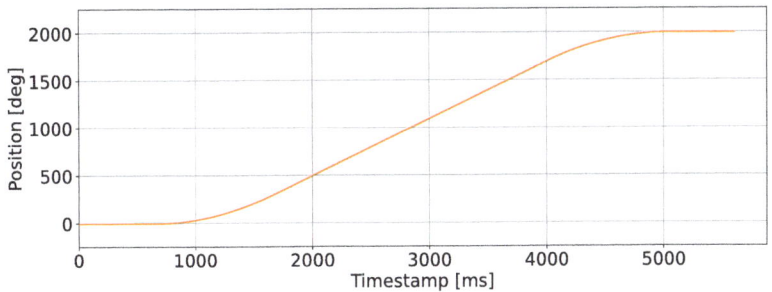

(a) Actual position (position command value: 2000°)

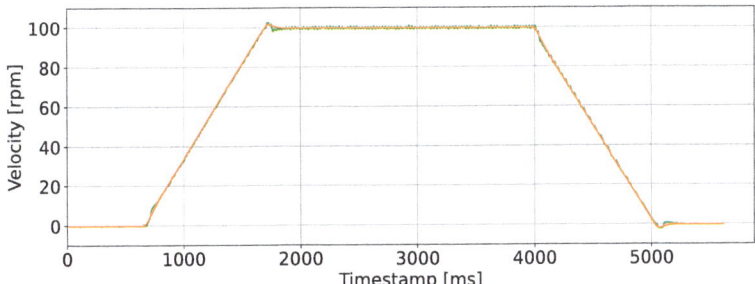

(b) Actual velocity (positioning velocity: 100 rpm, positioning acceleration: 10 rad/s^2)

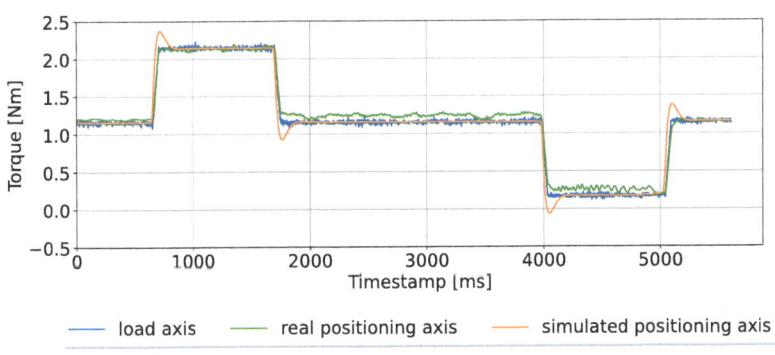

(c) Actual torques

Fig. 7. Resulting positioning behavior

The CPU load at the simulation device was about 7% and the memory usage was about 6%. This result is only partly applicable to real applications, as the chosen mechanical model is comparably simple and the imitated machine only contained one servo axis. Dynamic multi-body simulation with multiple axes would need significantly more CPU power. Nevertheless, this shows the real-time capability of the demonstrated technology.

4.2 Interpretation of Results

The torque discrepancy between load and positioning axis, which arise at constant speed and deceleration, may result from friction, axial displacement at the coupling and imprecise measurement. The torques are measured indirectly via the phase currents, which are transformed to a torque forming current, relying on several assumptions and a precise position feedback of the motor's rotor.

The remaining discrepancy of the simulated axis is resulting from the divergent control loops in the drive simulation model. In the real drive, the control loops are cascaded and run with a cycle time of 62.5 μs to 250 μs. In contrast, the simulated control loops are running with a cycle time of 1 ms. Simulating multiple small steps in each simulation cycle is possible in theory but circumstantial in practice, as complicated model structures are necessary to exchange the feedback values. Furthermore, running the simulation models with a step size of 62.5 μs is increasing the computational load significantly and therefore impairing the real-time capability. The resulting discrepancy can be further reduced by manual tuning of the control parameters in the simulation.

As the online simulation is integrated via the fieldbus and the drives run in a positioning mode, temporal synchronization is automatically handled within this demonstrator. If there were significant disturbances or modeling inaccuracies, discrepancies would arise that should be handled by a synchronization mechanism. Furthermore, if the considered drives ran in an open loop control or a velocity or torque control mode, the discrepancies would be integrated resulting in persistent position discrepancies, which are not compensated by the individual closed loop control and therefore requires all the more an adaption mechanism to reduce these discrepancies.

The demonstrator shows that a real-time capable online simulation integrated at field level could be realized by using components from industrial practice. The online simulation reflects the behavior of a real servo axis well and provides results within the cyclic fieldbus communication. Furthermore, the result confirms that virtual commissioning simulation models extended by dynamic models are suitable to reflect the dynamic behavior of a real application.

4.3 Remaining Challenges

A big challenge concerning online simulation is the initialization of the simulation, which has to reflect the current state of the manufacturing system. For this demonstrator, the real mechanics could be easily transferred to a stable initial state manually. However, for some machines there is no stable initial state and in industrial practice, the online simulation system should be fully automated and not rely on a manual initialization. Capturing the manufacturing system's state at any point of time and trying to apply this state to the online simulation is not working if comprehensive component models are used, as these usually are black box models, where the state could neither be fully captured from reality nor be set directly in the simulation environment.

Many contemplations on online simulation propose to use synchronization mechanisms [9,10,12,16]. Within this demonstrator, a synchronization is not necessary, as the main state variables of the manufacturing system are either discrete and stable, like configurations of the drive, or are controlled by the drive itself and therefore kept from diverging from the desired state, like the axis position or velocity. However, this does not apply generally, so the proposed relevance of synchronization mechanisms should not be queried by this work.

A disadvantage of the chosen interface is that it would be drastically more complicated to realize without a comprehensive drive model by the component manufacturer. Modeling the logic behavior of such an industrial component based on the fieldbus data and reflecting the numerous operation modes, configurations, and exceptions is too costly for a machine manufacturer to be a suitable solution for online simulation, so in this case, other interfaces and scopes for the simulation have to be chosen. In the ideal case, the component manufacturer provides comprehensive models of its components, as they know the particular system behavior the best and the modeling efforts are scaled.

5 Conclusion and Future Work

This article presents concepts and implementation approaches for real-time online simulation at field level in industrial automation. The main architectural aspect is the integration of a real-time simulation to an industrial control system at field level. Regarding this, different approaches are compared and a special interest is given on the data distribution within multi-functional edge devices.

Moreover, the concepts have been proven by the implementation of a demonstrator, which is realizing an online simulation for a conveyor belt application using real components from industrial practice. The demonstrator shows the feasibility of the proposed concepts for online simulation at field level and points out the drawbacks of the used methods.

For future work, special attention should be given to the aspects of initializing the online simulation with the current manufacturing system state and synchronizing the simulation to its real counterpart to prevent inaccurate reflection.

Acknowledgement. The work presented in this paper has been partly funded by the German Federal Ministry for Economic Affairs and Climate Action (BMWK) under the project 13IK001ZF "Software-Defined Manufacturing for the automotive and supplying industry https://www.sdm4fzi.de/".

References

1. Barth, M., Puntel-Schmidt, P., Hoernicke, M., Oppelt, M., Wolf, G., Stern, O., et al.: Methoden und Modelle der virtuellen Inbetriebnahme. In: Automation 2015 Benefits of Change-the Future of Automation, pp. 107–120. VDI Verlag (2015)
2. Basem, M.: Comparing PLC, software containers and edge computing for future industrial use: a literature review. Master's thesis, University of Skövde, School of Engineering Science (2022)

3. Deubert, D., Klingel, L., Selig, A.: Online simulation at machine level: a systematic review. Springer Int. J. Adv. Manuf. Technol. **131**(3), 977–998 (2024)
4. Kain, S., Heuschmann, C., Schiller, F.: Von der virtuellen Inbetriebnahme zur Betriebsparallelen Simulation. ATP Ed. **50**(08), 48–52 (2008)
5. Klingel, L., Heine, A., Acher, S., Dausend, N., Verl, A.: Simulation-based predictive real-time collision avoidance for automated production systems. In: 19th International Conference on Automation Science and Engineering (CASE) (2023)
6. Lechler, T., Fischer, E., Metzner, M., Mayr, A., Franke, J.: Virtual commissioning-scientific review and exploratory use cases in advanced production systems. Proc. CIRP **81**, 1125–1130 (2019)
7. Mellado, J., Núñez, F.: Design of an IoT-PLC: a containerized programmable logical controller for the industry 4.0. J. Ind. Inf. Integr. (2022)
8. Mihalič, F., Truntič, M., Hren, A.: Hardware-in-the-loop simulations: a historical overview of engineering challenges. Electronics **11**(15), 2462 (2022)
9. Nakaya, M., Fukano, G., Onoe, Y., Ohtani, T.: On-line simulator for plant operation. In: 2006 6th World Congress on Intelligent Control and Automation (2006)
10. Pantelides, C.C., Renfro, J.G.: The online use of first-principles models in process operations: review, current status and future needs. Comput. Chem. Eng. **51**, 136–148 (2013)
11. Pritschow, G., Röck, S.: Hardware in the Loop. Simulation of Machine Tools. CIRP Anna. **53**(1), 295–298 (2004). https://doi.org/10.1016/S0007-8506(07)60701-X
12. Santillán Martínez, G., Karhela, T., Ruusu, R., Lackman, T., Vyatkin, V.: Towards a systematic path for dynamic simulation to plant operation: OPC UA-enabled model adaptation method for tracking simulation. In: IECON 2017-43rd Annual Conference of the IEEE Industrial Electronics Society, pp. 5503–5508. IEEE (2017)
13. Scheifele, C., Scheifele, D., Eger, U., Daniel, C., Buchal, E., Röck, S.: ISG-virtuos-der Digitale Zwilling für die Praxis. In: Echtzeitsimulation in der Produktionsautomatisierung: Beiträge zu Virtueller Inbetriebnahme. Digitalem Engineering und Digitalen Zwillingen, pp. 61–74. Springer, Berlin Heidelberg Berlin, Heidelberg (2024)
14. Sekler, P.: Modellbasierte Berechnung der Systemeigenschaften von Maschinenstrukturen auf der Steuerung. PhD Thesis, University of Stuttgart (2012). ISW/IPA-Forschung und -Praxis;189
15. Stankovski, S., Ostojić, G., Baranovski, I., Tegeltija, S., Smirnov, V.: Robust automation with PLC/PAC and edge controllers. IFAC-PapersOnLine **55**(4) (2022)
16. Zipper, H., Auris, F., Strahilov, A., Paul, M.: Keeping the digital twin up-to-date-process monitoring to identify changes in a plant. In: 2018 IEEE International Conference on Industrial Technology (ICIT), pp. 1592–1597 (2018)

Open Access This chapter is licensed under the terms of the Creative Commons Attribution 4.0 International License (http://creativecommons.org/licenses/by/4.0/), which permits use, sharing, adaptation, distribution and reproduction in any medium or format, as long as you give appropriate credit to the original author(s) and the source, provide a link to the Creative Commons license and indicate if changes were made.

The images or other third party material in this chapter are included in the chapter's Creative Commons license, unless indicated otherwise in a credit line to the material. If material is not included in the chapter's Creative Commons license and your intended use is not permitted by statutory regulation or exceeds the permitted use, you will need to obtain permission directly from the copyright holder.

Comparative Analysis of Machine Learning Models in Production Environments Through Residual Distributions

Jan A. Zak[1(✉)] and Christian Weißenfels[2]

[1] Mercedes-Benz AG, Bela-Barenyi-Straße, 71059 Sindelfingen, Germany
jan_alexander.zak@mercedes-benz.com
[2] Augsburg University, Data-driven Computational Materials Science and Engineering, Universitätsstraße 2, 86159 Augsburg, Germany

Abstract. In the context of deploying robust solutions in production environments, evaluating the performance of different machine learning models on a specific application is crucial. While common metrics like the mean square error offer a concise summary of model accuracy, they may overlook nuanced differences in prediction performance. In this paper, we propose a visual comparison method based on analyzing the distribution of residuals across multiple machine learning models. By sorting, visualizing, and defining a cutoff value, we aim to provide machine learning engineers with a comprehensive yet practical approach to assess model performance, enabling informed decision-making for real-world applications. The corresponding code is published on GitHub.

Keywords: machine learning · residuals · model comparison · residual distributions

1 Introduction

Machine Learning (ML) engineers face numerous design choices when developing ML models for real-world applications.[1] These choices range from selecting an appropriate model type and tuning hyperparameters to feature engineering and managing outliers. This is a prominent issue in production environments. Engineers frequently search for a ML model that generalizes well on raw data or a specific application. The uncertainty in the search for a model is known by the no free lunch theorem and implies that a specific tune of a model may produce a good generalization for one application while producing an inferior generalization for another, as argued by Wolpert et al. [1,2]. Performance metrics like the mean square error (MSE) offer a standardized measure of model accuracy. A caveat of a single metric is that it intentionally simplifies the prediction error into one metric. Furthermore, outliers affect the MSE and skew the performance of certain models. Common visual approaches like the residual plot [3,4] and the partial residual plot over a predictor [5] are solutions of statisticians for an in-depth analysis and to visualize each prediction against its actual value, the residual. Thereby, they can

---
[1] https://github.com/JanAlexanderZak/SCAP_2024.

reveal uncaught relationships or outliers. The examination of residuals [6] is useful to identify: (a) outliers, (b) omitted effects, (c) heteroscedasticity, and (d) the error distribution. Single metrics, such as the MSE, are a kind of summary of the residual plot that strips away the details of the distribution. On the other hand, examining individual residual plots for each model can be tedious and time-consuming. The proposed residual distribution offers ML engineers the ability to compare multiple models, surpassing the limitations of a single metric. An illustrative example of this limitation is shown in Fig. 1.

2 Methodology

We define a feature matrix $\mathbf{X} \in \mathbb{R}^{n \times p}$ and a target vector $\mathbf{y} \in \mathbb{R}^n$, representing the input features and the corresponding true values, respectively. Given a regression model as a hypotheses $h : \mathbb{R}^{n \times p} \to \mathbb{R}^n$, we obtain the predicted values $\hat{\mathbf{y}} = h(\mathbf{X})$. In general, h can have three error types [7,8]: (1) bias, (2) variance, and (3) an irreducible error ε. Bias measures the deviance from the true mean y_0 and the expected value $\mathbb{E}[h(x_0)]$, and variance is the sensitivity to different data samples. Both can be found as the first and second terms on the right-hand side of Eq. 1. Together with ε, they compose the expected prediction error (EPE):

$$\text{EPE}(x_0) = (\mathbb{E}[h(x_0)] - y_0)^2 + \mathbb{E}\left[(h(x_0) - \mathbb{E}[h(x_0)])^2\right] + \varepsilon \qquad (1)$$

A more complex, low-bias hypotheses h typically fits the training data better, while a restrictive, low-variance h generalizes better [9]. Different algorithms have either more bias or variance, which results in a bias-variance trade-off. Flexible models have high variance and tend to overfit, whereas restrictive models have high bias and may underfit, e.g. by adhering to strong assumptions. The first term on the right-hand side in Eq. 1 is the squared error of the residuals e. It is defined by:

$$\ell(y, h(x)) = e^2, \quad \text{with} \quad e = y - h(x), \qquad (2)$$

where ℓ is a loss function. Its formulation as the MSE, noted in Eq. 3, is the most common loss function [7,10], partially because of its simplicity and differentiability. Alternatives that are less dependent on outliers are the mean absolute error (MAE) or the Huber loss [11].

$$\text{MSE} = \sum_{i=1}^{N} \frac{(y_i - \hat{y}_i)^2}{N} = \sum_{i=1}^{N} \frac{e_i^2}{N}. \qquad (3)$$

Here, the N is the total number of data points and is intentionally written inside the sum. The term within the sum may be summarized as an element e_i^2 of the residuals $\mathbf{e}$. The value of e_i depends on the accuracy of h for that particular instance. Per default, $\mathbf{e}$ is initially not sorted, but rather dependent on the instances of $\mathbf{y}$. For the purpose of this paper, the residuals are sorted by a function sort $: \mathbb{R}^n \to \mathbb{R}^n$ in monotone increasing order:

$$\mathbf{e}^{\text{sort}} = \text{sort}\{\mathbf{e}\} \qquad (4)$$

The sorted array $\mathbf{e}^{\text{sort}}$ has the distinctive property wherein each element $\mathbf{e}_i^{\text{sort}}$ is strictly less or equal than the succeeding element $\mathbf{e}_{i+1}^{\text{sort}}$. This property implies a progressive increase in the magnitude of residuals within the sorted array. The residual $\mathbf{e}_0^{\text{sort}}$ represents the smallest residual contribution to the MSE, while the largest contribution stems from $\mathbf{e}_N^{\text{sort}}$. Consequently, the former adds the least and the latter the most to the MSE. Thus, through the setup established in Eq. 3, the sum of the residuals $\mathbf{e}$ corresponds precisely to the MSE, providing a direct quantitative relationship between individual residuals and the aggregate error metric. This structured vector of residuals, $\mathbf{e}$, can be used for comparative analysis across multiple ML models. The vector can visually show the distribution of residuals across the data points, thereby ML engineers can discern patterns of model performance. Furthermore, a cutoff value may be defined to compare residuals of multiple models.

3 Application to Aluminum Resistance Spot Welding

We apply the residual distribution to a case study involving aluminum resistance spot welding (RSW). The basis of this investigation is a full factorial experiment [12,13], with the factors maximum current $I_{\max}$ [kA], maximum force $F_{\max}$ [kN], and electrode cap thickness l_{cap} [mm]. The response variable is the diameter d_n [mm] of the weld nugget. The room temperature T_{room} [°C] and the weld number w [−] have also been recorded as parameters. A two-sheet combination of 1.5-millimeter EN AW-6014 (AL6) has been tested with 176 experiments. Each experiment has been conducted with 30 repetitions such that the combination has been tested with $176 \times 30 = 5,280$ weld spots. The experiments have been conducted with a C-shaped transformer spot welding gun for aluminum alloys. The electric current is supplied and rectified with a medium-frequency welding transformer and controlled with a constant current control. The welding control is a Bosch Rexroth PRC7000. The sheets are pretreated with a titanium-zirconium passivation and a hotmelt [14]. A specific weld bonding adhesive BETAMATE^{TM} 1640 is automatically applied beforehand. The copper-chromium-zirconium electrode caps [15–17] are of the type A0 [18]. The cap geometry is defined in millimeters by: (a) cap thickness $l_{\text{cap}} \in \{7,9\}$, (b) cap radius $r_{\text{cap}} = 100$, and (c) cap diameter $d_{\text{cap}} = 20$. The sheets are peeled [19,20] after the weld process. Two nugget diameters $\{d_n^{(1)}, d_n^{(2)}\}$ are measured with a digital caliper at an angle of 90 degrees. The mean of both yields the weld nugget diameter d_n. The welding schedule consists of a preheat current of 12 kA over 200 milliseconds, a linear upslope of 70 milliseconds, and a main welding time of 90 milliseconds at $I_{\max}$. The electrode force $F_{\max}$ is held constant. Additional information may be found in Pestka et al. [21] investigating a similar material combination with an identical setup.

With this setup, the predictors and response take the shape of $\mathbf{X} \in \mathbb{R}^{5,280 \times 5}$ and $\mathbf{y} \in \mathbb{R}^{5,280}$:

$$\mathbf{X} = \begin{bmatrix} I_{\max} & F_{\max} & l_{\text{cap}} & T_{\text{room}} & w^1 \\ I_{\max} & F_{\max} & l_{\text{cap}} & T_{\text{room}} & w^2 \\ \ldots & \ldots & \ldots & \ldots & \ldots \\ I_{\max} & F_{\max} & l_{\text{cap}} & T_{\text{room}} & w^{30} \end{bmatrix}, \quad \text{and} \quad \mathbf{y} = \begin{bmatrix} d_n^1 \\ d_n^2 \\ \ldots \\ d_n^{5,280} \end{bmatrix}. \tag{5}$$

Here, superscripts indicate the weld number and identifier. The dataset is set up with $\mathscr{D} := \{(\mathbf{x}_n, \mathbf{y}_n)\}_{n=1}^{N}$. Here, $\mathscr{D}$ has $N = 5,280$ entries that are assumed to consist of i.i.d. samples from a predictor space $\mathbf{X} \in \mathscr{X}$ and a response space $\mathbf{y} \in \mathscr{Y}$. The learning task aims to generalize such that h performs well on unseen data. To that end, $\mathscr{D}$ is divided into a set for training and testing $\{\mathscr{D}^{\text{train}}, \mathscr{D}^{\text{test}}\}$ [22]. The objective is not to fully train a ML model but rather to demonstrate the proposed method. Consequently, default vanilla models such as linear regression and support vector regression [23] are selected. Ensemble models, as advocated by Schwartz et al. among others [24–26], are generally effective for tabular data. Therefore, a random forest regression, which reduces variance by averaging out numerous decorrelated trees [27], and XGBoost [28], which constructs a single tree with high variance to primarily reduce bias, are also chosen. The random forest is set up with 100 estimators, a minimum leaf sample of one, and no depth restriction, whereas the XGBoost model uses a maximum depth of six. Finally, a neural network model is included. All models except XGBoost are implemented in the Python library scikit-learn [29].

Table 1. Collection of the MSE, MAE, and R^2 for the chosen models

Vanilla Model	MSE	MAE	R^2
Linear regression	2.20	1.10	0.64
Support vector regression	1.86	0.94	0.70
Neural network	1.68	0.95	0.72
Random forest regression	0.78	0.65	0.90
XGBoost	0.76	0.65	0.91

The corresponding MSE values are summarized in Table 1, where it is observed that the ensemble models exhibit superior performance compared to the other models, with linear regression demonstrating the poorest performance. For example, the MSE of the linear regression and XGBoost model correspond to $R^2 = 0.64$ and $R^2 = 0.91$, respectively. The conventional residual plot may be utilized to compare the models. It is compared to the proposed residual distribution in Fig. 1. A more detailed depiction for the last 10% of data is shown in Fig. 2a. At first glance, the right-hand side of Fig. 1 is tidier and offers a succinct and intuitive interpretation. The trade-off is the loss of individual data instances. The chosen models generally fail to accommodate instances where zero-diameters are present within the dataset. These instances are situated along the x-axis in Fig. 1a. A zero-diameter is a notable example of values that are not considered outliers; they are physically plausible but contribute to an increase in the MSE.

Moreover, the benefit of the proposed method is demonstrated through the performance of the support vector regression (SVR) and the neural network (NN), as illustrated in Fig. 1b. In this context, the support vector regression surpasses the neural network in 97% of the dataset instances, yet it yields a higher overall MSE. When considering the presence of zero-diameters, the support vector regression may be a sufficient model, despite this not being evident solely through the evaluation of neither the

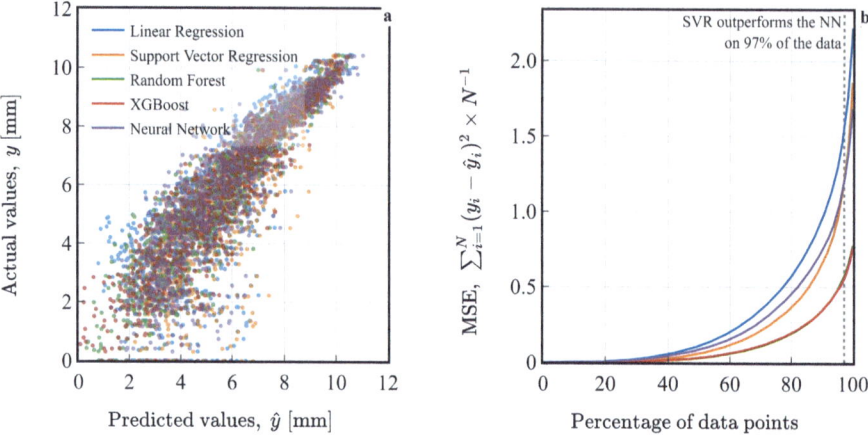

Fig. 1. Conventional residual plot and residual distribution of each model. The five vanilla ML models listed in Table 1 are compared. The figure highlights the challenge of effectively comparing these models using a residual plot. Notably, data points situated within the x-axis represent valid spot welds with zero-diameter; a feature that the models have not learned.

MSE nor Fig. 1a. Therefore, 97% may be defined as a cutoff value for this application. This distinction becomes apparent upon using Eq. 4 to address the zero-diameter phenomenon. Figure 2a focuses on the last 10% where the differences between the models become evident. Each model enters the plot at a different magnitude and depends differently on the last 10%. The different magnitude is explained by better or worse generalization across most instances. Figure 2b presents an alternative representation, demonstrating the contribution of each data point to the overall MSE. Here, the x-axis is identical to Fig. 2a. This illustration highlights a steep increase in the contribution of the last percentage points. Notably, the curves of the non-ensemble models begin to diverge after 95% due to the impact of the last $\sim 1\%$ of data points.

While the MSE is the most common loss function, by definition, it gives higher weight to outliers. This sensitivity to large deviations is advantageous in scenarios where significant errors are particularly undesirable, such as in production environments. However, when outliers are less of a concern or robustness to them is desired, alternatives are the aforementioned MAE or Huber loss. We compare the proposed method with the MAE, as it is less sensitive to outliers and preserves the unit of the data. The overall trends are similar to those obtained using the MSE, cf. Fig. 2, but the magnitude is approximately half as large. As presented in Table 1, the MAE suggests that the support vector regression model performs slightly better than the neural network. This is observed when examining the individual contribution of each data point, as evident in Fig. 3a, while, on the right-hand side in Fig. 3b, the support vector regression begins to diverge from the neural network towards the last percentage points. This indicates that support vector regression, despite being affected by high residuals, accumulates fewer residual differences in total, resulting in a lower overall MAE.

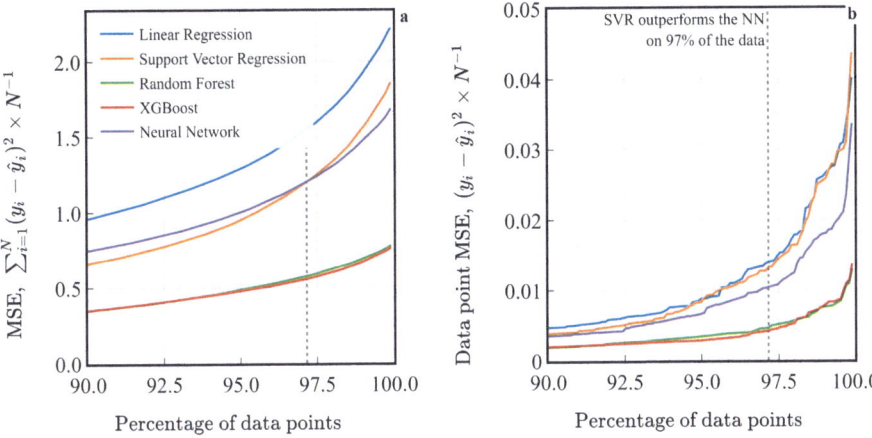

Fig. 2. Contribution of individual data points to the MSE. The left-hand side shows the cumulative and the right-hand side shows individual contributions and the real value contribution to the MSE. Both show the range between 90–100%. The steep increase toward the last percentage points is evident in any model.

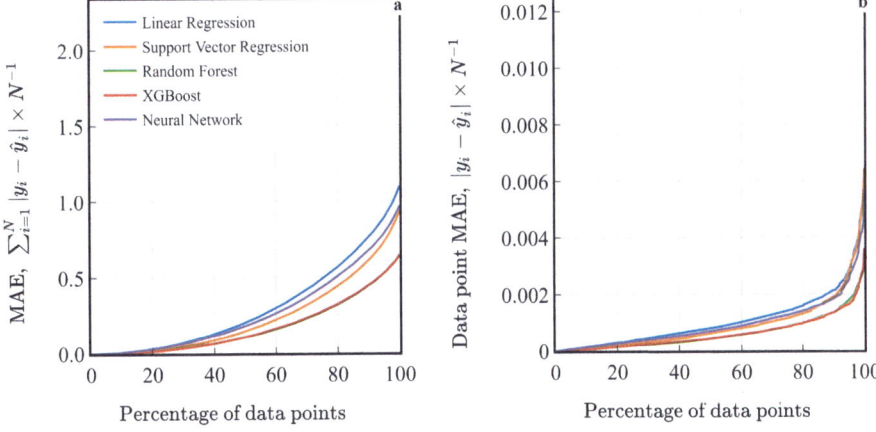

Fig. 3. Contribution of individual data points to the MAE. The left-hand side shows the cumulative error. The right-hand side shows the individual contributions. Note that the magnitude of the y-axis is a quarter of the previous figures.

4 Conclusion

This paper presents an approach to compare the performance of ML models based on the distribution of their residuals. Visualizing the sorted prediction errors offers a detailed yet accessible way to assess model performance across diverse datasets. Our approach is particularly valuable in production environments, where ML engineers need to make informed decisions about model selection and deployment. By considering not only aggregate metrics like the MSE but also the distribution of residuals, engineers can gain deeper insights into the behavior of different models and make more informed decisions about their suitability for real-world applications. Future research could explore extensions and refinements to our methodology, such as incorporating alternate computation of residuals. Moreover, applying our approach to a wider range of datasets and problem domains could further validate the effectiveness.

References

1. Wolpert, D.H., Macready, W.G.: No free lunch theorems for optimization. IEEE Trans. Evol. Comput. **1**(1), 67–82 (1997). https://doi.org/10.1109/4235.585893
2. Wolpert, D.H.: Ubiquity symposium: evolutionary computation and the processes of life: what the no free lunch theorems really mean: how to improve search algorithms. Ubiquity **2013**(December), 1–15 (2013). https://doi.org/10.1145/2555235.2555237
3. Anscombe, F.J., Tukey, J.W.: The examination and analysis of residuals. Technometrics **5**(2) (1963). https://doi.org/10.2307/1266059
4. Mansfield, E.R., Conerly, M.D.: Diagnostic value of residual and partial residual plots. Am. Stat. **41**(2), 107–116 (1987). https://doi.org/10.2307/2684221
5. Cook, R.D.: Exploring partial residual plots. Technometrics **35**(4) (1993). https://doi.org/10.2307/1270269
6. Cox, D.R., Snell, E.J.: A general definition of residuals. J. Royal Stat. Soc. B **30**(2), 248–275 (1968). http://www.jstor.org/stable/2984505
7. Hastie, T., Tibshirani, R., Friedman, J.: Additive models, trees, and related methods. In: The Elements of Statistical Learning. SSS, pp. 295–336. Springer, New York (2009). https://doi.org/10.1007/978-0-387-84858-7_9
8. James, G., Witten, D., Hastie, T., Tibshirani, R.: An Introduction to Statistical Learning, 2nd edn. Springer Texts in Statistics. Springer (2021). https://doi.org/10.1007/978-1-4614-7138-7
9. Russell, S., Norvig, P.: Artificial Intelligence, Global Edition, 4th edn. Pearson (2021). https://elibrary.pearson.de/book/99.150005/9781292401171
10. Murphy, K.P.: Probabilistic Machine Learning: An Introduction. Adaptive Computation and Machine Learning series. The MIT Press (2022). https://mitpress.mit.edu/books/probabilistic-machine-learning
11. Huber, P.J.: Robust estimation of a location parameter. Ann. Math. Stat. **35**(1) (1964). https://doi.org/10.1214/aoms/1177703732
12. Siebertz, K., van Bebber, D., Hochkirchen, T.: Statistische Versuchsplanung [Statistical design of experiments], 2nd edn. Springer Vieweg, Heidelberg (2017). https://doi.org/10.1007/978-3-662-55743-3
13. Cox, D.R., Reid, N.: Theory of the Design of Experiments. CRC Press (2000)
14. VDA: VDA 230-213: Prüfverfahren für die produktklassen prelube, prelube 2, hotmelt, spot lubricant [test methods for the product classes prelube, prelube 2, hotmelt, spot lubricant] (2022). https://webshop.vda.de/VDA/de/vda-230-213-082022
15. ISO: Din EN ISO 5182:2016-11: Resistance welding - Materials for electrodes and ancillary equipment (2016). https://doi.org/10.31030/2574960. Elektroden Werkstoff
16. DVS: DVS 2903:1998-10: Electrodes for resistance welding (1998). https://www.beuth.de/de/technische-regel/dvs-2903/12718553. Elektrodenkappen
17. Corp., M.M.: A-trode: the universal cap electrode for resistance welding of light-to-medium gauge coated and uncoated steels. https://cdn.luvata.com/docs/default-source/products/welding-products/a-trode/a-trode_a4_english.pdf. Accessed 07 Nov (2024). Luvata

18. ISO: Din EN ISO 5821:2010-04: Resistance welding - Spot welding electrode caps (2009). https://doi.org/10.31030/1560906. Kappen WPS
19. ISO: Din EN ISO 10447:2015-05: Resistance welding - Testing of welds - Peel and chisel testing of resistance spot and projection welds (2015). https://doi.org/10.31030/2307649. Abschälen
20. DVS: DVS 2916-1:2014-03: Testing of resistance welded joints - Destructive testing, quasi static (2014). https://www.beuth.de/de/technische-regel/dvs-2916-1/200030552. Abschälen
21. Pestka, J., Weihe, S.: Influence of the joining force on the nugget diameter during resistance spot welding of aluminum materials, book section 10, pp. 96–107. Arena2036. Springer (2023). https://doi.org/10.1007/978-3-031-27933-1_10
22. Chollet, F.: Deep Learning with Python. Manning Publications (2018)
23. Drucker, H., Burges, C.J.C., Kaufman, L., Smola, A., Vapnik, V.: Support vector regression machines. In: M. Mozer, M. Jordan, T. Petsche (eds.) Advances in Neural Information Processing Systems 9, pp. 155–161. The MIT Press, Denver, USA (1996). https://proceedings.neurips.cc/paper_files/paper/1996/file/d38901788c533e8286cb6400b40b386d-Paper.pdf
24. Shwartz-Ziv, R., Armon, A.: Tabular data: deep learning is not all you need. ArXiv.org (2021). https://arxiv.org/pdf/2106.03253.pdf
25. Grinsztajn, L., Oyallon, E., Varoquaux, G.: Why do tree-based models still outperform deep learning on tabular data? ArXiv.org (2022). https://arxiv.org/pdf/2207.08815.pdf
26. McElfresh, D., et al: When do neural nets outperform boosted trees on tabular data? ArXiv.org (2023). https://arxiv.org/pdf/2305.02997.pdf
27. Breiman, L.: Bagging predictors. Mach. Learn. **24**(2), 123–140 (1996). https://doi.org/10.1007/bf00058655
28. Chen, T., Guestrin, C.: XGBoost: a scalable tree boosting system (2016). https://doi.org/10.1145/2939672.2939785
29. Pedregosa, F., et al.: Scikit-learn: machine learning in Python. J. Mach. Learn. Res. **12**, 2825–2830 (2011). https://scikit-learn.org/

Open Access This chapter is licensed under the terms of the Creative Commons Attribution 4.0 International License (http://creativecommons.org/licenses/by/4.0/), which permits use, sharing, adaptation, distribution and reproduction in any medium or format, as long as you give appropriate credit to the original author(s) and the source, provide a link to the Creative Commons license and indicate if changes were made.

The images or other third party material in this chapter are included in the chapter's Creative Commons license, unless indicated otherwise in a credit line to the material. If material is not included in the chapter's Creative Commons license and your intended use is not permitted by statutory regulation or exceeds the permitted use, you will need to obtain permission directly from the copyright holder.

A Baseline Model for Nugget Diameter Prediction Based on Process Parameters for Aluminum Resistance Spot Welding

Jan A. Zak[1]([✉]), Jose M. Araya-Martinez[1], and Christian Weißenfels[2]

[1] Mercedes-Benz AG, Bela-Barenyi-Straße, 71059 Sindelfingen, Germany
jan_alexander.zak@mercedes-benz.com
[2] Data-driven Computational Materials Science and Engineering, Augsburg University, Universitätsstraße 2, 86159 Augsburg, Germany

Abstract. Resistance spot welding is a widely used joining method in automotive manufacturing, known for its efficiency and reliability. With growing regulatory scrutiny on vehicles' carbon footprint, the body-in-white faces higher weight requirements. Therefore, aluminum alloys, particularly 5xxx and 6xxx series aluminum, are increasingly favored over steel due to their strength-to-density ratio. Despite this shift, research on the resistance spot welding of aluminum in large-scale production remains limited compared to steel. Furthermore, with the advent of big data and industry 4.0, the development and optimization of aluminum resistance spot welding may be addressed as a machine learning problem. In this study, we establish a first baseline model through machine learning and big data to predict the nugget diameter. The model is trained on tens of thousands of weld spots for both mentioned alloys across a wide process parameter range. Moreover, we provide a workflow for identifying the optimal hyperparameters for a suitable training algorithm. We identify a reasonable algorithm, determine the most important process parameters, and visualize their effects with shapley additive explanations. The performance of the baseline model serves as a benchmark for future machine learning models and provides a cornerstone for the experimental design decisions within the field. To ensure the reproducibility of results, we make available our baseline models and code on GitHub (https://github.com/JanAlexanderZak/SCAP_2024).

Keywords: resistance spot welding · shapley additive explanations · industry 4.0 · body-in-white

1 Introduction

Resistance spot welding (RSW) is a widely used joining technique in the automotive industry due to its efficiency and reliability. In recent years, increased attention has been directed towards the body-in-white due to the increasing regulatory scrutiny on carbon dioxide emissions from vehicles [1,2]. Aluminum alloys often replace steel as the material of choice because of the ratio between its density and strength [3–7]. In distinction to other aluminum alloys, 5xxx and 6xxx series aluminum are primarily used for the body-in-white [8,9]. Both aluminum alloys are about a third lighter than steel while compromising only one-fifth of the strength. Multiple studies have extensively analyzed steel RSW and the effects of process parameters on the nugget diameter d_n, which is the prime quality metric for the process. However, comparatively limited research considers aluminum RSW, especially in large-scale production environments. Moreover, there is limited utilization of machine learning (ML) models for predicting d_n. From a ML perspective, one of the first steps in the ML workflow [10] is to establish a baseline model. In the context of regression learning tasks, a simplistic baseline entails using the mean value, linear or polynomial regression, while a more sophisticated alternative is employing a vanilla ensemble or deep learning model. Such an approach assesses the generalization capacity of a given model class in subsequent experiments. In a second step, the baseline model can be interpreted with an initial elucidation of the model. For the sake of interpretability, various avenues can be explored, including both model-agnostic or model-specific methods [11,12]. An example of a model-agnostic approach is the utilization of shapley additive explanations (SHAP) values [13] to represent feature effects on the response visually.

2 Dataset and Preprocessing

The basis of this investigation is a full factorial experiment [14,15]. We consider the factors: maximum current I_{max} [kA], maximum force F_{max} [kN], and electrode cap thickness l_{cap} [mm], as listed in Table 1. The response variable is the diameter d_n [mm] of the weld nugget. The room temperature T_{room} [°C] and the weld number $w \in \mathbb{N}$ have also been recorded as parameters. A two-sheet combination has been tested with 176 experiments by varying the factors. The upper and lower sheet thicknesses are between 1.0 and 3.0 mm. Each experiment has been conducted with 30 repetitions, leading to $176 \times 30 = 5,280$ weld spots per combination. The experiments have been conducted with a C-shaped transformer spot welding gun for aluminum alloys. The electric current is supplied and rectified with a medium-frequency welding transformer and controlled with a constant current control. The welding control is a Bosch Rexroth PRC7000. The specimen dimensions in the x-axis and y-axis are 88×500 mm. The welds are placed in two rows from the 1st to the 15th weld spot and from the 16th to the 30th weld spot, as shown in Fig. 1. The sheets are pretreated with a titanium-zirconium passivation and a hotmelt [16]. A specific weld bonding BETAMATE™ 1640 adhesive is automatically applied beforehand. The copper-chromium-zirconium electrode caps [17–19] are of the type A0 [20]. The cap geometry is defined in millimeters by: (a) cap thickness $l_{cap} \in \{7,9\}$, (b) cap radius $r_{cap} = 100$, and (c) cap diameter $d_{cap} = 20$. After welding, the sheets are peeled [21,22]. Two nugget diameters $\{d_n^1, d_n^2\}$ of one weld are measured with a digital caliper at an angle of 90°. The mean of both nugget diameters yields the weld nugget diameter d_n. The welding schedule consists of a preheat current of 12 kA over 200 ms, a linear upslope of 70 ms, and a main welding time of 90 ms at I_{max}. The electrode force F_{max} is held constant. Additional information may be found in Pestka et al. [23], who investigate a similar material combination with an identical setup. The dataset is set up with $\mathscr{D} := \{(\mathbf{x}_n, \mathbf{y}_n)\}_{n=1}^{N}$. Here, $\mathscr{D}$ has $N = 5,280 \times 23 = 121,440$ entries originating from 23 two-sheet combinations. The data are assumed to consist of

Table 1. Factor levels and controllable parameters

Factor/Parameter	Symbol	Min. (−)	Max. (+)	Unit	Value range
max. current	I_{max}	26	47	kA	$\{I_{max} \in \mathbb{N} \mid 26 \leq I_{max} \leq 47\}$
max. force	F_{max}	5	8	kN	$\{F_{max} \in \mathbb{N} \mid 5 \leq F_{max} \leq 8\}$
cap thk.	l_{cap}	7	9	mm	$\{7,9\}$
preheat	-		12	kA	$\{12\}$
squeeze time	-		200	ms	$\{t_{st} \in \mathbb{Z} \mid -200 < t_{st} \leq 0\}$
preheat time	-		400	ms	$\{t_{pt} \in \mathbb{N} \mid 0 < t_{pt} \leq 400\}$
upslope time	-		70	ms	$\{t_{ut} \in \mathbb{N} \mid 400 < t_{ut} \leq 470\}$
weld time	-		90	ms	$\{t_{wt} \in \mathbb{N} \mid 470 < t_{wt} \leq 560\}$

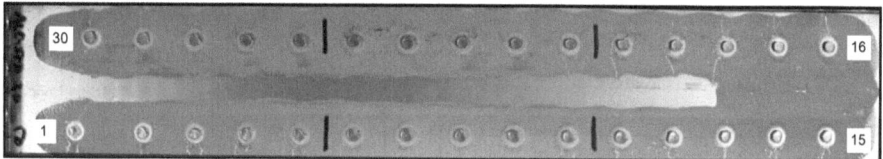

Fig. 1. Examplatory sheet with numbered weld nuggets and the sheet format

i.i.d. samples from a predictor space $\mathbf{X} \in \mathscr{X}$ and a response space $\mathbf{y} \in \mathscr{Y}$. With this setup, the predictors and response take a shape where $\mathbf{X} \in \mathbb{R}^{121,440 \times 5}$ and $\mathbf{y} \in \mathbb{R}^{121,440}$:

$$\mathbf{X} = \begin{bmatrix} I_{max} & F_{max} & l_{cap} & T_{room} & w^1 \\ I_{max} & F_{max} & l_{cap} & T_{room} & w^2 \\ \ldots & \ldots & \ldots & \ldots & \ldots \\ I_{max} & F_{max} & l_{cap} & T_{room} & w^{30} \end{bmatrix}, \quad \text{and} \quad \mathbf{y} = \begin{bmatrix} d_n^1 \\ d_n^2 \\ \ldots \\ d_n^{121,440} \end{bmatrix}. \quad (1)$$

Here, w is the weld number, and superscripts indicate the weld number and identifier. The learning task aims to generalize such that a model h performs well on unseen data. To that end, $\mathscr{D}$ is divided into a set for training, validation, and testing $\{\mathscr{D}^{train}, \mathscr{D}^{val}, \mathscr{D}^{test}\}$ [10]. Table 2 lists the available data per dataset and alloy. In general, the data has been acquired using industrial-quality sensors, necessitating minimal preprocessing with respect to data quality. We aggregate weld points across a single sheet with identical process parameters for the baseline model. This procedure averages out the experimental variance over one sheet but neglects electrode degradation over 30 welds. The average nugget diameter $\bar{d}_n$ is defined by:

$$\bar{d}_n = \frac{1}{30} \sum_{i=1}^{30} d_n^i, \quad (2)$$

where i references the i-th nugget diameter on the sheet. The resulting dataset $\mathscr{D}$ is equal to the dataset listed in Table 2 divided by a factor of 30. As each sheet is combined to calculate an average nugget diameter $\bar{d}_n$, the weld numbers w^i are aggregated accordingly into one value.

Table 2. Number of datapoints per dataset

	$\mathscr{D}^{train}$	$\mathscr{D}^{val}$	$\mathscr{D}^{test}$	Σ
EN AW-5182	41,184	13,728	13,728	68,640
EN AW-6014	31,770	10,590	10,590	52,800
Σ	72,864	24,288	24,288	121,440

3 Methodology

Ensemble models are generally effective for tabular data, as advocated by Schwartz et al. and others [24–26]. Therefore, XGBoost [27], which constructs a single regression tree with high variance to primarily reduce bias, is a popular model choice. We find that the alloys are distinct from one another because of different alloying elements and temper. Thus, two separate models are required, and we train a model for EN AW-5182 and one for EN AW-6014. Since it is unknown which hypotheses h generalizes best on the data, known as the no free lunch theorem [28,29], it is necessary to explore a wide range of hyperparameters. Hyperparameter optimization is treated as a formal process [30]. The search space is not explored manually nor by a grid but instead randomly. Bergstra et al. [31] experimentally show that this is a favorable approach compared to other mentioned methods. Floating point hyperparameters are defined on a logarithmic scale, and random configurations are suggested by hyperopt [32]. The hyperparameter setup is summarized in Table 3. The hyperparameter setup for the XGBoost model is summarized in Table 3. Here, alpha and lambda steer the L1 and L2 regularization of the model, respectively. The hyperparameter optimization is trained on $\mathscr{D}^{train}$ and evaluated on $\mathscr{D}^{val}$. For the sake of ML interpretability [33,34], the models are evaluated with the calculation of SHAP values [13]. This approach sets out that the optimal explanation of a given model h concerning the features $\mathbf{X}$, is the model itself. Since a flexible model becomes challenging to interpret, an explanatory model $\mathscr{E}$, on simplified features $\mathbf{X}'$, is defined to explain h. Lundberg et al. [13] assume that if $\mathbf{X} \approx \mathbf{X}'$, then the output of the

Table 3. Explored hyperparameter search space

Hyperparameter	Search space	Best value	
		AL5	AL6
number of estimators	[3, 200]	21	35
maximum depth	[3, 200]	8	8
maximum leaves	[3, 200]	53	61
maximum bins	[3, 200]	13	26
alpha	[0, 10]	4	4
lambda	[0, 10]	0.5	0.6
gamma	[0, 10]	1	1

original model h is similar to the explanatory model $h(\mathbf{X}) \approx \mathscr{E}(\mathbf{X'})$ with:

$$\mathscr{E}(\mathbf{X'}) = \varphi_0 + \sum_{i=1}^{N} \varphi_i \mathbf{X'}, \tag{3}$$

where, φ_0 is the baseline prediction and φ_i is the feature attribution of feature i. The summation estimates the original output $h(\mathbf{X})$ [11, 13]. The authors establish three desirable properties of SHAP: (a) local accuracy, (b) missingness, and (c) consistency. The SHAP values are calculated with the use of the Python library shap [13] on $\mathscr{D}^{\text{test}}$ as recommended by the authors.

4 Discussion

The two models achieve a $R^2 = 0.84$ and $R^2 = 0.89$ for EN AW-6014 and EN AW-5182, respectively. These results meet the prerequisite for SHAP values [13], as they are only reasonable with high model accuracy. Nevertheless, the results are not exceptional, consistent with expectations given a baseline model and limited feature engineering. The model for EN AW-5182 uses 21 estimators, a maximum depth parameter set to 8, an alpha value of 4, a lambda coefficient of 10.5, and a gamma parameter of 1. The model's configuration also includes maximum bins limited to 13 and a maximum number of 53 leaves. The model for EN AW-6014 uses 35 estimators, a maximum depth parameter set to 8, an alpha value of 4, a lambda coefficient of 0.6, and a gamma parameter of 4. The model's configuration includes maximum bins limited to 26 and a maximum number of 61 leaves. Figure 2 summarizes the SHAP values for EN AW-5182 on the left and EN AW-6014 on the right. The values quantify the deviation from a mean prediction, with color coding used to show an interaction effect. The figure depicts the effect of the maximum electric current I_{max}, maximum electrode force F_{max}, the combination thickness, and the room temperature T_{room} on the nugget diameter d_n. Those are the main factors and the parameter T_{room} for aluminum RSW that are known prior to the execution of the weld. In accordance with literature [35–40] the maximum current I_{max} shows the clearest effect. The absolute effect ranges over seven *millimeters*. Notably, the SHAP values begin to saturate with higher currents with EN AW-5182, which is not the case for EN AW-6014. The latter exhibits a higher variance but a similar trend. Regarding F_{max}, no discernible effect is visible for EN AW-5182, while EN AW-6014 exhibits a linear effect with high variance. Contrary to literature [41,42] decreased force does not result in higher d_n for EN AW-5182. The effect of the combination thickness shows a different effect between EN AW-5182 and EN AW-6014. In the former, both the absolute effect and variance are lower, yet both alloys appear to reach a saturation point at a thickness of 4.5 mm. Notably, two outliers are observed in asymmetric combination sheet thicknesses where the upper sheet is more than 2.5 times thinner than the lower sheet. Lastly, the room temperature does not show any effect on d_n.

A Baseline Model for Nugget Diameter Prediction 117

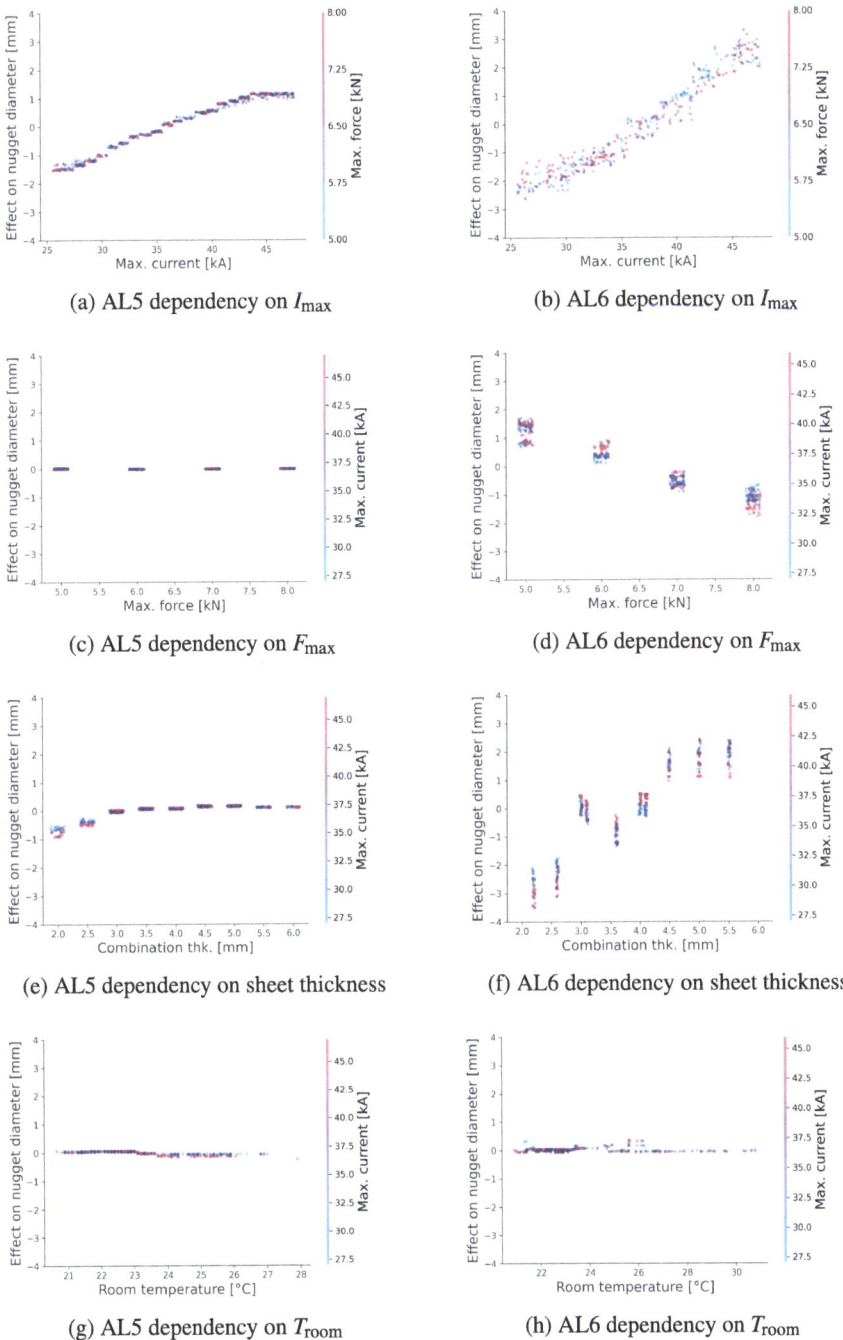

(a) AL5 dependency on I_{max}

(b) AL6 dependency on I_{max}

(c) AL5 dependency on F_{max}

(d) AL6 dependency on F_{max}

(e) AL5 dependency on sheet thickness

(f) AL6 dependency on sheet thickness

(g) AL5 dependency on T_{room}

(h) AL6 dependency on T_{room}

Fig. 2. SHAP value dependence plot of d_n on the features. The left shows EN AW-5182 (AL5) and right EN AW-6014 (AL6). The highest interacting feature is colored. The combined thickness equals that of the upper and lower sheet.

5 Conclusion

The paper presents an entirely data-driven approach for predicting the nugget diameter d_n for the RSW of two aluminum alloys, namely EN AW-5182 and EN AW-6014. The study establishes a baseline for future investigations by developing two separate and sufficiently accurate ML models. An interpretation of these models using SHAP values unveiled the feature attribution effects of electric current I_{max}, electrode force F_{max}, and the combination thickness between the alloys. While both alloys exhibit a clear dependency on I_{max}, the nature of the effect differs. Interestingly, the effect of F_{max} is minimal for EN AW-5182, contrary to expectations based on the effect of EN AW-6014. Future research avenues may include feature engineering of in-situ measurements to enhance model performance and explore alternative model classes.

References

1. EU: No. 443/2009 of the European parliament and of the council of 23 April 2009: setting emission performance standards for new passenger cars as part of the community's integrated approach to reduce CO2 emissions from light-duty vehicles (2009). https://eur-lex.europa.eu/legal-content/EN/ALL/?uri=celex%3A32009R0443
2. EU: No. 2019/631 of the European parliament and of the council of 17 April 2019: setting CO2 emission performance standards for new passenger cars and for new light commercial vehicles, and repealing regulations (EC) no 443/2009 and (EU) no 510/2011 (2019). https://eur-lex.europa.eu/legal-content/DE/ALL/?uri=CELEX:32019R0631
3. Hirsch, J.: Aluminium in innovative light-weight car design. Mater. Trans. **52**(5), 818–824 (2011). https://doi.org/10.2320/matertrans.L-MZ201132
4. Friedrich, H.E.: Leichtbau in der Fahrzeugtechnik [Lightweight Construction in Automotive Engineering], 1st edn. ATZ/MTZ-Fachbuch. Springer, Wiesbaden, Germany (2013). https://doi.org/10.1007/978-3-8348-2110-2
5. Pfestorf, M., van Rensburg, J.: Improving the functional properties of the body-in-white with lightweight solutions applying multiphase steels, aluminum and composites. SAE Technical Paper Series (2006). https://doi.org/10.4271/2006-01-1405
6. Tisza, M., Czinege, I.: Comparative study of the application of steels and aluminium in lightweight production of automotive parts. Int. J. Lightweight Mater. Manuf. **1**(4), 229–238 (2018). https://doi.org/10.1016/j.ijlmm.2018.09.001
7. Henriksson, F., Johansen, K.: On material substitution in automotive BIWs - from steel to aluminum body sides. Procedia CIRP **50**, 683–688 (2016). https://doi.org/10.1016/j.procir.2016.05.028
8. Gould, J.E.: Joining aluminum sheet in the automotive industry - a 30 year history. Weld. Res. **91**, 23–34 (2012)
9. Ostermann, F.: Anwendungstechnologie Aluminium [Application technology aluminum], 3rd edn. Springer Vieweg (2014). https://doi.org/10.1007/978-3-662-43807-7
10. Chollet, F.: Deep Learning with Python. Manning Publications (2018)
11. Molnar, C., Casalicchio, G., Bischl, B.: Interpretable machine learning – a brief history, state-of-the-art and challenges. In: Koprinska, I., et al. (eds.) ECML PKDD 2020. CCIS, vol. 1323, pp. 417–431. Springer, Cham (2020). https://doi.org/10.1007/978-3-030-65965-3_28
12. Ribeiro, M.T., Singh, S., Guestrin, C.: Why should i trust you? Explaining the predictions of any classifier (2016). https://doi.org/10.1145/2939672.2939778

13. Lundberg, S.M., Lee, S.I.: A unified approach to interpreting model predictions. In: Advances in Neural Information Processing Systems, pp. 4765–4774 (2017). https://proceedings.neurips.cc/paper_files/paper/2017/file/8a20a8621978632d76c43dfd28b67767-Paper.pdf
14. Siebertz, K., van Bebber, D., Hochkirchen, T.: Statistische Versuchsplanung [Statistical design of experiments], 2nd edn. Springer, Vieweg (2017). https://doi.org/10.1007/978-3-662-55743-3
15. Cox, D.R., Reid, N.: Theory of the Design of Experiments. CRC Press (2000)
16. VDA: VDA 230-213: Prüfverfahren für die produktklassen prelube, prelube 2, hotmelt, spot lubricant [test methods for the product classes prelube, prelube 2, hotmelt, spot lubricant] (2022). https://webshop.vda.de/VDA/de/vda-230-213-082022
17. ISO: Din EN ISO 5182:2016-11: resistance welding - materials for electrodes and ancillary equipment (2016). https://doi.org/10.31030/2574960
18. DVS: DVS 2903:1998-10: Electrodes for resistance welding (1998). https://www.beuth.de/de/technische-regel/dvs-2903/12718553
19. Mitsubishi Materials Corporation: A-Trode: the universal cap electrode for resistance welding of light-to-medium gauge coated and uncoated steels. https://cdn.luvata.com/docs/default-source/products/welding-products/a-trode/a-trode_a4_english.pdf. Accessed 07 Nov 2024
20. ISO: Din EN ISO 5821:2010-04: resistance welding - spot welding electrode caps (2009). https://doi.org/10.31030/1560906
21. ISO: Din EN ISO 10447:2015-05: resistance welding - testing of welds - peel and chisel testing of resistance spot and projection welds (2015). https://doi.org/10.31030/2307649
22. DVS: DVS 2916-1:2014-03: testing of resistance welded joints - destructive testing, quasi static (2014). https://www.beuth.de/de/technische-regel/dvs-2916-1/200030552
23. Pestka, J., Weihe, S.: Influence of the joining force on the nugget diameter during resistance spot welding of aluminum materials. In: SCAP 2022. Arena2036, pp. 96–107. Springer, Cham (2023). https://doi.org/10.1007/978-3-031-27933-1_10
24. Shwartz-Ziv, R., Armon, A.: Tabular data: deep learning is not all you need. arXiv preprint arXiv:2106.03253 (2021)
25. Grinsztajn, L., Oyallon, E., Varoquaux, G.: Why do tree-based models still outperform deep learning on tabular data? arXiv preprint arXiv:2207.08815 (2022)
26. McElfresh, D., et al.: When do neural nets outperform boosted trees on tabular data? arXiv preprint arXiv:2305.02997 (2023)
27. Chen, T., Guestrin, C.: XGBoost: a scalable tree boosting system (2016). https://doi.org/10.1145/2939672.2939785
28. Wolpert, D.H., Macready, W.G.: No free lunch theorems for optimization. IEEE Trans. Evol. Comput. **1**(1), 67–82 (1997). https://doi.org/10.1109/4235.585893
29. Wolpert, D.H.: Ubiquity symposium: evolutionary computation and the processes of life: what the no free lunch theorems really mean: how to improve search algorithms. Ubiquity **2013**(December), 1–15 (2013). https://doi.org/10.1145/2555235.2555237
30. Bergstra, J., Bardenet, R., Bengio, Y., Kegl, B.: Algorithms for hyper-parameter optimization (2011). https://proceedings.neurips.cc/paper/2011/file/86e8f7ab32cfd12577bc2619bc635690-Paper.pdf
31. Bergstra, J., Bengio, Y.: Random search for hyper-parameter optimization. J. Mach. Learn. Res. **13**, 281–305 (2012). https://www.jmlr.org/papers/volume13/bergstra12a/bergstra12a.pdf
32. Bergstra, J., Yamins, D., Cox, D.: Hyperopt: a python library for optimizing the hyperparameters of machine learning algorithms. In: van der Walt, S., Millman, J., Huff, K. (eds.) Proceedings of the Python in Science Conference, pp. 13–19. SciPy, Austin, USA (2013). https://doi.org/10.25080/majora-8b375195-003

33. Murdoch, W.J., Singh, C., Kumbier, K., Abbasi-Asl, R., Yu, B.: Definitions, methods, and applications in interpretable machine learning. Proc. Natl. Acad. Sci. **116**(44), 22071–22080 (2019). https://doi.org/10.1073/pnas.1900654116
34. Doshi-Velez, F., Kim, B.: Towards a rigorous science of interpretable machine learning. arXiv preprint arXiv:1702.08608 (2017)
35. Kim, E.W., Eagar, T.W.: Parametric analysis of resistance spot welding lobe curve. SAE technical paper series (1988). https://doi.org/10.4271/880278
36. Kim, E.W.: Analyses of resistance spot welding lobe curve. Dissertation (1989). https://dspace.mit.edu/handle/1721.1/100653
37. Florea, R.S., Solanki, K.N., Hammi, Y., Bammann, D.J., Castanier, M.P.: An experimental study of mechanical behavior of resistance spot welded aluminum 6061-T6 joints (2010). https://doi.org/10.1115/imece2010-39167
38. Florea, R.S., Bammann, D.J., Yeldell, A., Solanki, K.N., Hammi, Y.: Welding parameters influence on fatigue life and microstructure in resistance spot welding of 6061-T6 aluminum alloy. Mater. Des. **45**, 456–465 (2013). https://doi.org/10.1016/j.matdes.2012.08.053
39. Rashid, M., Medley, J.B., Zhou, Y.: Nugget formation and growth during resistance spot welding of aluminium alloy 5182. Can. Metall. Q. **50**(1), 61–71 (2011). https://doi.org/10.1179/000844311x552322
40. Kim, G.C., Hwang, I., Kang, M., Kim, D., Park, H., Kim, Y.M.: Effect of welding time on resistance spot weldability of aluminum 5052 alloy. Met. Mater. Int. **25**(1), 207–218 (2018). https://doi.org/10.1007/s12540-018-0179-3
41. Ao, S.S., et al.: Study on quality of resistance spot welded aluminum alloys under various electrode pressures. Front. Mater. Sci. Chin. **3**(1), 98–101 (2009). https://doi.org/10.1007/s11706-009-0004-5
42. Furukawa, K., Katoh, M., Nishio, K., Yamaguchi, T.: Influence of electrode pressure and welding conditions on the maximum tensile shear load. Q. J. Jpn. Weld. Soc. **24**(1), 10–16 (2006). https://doi.org/10.2207/qjjws.24.10

Open Access This chapter is licensed under the terms of the Creative Commons Attribution 4.0 International License (http://creativecommons.org/licenses/by/4.0/), which permits use, sharing, adaptation, distribution and reproduction in any medium or format, as long as you give appropriate credit to the original author(s) and the source, provide a link to the Creative Commons license and indicate if changes were made.

The images or other third party material in this chapter are included in the chapter's Creative Commons license, unless indicated otherwise in a credit line to the material. If material is not included in the chapter's Creative Commons license and your intended use is not permitted by statutory regulation or exceeds the permitted use, you will need to obtain permission directly from the copyright holder.

Modeling the Aging Behavior of the Catalyst Layer in PEM Fuel Cells

Theresa Uhlemayr[✉], Joachim Scholta, and Markus Hölzle

Zentrum für Sonnenenergie- und Wasserstoff-Forschung Baden-Württemberg (ZSW),
Lise-Meitner-Straße 24, 89081 Ulm, Germany
`theresa.uhlemayr@zsw-bw.de`

Abstract. Fuel cells are an environmentally friendly alternative to combustion engines since they use hydrogen to generate electric power.

To assure the fulfilment of the lifetime targets, extensive aging tests are required. In order to reduce the amount and costs of these tests, it is necessary to use a predictive simulation model.

A performance model has been developed by ZSW using Matlab®. This model is now being extended to include the degradation model.

There are many processes which cause degradation in the cell components. The focus of this work is on the electrochemical aging behavior of catalyst layer, including also first aspects of carbon corrosion and dehydrophobization effects.

The catalyst consists of the support material carbon and the catalyst material platinum. On the surface platinum and carbon oxidation and corrosion occurs, Pt ions dissolve and move into the membrane, particles detach and agglomerate. The result of these mechanisms is the reduction of the electrochemical active surface area which leads to performance loss. Additionally, the reduced ionomer coverage of the catalyst and carbon corrosion results in a decreasing contact angle, leading to dehydrophobization and thereby to a limited O_2 mass transport.

The degradation is calculated using an ODE system based on the particle radii and size distribution for every time step. At the beginning, the size of the particle radii and the distribution are defined. The change in these two variables is then calculated in the ODE system and the active surface area of platinum is computed as one influencing aging factor.

Keywords: PEMFC · Degradation · Modeling · Matlab · Catalyst Degradation

1 Introduction

Proton Exchange Membrane Fuel Cells (PEMFC) use hydrogen and oxygen to produce electrical energy through a chemical reaction. They can be used in both stationary and mobile applications, although research is mainly being carried out in mobile applications. The use in vehicles makes them an environmentally friendly alternative to combustion engines. Short charging times and long ranges make the fuel cell vehicles attractive compared to battery electric vehicles. In [1] battery electric vehicles (BEV) and fuel cell electric vehicles (FCEV) are compared. In average, BEVs achieve a range of 261 km with a charging time of >30 min up to 13 h, depending on the type of charging. FCEV double the range to 575 km with only 3 min to charge. On the other hand, the BEV has a higher efficiency and is cheaper to buy. It depends very much on the application which technology is better used where. The fuel cell electric vehicle is better suited for long distances, while short distances can be covered with a battery electric vehicle. This paper focuses on the fuel cell application.

Since electric vehicles are a very young technology, there is very few long-term experience available. For fuel cells mobile application the lifetime target is set to 8,000 h with a maximum performance drop of 10% [2]. Many endurance tests are necessary for cell development and to ensure that these goals are achieved. To reduce the time and cost of these tests, a mathematical model can be used to predict aging. This significantly speeds up development.

In the literature, 3-D and 1-D fuel cell simulations exist which different approached each [3–5]. In general, the 3-D models compute e.g. the current density distribution finely resolved based on the individual cell geometry. These tools are among other things used to improve the flow distribution during the geometry development.

1-D models with a clearly shorter computing time are used to predict e.g. the performance or calculate degradation mechanisms like in [6–8]. However, there is no 1-D performance model available, which contains completely all catalyst aging-mechanisms.

Ageing can be caused by various processes. For example, it can be influenced by the driving behaviour. Many start-ups and shut-downs have a negative effect on the lifetime. Environmental conditions such as temperature and humidity also have an impact - so it makes a difference whether the vehicle is used in Norway or Spain.

These influences (load cycling, temperature and humidity) lead to chemical aging processes within the cell components. On the other hand, there are mechanical influences which may lead to fractures in the bipolar plate, failure of the sealing gasket, deformation and thinning of the gas diffusion layer and crack and pinhole formation in the membrane.

Nguyen et al. [9] provide a detailed overview of the durability in PEM fuel cells and the aging mechanisms of each component.

In this work, the focus is on the electrochemical processes. In general, chemical processes can affect every component of the fuel cell. On the bipolar plate and gas diffusion layer, corrosion leads to thinning, conductivity loss and contact angle changes. The membrane loses conductivity, becomes leaky and perforated. In addition, the thickness is also reduced.

In the following, the focus is on catalyst related aging, which can be divided into various processes that have an impact on platinum (Pt), carbon (C) and the hydrophobicity.

The aim of the degradation model presented in this paper is to predict the performance based on all catalyst-aging mechanisms described below.

2 Theoretical Background

As mentioned above the chemical catalyst degradation consists of various processes that influence the properties of the platinum particles and the catalyst support material carbon. In the following, the mechanisms are explained.

2.1 Oxidation and Corrosion

The surface of the platinum and carbon particles can oxidize. The formation of the oxide layer on the surface of the Pt particles reduces the available active surface for reactions. If the surface of the catalyst support material carbon oxidizes, the platinum particles adhering to it detach, which also leads to a loss of the platinum surface.

Equations 1–3 describe the carbon oxidation process while the remaining three equations show the Pt oxidation. Carbon corrosion is a combination of reaction 3 and 6 [6, 7].

$$C^* + H_2O \leftrightarrow C-OH + H^+ + e^- \tag{1}$$

$$C-OH \leftrightarrow C=O + H^+ + e^- \tag{2}$$

$$C-OH + H_2O \rightarrow C^* + CO_2 + 3H^+ + 3e^- \tag{3}$$

$$Pt^* + H_2O \leftrightarrow Pt-OH + H^+ + e^- \tag{4}$$

$$Pt-OH \leftrightarrow Pt=O + H^+ + e^- \tag{5}$$

$$Pt-OH + C-OH \rightarrow C^* + Pt^* + CO_2 + 2H^+ + 2e^- \tag{6}$$

C* and Pt* stand for defect sided on the platinum and carbon surface [7].

2.2 Platinum Dissolution

Some detached particles dissolve in the ionomer and move into the membrane. In this case, we again have a reduced remaining Pt surface on the catalyst layer. The reaction can be represented as in Eq. 7 [6, 7].

$$Pt \leftrightarrow Pt^{2+} + 2e^- \tag{7}$$

2.3 Particle Size Redistribution

Pt particles can move freely. They can detach from the carrier material and change their position or deposit on other platinum particles.

Another mechanism (Ostwald ripening) occurs when two small platinum particles agglomerate to form a larger particle due to their surface tension.

2.4 Loss of Hydrophobicity

The catalyst has a hydrophobic property right from the start. However, if the amount of carbon or platinum changes over its lifetime, the material becomes more hydrophilic [10–12]. As described in [12], the composition of carbon, platinum and the ionomer specify the hydrophilic or hydrophobic property. While carbon alone is hydrophobic, the combination of carbon and platinum or carbon and ionomer is on the other hand hydrophilic. With the addition of platinum or ionomer, the complete structure of the three components becomes hydrophobic at the beginning.

During the lifetime, the amount of carbon and platinum changes, why the catalyst gets more hydrophilic.

Figure 1 shows the degradation mechanisms which occur on the catalyst surface. The top left shows the state of the catalyst at the beginning. Here there is a relatively even distribution of small platinum particles (purple) on the support material carbon (grey). The electrochemical aging processes described above are shown on the right-hand side.

At the end of the lifetime, fewer and larger particles are present (bottom left).

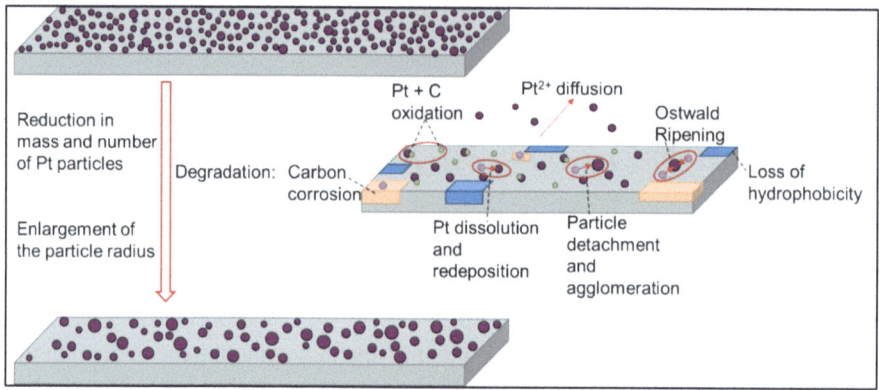

Fig. 1. Catalyst Degradation Mechanisms (own illustration)

2.5 Impact on Performance Degradation

All the processes described, except the loss of hydrophobicity, reduce the number of Pt particles and increase the size of those remaining. The electrochemical active surface area (ECSA) of the platinum is thus reduced, which leads to performance degradation.

The decrease in contact angle due to the loss of hydrophobicity causes water accumulation in the pores. This reduces the mass transport of oxygen, which also leads to a decreased cell performance.

3 Modeling

3.1 Basic Performance Model

A 1-D performance model has been developed in Matlab® as part of a doctoral thesis at the ZSW [8]. This model forms the basis for the extension to the aging model. The functionality of this existing model is explained below.

In the 1-D model, the channel is divided into a several number of segments $n_{segments}$ and it works in three loops: calculation of time step, cell voltage and segment (see the workflow in Fig. 2).

Within the model, a simple Springer membrane model for humidity and the sinus hyperbolicus Butler-Volmer model for anode and cathode is implemented. It can also model both co- and counterflow, has a transport term model for species and an extension to the contaminant model on the cathode side.

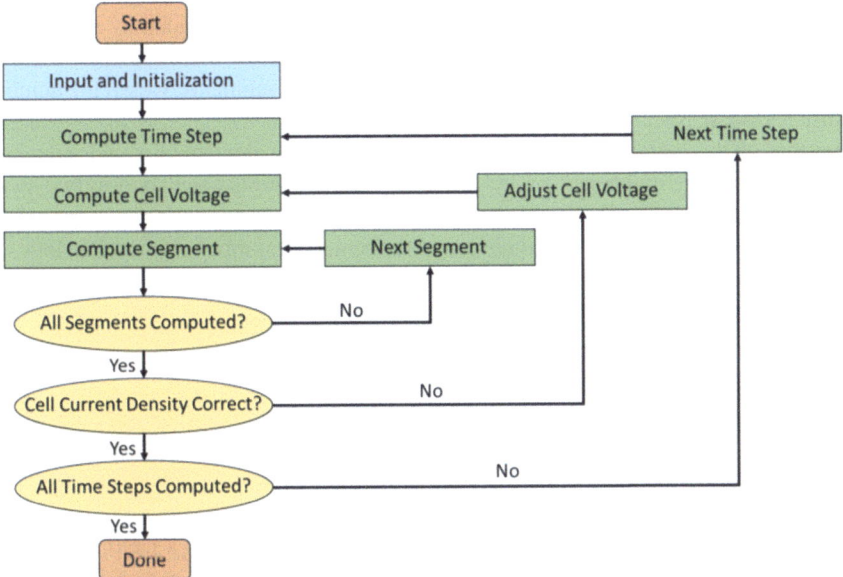

Fig. 2. Workflow of the basic performance model [8]

Equation 8 provides the formula for the cell voltage calculation.

$$U_{cell} = E - U_{act} - U_{ohm} - U_{trans} \tag{8}$$

From the theoretical maximum potential E, the voltage losses due to activation overpotential U_{act}, ohmic resistance U_{ohm} and mass transport losses U_{trans} are subtracted to get the cell voltage U_{cell}.

A more detailed description of this model can be found in [8].

The aging model is now placed in the inner loop. All degradation mechanisms are calculated for each segment and time step. In the following, this is described in more detail.

3.2 Extended Aging Model

Before the aging simulation, a number of radius size classes i is defined. In this model, the number of size classes is set to 80 and has an initial range of $R_{min} = 0.1$ nm to $R_{max} = 8$ nm with even spacing of $\Delta R = 0.1$ nm.

The initial number of platinum particle is calculated form the input parameter "Platinum Loading" according to Eq. 9.

$$n_{Pt\,particles} = \frac{Pt-loading * A}{\rho_{Pt} * \frac{4}{3} * \pi * R_0^3} \quad (9)$$

Here, in the numerator the *Pt-loading* in [kg/m^2] is multiplied by the active area A in [m^2]. In the denominator, the density ρ_{Pt} [kg/m^3] is multiplied by the volume of one particle [m^3], resulting in the weight of one particle [kg]. The volume is calculated based on the input parameter R_0, which is the average particle radius at the beginning. To get the number of particles, the total Pt mass on the catalyst is thus divided by the mass of a single platinum particle.

By dividing the number of Pt particles by the number of segments, we get the number of particles per segment $n_{Pt\,particles,i}$.

Further on the initial particle size distribution (PSD) is calculated for each segment. The PSD means the number of particles per size class i and is calculated in two steps according to Eq. 10 and 11 [6]. First, the standard distribution per size class N_i is calculated on the basis of R_0 and then scaled to the actual number of particles. Index 0 means the initial conditions before starting the simulation.

$$N_i = \frac{1}{0.31 * R_i * \sqrt{2\pi}} * e^{\frac{-(\log(R_i)-\log(R_0))^2}{2*0.31^2}} \quad (10)$$

$$PSD_{0,i} = n_{Pt\,particles,i} * \frac{N_i}{\sum N_i} \quad (11)$$

As mentioned in the theory part, the electrochemical active surface area is the parameter used to get the degradation-influence. For this, every time step a factor is built by the division of the initial ECSA by the ECSA of the respective time step. For the initial ECSA there are two possibilities: an input parameter or the calculation based on radius and distribution. In this model both variants are used. Equations 12–14 show the calculation of the initial ECSA in [m^2/kg], for which the particle surface area A and volume V must first be determined. For the comparison of the calculated and the "real" input value, the factor ζ is introduced according to [6] (see Eq. 15).

$$A_i = 4 * \pi * R_i^2 * PSD_{0,i} \quad (12)$$

$$V_i = \frac{4}{3} * \pi * R_i^3 * PSD_{0,i} \quad (13)$$

$$ECSA_0 = \frac{\sum A_i}{\rho_{Pt} * \sum V_i} \quad (14)$$

$$\zeta = \frac{ECSA_{input}}{ECSA_0} \tag{15}$$

In the following, the simulation process with the corresponding aging formulas are explained.

3.3 Oxidation and Corrosion

The oxide coverages of Pt and C are calculated as the first step. To reduce the time for the model to run, the reaction rates and coverages are not computed for each size class but as an average with a mean particle radius R_{mean}.

$$R_{mean} = \frac{\sum(R_i * PSD_i)}{\sum PSD_i} \tag{16}$$

The time dependent changes are calculated based on the chemical reaction rates taken with minor modifications from [7]. As a difference to [7], the high-temperature factor γ_{HT} is set to 1, since we focus on low temperature PEMFC in this work.

With the reaction rates, the time dependent changes of the coverages θ of Pt, PtOH, PtO, C, COH and CO are calculated based on [7].

Instead of computing the reaction rates and coverages for every size class, the average over all classes are calculated using R_{mean} instead of using R_i.

3.4 Pt Ion Concentration

After calculating the oxide coverages, the Pt ion concentration is computed. For this, the mean particle radius (Eq. 16) and additionally the mean surface area (Eq. 17) is used.

$$A_{mean} = 4 * \pi * R_{mean}^2 * \overline{PSD} \tag{17}$$

With $\overline{PSD}$ being the mean value across all size classes $\left(\overline{PSD} = \frac{\sum PSD}{i}\right)$.

Equations 18 and 19 present the calculation of the Pt^{2+}-concentration $c_{Pt^{2+}}$ based on [7].

$$\frac{dc_{Pt^{2+}}}{dt} = A_{mean} * k_{PtDiss} * \theta_{Pt}$$

$$* \left(e^{\frac{\alpha_{Diss}*(U-E_{Diss})}{R*T}} - \frac{c_{Pt^{2+}}}{c_{Pt^{2+}_{ref}}} * e^{-\frac{(1-\alpha_{Diss})*(U-E_{Diss})}{R*T}} \right) \tag{18}$$

$$E_{Diss} = 1.188 - \frac{1}{2*F} * \frac{\sigma_{Pt} * M_{Pt}}{\rho_{Pt} * R_{mean}} \tag{19}$$

Here k_{PtDiss} is the reaction rate constant for the Pt ion dissolution and θ_{Pt} means the platinum coverage that is left after the oxidation and corrosion processes. α represents the electron transfer coefficient, U the voltage and E the equilibrium potential for this reaction. The temperature dependency is considered in T. In this case, R is the universal gas constant and F the Faraday constant. σ means the surface tension and M the molar mass of platinum.

3.5 PSD and Radius Size Change

Using the previously calculated concentration and coverages, the change in distribution and radii is calculated for each size class i according to [7].

The change in PSD is made up of the mechanism of Ostwald ripening and detachment and agglomeration, whereby a distinction must be made here between three cases:

1. Calculation of the change in the 1st size class
2. Calculation of the change in size class 2–79
3. Calculation of the change in the 80th size class

The formulas for calculating the change in the radius size $\frac{dR}{dt}$ and the particle size distribution $\frac{dPSD}{dt}$ are taken from [7].

3.6 Carbon Corrosion

To get the impact of the carbon corrosion on the degradation, the following linear assumption is made. Equation 20 is taken from [7] with modifications.

$$\frac{dm_{C,corroded,i}}{dt} = M_C * \left(r_{C3,\,Rmean} + r_{PtC,\,Rmean}\right) * \sum R_i \tag{20}$$

$$m_{C,0,i} = A * d_{CL} * \rho_C \tag{21}$$

$$C_{loss,i} = \frac{\frac{dm_{C,corroded,i}}{dt}}{m_{C,0,i}} [\%] \; with \, Pt_{loss,i} = C_{loss,i}[\%] \tag{22}$$

$$PSD_i := PSD_i * \left(1 - Pt_{loss,i}\right) \tag{23}$$

With d_{CL} being the catalyst layer thickness, A the active area and M_C the molar mass of carbon. $r_{C3,\,rmean}$ and $r_{PtC,rmean}$ are the reaction rates for Eq. 3 and 6 taken from [7] with the modification of using R_{mean} instead of R_i.

3.7 ECSA

After the computation of the changes in the mentioned differential equations the ECSA is calculated for each segment and time step [6, 7]:

$$ECSA = \zeta * \frac{\sum 4 * \pi * R_i^2 * PSD_i}{\rho_{Pt} * \sum \frac{4}{3} * \pi * R_i^3 * PSD_i} \tag{24}$$

$$ECSA_{factor} = \frac{ECSA}{ECSA_0} \tag{25}$$

3.8 Loss of Hydrophobicity

In this model, the loss of hydrophobicity is dependent on the loss of carbon. The reduction of this increases the water coverage, as described in [10]. The contact angle is one of the measures of the loss of hydrophobicity. A completely hydrophobic catalyst surface would have a contact angle of 120°, while a completely hydrophilic surface would have a contact angle of 30°. Instead of the contact angle, a contact angle factor ΔDA is calculated (120° contact angle $\rightarrow \Delta DA = 0$; 30° $\rightarrow \Delta DA = 1$) [10–13].

$$\theta_{H_2O}(C_{loss}) = 2.271 * 10^{-3} * C^3_{loss[\%]} - 0.0003302 \\ * C^2_{loss[\%]} + 0.01575 * C_{loss[\%]} + 0.3095 \quad (26)$$

$$\Delta DA = \frac{\theta_{H_2O}}{\theta_{max}} \text{ with } \theta_{max} = \theta_{H2O}(C_{loss} = 100\%) \quad (27)$$

$$lwc = 0 \text{ for } p_{vapor,partial} < p_{sat} \\ lwc = \frac{p_{vapor,partial}}{p_{sat}} - 1 \text{ for } p_{sat} < p_{vapor,partial} < 2 * p_{sat} \quad (28) \\ lwc = 1 \text{ for } p_{vapor,partial} \geq 2 * p_{sat}$$

The contact angle factor ΔDA and liquid water content lwc are used to calculate the pore filling level.

If the water vapor partial pressure is above the saturation vapor pressure, water is present in liquid form. If H_2O is only present in gas form, the pores are not filled with water. The degree of pore filling is therefore zero. If liquid water is present, the following assumption is made to calculate the degree of pore filling.

$$pore \, filling \, level = lwc * \Delta DA \quad (29)$$

The pore filling level is used as a factor to compute the reduced oxygen transport.

4 Results

The results presented in this chapter are generated out of the parameter set shown in Table 1. Just the dwell time and repetitions are varied. In general the parameters in the equations above are taken from the corresponding reference. In case of modification the parameters are listed in the second part in Table 1.

Starting in chronological order, Fig. 3 shows the oxide coverages. Thereby the voltage changes every 3 s between 0.6 V and 0.9 V, starting with 0.9 V. Compared to [7] this plot shows a good agreement and the same characteristic of the single coverage curve shapes.

The following plots are the results of the $\frac{dR}{dt}$ and $\frac{dPSD}{dt}$ calculation. The plots in Fig. 4, 5 and Fig. 6 originated from the same simulation over 1,000 h with a voltage cycling between ~0.68 V and 0.93 V, a dwell time of 5min and 100 repetitions.

Figure 4 shows the change in the radii size over the 80 size classes. As shown, the radii of the larger classes increase while those of the smaller classes remain constant.

Table 1. Simulation Parameters

General Input			
Active area	300 cm²	Cathode stoichiometry	2.0
Coolant inlet temperature	72 °C	Anode stoichiometry	1.5
Coolant outlet temperature	83 °C	Cathode system in pressure	1.1 bar(g)
Cathode dew point	100 °C	Anode system in pressure	1.2 bar(g)
Anode dew point	100 °C	Cathode system out pressure	0.9 bar(g)
Co/counterflow?	Counterflow	Anode system out pressure	1.0 bar(g)
		H_2-fraction	0.7
		O_2-fraction	0.2095
Degradation Model Input and Parameter			
Pt loading	0.004 kg/m²	$k_{Pt,Diss}$	0.9999
ECSA input	40 m²/g	α_{Diss}	0.2702
R_0	1.58 nm	$c_{Pt2+,ref}$	26.7278
d_{CL}	20 µm		

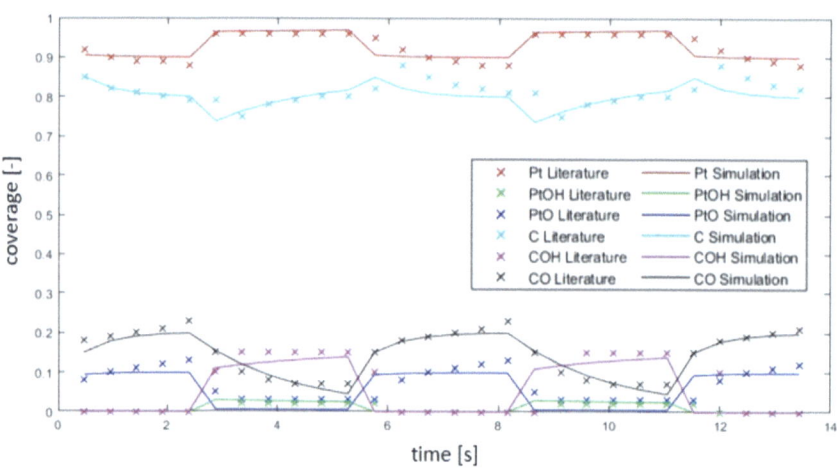

Fig. 3. Oxide coverage

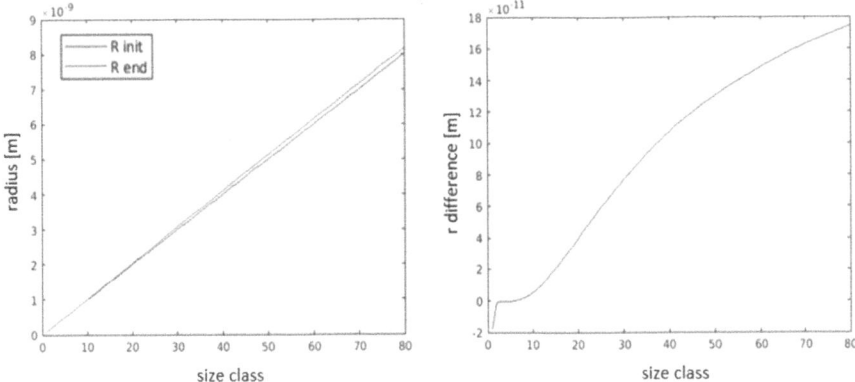

Fig. 4. Radius size change – total and difference

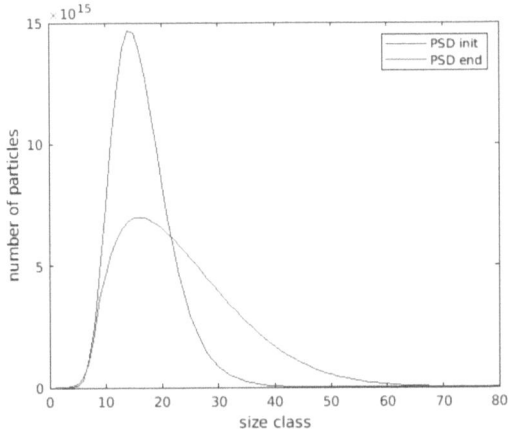

Fig. 5. Change in Particle size Distribution

Figure 5 shows the change in PSD over the 80 size classes. At the beginning, predominantly smaller particles are present. Over time, this distribution shifts to the right. After the 1000 simulated hours, fewer and larger particles are present.

As described above, the ECSA is calculated from the radius and the number of particles per class, from which the ECSA factor is then formed as the ratio of ECSA per time step and initial surface area. By increasing the radii (Fig. 4) and reducing the particles and shifting the distribution (Fig. 5), the active surface area is reduced, which can be observed in Fig. 6.

Figure 7 presents the total water coverage based on the carbon reduction. The more carbon gets lost the more water accumulates in the pores due to the loss of hydrophobicity according to [10]. In [10] Malek et al. present a microstructure-based model which predicts the water coverage on the whole catalyst layer based on the loss of the support

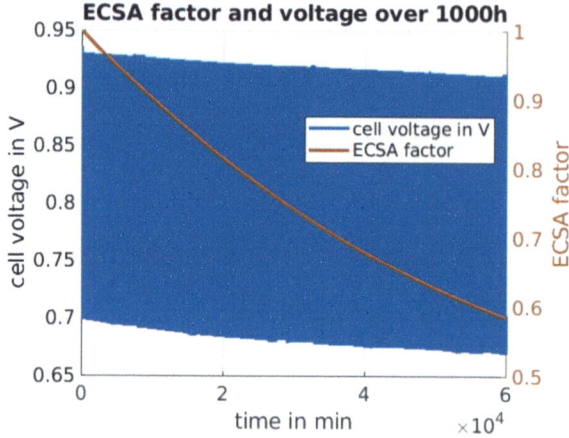

Fig. 6. ECSA-factor over 1000 h

material carbon. The literature data in Fig. 7 show the average values of the coverage on C and Pt. These data were used to create the corresponding formula (Eq. 26).

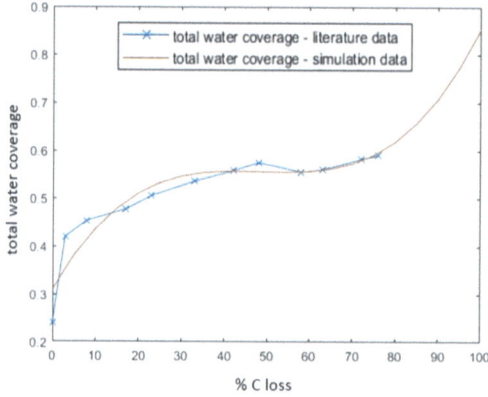

Fig. 7. Total water coverage. Comparison of simulation data with literature data in [10]

Based on the water accumulation and the loss of hydrophobicity the contact angle changes, which is shown in Fig. 8 The reduction of the contact angle is a function of carbon loss and follows the characteristic of the total water coverage. As described in the theory part the contact angle is 120° in the case of a fully hydrophobic material. A minimum contact angle of 30° represents a fully hydrophilic material. For the simulation the contact angle factor ΔDA is used. A factor of 0 means 120° contact angle (100% hydrophobic) while 1 is similar with 30° (100% hydrophilic). With more carbon loss the material gets more hydrophilic which leads to a reduced oxygen transport.

Figure 9 presents the Carbon loss in [%] and the change in DA over the 1,000 h-simulation. The less Carbon the more hydrophilic the catalyst becomes.

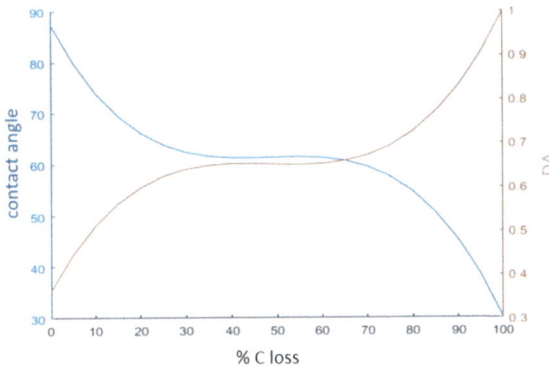

Fig. 8. Contact Angle reduction as a function of carbon loss

Figure 10 shows the pore filling level and DA over 1000 h. An increase in the degree of pore filling due to the very humid conditions can be seen.

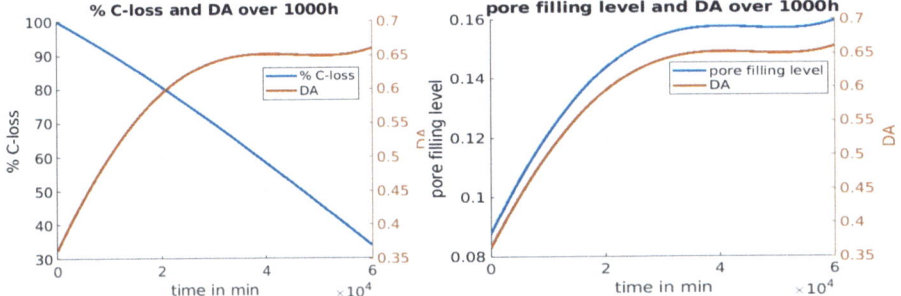

Fig. 9. Carbon loss and DA over 1,000 h   **Fig. 10.** Pore filling level over 1,000 h

In order to show the influence of the degree of pore filling level on the voltage, a comparison of four cases was made (see Fig. 11 and Fig. 12).

Case 1: Simulation of U without degradation (ECSA$_{factor}$ = 1, pore filling level = 0).
Case 2: Simulation of U with influence of ECSA and pore filling level.
Case 3: Simulation of U with pore filling level degradation only (ECSA$_{factor}$ = 1).
Case 4: Simulation of U with ECSA degradation only (pore filling level = 0).

Figure 12 shows the last five cycles as a zoom of Fig. 11.

As expected, the combination of both degradation factors (ECSA and pore filling level) leads to the highest decrease in cell voltage. The comparison of the cases three and four shows a slightly greater impact of the pore filling level of ~0.16 on the cell voltage than of the ECSA$_{factor}$ of ~0.58.

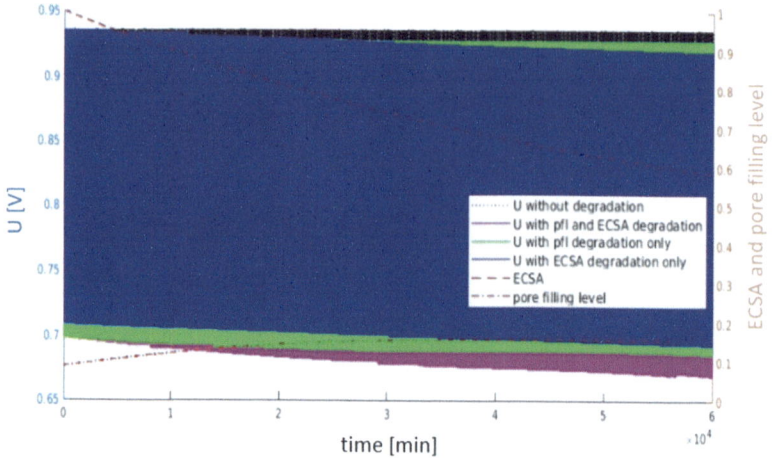

Fig. 11. Comparison of four cases

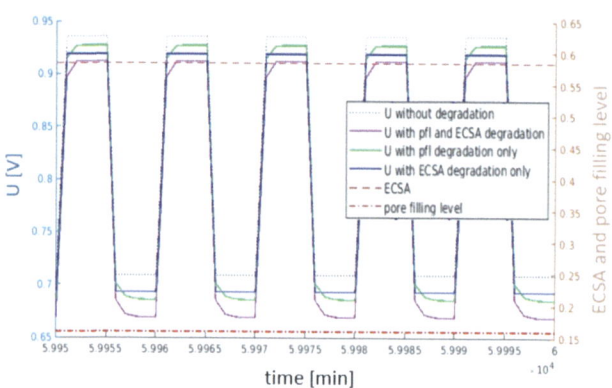

Fig. 12. Zoom of the last five cycles in Fig. 11

5 Conclusion and Future Work

This paper provides an overview of the electrochemical aging mechanisms in PEMFC catalyst layers. After the theoretical part, a modeling approach was presented. An existing 1D-performance model was extended to include the catalyst degradation model with all equations described. As one aging factor, the ECSA was calculated for each time step. The other impact on degradation was given by the loss of hydrophobicity and the reduced oxygen mass transport. All those mechanisms lead to performance loss, as presented in Fig. 3, 4, 5, 6, 7, 8, 9, 10 and Fig. 11. The results show a good agreement with literature data where available. Further comparison with experimental data is still ongoing.

An extension of this performance and catalyst degradation model to include the membrane aging mechanisms is planned as the next step.

Acknowledgment. The research was financially supported by the Federal Ministry of Economic Affairs and Climate Protection (BMWK) and the European Union as part of the economic stimulus package no. 35c in module b on the basis of a resolution of the German Bundestag under grant agreement no. 19S22006N ("Digitalization for Sustainability" – DigiTain).

Abbreviations

BEV	Battery electric vehicle
C	Carbon
ECSA	Electrochemical active surface area
FCEV	Fuel cell electric vehicle
PEMFC	Proton exchange membrane fuel cell
PSD	Particle size distribution
Pt	Platinum

References

1. de Wolf, D., Smeers, Y.: Comparison of battery electric vehicles and fuel cell vehicles. WEVJ **14**, 262 (2023). https://doi.org/10.3390/wevj14090262
2. Fuel Cell Technical Team Roadmap. https://www.energy.gov/sites/prod/files/2017/11/f46/FCTT_Roadmap_Nov_2017_FINAL.pdf. Accessed 21 Feb 2024
3. Fink, C., Fouquet, N.: Three-dimensional simulation of polymer electrolyte membrane fuel cells with experimental validation. Electrochim. Acta **56**, 10820–10831 (2011). https://doi.org/10.1016/j.electacta.2011.05.041
4. Schneider, S., Wilhelm, F., Scholta, J., Jörissen, L., Hunger, J.: Advancement of an OpenFOAM PEMFC toolbox and its validation on an automotive cell design (2023). https://github.com/EC-SIM/OpenFoamZSWPemfcToolbox. Presented at FDFC 2023, Ulm. Publication in progress
5. Sivertsen, B., Djilali, N.: CFD-based modelling of proton exchange membrane fuel cells. J. Power Sour. **141**, 65–78 (2005). https://doi.org/10.1016/j.jpowsour.2004.08.054
6. Jahnke, T., Futter, G.A., Baricci, A., Rabissi, C., Casalegno, A.: Physical modeling of catalyst degradation in low temperature fuel cells: platinum oxidation, dissolution, particle growth and platinum band formation. J. Electrochem. Soc. **167**, 13523 (2020). https://doi.org/10.1149/2.0232001JES
7. Kregar, A., Tavčar, G., Kravos, A., Katrašnik, T.: Predictive system-level modeling framework for transient operation and cathode platinum degradation of high temperature proton exchange membrane fuel cells. Appl. Energy **263**, 114547 (2020). https://doi.org/10.1016/j.apenergy.2020.114547
8. Wagner, T.: Experimental and model based analysis of LT PEM fuel cells exposed to gaseous impurities. Ph.D. Thesis, University of Ulm (2023)
9. Nguyen, H.L., Han, J., Nguyen, X.L., Yu, S., Goo, Y.-M., Le, D.D.: Review of the durability of polymer electrolyte membrane fuel cell in long-term operation: main influencing parameters and testing protocols. Energies **14**, 4048 (2021). https://doi.org/10.3390/en14134048
10. Malek, K., Franco, A.A.: Microstructure-based modeling of aging mechanisms in catalyst layers of polymer electrolyte fuel cells (2011)
11. Olbrich, W., Kadyk, T., Sauter, U., Eikerling, M.: Modeling of wetting phenomena in cathode catalyst layers for PEM fuel cells. Electrochim. Acta **431**, 140850 (2022). https://doi.org/10.1016/j.electacta.2022.140850

12. Olbrich, W., Kadyk, T., Sauter, U., Eikerling, M.: Review—wetting phenomena in catalyst layers of PEM fuel cells: novel approaches for modeling and materials research. J. Electrochem. Soc. **169**, 54521 (2022). https://doi.org/10.1149/1945-7111/ac6e8b
13. Leonard, N.D., Artyushkova, K., Halevi, B., Serov, A., Atanassov, P., Barton, S.C.: Modeling of low-temperature fuel cell electrodes using non-precious metal catalysts. J. Electrochem. Soc. (2015). https://doi.org/10.1149/2.0311510jes

Open Access This chapter is licensed under the terms of the Creative Commons Attribution 4.0 International License (http://creativecommons.org/licenses/by/4.0/), which permits use, sharing, adaptation, distribution and reproduction in any medium or format, as long as you give appropriate credit to the original author(s) and the source, provide a link to the Creative Commons license and indicate if changes were made.

The images or other third party material in this chapter are included in the chapter's Creative Commons license, unless indicated otherwise in a credit line to the material. If material is not included in the chapter's Creative Commons license and your intended use is not permitted by statutory regulation or exceeds the permitted use, you will need to obtain permission directly from the copyright holder.

Dynamic Process Reconfiguration Through Digital Product Passports: A Framework for Adaptive Production Control

Samed Ajdinović[✉], Moritz Walker, Rebekka Neumann, Nicolai Maisch, Michael Neubauer, Armin Lechler, and Oliver Riedel

Institute for Control Engineering of Machine Tools and Manufacturing Units, University of Stuttgart, Seidenstr. 36, 70174 Stuttgart, Germany
samed.ajdinovic@isw.uni-stuttgart.de

Abstract. In today's rapidly changing manufacturing environment, it is increasingly important to have dynamic process reconfiguration through adaptation control mechanisms. Digital Product Passports are also being developed and will become mandatory, such as for batteries in the European Union by 2027. Although this may initially require additional work, it can lead to synergies from the information obtained. By using CO_2 values, e.g., from a Digital Product Passport of components, you can optimize the CO_2 content of the end product to achieve a specific target and improve competitiveness. This can be achieved by adapting processes, such as choosing between high or low dynamics to influence overall energy consumption. A framework is necessary to extract specific information from DPP and make decisions for adaptation. This paper presents a framework architecture based on OPC UA and the AAS, accompanied by an illustrative example of a battery-packing handling process. The speed of the packing process is determined by the energy consumption values of the individual cells to reduce the total energy consumption value of the battery.

Keywords: Digital Product Passport · Reconfiguration · Asset Administration Shell · OPC UA

1 Introduction

In the context of the contemporary manufacturing environment, the imminent requirement for Digital Product Passports Digital Product Passports (DPPs), set to be implemented for products, such as batteries in Europe by 2027, prompts a critical examination of the advantages inherent in such systems. While mandatory adoption may initially seem daunting, it provides a unique opportunity to explore these passports' manifold benefits. Central to this exploration is the concept of process reconfiguration, which is an indispensable facet of modern manufacturing. In order to gain insight into the potential for information to drive transformative change, it is necessary to examine the ways in which DPPs can influence and enhance the reconfiguration process. The crux of this matter is the acknowledgment that the mandatory implementation of DPPs necessitates a comprehensive understanding of their potential contributions. This investigation considers not only the challenges posed by integration but also the opportunities presented by harnessing the data encapsulated in these DPP. This approach illuminates the pathway towards a more agile and responsive manufacturing ecosystem, which is empowered by the insights derived from DPPs.

The objective of this paper is to elucidate the intricate interrelationship between the mandatory adoption of DPP and the paradigm of process reconfiguration. The focus of the investigation is the development of an architectural framework that is designed to utilise DPP data as a catalyst for production optimisation. By utilising DPPs as inputs, this framework facilitates the implementation of nuanced decision-making, thereby enabling the deployment of adaptive control mechanisms that can drive efficiency and innovation. The utilisation of data from the DPP as closed-loop feedback represents a promising avenue for significantly enhancing the efficiency and sustainability of manufacturing processes. The mandatory implementation of DPP not only ensures transparency and traceability in the production chain but also offers substantial advantages that can be harnessed through such feedback mechanisms. By integrating data from DPPs, manufacturers can create a feedback loop that continuously monitors and adjusts production parameters. This dynamic approach ensures compliance with resource consumption limits while optimising the use of materials and energy throughout the production lifecycle. Consequently, the utilisation of DPP data for closed-loop feedback transforms regulatory compliance into an opportunity for the achievement of higher standards of environmental responsibility and operational excellence.

The following sections will present the fundamental concepts of DPP, interoperable semantic data modelling and communication and dynamic process reconfiguration in manufacturing. The subsequent section will investigate the state-of-the-art architectures for process reconfigurations based on digital twins, followed by the framework for adaptive control derived from DPP. Finally, these topics will be discussed before a conclusion is provided.

2 Fundamentals

The following section sets out the fundamental concepts that are essential for this work. At first, the concept of a DPP is explained. It then focuses on interoperable semantic data modelling and communication, as well as dynamic process reconfiguration in manufacturing.

2.1 Digital Product Passport

In various environmental initiatives, the DPP is regarded as a key promoter of sustainable value creation. The DPP is defined as a set of data that summarises the components, materials and chemical substances, as well as information on repairability, spare parts or proper disposal [1]. This is reflected in both the European Union's (EU) Green Deal [2] and the United Nation's 2030 Agenda [3]. The documentation of industrial product data plays a pivotal role in this system, enhancing transparency and sustainability use cases. A DPP serves as the central repository for the documentation of product properties, including the materials used, recycling instructions, and the carbon footprint. It will be used by both companies and customers, as well as by authorities to conduct targeted inspections and verify compliance with applicable legislation. An example of this is the EU Battery Regulation, which has passed a law that makes certain data readable through an interface [4]. This will be mandatory from February 2027 for manufacturers within the EU and for manufacturers supplying to the EU. The implementation of a DPP for batteries, referred to as the Battery Passport, is defined by the Battery Pass Consortium [5]. The general data attributes are defined into seven content clusters:

- General Information
- Labels and certifications
- Carbon footprint
- Supply chain due diligence
- Materials and composition
- Circularity and resource efficiency
- Performance and durability

When analysing research on DPPs, particular attention is paid to the definition of its content [6]. Another focus is on the information modelling of the DPPs and its implementation, including the integration of operational data. Initial approaches to implementation, which utiliseOpen Platform Communications Unified Architecture (OPC UA) at the operational technology level and the Asset Administration Shell (AAS) at the information technology level, have already been investigated in connection with the DPP [7]. This paper builds on these approaches and focuses on semantic data modelling and communication in the next chapter.

2.2 Interoperable Semantic Data Modeling and Communication

Automated data acquisition is one fundamental building block for the automated generation of DPPs. To achieve a high level of automatism, providing data originating at field devices, Programmable Logic Controllers (PLCs) and supervisory systems without context is insufficient. The data must be structured by using standardized metamodels and be enriched with widely recognised semantics [8]. OPC UA and AAS provide metamodels for the manufacturing industry and, in addition, the benefit of defined communication mechanisms.

Open Platform Communications Unified Architecture: In Industry 4.0, interoperability between components like sensors, control systems, and manufacturing execution systems is crucial.OPC UA supports standardized information exchange through a layered architecture, defining a standard information model, communication model, and conformance model [9]. OPC UA specifies base information models, such as Alarms and Conditions (AC), Data Access (DA), and Historical Access (HA), extended by Companion Specifications (CSs) to achieve vendor-independent unified interfaces and define OPC UA usage in specific environments [9]. One example is the CSs for Robotics, developed with the German Engineering Federation (VDMA) and various vendors [10], and the number of CSs is expected to grow, with VDMA coordinating over 53 working groups [11]. A CS consists of machine-readable NodeSet Extensible Markup Language (XML) files and a descriptive text document.

The central element of OPC UA, the OPC UA address space, is a graph of nodes connected by references. Nodes have attributes like *BrowseNames*, *DisplayNames*, and *NodeIds*. OPC UA distinguishes between hierarchical (e.g., *HasComponent*) and non-hierarchical (e.g., *HasTypeDefinition*) references. Optional and mandatory elements are defined using the *HasModellingRule ReferenceType* on *TypeDefinitionNodes* [12]. Services are defined ways for OPC UA clients to interact with servers in a remote procedure call-like fashion. E.g., the nodes' *BrowseName* attributes help find child nodes using the *TranslateBrowsePathToNodeId* service [12].

Furthermore, the information model OPC UA also defines the manner in which interactions with the address space, or communication, are conducted. There are two communication models: the server/client and the publish/subscribe model. Both mechanisms can be carried out over multiple transports. E.g., server/client can use Hypertext Transfer Protocol (HTTP) or Transmission Control Protocol (TCP), and publish/subscribe can use User Datagram Protocol (UDP), Advanced Message Queuing Protocol (AMQP), and Message Queue Telemetry Transport (MQTT).

Asset Administration Shell: While OPC UA is focused on data exchange on the shopfloor, flexible manufacturing systems can only be achieved by digitally representing the entirety of information required for manufacturing and covering the whole supply chain [13]. These higher levels are covered by the AAS.

The Industrial Digital Twin Association (IDTA) is in the process of creating standards for the implementation of the AAS. These standards encompass the metamodel, Application Programming Interfaces (APIs), and guidelines for data provision and semantics. The goal is to standardize the AAS, promoting seamless interoperability both within a single enterprise and across different enterprises. The AAS was conceived to provide a unified interface for communication and data access based on standardized data models tailored for the manufacturing industry. Essentially, an AAS acts as the digital twin of the associated asset, capturing and delivering all relevant information and functionalities throughout the asset's lifecycle [14]. Assets, which can be either physical or non-physical, are defined as any entities of value to an organization [15]. The AAS is structured into sub-models, each detailing specific aspects or perspectives of the asset's lifecycle [16].

The specification [15] outlines three distinct types of AAS, each fitting differently within the RAMI4.0 model [17]. The passive AAS is implemented as a file, typically in a standardized and serializable format such as XML or JSON. It enables the exchange of asset information between partners consistently, ensuring that information about the asset is directly accessible throughout its life cycle [18]. Reactive AASs are building on the passive AASs. This type provides information access through predefined APIs. The reactive AAS facilitates data exchange with users and application services via request-response mechanisms. Proactive AAS type incorporates logic to perceive and respond to its environment, executing actions autonomously based on its internal and external context [19].

2.3 Dynamic Process Reconfiguration in Manufacturing

A large body of research is regarding the software reconfiguration of real-time control systems. In DevOps in office IT, modular applications, e.g., microservices, are reconfigured by keeping them stateless. When replacing a service, only the underlying database needs to be migrated [20]. The components of a control architecture can be stateful. Thus, deterministic state transfer is needed to reconfigure an application at runtime.

Hofmeister et al. [21] present one of the first concepts for the dynamic reconfiguration of distributed systems. Here, the interfaces of software modules are extended to include the internal state, which can then be transferred to other components via messages. The determinism of the reconfiguration is not considered in the publication. Telschig et al. [20] present a concept for real-time updates in a container-based real-time architecture. The underlying cyclical execution model is divided into two phases: the configuration phase and the execution phase. During the configuration phase, different operations can exchange states between components. Since the state of a component can still change after the transfer, the Aggregator-Anticipator pattern is used under the assumption that the states can be computed from the inputs. The anticipator reconstructs the internal state from the messages received by the aggregator. Walker et al. [22] present another concept for reconfiguring real-time software at run-time. Each module is encapsulated into a container, and cohesive applications are formed by combining multiple real-time containers into directed acyclic execution graphs. These graphs can then be executed either event-based or cyclically. In [23], a concept for safeguarding such updates at runtime through additional update observers, e.g., based on online real-time simulation, is presented. The system uses a custom Kubernetes resource definition and controller to execute the observed update process. Although these concepts enable the reconfiguration of real-time software at runtime, they do not describe how such systems are integrated into higher-level systems. One such system is the adaptive control system proposed in this paper.

3 State of the Art Architectures for Process Reconfiguration Based on Digital Twins

Rapidly changing and customer-specific requirements for industrially manufactured products are forcing manufacturers to create flexible production lines that can be reconfigured for batch size one [24]. Mapping the relevant product characteristics in a machine-readable format in a digital twin is a key feature for automated, product-specific configuration of production processes. These characteristics can be used to derive the required capabilities and setting parameters for the production processes [25]. Iterative process planning at the workpiece level also makes it possible to use existing production resources as efficiently as possible [26]. It uses process and operational data from sub-processes to continuously optimise production planning. This concept can also be applied to the quality assurance of multi-stage manufactured parts. By using process parameters along the material flow, production deviations can be continuously recorded and conclusions drawn for further production steps [27]. Müller et al. [28] presents a concept for abstracting reconfiguration during production. A four-stage mechanism considers the identification of reconfiguration demands, the generation of alternative configurations, the evaluation of the identified alternatives, and the selection of a reconfiguration variant based on prioritisation. Depending on the defined production process requirements, optimised results can be generated.

The concepts presented can also be applied to information from DPPs. The information it contains can be used as a basis for repair, reuse, recycling or remanufacturing [24,29,30]. The Battery Passport is also intended to enable such use cases, to give batteries a second life or to be able to remanufacture them efficiently [5]. There is research on what data and information should be collected in DPPs in order to inform decision making for life cycle stage dependent process steps, e.g. [31]. Individual product data provides companies with information on economically and ecologically sensible next production steps for products which have already been used. However, the information from DPPs is not already used during the production process as a basis for individual process reconfiguration. Existing concepts from process configuration in industrial manufacturing can be combined with the information from DPPs to consider the DPP as a basis for product requirements.

4 Framework for Adaptive Control Out of Digital Product Passports

This chapter presents an architectural framework that enables the dynamic reconfiguration of processes based on the content of the DPP. First, the general framework is described in detail. The application of this framework is then illustrated using an example battery pack process.

4.1 Architecture

The general framework for enabling process reconfiguration based on the contents of the DPP is depicted in Fig. 1. The framework is exemplified by a process in which an order is available in the form of an AAS. Furthermore, the DPP of the product itself and the DPPs of the sub-components that interact with the actual product are available.

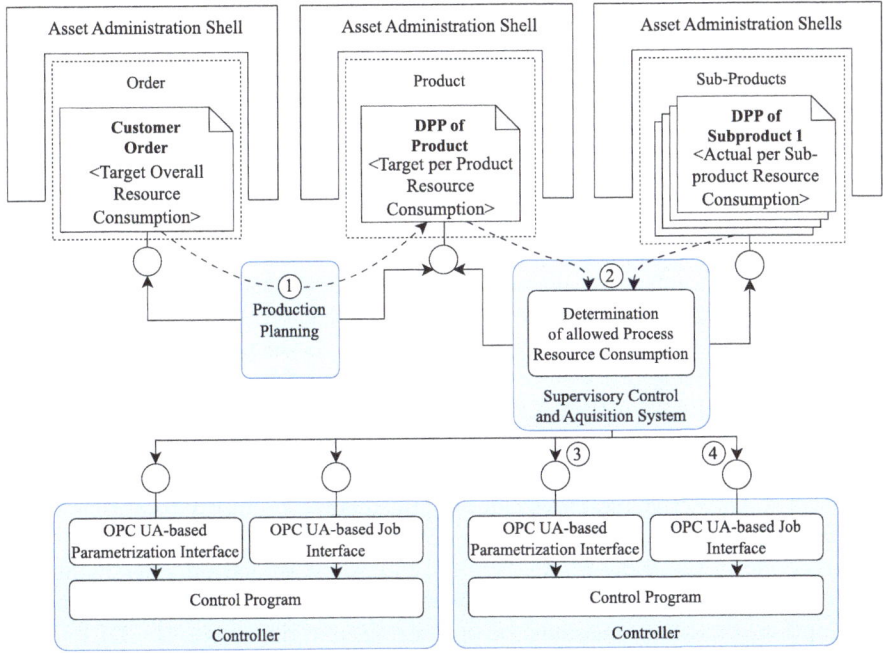

Fig. 1. Generic architecture for DDP-based process reconfiguration

Production planning (1) for the product commences upon receipt of the order. One of the specifications, which may originate from the customer or another source, concerns compliance with certain limits for resource consumption during the manufacturing process. This is where the data points of the product and its sub-products become relevant, as they contain the previous energy consumption of the upstream processes. It is crucial to acknowledge that the consumption of resources can exhibit considerable variability across instances, even when the underlying components remain identical. To illustrate, the proportion of renewable energy in the electricity mix of one sub-product may be greater than that of another.

By analysing the energy consumption of the individual sub-components, it becomes evident how much energy has been consumed thus far and how much remains until the limit value is reached. This allows the subsequent process's energy consumption to be determined in step (2). With this knowledge, the parameterisation can be carried out in step (3), in which the settings for precisely this product and this instance are adjusted in order to comply with the limit value. Once this has been completed, the process is initiated in step (4).

In order to facilitate seamless integration in higher-level systems, a north-bound interface for control applications must be based on open standards. This enables higher-level software components, such as manufacturing execution systems, to interact with the newly defined automation application. Standardised job interfaces, such as OPC UA-based job control, are employed as north-bound interfaces. These interfaces are compatible with ISA 95 [32] or alternatively, OPC UA-based job management can be used.

In addition to energy consumption, other factors such as material usage, emissions, and waste generation can be crucial for the sustainable manufacturing of the product. The DPPs contain comprehensive data on these aspects, allowing for a holistic assessment of the product's environmental impact. For example, the DPP may include information on the carbon footprint of the materials used, the efficiency of resource utilisation, and the methods employed in waste disposal. This detailed information facilitates more informed decision-making during the production planning phase, ensuring that not only energy consumption but all relevant environmental parameters are considered. As the production progresses, data from the ongoing processes can be continuously compared against the pre-defined limits and parameters. This feedback loop allows for immediate adjustments to be made, ensuring that the manufacturing process remains within the desired specifications. For instance, should a deviation in energy consumption be identified, corrective measures can be implemented promptly to mitigate any potential impact on the overall resource limits.

Furthermore, the framework's flexibility allows it to accommodate different production scenarios and constraints. Whether the focus is on minimising energy consumption, reducing emissions, or optimising material usage, the DPPs provide the necessary data to reconfigure the process accordingly. This adaptability is particularly valuable in dynamic manufacturing environments where conditions and requirements can change rapidly. By utilising the comprehensive data within the DPPs, manufacturers are able to rapidly modify their processes to align with new objectives or constraints, thereby maintaining efficiency and sustainability.

4.2 Use Case: Battery Handling and Packing

In order to illustrate the general framework, the example of battery production, or more precisely, the battery packing process, will be examined. In this process, a robot assembles the battery by precisely positioning the individual battery cells and integrating them into the battery module, as illustrated in Fig. 2.

The Fig. 2 illustrates an example of the general architecture, which includes a specification for a maximum energy consumption limit during battery production. This limit is set when an order is received and incorporated into the production planning process. The data points pertaining to the individual battery

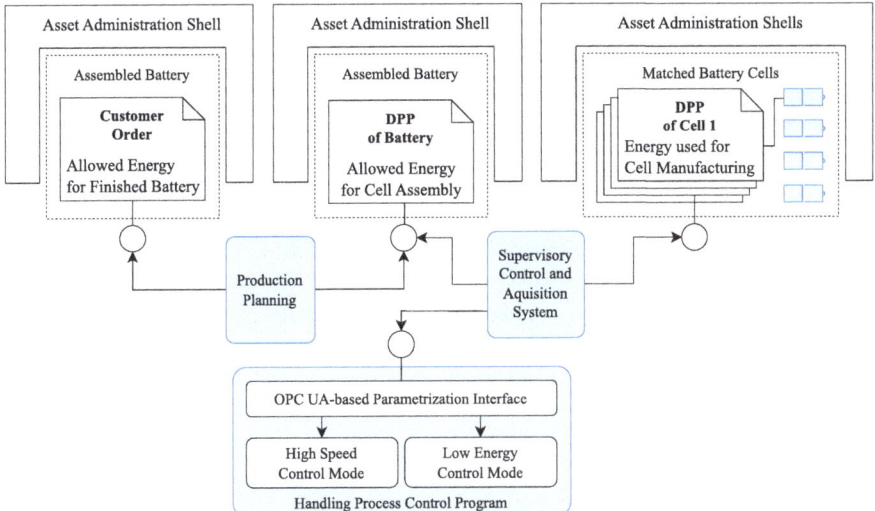

Fig. 2. Architecture for the reconfiguration of battery cell handling during packing to fulfill per cell and overall energy consumption limits

cell instances are analysed, and the production process is subsequently adjusted in accordance with the findings. Once the current energy balance of the individual cells has been determined, the process is parameterised using the OPC UA-based parametrisaton interface. The system then determines whether the robot tasked with battery packing should operate in maximum speed mode (if sufficient energy is still available up to the limit value) or in low-energy mode in order to conserve energy and not exceed the limit value, depending on the energy balance.

To illustrate this further, consider a scenario where an order specifies that the total energy consumption for the battery packing process must not exceed a certain limit. The data points of the battery cells indicate that the cells originate from distinct production batches, each exhibiting a disparate energy consumption profile. Batch A, for instance, may have cells produced using a greater proportion of renewable energy, resulting in a lower overall energy footprint. Conversely, Batch B may have cells produced with a greater reliance on conventional energy sources, resulting in a higher energy footprint.

The production planning system is designed to calculate the total energy consumption of the cells from Batch A and Batch B. In the event that the combined energy consumption of the cells from Batch A is significantly lower, the system may prioritize these cells for use in the battery modules in order to remain within the predefined limit. Nevertheless, in the event that the cells from Batch B are required due to constraints on the inventory, the system will adjust the robot's operation to a low-energy mode in order to compensate for the higher initial energy consumption.

During the packing process, real-time energy monitoring ensures that the total consumption remains within the predefined limits. Sensors and smart meters integrated with the OPC UA interface provide a continuous feed of data to the central control system. Should the energy consumption approach the critical threshold, the system is capable of dynamically adjusting the robot's speed and operational parameters in order to prevent exceeding the aforementioned limit.

For instance, should the energy consumption reach a certain critical threshold during the production process, the system may transition the robot to an low-energy mode, reducing its speed and optimising its movements in order to conserve energy. Furthermore, the system may implement temporary pauses or slowdowns during non-critical operations in order to further minimise consumption. The framework employs real-time data and adaptive control to ensure compliance with energy constraints while maintaining production efficiency.

5 Discussion and Outlook

In this study, we present an architectural framework that enables dynamic process reconfiguration based on the contents of Digital Product Passports. Our illustrative example of a battery-packing process demonstrates how DPPs can inform process adjustments to optimise energy consumption and meet specific targets. The ability to tailor production parameters based on data from DPPs highlights the potential for increased efficiency and competitiveness in manufacturing.

Nevertheless, the question remains as to whether the energy differences between fast and slow production speeds are significant enough to justify such reconfiguration. While slower processes might inherently use less energy, excessively reducing speed is not always economically viable. This balance between energy efficiency and productivity underscores the importance of finding a practical middle ground. Our framework demonstrates how DPP-based reconfiguration can work, but it is crucial to develop strategies that do not compromise overall production efficiency.

Furthermore, the economic implications of implementing DPP-driven reconfiguration require further investigation. Manufacturers must assess whether the benefits of reduced energy consumption outweigh the costs associated with slower production speeds. This assessment is essential to ensure that the reconfiguration strategy is both economically and operationally sustainable. Future research should focus on refining the framework to address these challenges. One area of interest could be the development of algorithms that optimise production parameters without significantly impacting throughput. Furthermore, the application of DPPs to other manufacturing processes could serve to further validate the framework's versatility and effectiveness.

6 Conclusion

This paper presents a novel architecture for dynamic process reconfiguration based on the contents of Digital Product Passports. By utilising data from DPPs, manufacturers can optimise energy consumption and meet specific environmental targets, thereby enhancing both sustainability and competitiveness. The battery-packing process serves as an illustrative example of how DPPs can be employed to adjust production parameters in real-time, thereby ensuring compliance with resource consumption limits.

Our findings indicate the potential benefits of DPP-based reconfiguration, yet also emphasise the necessity for a balanced approach. While slower production speeds can reduce energy consumption, they may not always be economically viable. Consequently, it is of paramount importance to devise strategies that optimise both energy efficiency and productivity.

The framework proposed in this paper offers a foundational approach to integrating DPPs into manufacturing processes. Future work should concentrate on further developing this framework, creating algorithms that balance energy savings with throughput, and extending the framework to other manufacturing scenarios.

In conclusion, while the implementation of DPP-driven process reconfiguration presents challenges, it also offers significant opportunities for advancing sustainable manufacturing practices. Continued research and development in this area will be essential to fully harness the potential of DPPs and achieve more efficient and eco-friendly production processes.

Acknowledgement. The authors would like to thank the Federal Ministry for Economic Affairs and Climate Action (BMWK) for funding the joint projects: SDM4FZI and SFTwin as part of the "Future Investments in the Automotive Industry" program.

References

1. Bundesministerium für Umwelt, Naturschutz, nukleare Sicherheit und Verbraucherschutz, Was ist ein digitaler produktpass? https://www.bmuv.de/faq/was-ist-ein-digitaler-produktpass#:~:text=Der%20digitale%20Produktpass%20ist%20ein,Entsorgung%20f%C3%BCr%20ein%20Produkt%20zusammenfasst
2. European Comission: The European green deal (2019). https://eur-lex.europa.eu/legal-content/EN/TXT/HTML/?uri=CELEX:52019DC0640
3. United Nations: Transforming our world: the 2030 agenda for sustainable development (2015). https://www.un.org/en/development/desa/population/migration/generalassembly/docs/globalcompact/A_RES_70_1_E.pdf
4. European Parliament and EU Council: Regulation (eu) 2023/1542 of the European parliament and of the council of 12 july 2023 concerning batteries and waste batteries, amending directive 2008/98/ec and regulation (eu) 2019/1020 and repealing directive 2006/66/ec (2023). https://eur-lex.europa.eu/legal-content/DE/TXT/PDF/?uri=CELEX:32023R1542

5. Battery Pass Consortium: Battery passport content guidance: achieving compliance with the eu battery regulation and increasing sustainability and circularity (2023). https://thebatterypass.eu/assets/images/content-guidance/pdf/2023_Battery_Passport_Content_Guidance.pdf
6. Stratmann, L., Hoeborn, G., Pahl, C., Schuh, G.: Classification of product data for a digital product passport in the manufacturing industry (2023). https://www.repo.uni-hannover.de/handle/123456789/13573
7. Ajdinović, S., Strljic, M., Lechler, A., Riedel, O.: Interoperable digital product passports: an event-based approach to aggregate production data to improve sustainability and transparency in the manufacturing industry. In: 2024 IEEE/SICE International Symposium on System Integration (SII), pp. 729–734. IEEE (2024)
8. Neumann, R., von Arnim, C., Neubauer, M., Lechler, A., Verl, A.: Requirements and challenges in the configuration of a real-time node for OPC UA publish-subscribe communication. In: International Conference on Mechatronics and Machine Vision in Practice (2023)
9. OPC Foundation: OPC UA Part 1 - Overview and Concepts Release V 1.05.02 (2022)
10. OPC Foundation: OPC 40010-1: Robotics - Vertical Integration Release V 1.0 (2019)
11. Herden, H.: Which OPC UA working groups exist in the VDMA? (2022). https://www.vdma.org/viewer/-/v2article/render/1248676
12. OPC Foundation: OPC UA Part 3 - Address Space Model Release V 1.05.01 (2022)
13. Ellwein, C., Neumann, R., Verl, A.: Software-defined manufacturing: data representation. In: 16th CIRP Conference on Intelligent Computation in Manufacturing Engineering, CIRP ICME 22, Italy (2022)
14. Bundesministerium für Wirtschaft und Energie (BMWi) Industrie 4.0. Details of the Asset Administration Shell - Part 1 (2020)
15. Bader, S., Barnstedt, E., Bedenbender, H., Berres, B., Billmann, M., Ristin, M.: Details of the asset administration shell - part 1 : the exchange of information between partners in the value chain of Industrie 4.0 (version 3.0rc02). https://www.plattform-i40.de/IP/Redaktion/EN/Downloads/Publikation/Details_of_the_Asset_Administration_Shell_Part1_V3.html
16. Miny, T., Thies, M., Epple, U., Diedrich, C.: Model transformation for asset administration shells. In: IECON 2020 The 46th Annual Conference of the IEEE Industrial Electronics Society, pp. 2207–2212 (2020)
17. Miny, T., et al.: Semi-automatic testing of data-focused software development kits for Industrie 4.0. In: 2022 IEEE 20th International Conference on Industrial Informatics (INDIN), pp. 269–274. IEEE (2022)
18. Fuchs, J., Schmidt, J., Franke, J., Rehman, K., Sauer, M., Karnouskos, S.: 2019 24th IEEE International Conference on Emerging Technologies and Factory Automation (ETFA): Paraninfo Building, University of Zaragoza, Zaragoza, Spain, 10-13 September, 2019 : Proceedings. Piscataway, NJ. IEEE (2019). https://ieeexplore.ieee.org/servlet/opac?punumber=8851311
19. Weiss, M., Wicke, K., Wende, G., Pakala, H., Gill, M.S.: Masimo - development and research of industry 4.0 components with a focus on experimental applications of proactive asset administration shells in data-driven maintenance environments
20. Telschig, K., Knapp, A.: Time-critical state transfer during operation of distributed embedded applications, vol. 1, pp. 516–523 (2019)
21. Hofmeister, C., Purtilo, J.: Dynamic reconfiguration in distributed systems: adapting software modules for replacement. In: 1993 Proceedings. The 13th International Conference on Distributed Computing Systems, pp. 101–110 (1993)

22. Walker, M., Tasci, T., Lechler, A., Verl, A.: Analysis of real-time execution models for container-based control applications, pp. 14–24. Springer, Cham (2023)
23. Walker, M., Klingel, L., Oechsle, S., Neubauer, M., Lechler, A., Verl, A.: Safeguarded continuous deployment of control containers through real-time simulation. In: 2023 IEEE 28th International Conference on Emerging Technologies and Factory Automation (ETFA), pp. 1–8 (2023)
24. Plattform Industrie 4.0: Fortschreibung der anwendungsszenarien der plattform industrie 4.0 (2016). https://www.plattform-i40.de/IP/Redaktion/DE/Downloads/Publikation/fortschreibung-anwendungsszenarien.pdf?__blob=publicationFile&v=7
25. Backhaus, J., Reinhart, G.: Digital description of products, processes and resources for task-oriented programming of assembly systems. J. Intell. Manuf. **28**(8), 1787–1800 (2017)
26. Dietrich, D., Neubauer, M., Lechler, A., Verl, A.: Iterative planning as a holistic framework for production system-wide optimization control loops. In: Silva, F., Pereira, A.B., Campilho, R. (eds.) Flexible Automation and Intelligent Manufacturing: Establishing Bridges for More Sustainable Manufacturing Systems, pp. 611–621. Springer, Cham (2024)
27. Dietrich, D., Neubauer, M., Lechler, A., Verl, A.: Fehlerfreie produktion durch flexibilisierte prozesse: Wie sich planung und fehlerkompensation durch eine softwaredefinierte fertigung verändern. wt Werkstattstechnik online, vol. Bd 112, no. 5 (2022)
28. Müller, T., Jazdi, N., Schmidt, J.-P., Weyrich, M.: Cyber-physical production systems: enhancement with a self-organized reconfiguration management. Procedia CIRP **99**, 549–554 (2021)
29. Jansen, M., Meisen, T., Plociennik, C., Berg, H., Pomp, A., Windholz, W.: Stop guessing in the dark: Identified requirements for digital product passport systems. Systems **11**(3) (2023). https://www.mdpi.com/2079-8954/11/3/123
30. EUROPÄISCHE KOMMISSION: Mitteilung der kommission an das europäische parlament, den europäischen rat, den rat, den europäischen wirtschafts- und sozialausschuss und den ausschuss der regionen der europäische grüne deal (2019)
31. Jensen, S.F., Kristensen, J.H., Adamsen, S., Christensen, A., Waehrens, B.V.: Digital product passports for a circular economy: data needs for product life cycle decision-making. Sustain. Prod. Consumption **37**, 242–255 (2023)
32. OPC Foundation: OPC 10031-4. OPC UA for ISA-95: Part 4: Job control (2021)

Open Access This chapter is licensed under the terms of the Creative Commons Attribution 4.0 International License (http://creativecommons.org/licenses/by/4.0/), which permits use, sharing, adaptation, distribution and reproduction in any medium or format, as long as you give appropriate credit to the original author(s) and the source, provide a link to the Creative Commons license and indicate if changes were made.

The images or other third party material in this chapter are included in the chapter's Creative Commons license, unless indicated otherwise in a credit line to the material. If material is not included in the chapter's Creative Commons license and your intended use is not permitted by statutory regulation or exceeds the permitted use, you will need to obtain permission directly from the copyright holder.

EtherCAT Tunneling Through Time-Sensitive Networks: An Experimental Evaluation

Marc Fischer[✉], Moritz Walker, Philipp A. Neher, Michael Neubauer, Armin Lechler, and Alexander Verl

Institute for Control Engineering of Machine Tools and Manufacturing Units, University of Stuttgart, Seidenstr. 36, 70174 Stuttgart, Germany
`marc.fischer@isw.uni-stuttgart.de`

Abstract. Due to flexibility requirements, the strictly horizontal communication of the automation pyramid is converging. This includes communication at the field level. Due to the long lifetime of machines in manufacturing, it is important to support converged communication for existing fieldbuses such as EtherCAT. This enables the implementation of brownfield approaches for gradually implementing converged networks in existing plants. EtherCAT is a widely used industrial Ethernet-based fieldbus protocol for communication between programmable logic controllers and field devices. This work analyzes the tunneling concept of EtherCAT through a Time-Sensitive Networking (TSN) network from the literature and contributes an empirical evaluation based on a test setup with multiple EtherCAT networks and EtherCAT slaves. The tunneling of EtherCAT through TSN based on Virtual Local Area Networks (VLANs) is demonstrated to be a viable option, allowing the utilization of existing EtherCAT devices without the necessity for adaptation. A comparison of the Linux features SO_TXTIME and a simple RAW_SOCKET reveals that both introduce jitter, which is compensated by the EtherCAT slaves to a few microseconds.

Keywords: Fieldbus · Network convergence · EtherCAT · Time-sensitive networking · TSN · Real-time

1 Introduction

The multifaceted developments of globalization, regionalization, and personalization of markets have a profound impact on manufacturing companies, yet the prevailing approaches to production technology, are inadequate to address these challenges, as outlined in [3]. In light of the comprehensive nature of the requirements, various technologies and paradigms are regarded as enablers for flexibility and changeability [15].

The enablers share a commonality in their need for increasing communication of the various systems utilized in manufacturing. This is because the ability to be flexible depends on the transparency of information being made available to all systems involved at all times. From the perspective of the automation pyramid, communication is becoming increasingly integrated, with horizontal and vertical streams of communication occurring within the same network [9]. This implies that communication streams with disparate timing and throughput requirements are accommodated by the same network. Historically, communication in manufacturing has been severely divided between the shop floor, known as operations technology (OT), and the higher levels, known as information technology (IT). The implementation of OT communications was achieved through the use of fieldbuses to meet the real-time requirements. Given the diversity of approaches to real-time implementation, a multitude of fieldbuses emerged over time. In the present era, the majority of fieldbuses, including EtherCAT, Profinet, and Sercos III, are Ethernet-based. This same layer is also commonly utilized in the field of IT. The higher communication layers of the Open Systems Interconnection (OSI) model, such as routing, diverge from those of IT due to the necessity of real-time functionality. In addition, some fieldbuses extend Ethernet in incompatible ways. Consequently, fieldbuses are only compatible with IT-based Ethernet networks to a limited extent [17].

Time-Sensitive Networking (TSN) facilitates the coexistence of OT and IT traffic by standardizing a real-time capable data link layer by extending the Ethernet standards. This enables the preservation of the real-time requirements of OT traffic while enabeling the convergence of communication in manufacturing. TSN is regarded as a key enabler technology for this convergence [16]. The Institute of Electrical and Electronics Engineers (IEEE) is responsible for the standardization of TSN, which is a collection of different standards.

Because TSN covers only the lower two communication layers, the higher layers are implementation specific. OPC UA with the Pub/Sub extension for real-time communication, is a promising candidate to fill these layers on the green field [13,16]. However, the transition to new technologies will not be seamless due to its complexity, unresolved dependencies, and the network configuration [1]. Therefore, the interoperability with existing technology is an important step towards a convergent communication based on TSN. Thus, many fieldbuses try to achieve compatibility by using TSN on the lower two layers, like EtherCAT [7] or Profinet [14]. This is called tunneling.

Although the theoretical foundations have been laid, these approaches need to be verified trough latency measurements in realistic scenarios, as real-time systems can be disrupted by numerous factors that are often overlooked in the context of theoretical assumptions. Thus, this paper presents an analysis and empirical evaluation of the compatibility of EtherCAT in TSN networks. In contrast to previous literature [2,11], which has primarily focused on simulative evaluation, this study provides empirical evaluation with multiple EtherCAT networks and commercial off-the-shelf EtherCAT slaves in a cloud-based control scenario. The remainder of this work is structured as follows. Section 2 provides an overview of existing architectures. Section 3 provides a short introduction to EtherCAT and TSN. In Sect. 4, we present our architecture and explain the empirical evaluation method that we have employed. The results of this evaluation are presented and discussed in Sect. 5. In Sect. 6 a conclusion is drawn.

2 Related Works

The EtherCAT Technology Group (ETG), which is responsible for the standardization of EtherCAT, has published the TSN communication profile ETG.1700 [8]. The primary contribution of this profile is the description of two unidirectional streams through the use of the Time-Aware Shaper originally defined in IEEE 802.1Qbv-2015 and now defined in IEEE 802.1Q-Amd3-2017 and the Stream Reservation Protocol originally defined in IEEE 802.1Qcc-2018 and now part of IEEE 802.1Q, to which the EtherCAT traffic is mapped. This is referred to as stream adaptation, which necessitates the implementation of an adaptation layer in the EtherCAT master and slaves. The profile stipulates that the time of the EtherCAT network must be synchronized to the TSN network time. To achieve this, the EtherCAT master must be synchronized to the TSN network time via the Precision Time Protocol (PTP). The master then synchronizes the slaves via the distributed clocks (DC) synchronisation mechanism of EtherCAT.

In [2], a simulation of clock synchronisation is conducted using the OMNeT++ network simulation framework. The authors examine a range of test cases to ascertain the feasibility of achieving a target cycle time of 100 μs and a target jitter of less than $1\mu s$.

In [11], the authors analyze the clock synchronization of the EtherCAT master and slaves and the TSN network. They formally verify the correctness and precision of EtherCAT DC synchronization through a TSN network, where the EtherCAT master acts as a reference clock and is synchronized to the TSN clock via gPTP.

3 Introduction to EtherCAT and TSN

An EtherCAT network consists of a single master and several slaves, whose physical wiring can be implemented as a combination of star, tree, and line topologies. The logical topology is always a ring. The master is the sole communication device that creates EtherCAT frames. This collective telegram contains the EtherCAT data of all slaves and traverses all communication devices, which receive the output process data, enter the input data, and forward the telegram. The telegram is returned to the master by the final device. In addition to the cyclical process data, data is exchanged acyclically via mailboxes. Synchronization between devices in an EtherCAT network is based on the DC mechanism. Typically, the clocks of all DC-capable devices are synchronized to the clock of the first DC-capable slave with an accuracy of around $100\,ns$. This enables the assignment of high-precision time stamps to recorded data and the achievement of synchronization, which is essential for motion control applications, among others [6].

The TSN Task Group, which is part of the IEEE 802.1 Working Group, is responsible for defining the mechanisms and protocols that enable real-time critical communication via Ethernet without the need for proprietary adaptations. TSN is independent of the communication medium used and consists of a number of different protocols and procedures defined in different standards. A key protocol of TSN is the Stream Reservation Protocol (SRP), which was originally defined in IEEE 802.1Qcc. SRP is employed to reserve transmission capacities within a network comprising talkers, listeners, and switches. These streams must be forwarded by the switches in a manner that preserves the real-time properties of the communication. Prioritizing messages is insufficient for this purpose. For instance, the forwarding of a message with a higher priority could be delayed if the destination port is occupied with a lower-priority message. Consequently, TSN switches employ the Time-Aware Shaper (TAS) methodology for time-controlled message forwarding in accordance with IEEE 802.1Qbv. This can be understood as a time window procedure [5].

4 Evaluation Scenario and Empirical Evaluation Method

Figure 1 depicts the test setup for our evaluation in a cloud-based scenario. This scenario is derived from the Software-defined manufacturing vision, which advocates for the separation of hardware and software [10]. Building upon this vision, we have developed a scenario wherein multiple machines are controlled by a single industrial personal computer (IPC) [12]. This scenario is represented by three EtherCAT networks, with their three EtherCAT masters deployed on the single IPC. Each network is connected to multiple slaves.

EtherCAT employs broadcast Media-Access-Control-Addresses (MAC-Addresses) to deliver frames to any slave. Consequently, the use of the same wire for multiple EtherCAT networks necessitates the implementation of additional mechanisms to keep the networks separated, since the broadcast address cannot be configured. Instead of utilising the stream adaptation described in the ETG.1700 profile, which maps broadcast traffic to specific MAC-Addresses, we employ multiple Virtual Local Area Networks (VLANs) defined in the IEEE 802.1Q standard. This allows the use of the conventional EtherCAT bus coupler, such as the Beckhoff EK1100, in place of the specialized TSN-capable couplers, such as the Beckhoff EK1000.

Two measurement methods are employed to assess the real-time performance of the system. First, the traffic traversing each cable is monitored and measured with a network test access point (TAP). To quantify real-time performance, a TAP with deterministic monitoring latency is utilized, which mirrors the traffic (PROFITAP C8-1G). The mirrored traffic is then analyzed using a separate evaluation IPC by taking the receive time stamp of the network card (Intel i350). The clocks of the network card are synchronized to the TSN network time domain using linuxptp. This methodology enables the determination of network packet latencies according to the TSN network time domain. Secondly, a periodic rectangular signal is generated with a digital output device by toggling the output

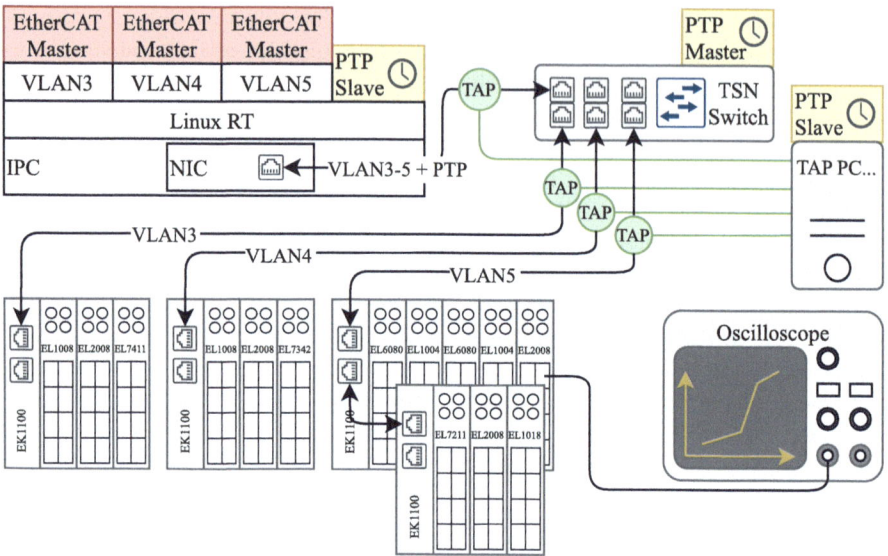

Fig. 1. Test setup and evaluation method.

in each cycle via the EtherCAT Master. The signal is then measured with an oscilloscope by setting the oscilloscope trigger to the falling edge and measuring the time distance to the rising edge. This enables the determination of the signal jitter in the EtherCAT slaves. A typical application in which a cyclical signal change is required is the setpoint generation of a drive. This typically takes place in every cycle.

The TSN schedule depicted in Fig. 2 is employed for the execution of the described scenario. Three distinct traffic classes are utilized for the *(i)* real-time, *(ii)* PTP time synchronization, and *(iii)* best effort traffic. A time slot of 100 µs is assigned to each stream by configuring the EtherCAT master's transmission behavior accordingly.

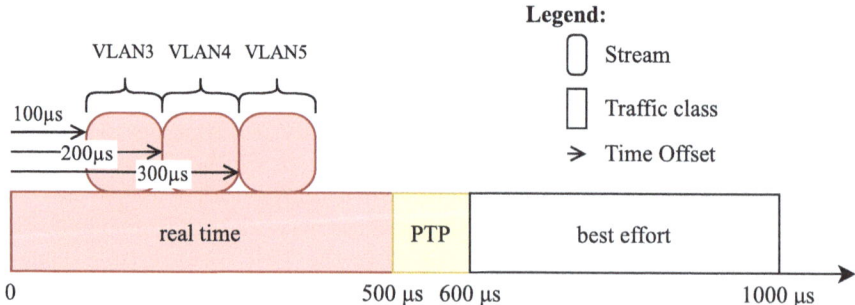

Fig. 2. TSN schedule with three classes and three streams

The EtherCAT masters are implemented using the libethercat stack [4], which is developed by the German Aerospace Center (DLR). This stack runs on an Intel Xeon W-11555MRE CPU with 16 GB of memory, Linux 5.15 PRE-EMPT_RT with real-time tuning, and a Intel i225 network card. The packet sending is implemented via a Linux RAW_SOCKET with two configurations. The first configuration leverages the launch time feature SO_TXTIME of network cards, while the second configuration does not.

5 Results and Discussion

The network packet latencies are evaluated as follows: Only the cyclic process data packets are captured. Based on the planned start time of the transmission $t_{\text{txplanned}}$, which is the offset of a stream in a cycle, the latencies are calculated with the time when a packet arrives at the TAP t_{meas} evaluation PC with t_{meas} with $\Delta t = t_{\text{meas}} - t_{\text{txplanned}}$. The Jitter J is calculated by $J = \Delta t_{\max} - \Delta t_{\min}$. The two packet directions are monitored separately by the TAP. Therefore, the four monitoring points Master → Switch, Switch → Slave, Slave → Switch, and Switch → Master are used.

In the first measurement, all EtherCAT masters operate in parallel for one hour. The SO_TXTIME feature is employed with the delta property set to 40 μs. This implies that the kernel offloads the packets to the network card 40 μs before the scheduled transmission time. The offset of the three EtherCAT streams is selected in accordance with the configuration depicted in Fig. 2.

Figure 3 illustrates the latencies for VLAN 4. It can be observed that the latencies exhibit outliers, with a jitter of approximately 23 μs. These outliers are present in all measurement points, indicating that the source of jitter is the IPC. The packet distance should be $1\,ms$, but higher distances occur in all measurement points.

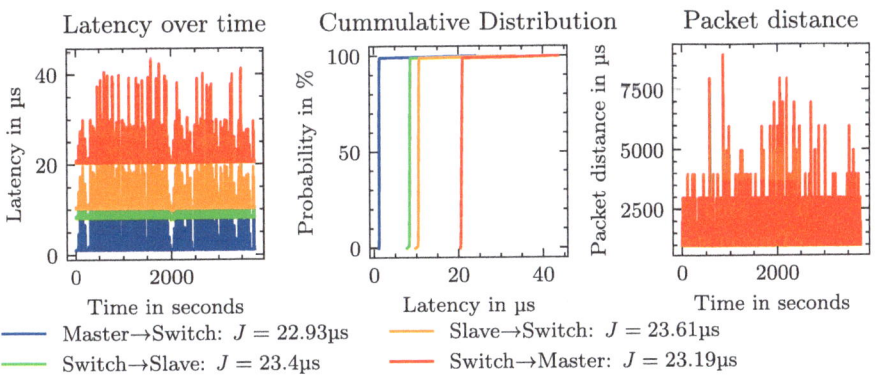

Fig. 3. Latency with SO_TXTIME for VLAN 4

This implies that not every cycle results in the transmission of a packet, which is attributable to the IPC. The packet loss is illustrated in Table 1. Assuming the SO_TXTIME in the network card is functioning correctly, the jitter and packet loss are attributed to the Linux network stack. The measured latencies permit the calculation of the processing latency in the switch and in the slaves, as illustrated in Table 1.

Table 1. Median frame processing latency

	Median frame processing latency			Packet loss
	Switch (to slave direction)	Slaves	Switch (from slave direction)	
VLAN 3	7.24 µs	2.04 µs	10.01 µs	$9.1 \cdot 10^{-2}\%$ (3410 Packets)
VLAN 4	7.26 µs	2.09 µs	10.22 µs	$1.6 \cdot 10^{-1}\%$ (6343 Packets)
VLAN 5	5.87 µs	5.07 µs	8.28 µs	$6.6 \cdot 10^{-2}\%$ (2490 Packets)

The lower processing latency of VLAN 5 disappears when a different port of the switch is used. Therefore, we assume that the ports of the switch have different processing latencies. The higher latencies for the "from slave direction" are probably caused by the lower data rate of 100 Mbps for the EtherCAT slaves compared to the 1000 Mbps connection to the IPC.

In order to assess the performance of SO_TXTIME in comparison to a classical RAW_SOCKET, an additional measurement under identical conditions is provided in Fig. 4. It can be observed a jitter of $J = 30 \mu s$, yet no packet loss is evident.

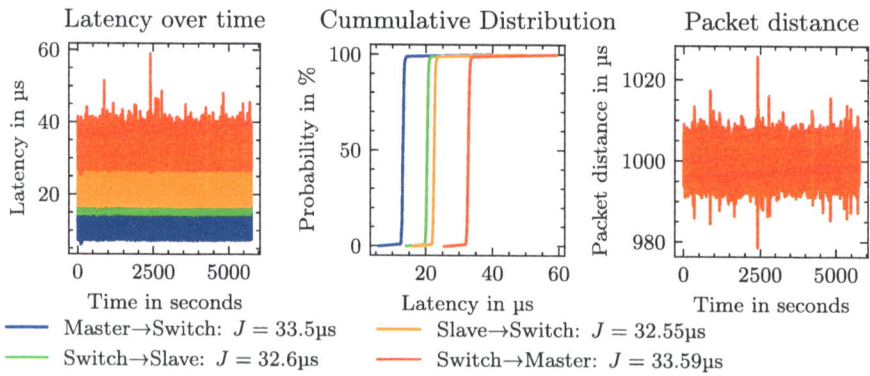

Fig. 4. Latency with RAW SOCKET for VLAN 4

The maximum, minimum, and mean values of the periodic rectangle signal is depicted in Fig. 5. The measurements with the oscilloscope are taken for the duration of one minute. For SO_TXTIME, a timespan without packet loss is chosen. The jitter using SO_TXTIME is approximately 2 μs, which is significantly lower than that of RAW_SOCKET, which is approximately 7 μs. It can be observed that the signal jitter is lower than the packet jitter of the periodic EtherCAT packet. This is due to a PI-controller, which is used to synchronize the time synchronization signals from the master to the slave clocks inside the slaves.

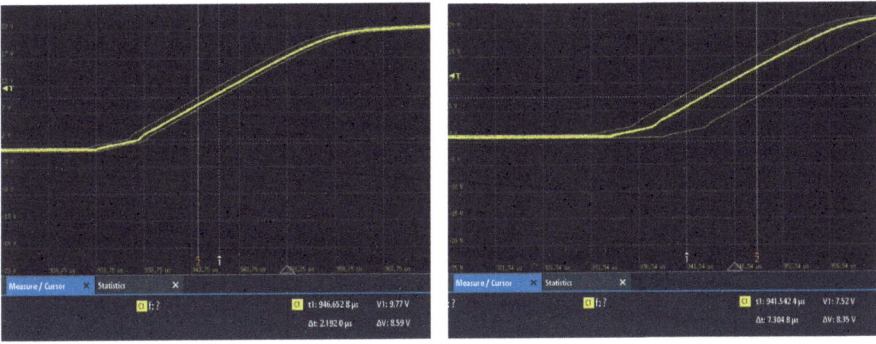

(a) Rectangle signal with SO_TXTIME (b) Rectangle signal with RAW_SOCKET

Fig. 5. Jitter of periodic rectangle signal

It can be concluded that SO_TXTIME improves jitter and enables greater precision in the operation of EtherCAT networks. Further optimization of jitter can be achieved by optimizing the parameters. However, the current level of packet loss does not make the feature attractive. It can also be observed that the jitter in the slaves is reduced to the order of a few microseconds, which is sufficient for many applications.

6 Conclusion

It can be concluded that a tunneling of EtherCAT through TSN based on VLANs is possible and allows the usage of existing EtherCAT devices without any adaptation. Therefore, the effort is reduced in comparison to the draft ETG.1700 of the ETG, where both the EtherCAT master and slaves must be adapted. Moreover, the literature indicates that the time synchronization, which has been analyzed in detail, behaves in a manner consistent with the predictions made. This allows the synchronization of EtherCAT slaves with the DC by the EtherCAT master. Furthermore, it has been demonstrated that the operation of multiple EtherCAT networks through a single cable is possible with the presented architecture. Using the presented evaluation method, similar setups can be tested.

In addition, the setup and configuration of a TSN network is possible with commercial off-the-shelf hardware. However, the Linux network stack caused jitter around 30 μs on our test system. Another Linux network feature is the eXpress Data Path (XDP) which allows the transmitting and receiving of packets while bypassing the kernel network stack and reduces the jitter. Currently, the SO_TXTIME cannot be used in combination with XDP, but patches[1] with the aim to bring this feature into mainline Linux exist. Thus, XDP is an open topic for future works.

Acknowledgment.. The authors would like to thank the BMWK for partly funding the SDM4FZI joint project as part of the"Future Investments in the Automotive Industry (german: Zukunftsinvestitionen in der Fahrzeugindustrie)" funding program, which is funded by the European Union - NextGenerationEU. This work was partly supported by the Deutsche Forschungsgemeinschaft (DFG, German Research Foundation) - Project-ID 420528256.

References

1. von Arnim, C., Dragan, M., Frick, F., Lechler, A., Riedel, O., Verl, A.: TSN-based converged industrial networks: evolutionary steps and migration paths. In: 2020 25th IEEE International Conference on Emerging Technologies and Factory Automation (ETFA), pp. 294–301. IEEE (2020). https://doi.org/10.1109/ETFA46521.2020.9212057
2. Balakrishna, B., Meinardus, B., Kontopoulos, L.: Simulation framework for EtherCAT over TSN. In: 2021 IFIP Networking Conference (IFIP Networking), pp. 1–6. IEEE (2021). https://doi.org/10.23919/IFIPNetworking52078.2021.9472810
3. Bauernhansl, T.: Die vierte industrielle revolution – der weg in ein wertschaffendes produktionsparadigma. In: Vogel-Heuser, B., Bauernhansl, T., ten Hompel, M. (eds.) Handbuch Industrie 4.0 Bd.4, pp. 1–31. Springer, Berlin, Heidelberg (2016). https://doi.org/10.1007/978-3-662-53254-6_1
4. Burger, R., Fischer, M., Da-LiFe: Robert-Burger/libethercat: 0.5.1 (2024). https://doi.org/10.5281/zenodo.11170808
5. Dürkop, L.: Automatische Konfiguration von Echtzeit-Ethernet: Dissertation, Technologien für die Intelligente Automation, vol. Band 5. Springer, Berlin, Heidelberg (2016). https://doi.org/10.1007/978-3-662-54125-8
6. EtherCAT Technology Group: Ethercat - the ethernet fieldbus. https://www.ethercat.org/en/technology.html
7. EtherCAT Technology Group, Weber, K.: EtherCAT and TSN - best practices for industrial ethernet system architectures (2018). https://www.ethercat.org/download/documents/Whitepaper_EtherCAT_and_TSN.pdf
8. EtherCAT Technology Group, Weber, K.: Ethercat TSN communication profile: ETG.1700 s (d) v0.9.1 (2018). https://www.ethercat.org/de/downloads/downloads_254672A7CED54910B655F565B974F5AD.htm
9. Kleinemeier, M.: Von der automatisierungspyramide zu unternehmenssteuerungsnetzwerken. In: Bauernhansl, T., ten Hompel, M., Vogel-Heuser, B. (eds.) Industrie 4.0 in Produktion, Automatisierung und Logistik, pp. 571–579. Springer, Vieweg, Wiesbaden (2014). https://doi.org/10.1007/978-3-658-04682-8_29

---

[1] https://lore.kernel.org/bpf/20240325020928.1987947-1-yoong.siang.song@intel.com/.

10. Lechler, A., Riedel, O., Coupek, D.: Virtual representation of physical objects for software defined manufacturing. DEStech Trans. Eng. Technol. Res. (ICPR) (2018). https://doi.org/10.12783/dtetr/icpr2017/17652
11. Mateu, D.B., Hallmans, D., Ashjaei, M., Papadopoulos, A.V., Proenza, J., Nolte, T.: Clock synchronization in integrated TSN-EtherCAT networks. In: 2020 25th IEEE International Conference on Emerging Technologies and Factory Automation (ETFA), pp. 214–221. IEEE (2020). https://doi.org/10.1109/ETFA46521.2020.9212153
12. Neubauer, M., Reiff, C., Walker, M., Oechsle, S., Lechler, A., Verl, A.: Cloud-based evaluation platform for software-defined manufacturing. at - Automatisierungstechnik **71**(5), 351–363 (2023). https://doi.org/10.1515/auto-2022-0137
13. Pfrommer, J., Ebner, A., Ravikumar, S., Karunakaran, B.: Open source OPC UA PubSub over TSN for realtime industrial communication. In: 2018 IEEE 23rd International Conference on Emerging Technologies and Factory Automation (ETFA), pp. 1087–1090. IEEE, Piscataway, NJ (2018). https://doi.org/10.1109/ETFA.2018.8502479
14. PROFIBUS Nutzerorganisation e.V.: Profinet over TSN: Guidline (2021)
15. Thames, L., Schaefer, D.: Software-defined cloud manufacturing for industry 4.0. Procedia CIRP **52**, 12–17 (2016). https://doi.org/10.1016/j.procir.2016.07.041
16. Trifonov, H., Heffernan, D.: OPC UA TSN: a next-generation network for industry 4.0 and iiot. Int. J. Pervasive Comput. Commun. **19**(3), 386–411 (2023). https://doi.org/10.1108/IJPCC-07-2021-0160
17. Wollschlaeger, M., Sauter, T., Jasperneite, J.: The future of industrial communication: automation networks in the era of the internet of things and industry 4.0. IEEE Ind. Electron. Mag. **11**(1), 17–27 (2017). https://doi.org/10.1109/MIE.2017.2649104

Open Access This chapter is licensed under the terms of the Creative Commons Attribution 4.0 International License (http://creativecommons.org/licenses/by/4.0/), which permits use, sharing, adaptation, distribution and reproduction in any medium or format, as long as you give appropriate credit to the original author(s) and the source, provide a link to the Creative Commons license and indicate if changes were made.

The images or other third party material in this chapter are included in the chapter's Creative Commons license, unless indicated otherwise in a credit line to the material. If material is not included in the chapter's Creative Commons license and your intended use is not permitted by statutory regulation or exceeds the permitted use, you will need to obtain permission directly from the copyright holder.

Dependable Cyber-Physical Matrix Production Systems Utilizing Holonic Multi-agent Systems

Jonathan Bartels[1,2](✉), Simon Komesker[1,2], William Motsch[3], Katharina Hengel[3], Achim Wagner[1,3], and Martin Ruskowski[1,4]

[1] RPTU Kaiserslautern-Landau, Gottlieb-Daimler-Str. 47, 67663 Kaiserslautern, Germany
jonathan.bartels@volkswagen.de
[2] Volkswagen AG, Berliner Ring 2, 38440 Wolfsburg, Germany
[3] DFKI GmbH, Trippstadter Str. 122, 67663 Kaiserslautern, Germany
[4] Technologie-Initiative SmartFactory KL e.V, Trippstadter Str. 122, 67663 Kaiserslautern, Germany

Abstract. In the domain of reconfigurable production systems, Cyber-Physical Matrix Production Systems (CPMPS) are recognized for their advanced levels of operational flexibility. Given the inherent flexible material flow, these loosely coupled systems are characterized by dynamic interdependencies and rapid changes in order sequencing and allocation. This leads to major challenges in production flow control including the emergence of instable behaviors decreasing robustness and threatening overall performance.

Traditional methodologies for assessing and enhancing the reliability and ensuring the robustness of the system do not tackle the dynamic behavior of re-configurable production systems. Due to rigid probabilistic assumptions, efficiency decreases and reasoning in fault propagation is not apparent. For this reason, dependable systems engineering embraces formal descriptions of the systems' dynamical behaviors and continuous monitoring of system properties.

This paper proposes the application of distributed artificial intelligences in the form of holonic multi-agent system (MAS) that integrate the concepts of dependability as part of the system design. Multi-level monitoring of state properties and fault-tolerant control mechanisms are used to minimize deviation between the modelled and observed behavior, therefore ensuring robustness and securing the system's intended operation. The presented framework demonstrates feasibility by first implementations of dynamic interaction mechanisms for subsidiary decision improving makespan while remaining flexible.

Keywords: reconfigurable production systems · fault-tolerant production flow control · holonic multi-agent systems · dependable systems engineering

1 Introduction

In the rapidly evolving landscape of manufacturing technologies, reconfigurable production systems stand out for their ability to adapt to changing product demands and production conditions rapidly. Rigid linear assembly systems struggle to adapt to the current demand for flexibility and can only be reconfigured with significant down-time. Therefore, service-oriented and reconfigurable production systems emerge as suitable alternatives to existing production systems [1]. In the case of Cyber-Physical Matrix Production System (CPMPS), the inherent modular structure and flexible transportation between production entities leverages the ability to anticipate any deviations dynamically through reconfiguration and therefore minimizing throughput loss caused by the linkage of failure rates [2, 3].

The high degree of freedom in CPMPS introduces significant complexity, making it challenging for traditional centralized control approaches to calculate optimal solutions in a reasonable timeframe [4]. Instead, agent-based control systems, which utilize multiple distributed autonomous agents, have emerged as suitable solutions and been successfully implemented in various domains [5, 6]. These Multi-Agent Systems (MAS) can divide and conquer planning, scheduling, and control problems through negotiation and learning mechanisms [6, 7].

Nonetheless, as highlighted by [7–9] complex networks of loosely coupled systems such as MAS, confront new threats through unpredictable interdependencies. The newly acquired degrees of freedom and flexibility lead to complexities with unknown effects on the systems behavior. Purely data-driven approaches such as reward-based deep reinforcement learning (DRL) may solve the problem of complexity, yet leave the effect of unknown emerging system behaviors due to possible non-linear feedback and local minima. This threatens reliability of the system exposing the gap of combining this emerging technology with more reliable control methods [8–10].

Therefore, the main objective of this research is based on MAS based production control of CPMPS and the integration of the principles of robustness and resilience, which are closely tied to the fundamentals of dependability [10]. This leads to the central research question: How can MAS be designed to integrate mechanisms that asses and enhance the dependability of CPMPS? This addresses the gap in the understanding of flexible agent-based control in combination with required formal assessment of dependability. To address these issues, the state of the art is examined for dependable agent based solutions. Then, a generic methodology for applying principles of dependability is utilized and adapted to design a holonic multi-agent systems (HMAS). Lastly, components and interactions are designed to enhance the overall robustness against failures and resilience of CPMPS and validated in experiments.

2 State of the Art for Dependable MAS

This section will introduce the fundamentals of MAS and the connection to the principles of the discipline of Dependable Systems Engineering. The final sub-section will provide an overview of existing approaches and their advantages as well as possible limitations to the system application.

2.1 Multi-agent Systems for Planning, Scheduling and Control

MAS consist of multiple interacting agents, each with distinct behaviors, collectively working towards defined objectives. MAS offer a flexible and scalable approach for control and coordination, essential for handling high degrees of freedom and dynamic nature of systems such as CPMPS [11]. While MAS are effective for managing complex systems, the choice of control mechanisms significantly impacts scalability and system performance [7, 12]. MAS can be organized in centralized, decentralized, or hybrid forms representing entities in the production. Centralized MAS feature hierarchical agents overseeing the entire system, making global decisions. While providing optimal solutions, they suffer from scalability issues and single points of failure, making them less suitable for dynamic environments like CPMPS. Decentralized MAS allow to operate autonomously with local context, enhancing scalability and response-time but potentially resulting in suboptimal global performance due to instable behavior. Hybrid MAS balance global optimization with local autonomy, where agents operate independently but are coordinated by higher-level agents [12].

However, when considering the integration of agent systems into industrial multi-level environments, the robust and resilient control of the intended operation of the system is of uppermost priority [7, 8]. To achieve this goal, the HMAS approach of encapsulated systems of systems is suitable as it can solve a certain degree of global optimization combined with fast-responding local interaction [10, 13]. Nonetheless, despite the flexibility offered by MAS, independent from the control paradigm, their perceived intransparency and potential unreliability raise concerns about their dependability [8]. As environments such as production are safety-critical, they demand full consideration of attributes like safety, reliability, maintainability and availability – dimensions that are essential for a system to be considered dependable [14].

2.2 Dependable Systems Engineering

The discipline of Dependable Systems Engineering includes the Design and Implementation of highly-robust and resilient systems. The concept of dependability emerged as part of the need for a methodological approach of assessing a systems property and evaluate it's intended and actual behavior. The focus is on model-based systems engineering to enable a formal and safe integration [14]. This differs fundamentally from reliability as depicted in the comparison in Table 1.

In order to maintain the aforementioned system attributes threatened by faults, errors and failures, several measures are employed. According to the general taxonomy by Laprie [14], fault prevention aims to prevent the occurrence of faults; fault tolerance ensures that the system continues to operate correctly despite faults; fault removal involves reducing the number of existing faults; and fault forecasting estimates and predicts future faults and assesses their potential impact. Implementing these mechanisms is crucial for preventing fault propagation, thereby maintaining an acceptable level of performance and safety [14, 15].

Traditional risk assessment methodologies in reliability engineering rely on probabilistic assumptions to generate fault rates, leading to extensive experimentation, leaving Boolean fault modeling inadequate for unexpected changes. Similarly, Markov Chains, which use probabilistic features to predict future system states, can lead to a combinatorial explosion of states when dealing with high degrees of freedom, thus contradicting the dynamic nature of reconfigurable production systems [15].

Table 1. Difference between Reliability and Dependability [15]

Subject	Reliability	Dependability
Modelling	Probabilistic, based on random processes	Deterministic, based on dynamical behavior
Assessment	Function of failure probabilities	Functional in the state deviation
Means	Decrease of failure probability by redundancies and diversity	Decrease of state deviation by fault-tolerant control

2.3 Related Work for Dependability with MAS

When looking at the concept of reliability and dependability within production control with MAS, often the concept of holonic systems is found. One of the most prominent approaches tackling the challenge of chaotic behavior is found in the ADACOR^2 system by Barbosa et al. [16]. Here, the holonic approach is used and an advanced PID controller for stabilization in the production control is integrated. Inspired by classical control theory this leverages adaptive self-organization of the holons, yet does not fully examine the dimensions of dependability as such. In contrast, this leaves room for research on the design and training of controllers to enhance system dependability.

Similarly, Heid et al. [17] shows a significant step forward in dependable production systems by incorporating flexibility to manage evolving risk assessment effectively. Here the multi-level design introduces a safety agent with pre-configured hazard rules for strategical, tactical and operational levels. Nonetheless, the focus is on control-device level and lacks fault-tolerant implementation for other levels. Komesker et al. [4] developed a hybrid planning, scheduling and control architecture which serves as an enabler for multi-level mechanisms due to its fractal approach. Yet, it is noted that there is need for further exploration of fault tolerance mechanisms as the approach predominantly focuses on the underlying system architecture. With similar focus Bayanifar [18] developed a Failure Mode and Effects Analysis method. Nonetheless, uncategorized errors are neglected which especially in multi-level environments is dangerous as emerging behavior can't all be categorized beforehand.

Wannagat and Vogel-Heuser [19] apply Fault Tree Analysis (FTA) to enhance fault diagnosis, offering a structured method to manage system failures. Here, especially the dependability requirements are examined and results show stabilization effects on the systems behavior in case of a physical fault. This approach focuses on control device level only and does not consider predictive dimension since it is based on predefined faults in FTA. Rehberger et al. [20] emphasizes the dimension of dependability by aiming to improve planning processes and ensure right timing for real-time scheduling, yet does not consider runtime control and emerging behaviors.

As [8–10, 21] conclude that when it comes to complex networks of units, most existing safety-related approaches are not dynamic enough. They emphasize the need for safety-focused strategies for the control design. While this can set limitations on flexibility, they must be balanced out to enable dependable control. This comes together with the need for concepts to tackle emerging behaviors and not just categorized faults to make these systems applicable for large-scale modular production systems [21]. The related work reveals possibilities and requirements for a combination of adaptive control through MAS considering principles of dependability. However, generic holonic structures and formal methods for dependability assessment as well as unified reconfiguration are still not fully integrated.

3 Approach for Integrating Principles of Dependable Systems Engineering into MAS for CPMPS

One approach to cope with complexity and flexibility at the same time is the combination of model-based techniques for accessing the deterministic behavior attributes of the controlled system and the integration of data-driven methods for more flexibility and automatic model generation. While the latter can be used to quickly find new solutions such as production order reconfigurations, these solutions can be checked for feasibility and predictability by deterministic models. A generic system architecture (GSA) for the assessment of dependability in continuous and discrete systems has been developed in [15] which is based on a recursive-nested behavior control (RNBC) introduced in [22]. This formal model-based methodology aims at controlling complex systems through behavior-based models, and therefore is chosen and examined for applicability to HMAS with Active Fault-Tolerant Control (AFTC).

To enable the developed system for large-scale CPMPS, this framework is based on the holonic architecture by [4, 23] and will be extended to assess and leverage dependability by using model-based behavior control while also enable data-driven flexibility inside the holons. Based on the GSA, the methodology starts with a Behavior-Based System Decomposition and Description, to assess system properties and composition, ensuring scalability by decomposing the system into controllable subsystems. Defining system behavior is crucial for dependability, allowing deviation assessment and ensuring intended behavior. The System Architecture is then designed for control of the production, deriving a generic fractal architecture by matching components with the HMAS. In the subsequent sections, the Interaction Design for Multi-Level Monitoring and AFTC for dependable control is derived.

3.1 Behavior-Based System Decomposition and Description

Since the ISA 95 layered architecture for automation systems struggles with the flexibility accompanied by new modular production systems, the Reference Architecture Model Industry 4.0 (RAMI 4.0) is proposed for modern production systems [11, 24]. It has been demonstrated before that the nested behavior-based control structure, original developed for autonomous mobile robots, can be applied to other types of semi-autonomous systems such as rehabilitation systems, medical robots and unmanned air vehicles. In the case of RAMI 4.0, it is possible to use different adopted model types for every hierarchy level, e.g. continuous state space models on the motion levels, hybrid models on the machine level and discrete event systems on the planning and scheduling levels as the environment in this work. These can be trajectories for robotics or composite Key Performance Indicators as defined in ISO 22400-1 such as ratios, efficiency, effectiveness or rates. Examples for behavioral information and time increments for state updates can be cycle times for control devices in ms, makespan of products in hours or supply chain metrics on factory level, as sets of trajectories.

Therefore, it is proposed to define the mission of a system, which represents the intended behavior of the system in a limited set of reference trajectories. Such trajectories are common practice in many engineering fields and similar forms such as trust vectors exist, yet mainly on control device level not discrete event levels [15, 17]. In the case of CPMPS these missions can be represented as trajectories of production metrics over time derived from production plans and orders by solving methods such as heuristics. Sub-missions can then be derived for further granularity and be used for dependability assessment on lower levels. Plans will be formulated at top level and then divided and conquered defining the intended behavior b_{ref} of the system [4]. According to Wagner [15], the deviation from this mission can be defined as Dynamic Performance δ_p as depicted in Fig. 1. The deviation from the defined safety margins $s_{high/low}$ is defined as Dynamic Safety δ_s.

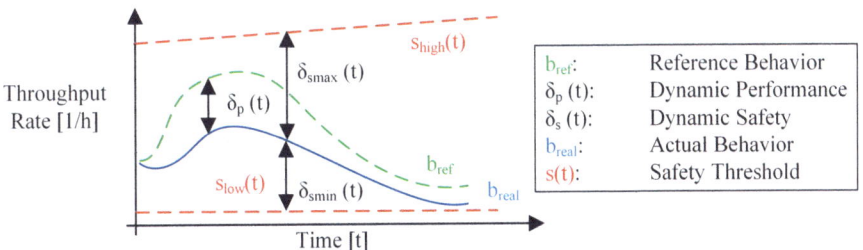

Fig. 1. Exemplary Schematic Overview of Behavior Metrics adapted from [15]

This integrated approach to describe the behavior and define thresholds to use for evaluation functions and triggers accordingly. The corresponding metrics can be combined to serve as a unified measure for dependability depicted in Eq. 1. This allows to compare control designs regarding a combined normalized measure of an integrated dependability measure (IDM). Given the event-based updates in the production environment, the time-discrete version over n samples is defined as follows.

$$D(n) = 1 - \frac{1}{m}\sum_{k=1}^{n}\sum_{j=1}^{d} a_j[1 - A_j(u(k), y_r(k), y(k), \theta_j)] mit \sum_{j=1}^{d} a_j = 1. \quad (1)$$

The IDM $D(n)$ shows the weighted discrepancy between the wanted and the actual behavior in terms of Dynamic Performance and Dynamic Safety. The IDM coefficients (acceptance functions) $A_j(k)$ in the integral are time dependent functions of defined dynamic properties, which are normalized in the interval [0, 1] and weighted by the constant factors a_j, while the sum over all $j = 1\ldots d$ factors is unity. Further variables are the mission input trajectory $u(k)$, the reference output trajectory $y_r(k)$, actual system output $y(k)$ and the overall mission time m. The vector θ_j includes performance parameters [15].

3.2 Dependable System Architecture Utilizing HMAS

As the examination of the state of the art and the related work suggests, the holonic system architecture is suitable for being established over several production layers. The system decomposition as suggested by the generic methodology matches the system decomposition from the level specific control in the HMAS by [23], as it done according to RAMI 4.0. The generic system architecture based on RNBC by Wagner is adapted to the principles of HMAS and depicted in the following architecture model in Fig. 2. Figure 1 the RNBC of a system requires monitoring and reconfiguration components. If we match these controller elements with the agents from [23], the functionalities can be found in the Data Agent (DA), the Deviation Agent (DevA), the Process Orchestration Agent (POA) and the Production Flow Agent (PFA) which are extended by services. Corresponding to the RNBC and the HMAS, the following generic component diagram results with their respective levels for $Holon_{Ln}$ and two $Holon_{Ln-1}$.

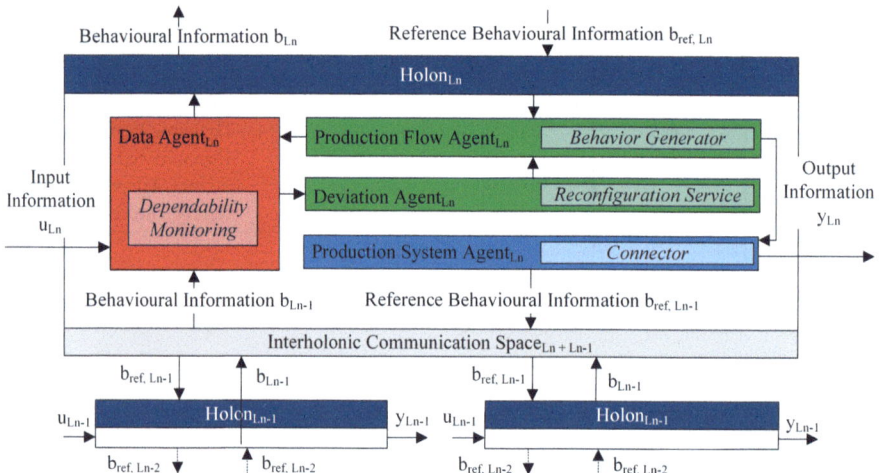

Fig. 2. Components of the HMAS for dependability integration: Monitoring (red), Reconfiguration (green) and Control (blue)

In the initial step the intended behavior $b_{ref,\,Ln}$ is defined. According to the definition of missions and sub-mission, these are distributed over the several layers of the architecture. The PFA_{Ln} is used to split an assigned mission as part of the reference behavior $b_{ref,Ln}$ to the holon and a reference $b_{ref,\,Ln-1}$ into several sub-reference behaviors. These are passed on to the corresponding holon via the POA_{Ln} and are further processed according to the fractal concept. The corresponding sub-mission $b_{ref,\,Ln-1}$ are forwarded to sub-holons or same-level commands send as output information y_{Ln}. The sampling rates for monitoring correspond to the levels and thus lead to fast inner feedback control loops and superordinate slower loops as depicted in Sect. 4.

The DA_{Ln} is subscribed to the Interholonic Communication Space with subordinate holons and thus takes over the monitoring task by receiving externally sensed input information u_{Ln} from the same level and behavioral information $b_{ref,\,Ln-1}$ from sub-holons.

Here, the DA_{Ln} receives all information on the production and transportation entities of the subordinate level. The events contain formatted information and are pre-processed and used for metrics calculation such as the Dynamic Performance and Dynamic Safety for IDM and state estimations. Following the GSA and HMAS concept, the information is evaluated by the $DevA_{Ln}$ with the help of its Reconfiguration $Service_{Ln}$ for minimizing behavior deviation to restore measurable dependability.

4 System Design for Dependable Control of CPMPS

The following section provides an overview over the production use case and the missions to be fulfilled. Based on the problem description the interactions and signals for the agents are defined, implemented and validated in a simulation in Sect. 5.

4.1 Use Case Description

The reconfigurable production system consists of a large-scale CPMPS with skill-based resources, grouped into stations and lane-based AGV for inter-station and smaller free-roaming AGV for intra-station transfer of parts. Each resource includes multiple skills for production orders and has defined failure rates. According to the system architecture, the developed HMAS utilizes services for solving the production mission which is the daily production program provided by the ERP and preceding planning. The subsequent problem for the factory and the system of interest is defined as an np-hard flexible job shop scheduling (FJSS) problem which optimizes order allocation to resources and sequencing on multiple levels for makespan. The goal is to secure the overall production mission by the heuristic and the metrics defined within. To achieve this, the system will divide the initial reference plan from the ERP System to generate sub-missions for station and resource level and will use reconfiguration such as level-specific order reallocation and sequencing to adapt to deviations locally.

4.2 Interaction Concept for Multi-level Monitoring and Active Fault-Tolerant Control

As previously mentioned, the interaction mechanisms are the heart of MAS and predominantly influence the performance of the overall system [25]. The resulting signal diagram depicted in Fig. 3 illustrates the coordination among agents in a holonic multi-agent system for controlling the CPMPS. This system consists of the Process Orchestration Agent (POA) for control, the Data Agent (DA) for monitoring and the Deviation Agent (DevA) plus Production Flow Agent (PFA) for reconfiguration of the system. Moreover, the Holonic Levels are now matched to the RAMI 4.0 resource, station and factory for the use case problem from the previous Sect. 4.1 encapsulating the introduced agents.

The sequence begins with the daily production plan ERP $Data_{L1, ref}$ of the ERP System being send to the $Holon_{Factory}$ transmitting the first reference information. The metric of products for the day is provided by a higher hierarchy of the ERP system that is not observed in the system but provides the production mission for the overall factory. Using the skill-based decomposition the PFA will generate suborders which will serve as production missions for lower levels as part of a Backlog$_{L2,\,ref}$ for multiple Holon$_{Station}$. As this is the first reference, the PFA will generate sub-behaviors by using a heuristic service to solve the FJSS problem optimizing the overall makespan and get a first estimation of the model behavior. The DA of the Holon$_{Factory}$ will store the corresponding Trajectories$_{L1,\,ref}$ in form of aggregated metrics over time and scheduled orders for continuous monitoring. The sub-orders are sent as a command collection of orders $O_{L2,\,ref}$ via the POA controller and the Interholonic Communication Space to its loosely coupled child-holons. This down-stream information flow is continued inside the station level where the sub-orders of the orders are allocated and sequenced to the resources which will then executed the corresponding Process$_{L3,\,ref}$.

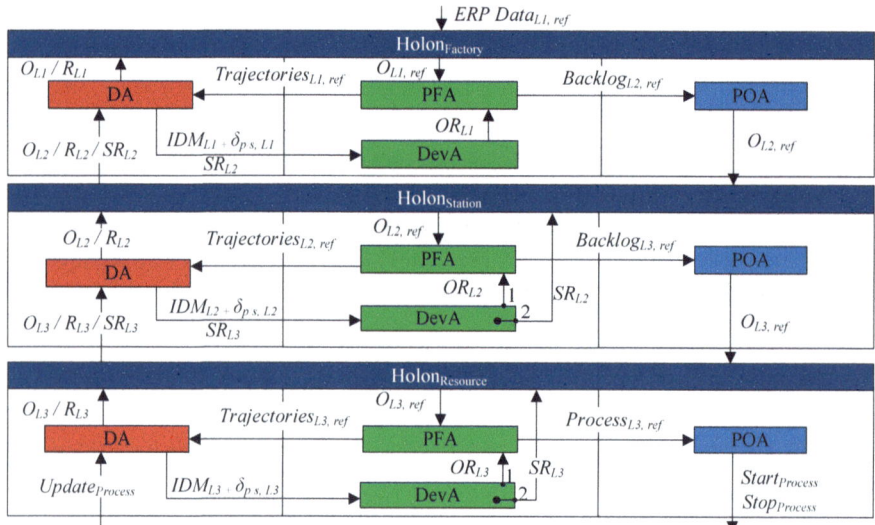

Fig. 3. Signal schemata for fast inner feedback control loops and slower outside feedback control loops with Monitoring (red), Reconfiguration (green) and Control (blue)

Given that the RNBC is a feedback control, the process values Update$_{Process}$ are collected from the field devices to the resource level and forwarded bottom-up. The relevant information for orders O_{L3} and resource states R_{L3} are processed by the DA$_{L3}$ of the Holon$_{Resource}$. After pre-processing and monitoring the relevant information such as the IDM, these are forwarded to the internal DevA for reconfiguration activities which is a resequencing of orders O_{L3} as part of the internal Backlog$_{L3,\,ref}$ based on the evaluation of the δ_p and δ_s. If another sub-level would be integrated an order reallocation to field devices would be a suitable action here as well. State deviations occur in case of break-downs, late orders or full buffers due to uncertainties in the planning or non-linear feedback as part of emerging system behavior.

The DevA$_{Ln}$ on every level plays a crucial role by assessing the behavior information to determine the necessary actions to reestablish reference behavior. Based on decision rules regarding the IDM including the δ_p and δ_s three potential outcomes can arise from the DevA$_{Ln}$ evaluation on every level: (1) Additional support is required beyond the local capabilities of the holon as the local production mission is not retrievable and no reconfiguration is feasible, a support request SR_{LN} is sent to a higher-level holon. (2) If the system decides for a feasible internal order reconfiguration to minimize deviation to the intended behavior, this information is forwarded to the decision support services on the level for reconfiguration such as DRL agents or heuristics. (3) No action is taken as the deviation is in acceptable boundaries. Each reconfiguration service is provided with information about the local context of the system. The services for order allocation and sequencing differ in their horizon, speed and performance corresponding to the slow outer and fast inner control loops and will be presented in Sect. 5.1.

5 Simulation Study

The developed HMAS is connected to a discrete-event simulation for multi-method simulation to test the mechanisms in cooperation with the services. The setup for proving the feasibility is described in Sect. 5.1 and the results are discussed after.

5.1 Setup

The simulation framework consists of the industrial-grade discrete event simulation tool *Plant Simulation* connected to the MAS framework *Janus* [26]. The connection is established via asynchronous OPC UA to account for delay in the subscription interval. Containerized and pre-trained DRL agents are spawned for each production entity reconfiguration commands and are continuously trained during simulation.

For the simulation three different scenarios were compared: (1) Central: Fulfilling strictly the central reference schedule with minimum deadlock-avoidance techniques, (2) Active DRL: Using station-level reconfiguration for resource allocation and sequencing with higher-priority than initial schedule, (3) Passive DRL: Using station-level reconfiguration for resource allocation and sequencing with lower priority than initial schedule. Each scenario works with the same initial schedule and will have the same failure profiles for machines given the individual technical availability of 95%. The ERP Data provides the initial daily plan for 200 products. The reconfiguration services in the DRL scenarios are configured with the same decision rules and allow deviation of 30 min per order when it comes to tardiness. The overall schema for the HMAS and DES integration is depicted in Fig. 4 showing the system boundaries for an example on each level. On the outer slower loop of the $Holon_{Factory}$ the overall throughput is continuously compared to the initial reference.

In case of large deviations as the systems runs empty, the reconfiguration is triggered releasing new orders from its backlog into the systems and optimizing the allocation and sequence to achieve the intended production mission of several products.

For the reconfiguration for $Holon_{Station}$ an existing approach called auction-based online scheduling with reinforcement learning (ABOS RL) from [27] is used to manage local deviations. The ABOS RL is pre-trained on optimizing the makespan and minimizing tardiness within individual stations, ensuring the throughput is maintained even if there are disruptions. The DevA of the lower $Holon_{Station}$ will gain access to local products and orders as well as neighboring products that are within reach of the station timely

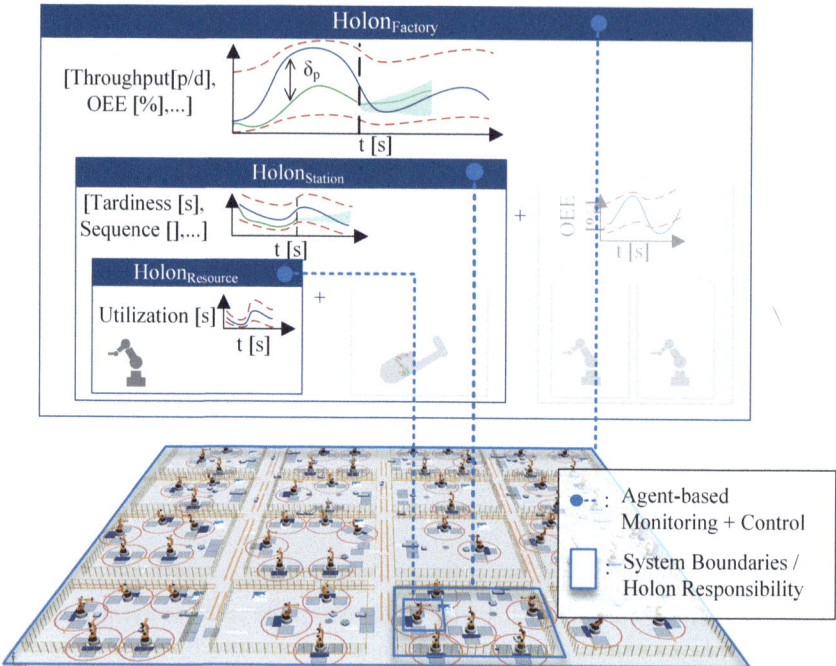

Fig. 4. Schematic overview of the developed HMAS (top) and DES (bottom)

and spatially allowing the $\text{Holon}_{\text{Station}}$ to reconfigure in its system boundaries with little horizontal overlap. For $\text{Holon}_{\text{Resource}}$ mechanisms according to FIFO decisions are used to reconfigure its own $\text{Backlog}_{\text{L3, ref}}$ sequence locally to improve local utilization for the intended production mission as long as the $\text{Holon}_{\text{Station}}$ is not intervening. The IDM for triggering the reconfigurations is solely evaluated based on the delta in performance in a one-dimensional trajectory in current implementations.

5.2 Results and Discussion

The results as depicted Table 2 show a clear picture when it comes to overall makespan. The initial plan with no alternative paths has a throughput time of 13710 s for the set number of products. The Active DRL can produce the same number of products in 12877 s and therefore needs more than 6% less than without deviation detection. The Passive DRL has a restrictive operation area and will have lower priorities than heuristic based orders showing negative effects in the makespan.

Average tardiness, describing the delay of orders, is a factor to be evaluated and served as the main reference trajectory in the framework. Therefore, it shows the difference between the planned finish time provided by a heuristic and the actual time for an order and its suborders accordingly. The results are to be expected as the Active DRL reduces tardiness through reconfiguration, as illustrated in Fig. 5 (left). This can help for short-term tardiness but can also eventually have negative consequences in the long term,

Table 2. Simulation results for different deviation strategies

Metric	Central FIFO	Active DRL	Passive DRL
Makespan [s]	13710	12877	13757
Average throughput time [s]	6033.7	5748.0	5804.4
Average tardiness [s]	1524.0	916.3	906.9
AGV utilization (lane-based) [s]	419325	612134	666099
σ of orders per resource [pc.]	234.7	175.3	157.3

as seen in the overall makespan for the Passive DRL. Metrics related to transportation via AGV usage show increased activity for the DRL strategies. These strategies show a lane-based AGV utilization significantly higher than the central strategy increasing transportation effort yet leveraging makespan in the long-term.

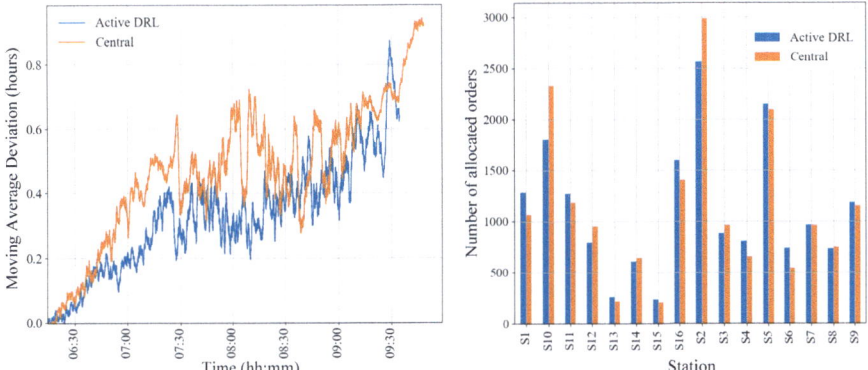

Fig. 5. Dynamic Performance on factory level for throughput (left) effects of order reconfiguration on station-level between Active DRL (blue) and Central strategy (orange)

Figure 5 shows the moving average deviation of the orders collected over all stations for the factory for the two main strategies of the Central FIFO and Active DRL. Since all machines are equipped with availabilities of over 95% and the initial plan is optimized for an ideal production, the loss is inevitable and increases over time. As is evident, the Active DRL manages to minimize the deviation of the individual orders from their intended start. The Central FIFO shows a higher variance in the tardiness of individual orders due to higher vulnerability due to being unable to dodge failures.

The standard deviation in order allocation per resource indicates workload distribution balance and reconfiguration activities. The central control strategy has the highest deviation at 234.7 per resource which was the initial global optimum for a failure free production for the overall makespan of 200 products. The Active DRL reduces it significantly, demonstrating more efficient workload distribution necessary as part of the deviations. This is shown in Fig. 5 (right) aggregated to stations.

In conclusion, it is evident that the DRL strategies successfully minimize deviations to predicted performance for the observed period. This indicates that the combination of model-based behavior generation and evaluation with data-driven flexibility and speed can minimize delays, maintaining performance within acceptable limits under varying conditions and potential disruptions over multiple-levels. The implemented DRL strategies enhance the system's ability to adapt to changing conditions, ensuring that performance remains within limits defined by configurable behaviors.

6 Conclusion and Further Research

In conclusion, the approach presented in this paper offers a promising solution to addressing the challenges faced in reconfigurable production systems, giving transparency and accessibility to the dimensions of dependability. By establishing a holonic control design, a nested control over the system dynamics is achieved, serving the multi-level CPMPS control. The fusion of HMAS with dependability principles creates a fault-tolerant system architecture, enhancing overall performance. The interactions allow continuous IDM monitoring and control of the holonic systems and future extensive analysis on combinations of trajectories and reconfiguration services. Divided in fast inner loops and slow outer loops, this enables quick adjustments to deviations while also handling slower trends operating on larger time increments. Furthermore, the integration of self-similar components ensures reusability within the system with reconfiguration services to minimize deviations of the reference the future. Using the heuristic for reference behaviors grants transparency and therefore control over the system's operations.

Future research should explore the multi-dimensional description of trajectory sets to provide a extensive understanding and forecast of the behaviors and fault-propagation across levels. Another focus in future work will be the trade-off between model-based restrictions and data-driven actions provided by DRL shown in the DRL configurations in this work. Balancing these aspects is essential to maximize the system's performance while maintaining flexibility, yet always assuring dependability. This will benefit from the IDM which can be used to optimize the DRL policies and evaluate actions while maintaining model-based formalism for transparency.

Acknowledgements. This work was supported by the European Union under the "Horizon 2020" research program (Grant no. 957204) within the project "Multi Agent Systems for Artificial Intelligence" (MAS4AI).

References

1. Colombo, A.W., Jammes, F., Smit, H., Harrison, R., Lastra, J.L.M., Delamer, I.M.: Service-oriented architectures for collaborative automation. In: 31st Annual Conference of IEEE Industrial Electronics Society, 2005, Raleigh, NC, USA, pp. 2649–2654. IEEE (2005). https://doi.org/10.1109/IECON.2005.1569325
2. Kern, W., Rusitschka, F., Kopytynski, W., Keckl, S., Bauernhansl, T.: Alternatives to assembly line production in the automotive industry. In: The 23rd International Conference on Production Research (2015)

3. Schönemann, M., Herrmann, C., Greschke, P., Thiede, S.: Simulation of matrix-structured manufacturing systems. J. Manuf. Syst. **37**, 104–112 (2015). https://doi.org/10.1016/j.jmsy.2015.09.002
4. Komesker, S., Motsch, W., Popper, J., Sidorenko, A., Wagner, A., Ruskowski, M.: Enabling a multi-agent system for resilient production flow in modular production systems. Procedia CIRP **107**, 991–998 (2022). https://doi.org/10.1016/j.procir.2022.05.097
5. Vrba, P., et al.: Rockwell automation's Holonic and multiagent control systems. IEEE Trans. Syst. Man. Cybern. C. **41**, 14–30 (2011). https://doi.org/10.1109/TSMCC.2010.2055852
6. Cruz Salazar, L.A., Ryashentseva, D., Lüder, A., Vogel-Heuser, B.: Cyber-physical production systems architecture based on multi-agent's design pattern—comparison of selected approaches mapping four agent patterns. Int. J. Adv. Manuf. Technol. **105**, 4005–4034 (2019). https://doi.org/10.1007/s00170-019-03800-4
7. Leitao, P., Mařík, V., Vrba, P.: Past, present, and future of industrial agent applications. IEEE Trans. Ind. Inf. **9**, 2360–2372 (2013). https://doi.org/10.1109/TII.2012.2222034
8. Vogel-Heuser, B., Lüder, A.: Agenten und Zuverlässigkeit – ein Widerspruch? at - Automatisierungstechnik **65**, 719–720 (2017). https://doi.org/10.1515/auto-2017-0102
9. Piardi, L., Leitao, P., de Oliveira, A.S.: Fault-tolerance in cyber-physical systems: literature review and challenges. In: 2020 IEEE 18th International Conference on Industrial Informatics (INDIN), pp. 29–34. IEEE (2020). https://doi.org/10.1109/INDIN45582.2020.9442209
10. Ehrlich, M., et al.: Alignment of safety and security risk assessments for modular production systems. Elektrotech. Inftech. **138**, 454–461 (2021). https://doi.org/10.1007/s00502-021-00927-9
11. Sidorenko, A., Motsch, W., Van Bekkum, M., Nikolakis, N., Alexopoulos, K., Wagner, A.: The MAS4AI framework for human-centered agile and smart manufacturing. Front. Artif. Intell. **6** (2023). https://doi.org/10.3389/frai.2023.1241522
12. Rieger, C.G., Gertman, D.I., McQueen, M.A.: Resilient control systems: next generation design research. In: 2009 2nd Conference on Human System Interactions, Catania, Italy, pp. 632–636. IEEE (2009). https://doi.org/10.1109/HSI.2009.5091051
13. Unland, R.: Chapter 2 - industrial agents. In: Leitão, P., Karnouskos, S. (eds.) Industrial Agents, pp. 23–44. Morgan Kaufmann, Boston (2015). https://doi.org/10.1016/B978-0-12-800341-1.00002-4
14. Laprie, J.C.: Dependability: basic concepts and terminology. In: Laprie, J.C. (ed.) Dependability: Basic Concepts and Terminology. Dependable Computing and Fault-Tolerant Systems, vol. 5, pp. 3–245. Springer, Vienna (1992). https://doi.org/10.1007/978-3-7091-9170-5_1
15. Wagner, A.: Modeling, analysis, and design of dependable systems with application to robotics and assistance technology. Habilitationsschrift, Universität Mannheim (2018)
16. Barbosa, J., Leitão, P., Adam, E., Trentesaux, D.: Dynamic self-organization in holonic multi-agent manufacturing systems: the ADACOR evolution. Comput. Ind. **66**, 99–111 (2015). https://doi.org/10.1016/j.compind.2014.10.011
17. Heid, M., et al.: Sichere und effiziente prod. durch agentensysteme. atp. **63**, 60–67 (2022). https://doi.org/10.17560/atp.v63i11-12.2617
18. Bayanifar, H.: Agent-based mechanism for smart distributed dependability and security supervision and control of Cyber-Physical Production Systems. Dissertation, Otto-von-Guericke-Universität Magdeburg (2017)
19. Wannagat, A., Vogel-Heuser, B.: Agent oriented software-development for networked embedded systems with real time and dependability requirements in the domain of automation. IFAC Proc. Vol. **41**, 4144–4149 (2008). https://doi.org/10.3182/20080706-5-KR-1001.00697
20. Rehberger, S., Spreiter, L., Vogel-Heuser, B.: An agent-based approach for dependable planning of production sequences in automated production systems. at - Automatisierungstechnik **65**, 766–778 (2017). https://doi.org/10.1515/auto-2017-0040

21. Karnouskos, S., Ribeiro, L., Leitao, P., Luder, A., Vogel-Heuser, B.: Key directions for industrial agent based cyber-physical production systems. In: 2019 IEEE International Conference on Industrial Cyber Physical Systems (ICPS), Taipei, Taiwan, pp. 17–22 (2019). https://doi.org/10.1109/ICPHYS.2019.8780360
22. Badreddin, E.: Recursive nested behavior control structure for mobile robots. In: Intelligent Autonomous Systems 2, An International Conference, pp. 586–596. IOS Press (1989). https://doi.org/10.1016/0921-8890(91)90002-3
23. Komesker, S., Bartels, J., Wagner, A., Ruskowski, M.: Multi-agent interaction structure for enabling subsidiary planning and control in modular production systems. In: Kiefl, N., Wulle, F., Ackermann, C., Holder, D. (eds.) SCAP 2022. ARENA2036, pp. 354–364. Springer, Cham (2023). https://doi.org/10.1007/978-3-031-27933-1_33
24. Trunzer, E., et al.: System architectures for Industrie 4.0 applications: derivation of a generic architecture proposal. Prod. Eng. Res. Dev. **13**, 247–257 (2019). https://doi.org/10.1007/s11740-019-00902-6
25. Bandini, S., Manzoni, S., Vizzari, G.: Agent based modeling and simulation: an informatics perspective. J. Artif. Soc. Soc. Simul. **12** (2009)
26. Rodriguez, S., Gaud, N., Galland, S.: SARL: a general-purpose agent-oriented programming language. In: 2014 IEEE/WIC/ACM International Joint Conferences on Web Intelligence and Intelligent Agent Technologies, pp. 103–110. IEEE (2014). https://doi.org/10.1109/WI-IAT.2014.156
27. Hengel, K., Wagner, A., Ruskowski, M.: A dynamic multi-objective scheduling approach for gradient-based reinforcement learning. In: 18th IFAC Symposium on Information Control Problems in Manufacturing 2024, Vienna, Austria. Elsevier (2024)

Open Access This chapter is licensed under the terms of the Creative Commons Attribution 4.0 International License (http://creativecommons.org/licenses/by/4.0/), which permits use, sharing, adaptation, distribution and reproduction in any medium or format, as long as you give appropriate credit to the original author(s) and the source, provide a link to the Creative Commons license and indicate if changes were made.

The images or other third party material in this chapter are included in the chapter's Creative Commons license, unless indicated otherwise in a credit line to the material. If material is not included in the chapter's Creative Commons license and your intended use is not permitted by statutory regulation or exceeds the permitted use, you will need to obtain permission directly from the copyright holder.

Improving Automated Manufacturing Processes by Applying Agent-Based Planning and Control

Simon Komesker[1,2(✉)], Jonathan Bartels[1,2], Bastian Lang[3], Achim Wagner[1,4], and Martin Ruskowski[1,5]

[1] Department of Machine Tools and Control Systems (WSKL), RPTU Kaiserslautern-Landau, Gottlieb-Daimler-Street 47, 67663 Kaiserslautern, Germany
simon.komesker@volkswagen.de
[2] Volkswagen AG, Berliner Ring 2, 38440 Wolfsburg, Germany
[3] Flexis AG, Industriestraße 6, 70565 Stuttgart, Germany
[4] DFKI GmbH, Kaiserslautern, Trippstadter Street 122, 67663 Kaiserslautern, Germany
[5] Technologie-Initiative SmartFactory KL E.V., Trippstadter Street 122, 67663 Kaiserslautern, Germany

Abstract. In automated production lines, process automation devices with high individual availability are rigidly interlinked. The resulting risk of interlinking losses is minimized by decoupling buffers, offering more flexibility by the cost of additional space. Instead, flexible production concepts with more degrees of freedom could be enabled by applying intelligent control systems.

To optimize and control industrial autonomous production systems with different classes of changeability, IT system architectures are required that can support strategic design as well as tactical and operational control decisions. Agent-based modeling and simulation can be used to implement the concept of a digital twin for applying intelligent planning and control. The findings on the interactions of individual system components can support planners in the design of production systems and can also be used as a basis for optimizing decision-making behavior in operations.

In this paper a simulation study is conducted, in which an agent-based system architecture is applied to control an automated production process in a body shop embracing the different degrees of freedom for altering scenarios. The results show a robust production by enabling a different production concept with reconfigurable static robot cells connected via flexible mobile robots. Whilst reducing the number of production resources and buffer capacities in comparison to rigidly linked scenarios, productivity could be improved. In addition, the integration of the reconfigurable system with more flexibility into an automated production line increases the availability and thus the productivity of the overall production line.

Keywords: systems architecture · autonomous decision making · digital methods and models · reconfiguration · flexibility

1 Introduction

In recent years, modular production concepts have proven to be a resilient alternative to purely line-managed production systems due to their classes of change capabilities. In this article, CPMPS (cyber-physical matrix production systems) are referred to as modular hierarchical systems of systems and, accordingly, CPMPSoS. Based on the modular product structure and the flexible process sequences, measures are sought to coordinate the changeable overall system with reconfigurable resources and flexible process sequences, which can be provided by decision-supporting system structures. Both the design-planning aspects (layout and resource planning) and the operational-controlling aspects (material flow, process sequence) must be considered. A framework that supports a production system from strategic design to operational control and should be able to coordinate the production and material flow as an integrated overall system to determine the appropriate production concept. This paper will combine various approaches for holistic planning, scheduling and control.

2 Planning, Scheduling and Control of Modular Production Systems (CPMPS)

The level of automation in industry has risen steadily since the introduction of PLCs in the 1980s. The current need for more resilience and the associated use of flexible and reconfigurable operating resources requires a move away from rigidly automated process sequences and unchanging batch sizes across the entire product and production life cycle. A modular production system as an integrated overall system has various classes of adaptability for better resilience [1]. To cope with the complexity of decision-making in these production systems, different time horizons must be considered for planning, scheduling and control. Planning describes strategic order planning, while scheduling describes tactical order planning, and control or dispatching is the short-term release and operative order processing [2]. The individual types of planning, scheduling and control are increasingly being combined with the ability to change the scheduling of line-managed production systems and have been a well-researched field for decades [3]. Large sequencing problems are usually solved with metaheuristics and CP technologies. The challenge here is to efficiently solve the respective planning problem by applying appropriate methods and procedures at the corresponding hierarchy level [5].

The flexible job shop problem is suitable for scheduling production systems with various methods and procedures have been researched in recent years [5]. However, the decision-making behavior in the event of a disruption poses a challenge for overall planning and control, regardless of the production concept. One way of simulating complex decision-making situations is the method of agent-based modelling and simulation [6]. This makes it possible to consider both design problems relating to the adaptability of the system and the short-term decision-making behavior for immediate operation in terms of flexibility and reconfigurability. MAS are recently being used to respond to the demand of reconfigurable and flexible planning and control for CPPS [7]. Various projects are also already investigating the use of agent-based and holonic system architectures for the control of modular production systems [8, 9]. RAMI4.0 can be extended using MAS and enables autonomous and holistic decision-making [10, 11]. To realize this, an agent-based system architecture and structured interaction are required [10, 12]. The objects of the production system can be described as skills/capabilities using standardized information models [13]. This forms the basis for skill-based shared production in the sense of manufacturer-independent process execution and order processing [14].

So far, an integrated approach for planning, scheduling and control of changeable industrial production systems does not exist [12]. In the automotive sector resequencing only occurs in storage buffers on factory level e.g. between the body shop and paint-shop to find the most efficient sequence for the following area [15]. By allowing more flexibility, the sequences could be altered and adapted to current events. This allows leveraging the degrees of freedom of flexible and reconfigurable process devices. Production systems can be considered as CPPSoS or CPMPSoS and should be designed by systems engineering design principles [12].

3 Applying an Agent-Based Framework for Improving an Automated Production Process by Integrated Planning, Scheduling and Control

To meet the upcoming demand of changeability in flexible and reconfigurable production systems the application of a framework that combines the different concepts of skill-based information modelling, service-orientation and agent-based interaction is required [12]. Key concept of agent-based control is the interaction modelling which will define the performance of the overall system and target KPI achievement [6].

The approaches to agent-based control were hierarchically structured by Komesker et al. using holons and implemented in an agent-based service-orientated framework [10, 12]. When designing the production layout, both process- and product-oriented structuring methods can be applied and abstracted on a capability basis. The application scenario originates from the production industry and is applicable to several hierarchy layers as per ISA 95, the system architecture according to [10] is used for the system design. Specified versions of the generic roles and interactions will be tested for the design of the control algorithms with global KPI optimization. The basic architecture from [10] enables multi-level control loops by implementing them using holonic agents. The $Holon_R$ represents the corresponding resources, $Holon_S$ the station and $Holon_F$ the factory. The AGV are represented by $Holon_T$ and follow the same fractal approach. Data is exchanged across levels using data agents (DA) while control commands are communicated via the process orchestration agent (POA). The production flow agent (PFA) optimizes order sequencing and allocation, which is continuously checked for deviations with the help of the deviation agent (DevA) and the incorporated quality agent (QA). The material agent (MA) will provide transport orders to multiple $Holon_T$. Although the architecture was initially developed for modular production systems, the encapsulation allows the aggregation of neighboring line-assembly stations as holons ($Holon_{S1, LIN}$ and $Holon_{S3, LIN}$) and keep them as black boxes. As the architecture proposes, each production entity is now encapsulating complexity and sub-systems in their holonic form as well. Based on cross-level interaction and the encapsulation in holons, the interaction scheme as shown in Fig. 1 was developed.

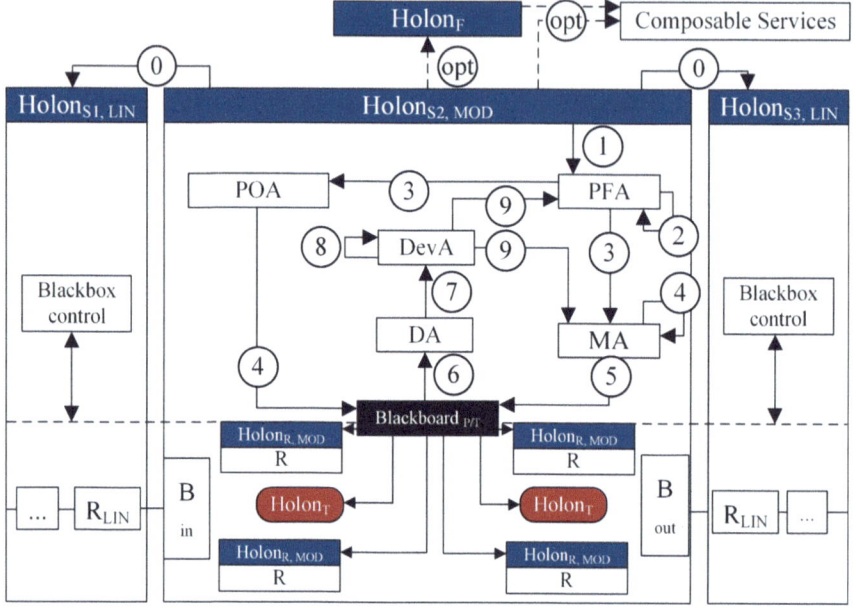

Fig. 1. Interaction Schemata for the integrated Scheduling and Deviation Detection

The center $Holon_{S2,\,MOD}$ gets a preview of upcoming products and remaining items of the batch following a certain distribution and are accessed via horizontal communication between the $Holon_S$ in step (0). When a product enters the input buffer B_{in} and the product data is read (1), the $Order_P$ allocation is triggered inside the PFA (2). The next station is selected according to the shortest waiting time and matched with the required skill which can be optimized with composable services (optional). Then in (3), the MA is informed and can choose between an ad-hoc allocation to the loading bay vehicles if available or a transport blackboard visible to all $Holon_T$ in (3). In step (4), Then, sorted by priority each $Holon_T$ receives the first available $Order_T$ via the blackboard mechanism (5). After successful transportation of the product the process is started. During gathering of $Order_{T/P}$ data in step (6, 7), resource and order data is analyzed and if a deviation is detected (8) a measure is derived. These machine failures are first detected by the DevA inside the sub-system $Holon_{R,\,MOD}$ and forwarded accordingly to the DA in $Holon_S$ as part of (5, 6). With the estimated process deviation, the Station DevA informs the MA and PFA about reallocation of the $Order_{T/P}$ (9). If a rerouting is necessary due to the dynamic batch sizes the DevA informs the MA, that reassigns and releases a new $Order_T$ onto the blackboard. The agent-based framework integrates services that can handle the scheduling of a FJSSP with different degrees of freedom on different hierarchy levels. Each functionality is deployed as a composable service, and the agents orchestrate the deployment of a suitable algorithm.

4 Simulation Study for Implementing Modular Production Structures in a Highly Automated Production Process

To show the applicability of the applied model in Sect. 3, a simulation study is conducted. Basis for the integrated planning, scheduling and control is a linear structured production system that shall be evaluated towards a different production concept.

The simulation study is done with an industrial grade discrete event simulation and contains an area in the body shop including 64 robots, deployed in a line. To identify a suitable production layout for a resilient production flow, the production system is decomposed in subareas and modularized in process clusters. As a result, 12 of the 64 robots are grouped in a subarea and offer a joining operation for the product modules depicted as blue area in Fig. 2. The joining operation can be regarded as a service of that subarea. This subarea and the processes of robots $R_{S2,1}$ to $R_{S2,12}$ are being decomposed and analyzed to find the ideal production concept. These robots were examined regarding the application of skill-based production and were found to be suitable for a unified $Skill_X$ combining previous skills. The line-based and the modular assembly are examined as integrated part into the overall production.

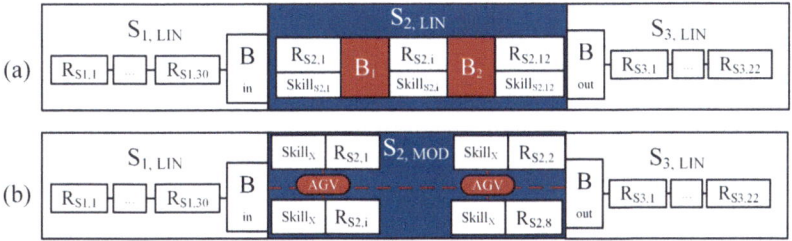

Fig. 2. Full rigid line-assembly (a) and parallelized assembly resources with AGV (b)

For the industrial case a rigid line-assembly for multiple product variants was chosen. The overall system has a total of 30 robots before the system of interest (S_2) encapsulated in station S_1 and 22 subsequent robots in S_3. The initial goal was to optimize throughput of the overall line (S_1 to S_3) without increasing buffer space B_{in} and B_{out}. Given the sequential loss due to failures in a line assembly, 12 separate skills ($Skill_1$ to $Skill_{12}$) were combined to form one $Skill_x$ summarizing the 12 similar process steps depicted in Fig. 2. This leads to station $S_{2,\ MOD}$ with up to 8 robots and 7 AGV. The full line-assembly station has fixed 12 robots with two large buffer spaces B_1 and B_2 inside station S_2 to achieve the required utilization in the first place. Every robot and buffer has a calculated failure rate based on their composition with a defined mean time to repair of 300 s. Reasonable buffer sizes and unlimited buffers were checked initially and compared to the modular integrated line. The utilization rate is based on a predefined 70.58 products per hour being the maximum possible throughput. For the study both versions were compared for 30 days until the standard-deviation between the runs of the same configuration was low enough. The control systems of the neighboring stations were never touched. The configurations with linear station $S_{2,\ LIN}$ were identified in preceding examinations by planning experts, whereas $S_{2,\ MOD}$ configurations were identified in stand-alone experiments as well as checked for feasibility regarding pre-defined dimensions and layout restrictions. LIN_3 corresponds to an unlimited non-feasible buffer size for comparison. The average results are in Table 1.

Table 1. Comparison of integrated line-assembly (LIN) and modular assembly stations (MOD)

ID	Rob	AGV	B_{in}	B_1	B_2	B_{out}	Sum Buffer	Throughput [1/h]	Utilization [%]
LIN_1	12	-	4	4	4	4	16	61.12	86.59
LIN_2	12	-	8	8	8	8	32	61.85	87.62
LIN_3	12	-	16	16	16	16	64	62.50	88.54
MOD_1	7	6	8	-	-	8	16	61.82	87.57
MOD_2	8	7	8	-	-	8	16	62.09	87.97
MOD_3	8	7	16	-	-	16	32	62.38	88.39

The modular assembly outperforms the line in comparable configurations as evident in the experiments although the substituted number of resources is a fraction of the overall system. LIN_2 with MOD_2 have the same size of buffers B_{in} and B_{out}. Nevertheless, the buffers B_1 and B_2 inside S_2 are eliminated and a better result with only 7 AGV is achieved. The internal buffers are substituted by using half the amount of AGV. The throughput increases by 0.25 products per hour, which leads to an increase of 5 products in 24 h and an increase of 172 in the observation period of 30 working days. If the AGVs are left out and the total number of buffers is compared (LIN_2 vs. MOD_3 or LIN_1 vs. MOD_2) a performance increase of 0.50% to 1.28% utilization for the line is seen. Deviations from a calculated batch size of 12 were allowed up to 3 pieces including the effects on changeover time for subsequent stations. The distribution revealed 91% of batch sizes were 12 products, whereas over 8% had 11 respectively 13 as batch size.

The results of the simulation study show potential for resource reduction while also improving overall performance. In previous stand-alone simulations for Station S_2 with unlimited input flow, the modular outperforms the linear concept with over 11% more output due to the avoidance of the multiplication of resource failures. Nonetheless, even in the integrated approach, the overall performance for the system consisting of 64 robots can be leveraged by 24 products per day.

5 Conclusion and Outlook

In this paper, it was shown that the application of suitable methods and models in an agent-based framework with service-orientated scheduling enables the integrated planning and control of a production system. Different variants of a production system could be simulated, considering different classes of change capabilities of a CPMPS on different hierarchy levels. With that, it offers decision support by means of agent-based interaction and service-oriented scheduling for different layout variants and process devices. The integration of flexible and reconfigurable process devices demonstrated an increase in the availability of the overall system with a reduced number of production resources. For further application, the various scenarios can be expanded and prepared for practical use with an investment and operating cost analysis.

Acknowledgements. This work was supported by the European Union under the "Horizon 2020" research program (Grant no. 957204) within the project "Multi Agent Systems for Artificial Intelligence" (MAS4AI).

References

1. Plattform Industrie 4.0, Umsetzung von cyber-physischen Matrixproduktionssystemen, Forschungsbeirat Industrie 4.0/acatech – Deutsche Akademie d. Technikwissenschaften (2022)
2. Wu, K., Shen, Y.: A unified view on planning, scheduling and dispatching for a factory. In: Advanced Engineering Informatics, **46** (2020)
3. Solnon, C., Cung, V.-D., Nguyen, A., Artigues, C.: The car sequencing problem: Overview of state-of-the-art methods and industrial case-study of the ROADEF'2005 challenge problem. In: Eur. J. Oper. Res. **191**, pp. 912–927 (2008)
4. Li, X., et al.: Survey of integrated flexible job shop scheduling problems. In: Computers & Industrial Engineering (2022)
5. Bandini, S., Manzoni, S., Vizzari, G.: Agent based modeling and simulation: an informatics perspective. In: J. Artif. Soc. Soc. Simul. **12** (2009)
6. Huckert, J.L., Sidorenko, A., Wagner, A.: Analysis and assessment of multi-agent systems for production planning and control. In: Flexible Automation and Intelligent Manufacturing: Establishing Bridges for More Sustainable Manufacturing Systems. FAIM 2023., Silva, F.J.G.; Pereira, A.B.; Campilho, R.D.S.G., pp. 687–698 (2024)
7. Nielsen, C., Avhad, A., Schou, C., Ribeiro da Silva, E.: Control system architecture for matrix-structured manufacturing. In: Computers in Industry (2023)
8. Buckhorst, A., do Canto, M., Rabelo, R., Schmitt, R.: A holonic control system approach for line-less mobile assembly. In: Procedia CIRP **107**, pp. 1449–1454 (2022)
9. Komesker, S., Bartels, J., Wagner, A., Ruskowski, M.: Multi-agent interaction structure for enabling subsidiary planning and control in modular production systems. In: Advances in Automotive Production Technology – Towards Software-Defined Manufacturing and Resilient Supply Chains, Springer International Publishing, pp. 354–364 (2023)
10. Sidorenko, A., Motsch, W., van Bekkum, M., Nikolakis, N., Alexopoulos, K., Wagner, A.: The MAS4AI framework for human-centered agile and smart manufacturing. In: Front. Artif. Intell. **6**, p. 1241522 (2023)
11. Komesker, S., Motsch, W., Popper, J., Sidorenko, A., Wagner, A., Ruskowski, M.: Enabling a multi-agent system for resilient production flow in modular production systems. In: PROCEDIA CIRP **107**, pp. 991–998 (2022)
12. Köcher, A., et al.: A reference model for common understanding of capabilities and skills in manufacturing at -. Automatisierungstechnik **71**(2), 94–104 (2023)
13. Bergweiler, S., Hamm, S., Hermann, J., Plociennik, C., Ruskowski, M., Wagner, A.: Produktion Level 4 - Der Weg zur zukunftssicheren und verlässlichen Produktion, Kaiserslautern: Whitepaper SmartFactory KL (2022)
14. Stauder, M., Kühl, N.: AI for in-line vehicle sequence controlling: development and evaluation of an adaptive machine learning artifact to predict sequence deviations in a mixed-model production line. In: Flexible Services and Manuf. Journal, **34**, p. 709–747 (2022)

Open Access This chapter is licensed under the terms of the Creative Commons Attribution 4.0 International License (http://creativecommons.org/licenses/by/4.0/), which permits use, sharing, adaptation, distribution and reproduction in any medium or format, as long as you give appropriate credit to the original author(s) and the source, provide a link to the Creative Commons license and indicate if changes were made.

The images or other third party material in this chapter are included in the chapter's Creative Commons license, unless indicated otherwise in a credit line to the material. If material is not included in the chapter's Creative Commons license and your intended use is not permitted by statutory regulation or exceeds the permitted use, you will need to obtain permission directly from the copyright holder.

Measuring Resilience: A New Perspective on Assessing Production Facilities Through an OEE Based Resilience Metric

Jonas Knüpper[1]([✉]), Alexander Haas[1], Bernd Lüdemann-Ravit[1], Fabian Schmitt[2], Pascal Grieser[3], Andreas Hanzelmann[3], Miriam Schleipen[4], and Dimitrios Genikomsidis[5]

[1] University of Applied Sciences Kempten, Institute of Production and Informatics, Mittagstraße 28a, 87527 Sonthofen, Germany
jonas.knuepper@hs-kempten.de
[2] Mercedes-Benz AG, Bela-Barenyi-Straße, 71059 Sindelfingen, Germany
[3] Mercedes-Benz Tech Motion GmbH, Gutenbergstr. 20, 70771 Leinfelden-Echterdingen, Germany
[4] EKS InTec GmbH, Danziger Straße 3, 88250 Weingarten, Germany
[5] ASCon Systems Holding GmbH, Curiestraße 5, 70563 Stuttgart, Germany

Abstract. Measuring the resilience of industrial production facilities during the production is very complex due to the variety of possible influences and interactions. In addition, production systems are characterized by a dynamic environment with increasing adjustments and changes. A metric for measuring the resilience of individual production facilities during usage phase is intended to help overcome these challenges while meeting requirements such as applicability and comparability. The objective of the metric focuses on the evaluation of measures to remedy resilience relevant events. For this purpose, classified events are considered as a total quantity of changes, adjustments and disruptions together with the measures taken, each in isolation with regard to the productivity of a production facility. An adapted OEE approach with the three weighted influencing factors of quality, performance and availability is developed as the basis for evaluating productivity. This approach evaluates quality based on the direct and indirect costs incurred for rework and post-production. Performance is determined by the cycle time of the facility. The time share of value-adding processes in the period under review is evaluated as a measure of availability. Including the time parameter, which is recorded alongside the resilience event, the metric results in a comparable resilience value. By classifying the event in terms of speed of occurrence and degree of awareness, it is possible to derive the characteristics of adaptability, innovation capability, robustness and improvisation capability of a production facility. In addition, the resilience value determined enables further analysis and the application of resilience strategies.

Keywords: resilience measurement · resilience quantification · production system evaluation · OEE evaluation · production KPI · comparison parameter for production

1 Introduction

Resilience is an increasingly used terminology in the context of Industry 4.0 and digital production. Uncertainties, e.g. in planning due to supply bottlenecks, lack of specialized personnel or other restrictions require resilient production systems. Resilience describes the ability of a technical system to recognize disruptions and subsequently reach a new state of stability again [1, 2]. As resilience capabilities can be analyzed at multiple levels of the value-added chain, an overall assessment is very complex. While indices already exist for supply chains [3], research into assessing the resilience of production facilities is still in the starting blocks. This article examines the determination of the resilience of individual production facilities during production phase and proposes a metric for calculating a resilience index. In order to develop a new concept for measuring the resilience of a production plant, requirements for an evaluation metric are formulated. Key performance indicators (KPIs) are then described that allow conclusions to be drawn about resilience. Finally, the resilience metric is presented and evaluated using an exemplary calculation of the resilience value in production scenarios.

2 Requirements and Application Area

The aim of the resilience metric is to create an evaluation system to determine a reliable resilience index. We propose the following requirements for a metric that makes such an index calculable.

For comparability and further evaluations such as trend analyses, the index must be **quantifiable**. This also enables cross-plant comparability of resilience responses, benchmarks and knowledge sharing. For integration and acceptance in industrial production facilities **industrial relevance** is another requirement. Key figures for data acquisition, classification and evaluation should be standard industry practice. In addition to the assessment, the index should also be able to show **potential for improvement**. The **useability** of the resilience metric extends to various stakeholders, with the operator of the production facility being the primary user. By actively applying the metric, the operator gains insights into the resilience of the facility, which is the starting point for targeted improvements. **Increasing reliability** as the aim of the resilience index is a further requirement. The index is intended as an effective tool for risk mitigation and enhancing reliability, robustness and efficiency. It is supposed to be a significant quality criterion for equipment suppliers, that is not only used internally but also can be used across the value added chain.

3 Influencing Factors of Resilience

A main characteristic of a production facility is the productivity. Productivity outages are usually linked to events such as adjustments, failures or specific changes to a facility. These events can be classified according to speed of occurrence and degree of recognition [4]. The classification is used to derive different resilience capabilities in the event reaction such as adaptability, innovation capability, robustness or improvisation capability.

Resilience events can be defined as events, that have an impact on productivity criteria of a production facility. An event starts when its entry is triggered and signaled. After event classification a corrective measure is defined for resilience relevant events and implemented with a possible delay. The event ends when the balance state or a target state is reached [5]. Figure 1 shows the event sequence.

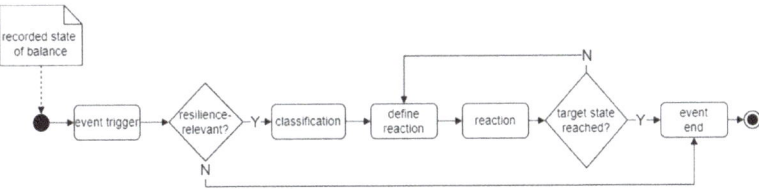

Fig. 1. Resilience event sequence

Several decisive time periods during the course of an event can be identified as shown in Fig. 2. An exemplary resilience curve is shown with increased value due to effort triggered by the event. t_B as confirmation period to clarify resilience relevance between event start T_E and time of resilience clarification T_{IR}. The subsequent period t_L for finding a measure until T_{IB} when the measure is defined. After a possible delay t_{BV} the measure starts at T_{AEB}. With the end of the measure T_{EEB} the recover period t_{WA} until reaching the target state at T_{ZE} begins.

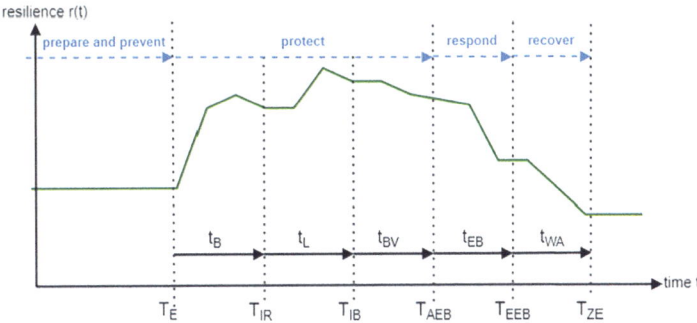

Fig. 2. Resilience event time periods

The five phases of resilience [6] can be seen in the sequence of an event. If disruptions occur despite preparation strategies (prepare) and preventative avoidance measures (prevent), protective mechanisms are initiated (protect). In the following reaction phase, an attempt is made to remedy the drop in performance (respond) in order to subsequently reach the state of stability (recover).

4 Design of an OEE Based Resilience Metric

The effects of external influences are reflected in the key indicators of a production facility. In addition to typical key figures such as MTBF (*mean time between failures*) and MTTR (*mean time to repair*). The OEE (*overall equipment effectiveness*) is an important indicator of productivity [5,7]. OEE is already widely used in industry to evaluate productivity [8] and is utilized to analyze production losses [9]. It encompasses the three main sources of productivity losses: $OEE = availability \cdot performance \cdot quality$. These features enables its utilization as KPI for the development of a resilience metric.

$$OEE = \frac{operating\ time}{scheduled\ time} \cdot \frac{produced\ parts \cdot cycle\ time}{run\ time} \cdot \frac{good\ parts}{produced\ parts} \quad (1)$$

With OEE as key indicator for the balance state of the production facility, availability $v(t)$, performance $l(t)$ and quality $q(t)$ can be derived as criteria for the resilience metric. As they can change during an event, they are formulated as functions of time. Weighting the criteria $(\alpha_V, \alpha_L, \alpha_Q)$ enables individual evaluation and resilience strategies. The proposed metric considers each resilience relevant event individually. The time dependent resilience indicator is called r. The lower the value of r, the more resilient the event reaction was.

$$r(t) = \alpha_Q \cdot q(t) + \alpha_V \cdot v(t) \cdot \alpha_L \cdot l(t) \quad (2)$$

The resulting effect after an event also must be considered. Improvement caused by the measure should have a positive effect on the additional improvement indicator R_I. This can be assessed by comparing the criteria in a period before and after the event.

$$R_I = \alpha_Q \cdot \frac{Q_b - Q_a}{Q_b} + \alpha_V \cdot \frac{V_b - V_a}{V_b} + \alpha_L \cdot \frac{L_b - L_a}{L_b} \quad (3)$$

with Q_b as average quality, V_b as average availability, L_b as average performance in balance state before and Q_a, V_a, L_b similarly after the event.

A suitable evaluation methodology needs to be selected for each criterion. To analyze the **quality** of a facility, direct and indirect costs can be used as an indicator [10]. Examples of direct costs for maintaining the production are required resources such as personnel and materials. Indirect costs with no direct impact on production arise for error reporting or storage costs. In contrast to the original OEE approach, which is based on the proportion of good parts to all produced parts or the FPY (*first pass yield*), costs provide information about how much rework or effort was required to produce the parts.

The **performance** of a production facility is often defined by its output speed. This indicator records the quantity of produced parts in a defined production period. The output speed is the time it takes for one part to be produced including idle and waiting times. A lower output speed leads to higher performance. As the indicator can fluctuate, it is advisable to use the average output speed in a period. It is assumed that production is continuous during a resilience event so the specified quantity of parts to be produced is irrelevant.

To assess the **availability** the time of non-value-adding processes is compared with the total time since event start. Non-value-adding processes that have a negative impact on the availability include set-up processes, maintenance, disruptions and specified production breaks. In contrast to the general OEE definition, the scheduled production time is omitted to avoid the influence of production management. The consideration is made with the reciprocal value of availability (non-availability) to ensure that the resilience indicator increases in line with performance and quality as the availability indicator decreases.
To calculate resilience indicator R_t for one period, all criteria in the event period are considered and integrated over time:

$$R_t = \int^t \alpha_Q \cdot q(t) + \alpha_V \cdot v(t) + \alpha_L \cdot l(t) \, dt \tag{4}$$

Costs for quality consideration are divided into basic costs K_b taken from the balance state before the event, and additional costs, which occur during the resilience periods. The summed total costs K_t are normalized to the period under review using the basic costs. Availability value is assumed to be constant in the time periods of the resilience event and is also normalized with the availability of the balanced state. The average output speed is normalized in the same way. This leads to following calculation formula for the individual time periods:

$$R_t|_{t_1}^{t_2} = \alpha_Q \cdot \left(\frac{K_t - K_b}{K_b} \cdot \frac{t_2 - t_1}{t_m}\right) + \alpha_V \frac{V_t - V_b}{V_b} + \alpha_L \cdot \frac{L_t - L_b}{L_b} \tag{5}$$

with t_1 as start and t_2 as end of time period, t_m as time period for recording the balance state, V_t as average availability and L_t as average output speed in the considered time period. The assessment must be carried out for the individual resilience event periods and results in overall resilience indicator R:

$$R = R_t(t_B) + R_t(t_L) + R_t(t_{BV}) + R_t(t_{EB}) + R_t(t_{WA}) \tag{6}$$

The resilience index for the event under consideration can thus be calculated.

5 Verification with Selected Change Scenarios

For the application of the resilience metric, two selected change scenarios classified as adaptability events are evaluated regarding quality, performance and availability. During this phase of elaboration, the resilience metric provides comparative indicators.

Change of Process After Infiltration into the Line: A quality-related change of process is implemented through adding a measuring station after the insertion phase. This leads to direct and indirect adjustment costs and higher cycle times. The quality increases after the event. An exemplary progression with costs and performance over time is shown in Fig. 3.

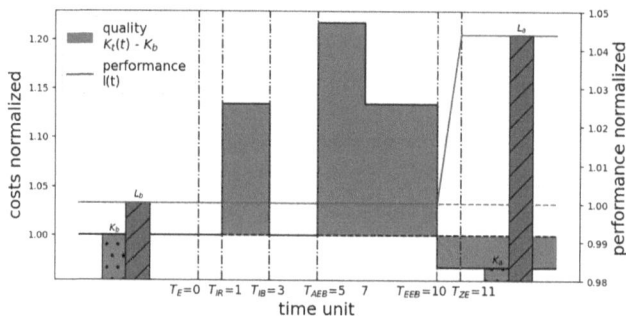

Fig. 3. Costs and performance progression

Exemplary calculation for t_L with $K_b = 1$, $K_t = 1.13$ and $K_a = 0.97$:

$$R_t(t_L) = \alpha_Q \cdot (\frac{1.13 - 1}{1} \cdot \frac{3 - 1}{1}) = \alpha_Q \cdot 0.26 \tag{7}$$

Following 6, the overall resilience can be calculated: $R_t = \alpha_Q \cdot 1,07 + \alpha_L \cdot 0,04$. The improvement rate is $R_I = -\alpha_L \cdot 0.044 + \alpha_Q \cdot 0,033$. The higher cycle time results in negative performance improvement.

Change of Process to Prevent Adhesive Curing: To improve quality and reduce rework, a change of process involves removing components from several supply stations, resulting in shorter dwell times. The event results in an adaptation investment with cost improvement after the modification. The resilience is determined $R_t = \alpha_Q \cdot 0,73$ with an improvement rate of $R_I = \alpha_Q \cdot 0,042$.

It can be noted that the resilience values depend on the weighting factors α_Q and α_L. With both set to 1 it can be seen, that in the first scenario the reaction has a higher index (1.11 vs 0.73) while the improvement rate is lower (-0.011 vs 0.042). This suggests a more effective measure in scenario 2.

6 Conclusion

The proposed new approach of a resilience metric demonstrates how resilience can be quantified. The calculated indices R and R_I can be used to evaluate actions taken to recover from resilience events, not just in terms of good or bad. Using OEE as a basis can lead to acceptance and reduce implementation barriers. By weighting the criteria, different strategies can be used to assess resilience. A next step is to create a model to determine appropriate weighting factors. The application of this metric requires a large dataset for extended validation and observation of the resilience index over comparable events. The metric is designed for production phase events and could be expanded to the engineering phase.

Acknowledgement. The results of this article were produced in the Werk 4.0 research project. The Werk 4.0 research project is funded by the European Union and the Federal Ministry for Economic Affairs and Climate Protection (BMWK).

References

1. Moghaddam, M., Deshmukh, A.: Resilience of cyber-physical manufacturing control systems. Manuf. Lett. **20** (2019). https://doi.org/10.1016/j.mfglet.2019.05.002
2. Heinicke, M.: Resilienzorientierte Beurteilung von Produktionsstrukturen, Otto von Guericke University Library, Magdeburg, Germany (2017). dx.doi.org/10.25673/5255
3. Ivanov, D.: Management der Resilienz in Lieferketten. In: Einführung in die Widerstandsfähigkeit der Lieferkette. Springer Gabler, Cham (2023). https://doi.org/10.1007/978-3-031-25186-3_2
4. Gößling-Reisemann, S. et al.: Vulnerabilität und Resilienz von Energiesystemen. In: Radtke, J., Hennig, B. (eds.), Energiewende - Beiträge der Wissenschaft. Metropolis-Verlag (2013)
5. Sielaff, L., et al.: Resilienzmessgrößen fur Produktionssysteme/Resilience Metrics for Production Systems. wt Werkstattstechnik online **113** (2023). https://doi.org/10.37544/1436-4980-2023-06-58
6. Buß, D., et al.: Resiliente Wertschöpfung in der produzierenden Industrie - innovativ, erfolgreich, krisenfest (2001). https://doi.org/10.13140/RG.2.2.24599.44964
7. Focke, M., Steinbeck, J.: Steigerung der Anlagenproduktivität durch OEE-Management Definitionen, Vorgehen und Methoden - von manuell bis Industrie 4.0. Springer Gabler (2018). https://doi.org/10.1007/978-3-658-21456-2
8. Ng Corrales, L., Lambán, M., Hernandez Korner, M., Royo, J.: Overall equipment effectiveness: systematic literature review and overview of different approaches. Appl. Sci. 10, 6469 (2020). https://doi.org/10.3390/app10186469
9. Gendre, Y., Waridel, G., Guyon, M., Demuth, J., Guelpa, H., Humbert, T.: Manufacturing execution systems: examples of performance indicator and operational robustness tools. CHIMIA Int. J. Chem. **70**. 616–620. (2016). https://doi.org/10.2533/chimia.2016.616
10. Kletti, J., Schumacher, J.: Die Schwachstellen der klassischen Produktion. In: Die perfekte Produktion. Springer Vieweg, Heidelberg. (2014). https://doi.org/10.1007/978-3-662-45441-1_3

Open Access This chapter is licensed under the terms of the Creative Commons Attribution 4.0 International License (http://creativecommons.org/licenses/by/4.0/), which permits use, sharing, adaptation, distribution and reproduction in any medium or format, as long as you give appropriate credit to the original author(s) and the source, provide a link to the Creative Commons license and indicate if changes were made.

The images or other third party material in this chapter are included in the chapter's Creative Commons license, unless indicated otherwise in a credit line to the material. If material is not included in the chapter's Creative Commons license and your intended use is not permitted by statutory regulation or exceeds the permitted use, you will need to obtain permission directly from the copyright holder.

From Market Research to Manufacturing: A Conjoint Analysis and Reconfigurable Manufacturing System Framework for Product-Line Optimization

Sascha Voekler and Ulrich Berger[✉]

Chair of Automation Technology, Brandenburg University of Technology Cottbus,
Platz der Deutschen Einheit 1, Senftenberg, 03046 Cottbus, Germany
sascha.voekler@b-tu.de

Abstract. In an era characterized by the rapid evolution of consumer preferences and diminishing product life cycles, the significance of reconfigurable manufacturing systems (RMS) has surged. For businesses, swiftly adapting to these fluctuating demands is imperative. Conjoint analysis has emerged as a prevalent method for capturing consumer preferences within market research. Subsequently, to delineate an optimal product-line that aligns with these preferences, product-line optimization strategies, such as profit maximization models, are employed. These models comprehensively consider various costs associated with the product line. RMS endeavors to minimize expenses, duration, waste, CO_2 emissions, and other objectives within the production process through a multi-objective framework. This study introduces a novel approach by integrating a profit maximization model from product-line optimization with the cost function of RMS, facilitating a multi-objective optimization framework that adeptly accommodates customer preferences within RMS. While not aiming to delineate the Pareto frontier, this paper elucidates the potential implications of the new framework on the configuration of the Pareto frontier. Additionally, it provides a comprehensive overview of RMS applications in the automotive industry.

Keywords: Reconfigurable Manufacturing Systems · Product-line Optimization · Conjoint Analysis · Multi-objective Optimization

1 Introduction

In today's manufacturing landscape, the ability to rapidly adapt to market changes and evolving consumer preferences is crucial for maintaining competitive advantage. This dynamic environment, characterized by short product life cycles and a demand for high customization, poses significant challenges for manufacturers. To navigate these complexities, it is essential to leverage cutting-edge strategies that align production capabilities with market needs efficiently and sustainably. Among these strategies, reconfigurable manufacturing systems (RMS) and product-line optimization (PLOpt) based on consumer preference data obtained by conjoint analysis (CA) stand out due to their potential to significantly enhance manufacturing responsiveness and economic performance. The integration of these systems promises a novel approach to manufacturing that optimally aligns product features with consumer preferences while ensuring production flexibility and cost-effectiveness. The integration of these two models yields a multi-objective optimization framework.

1.1 Motivation

The rapid transformation of consumer preferences and the decreasing lifespans of products necessitate innovative approaches in manufacturing that can swiftly adapt and respond to market dynamics. RMS have gained prominence due to their flexibility and efficiency in addressing the demand for customized and high-quality products. Concurrently, PLOpt, particularly when combined with Conjoint Analysis (CA), has proven indispensable in aligning product offerings with consumer desires through profit maximization models that factor in various costs and preferences.

The motivation for this research stems from the critical need to integrate RMS with PLOpt to enhance both the adaptability and economic efficiency of manufacturing operations. By combining the robust flexibility of RMS with the strategic acumen of PLOpt, the proposed multi-objective optimization framework seeks to optimize product lines in a manner that not only maximizes profitability but also minimizes costs and environmental impacts. This integration promises a more dynamic approach to manufacturing that can effectively cater to consumer preferences, reduce time and waste, and lower CO_2 emissions, thereby redefining the paradigms of sustainable and responsive manufacturing. Therefore, we employ a novel optimization framework shown in Fig. 1.

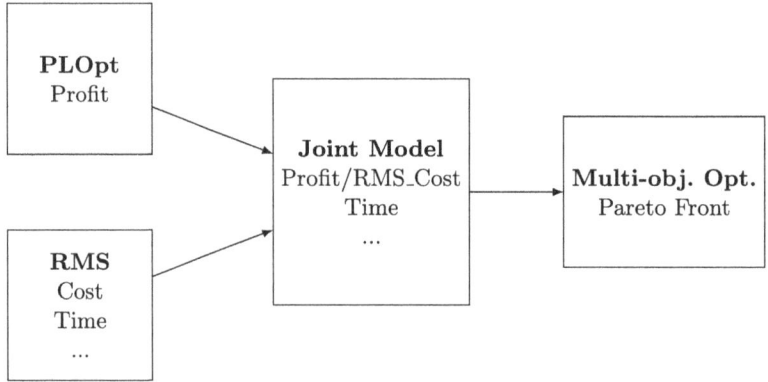

Fig. 1. Illustration of our novel Optimization Framework. (Source: Own representation)

Moreover, the automotive industry, with its complex supply chains and stringent quality demands, serves as an ideal backdrop for illustrating the transformative potential of this integrated approach. By implementing this novel framework, manufacturers can expect not only to meet but anticipate market changes, ensuring continued competitiveness and innovation in a rapidly evolving economic landscape. Thus, this study does not merely propose a theoretical model but addresses a tangible industry need, marking a significant step forward in the synthesis of consumer-oriented product development and adaptable manufacturing systems.

1.2 Product-Line Optimization (PLOpt) with Conjoint Analysis (CA)

Building on the work of [49], product line optimization based on conjoint analysis has rapidly evolved over the ensuing years and decades. Conjoint analysis, which forms the gold standard for measuring consumer preferences in marketing research, has been extensively developed and applied [12, 17, 35]. To conduct a conjoint analysis, marketing researchers predefine the attributes of the product or service of interest (such as color, price, etc.) and their respective levels (e.g., green, blue, 5 EUR, 10 EUR, etc.). Various types of conjoint analysis are available, with the most significant being traditional conjoint analysis, adaptive conjoint analysis, and particularly Choice-based Conjoint Analysis, which accounts for 94% of implementations [42]. The introduction of Choice-based Conjoint Analysis [37] placed consumers in realistic purchasing decisions by que'rying complete product profiles. The breakthrough of Choice-based Conjoint Analysis as the dominant method occurred in the 1990s with the advent of Hierarchical Bayes Analysis [4, 5, 34], enabling the estimation of consumer preferences at an individual level. This means there is a preference value for each attribute level of a product or service, known as a part-worth, for each consumer. A definition originating from [36] describes a product-line as a spectrum of goods of the same type, which differ in the quality of the individual goods [36]. Referring to this definition, [6] see a product-line as a bundle of products that essentially address the same customer needs (see Fig. 2). These products differ in terms of the presence or absence of certain features.

Based on these preference values, whether at an aggregated level or from other types of conjoint analysis, products and services can be optimized and optimal product lines can be determined under certain constraints. Models are needed to define which goals should be achieved under which conditions for determining optimal products, services, and product lines. For example, a company's goal may be to maximize its market share for a specific product group, considering consumer preferences and the status quo market.

Product-line optimization models from the literature generally pursue two goals: to maximize the market share or profit of a product or product line, taking into account consumer preferences and the status quo market. Market share maximization models include the Buyers' Problem and the Share-of-Choice Problem [18, 28].

Fig. 2. Illustrative example of a car product-line. (Source: Created with DALL-E in ChatGPT from OpenAI)

Profit maximization models are primarily represented by the Seller's Problem [18,28] and the model by [16]. Depending on the specific formulation of the problem, the Seller's Problem takes into account individual contribution margins for the attributes of a product or a specific price for a particular product [8].

For modeling the consumer's decision-making process when choosing a specific product, there are several decision rules regarding the preference for a product. These are divided into deterministic and probabilistic decision rules. For the profit maximization objective function in this study, we use a deterministic decision rule that will be described in the next sections.

1.3 Reconfigurable Manufacturing Systems (RMS)

Reconfigurable Manufacturing Systems (RMS) represent a significant advancement in manufacturing technology, aligning with the principal objectives of enhancing productivity, reducing inaccuracy, and minimizing time wastage during changeovers [21]. These systems are recognized for their quick responsiveness to market fluctuations, such as the introduction of new products, combining high quality with cost efficiency [3]. RMS are designed to produce complex, highly customized products in any quantity, merging the benefits of mass production with those of customization [44]. This is achieved through a flexible structure that includes characteristics such as customization, convertibility, modularity, diagnosability, scalability, and integrability [31–33]. Modularity allows for interchangeable hardware and software components, facilitating easy upgrades. Integrability ensures smooth integration of various modules through a common interface. Customization provides the necessary flexibility within a product family, while convertibility enables systems to evolve their functionality over time. These features not only support the system's ability to adjust rapidly to changes but also facilitate efficient reconfiguration in terms of time and effort [22].

The design of RMS is centered around product families, grouping products based on geometric similarities which dictate the system's configuration. This approach enables the production system to be reconfigured for each product family, allowing for a continuous adaptation to production demands [15]. Moreover, the modular nature of RMS promotes the reuse and interchangeability of components, enhancing the system's ability to introduce new technologies and maintain flexible production capabilities.

Mass customization has emerged as a pivotal strategy in modern manufacturing, addressing customer satisfaction by providing products that combine extensive variation and adaptation with competitive pricing [43]. This concept places considerable pressure on manufacturers to incorporate new and complex functionalities continuously, challenging their operational capacities [13]. Traditional systems such as Dedicated Manufacturing Lines (DML) and Flexible Manufacturing Systems (FMS) have their limitations. DML, while cost-effective for bulk production, lacks variety and may underutilize capacity under competitive pressure, leading to losses [31]. FMS, although versatile, require substantial initial investments and are not designed for structural changes [29]. RMS addresses these issues by combining the high production rate of DML with the flexibility of FMS, offering a cost-effective and responsive solution to rapid market changes and regulatory fluctuations.

The increasing global demand for both capital and consumer goods necessitates the development of new manufacturing systems that consider economic, environmental, and social factors. RMS enhances system sustainability and market responsiveness, making it a preferable choice in the modern manufacturing landscape [24]. For researchers and practitioners looking to optimize RMS further, various objective functions have been identified, including cost considerations for sustainability, time efficiency, and process planning. These functions address environmental impacts, such as greenhouse gas emissions and waste, and are typically optimized using advanced algorithms like Genetic Algorithms and Simulated Annealing [24,48]. This comprehensive approach underscores the transformative potential of RMS in meeting the evolving needs of the market while prioritizing sustainability.

2 Literature Review for Automotive Industry

In response to the increasing complexity and dynamic demands of the automotive industry, both RMS and product-lines (PL) have emerged as critical strategies for maintaining competitiveness. To illustrate a selection of applications and developments in these areas, Table 1 provides an overview of seminal works that have shaped current practices and methodologies, although it is not exhaustive.

Table 1. Applications of RMS and Product-lines (PL) in the Automotive Industry. (Source: Own representation)

Author	Topic	Year	Description
[14]	RMS	2015	Focuses on planning changeable and reconfigurable assembly systems for automotive manufacturing, utilizing process modules tailored to automotive assembly parts characteristics.
[39]	RMS	2017	Develops a tool and methodology for assessing adaptability of RMS in the automotive industry, crucial for the transition to Industry 4.0 smart factories.
[40]	RMS	2019	Applies a hybrid simulation and optimization approach to resource allocation in RMS, specifically in the context of automotive production planning.
[41]	RMS	2019	Proposes a criterion to assess the reconfigurability of existing automotive manufacturing systems, analyzing their capacity to adapt to product and volume variations.
[30]	RMS	2020	Discusses RMS used in manufacturing automotive engines and powertrain components, highlighting characteristics essential for automotive industry needs like scalability and modularity.
[2]	RMS	2024	Suggests a methodology for developing RMS focusing on adaptability and reconfigurability throughout the life cycle of automotive manufacturing systems.
[46]	PL	2015	Discusses how automotive engineering exemplifies extreme product line engineering (PLE), with implications including efficient manufacturing, complexity management, and automated generation of calibration parameters. Focuses on General Motors as a case study for large-scale automotive PLE.
[38]	PL	2017	Explores the operational effects of product line breadth in the automotive industry, emphasizing the balance between mismatch costs and production efficiencies through product proliferation.
[9]	PL	2017	Analyzes the impact of product line design in automotive car sharing models, highlighting trade-offs between driving performance and fuel efficiency, and its effect on OEM profits and environmental standards.
[45]	PL	2018	Focuses on optimizing product line architectures in the automotive industry, utilizing multi-objective optimization techniques to manage system design complexities.
[25]	PL	2022	Reviews methods and techniques in automotive software product line development, stressing the importance of managed evolution and agile methods in adapting software for various vehicle generations.

The table is segmented into two major topics: RMS, which focuses on the adaptability and efficiency of manufacturing processes, and PL, which deals with the strategic design and optimization of product lines to maximize operational effectiveness and market responsiveness. Each entry in the table lists the key authors, the year of publication, and a concise description of the study's contribution to the automotive industry, highlighting specific innovations or findings relevant to either RMS or PL. This overview facilitates a deeper understanding of how these methodologies are implemented to tackle the challenges of automotive manufacturing and design, underscoring their importance in driving technological advancement and operational efficiency in the sector.

3 Problem Description

In this section, we present two foundational models, PLOpt and RMS, which are central to our study. Initially, each model is introduced individually, detailing their respective methodologies and unique contributions to the optimization process. Subsequently, we integrate these models to formulate a comprehensive multi-objective optimization framework. This combined model is designed to compute the Pareto front, enabling the identification of optimal solutions that balance the objectives encapsulated by PLOpt and RMS. For a clearer understanding of the numerous parameters and variables involved in the optimization problems, please refer to the conventions outlined in Table 2.

3.1 Model for PLOpt

Product-line optimization models based on conjoint analysis are combinatorial optimization problems and therefore NP-hard [28]. For a single optimal product, there are $J = \prod_{k=1}^{K} L_k$ possible product combinations. Let R be the number of products to be added to a product line. The number of possible product-lines to be calculated would be $\binom{J}{R} = \binom{\prod_{k=1}^{K} L_k}{R}$.

For modeling a consumer's choice decision for a specific product, there are several decision rules regarding the preference of a product. These are divided into deterministic and probabilistic decision rules. For the profit maximization objective function in this work, we use a deterministic decision rule. The utility of a product $j \in J, j = 1, 2, \ldots, J$, for an individual $i \in I$ is denoted by $u_{ij}, i \in I, j \in J$. This is derived from the product of the part-worth values β_{ikl} with the binary variable x_{jkl} over the sum of all attributes K and all attribute levels L_k, where $x_{jkl} = 1$ if product $j \in J$ has attribute level $l \in L_k$ of attribute $k \in K$. If this is not the case, then $x_{jkl} = 0$. The total utility of a product j for a consumer i is given by:

$$u_{ij} = \sum_{k=1}^{K} \sum_{l=1}^{L_k} \beta_{ikl} x_{jkl} . \tag{1}$$

Once the aggregated or individual utility values of the products u_{ij} are calculated, the choice probabilities of the products of interest can be determined

Table 2. Description of Parameters (Params) and Decision Variables (Vars) for PLOpt and RMS.

Params/Vars	Description
PLOpt Params	
I	Set of consumers with $i \in I$
K	Set of product attributes with $k = 1, 2, ..., K$,
L_k	Set of attribute levels with $l = 1, 2, ..., L_k, k = 1, 2, ..., K$,
J	Set of all permissible products with $j = 1, 2, ..., J$,
R	Set of products to be included in the product-line with $j = 1, 2, ..., R$,
β_{ikl}	Part-worth utility of the i-th consumer/segment for the l-th attribute level of the k-th attribute,
u_{ij}	Utility of product j for an consumer i,
$prob_{ij}$	Choice probability of the j-th product for the i-th consumer,
PLOpt Vars	
x_{jkl}	Binary decision variable of the j-th product for the l-th attribute level of the k-th attribute ($= 1$, if the j-th product has the l-th attribute level of the k-th attribute; $= 0$, otherwise),
x_{ijkl}	Binary decision variable of the i-th consumer for the j-th product for the l-th attribute level of the k-th attribute ($= 1$, if the product candidate j assigned to consumer i has the l-th attribute level of the k-th attribute; $= 0$, otherwise),
x_{ij}	$= 1$, if a product j from the product-line provides consumer i with equal or lower utility than their status quo product; $= 0$ otherwise,
y_i	$= 1$, if firm decides to produce product j; $= 0$, otherwise.
RMS Params	
n	Number of operations
OP	Set of operations
d, d'	Index of operations
PR_d	Set of predecessors of operation OP_d
m	Number of machines
M	Set of machines
g, g'	Index of machines
t, t'	Index of triplets
TO_d	Set of available triplets for operation OP_d
TM_g	Set of available triplets using machine M_g
T	Set of triplets, where $T = TO_d \cup TM_g$
c, c'	Index of configurations
tl, tl'	Index of tools
p, p'	Index of positions in the sequence
$CCM_{g,g'}$	Machine changeover cost per time unit
$CCC_{c,c'}$	Configuration changeover cost per time unit
$CCT_{tl,tl'}$	Tool changeover cost per time unit
$Pc_{d,t}$	Operation OP_d processing cost when using triplet t per time unit
$TCM_{g,g'}$	Machine changeover time
$TCC_{c,c'}$	Configuration changeover time
$TCT_{tl,tl'}$	Tool changeover time
$Pt_{d,t}$	Operation OP_d processing time when using triplet t
RMS Vars	
$x_{d,p}^t$	$= 1$ if operation OP_d is using triplet t at the pth position, 0 otherwise.
$y_{p,t}^g$	$= 1$ if machine M_g is using triplet t at the pth position, 0 otherwise.
$MC_p^{p-1}(g, g')$	$= 1$ if there has been a change from machine M_g to machine $M_{g'}$ between positions $p-1$ and p, 0 otherwise.
$TC_p^{g,p-1}(t, t')$	$= 1$ if there has been a change from triplet t to triplet t' of machine M_g between positions $p-1$ and p, 0 otherwise.

using a deterministic decision rule, the so-called first choice decision rule [19,20]. A consumer chooses the product alternative that provides the highest utility, assuming that customers behave in a utility-maximizing manner.

$$prob_{ij} = \begin{cases} 1, & \text{if } u_{ij} \geq u_{ij'} \quad \forall j', \\ 0, & \text{otherwise.} \end{cases} \quad (2)$$

Here, $prob_{ij}$ denotes the probability that a consumer $i \in I$ chooses a particular product $j \in J$ over product j'.

We now present our mathematical programming formulation for the product line design problem using the profit maximization model of [8]. Let J represent the set of all possible products and I represent the set of consumers. For each consumer $i \in I$, her preference is indicated by a mapping $prob_{ij}$, where $j_2 \prec_i j_1$ signifies that consumer i prefers product j_1 over j_2 representing the first choice decision rule (see equation (2)). The conjoint model assumes that all products can be linearly ordered by each consumer, meaning that preferences are transitive. Let $Profit_j$ denote the profit from selling one unit of product j for $j \in J$, and let R denote the size of the product line to be designed. We define a binary variable x_{ij} to represent consumer i's decision to buy product j, and another binary variable y_j to represent the firm's decision to produce product j. The mathematical programming formulation of the product line design problem is then given as follows:

$$\max_{x,y} \sum_{j \in J} \sum_{i \in I} Profit_j \cdot x_{ij} \quad (3)$$

subject to (4)

$$\sum_{j \in J} y_j \leq R \quad \text{(capacity)} \quad (5)$$

$$\sum_{j \in J} x_{ij} \leq 1 \quad \forall i \in I \quad \text{(consumption)} \quad (6)$$

$$x_{ij} \leq y_j \quad \forall (i,j) \in I \times J \quad \text{(availability)} \quad (7)$$

$$x_{ij} \leq 1 - y_{j'} \quad \forall (i,j) \in I \times J, j \prec_i j' \quad \text{(preferences)} \quad (8)$$

$$x_{ij} \in \{0,1\}, y_j \in \{0,1\} \quad \forall (i,j) \in I \times J \quad \text{(binary)} \quad (9)$$

There are four types of constraints in the model: the firm's capacity, consumers' consumption, product availability, and consumers' preferences. The capacity constraint (5) ensures that the size of the product line is at most R. The consumption constraint (6) ensures that each consumer will choose at most one product. Moreover, consumers can choose product j only if it is available, which is reflected in the availability constraint (7). Lastly, each consumer must choose according to her preferences among the existing products, captured by

$$\min\ f_c = \sum_{p=1}^{n} \sum_{d=1}^{n} \sum_{t \in TO_i} x_{d,p}^{t} \cdot Pc_{d,t} \cdot Pt_{d,t}$$

$$+ \sum_{p=2}^{n} \sum_{g=1}^{m} \sum_{g'=1}^{m} MC_p^{p-1}(g,g') \cdot CCM_{g,g'} \cdot TCM_{g,g'}$$

$$+ \sum_{p=2}^{n} \sum_{g=1}^{m} \sum_{t \in M_g} \sum_{t' \in M_g} TC_p^{g,p-1}(t,t') \cdot (CCT_{tl,tl'} \cdot TCT_{tl,tl'} + CCC_{c,c'} \cdot TCC_{c,c'})$$

$$+ \text{Additional\_Costs} \qquad (10)$$

$$\min\ f_t = \sum_{p=1}^{n} \sum_{d=1}^{n} \sum_{t \in TO_d} x_{d,p}^{t} \cdot Pt_{d,t}$$

$$+ \sum_{p=2}^{n} \sum_{g=1}^{m} \sum_{g'=1}^{m} MC_p^{p-1}(g,g') \cdot TCM_{g,g'}$$

$$+ \sum_{p=2}^{n} \sum_{g=1}^{m} \sum_{t \in M_g} \sum_{t' \in M_g} TC_p^{g,p-1}(t,t') \cdot (TCC_{c,c'} + TCT_{tl,tl'}) \qquad (11)$$

$$\min\ f_a = f_a(x_{d,p}^{t}, y_{p,t}^{g}, MC_p^{p-1}(g,g'), TC_p^{g,p-1}(t,t'), Z_b) \qquad (12)$$

subject to

$$\sum_{i=d}^{n} \sum_{t \in TO_d} x_{d,p}^{t} = 1 \forall p = 1, \ldots, n \qquad (13)$$

$$\sum_{p=1}^{n} \sum_{t \in TO_d} x_{d,p}^{t} = 1 \forall d = 1, \ldots, n \qquad (14)$$

$$\sum_{t \in TO_d} x_{d,p}^{t} \cdot |PR_d| \le \sum_{p'=1}^{p-1} \sum_{t' \in TO_{d'}} x_{d',p'}^{t'} \forall i = 1, \ldots, n, \forall p = 2, \ldots, n \qquad (15)$$

$$\sum_{g=1}^{m} \sum_{t \in TM_g} y_{p,t}^{g} = 1 \forall p = 1, \ldots, n \qquad (16)$$

$$y_{p,t}^{j} \ge x_{d,p}^{t} \forall d = 1, \ldots, n, \forall p = 1, \ldots, n, \forall g = 1, \ldots, m, \forall t \in TM_g \qquad (17)$$

$$\sum_{d=1}^{n} x_{d,p}^{t} + \sum_{d=1}^{n} x_{d,p-1}^{t'} \le MC_p^{p-1}(g,g') + 1 \forall p = 2, \ldots, n, \forall t, t' \in T \qquad (18)$$

$$y_{p,t}^{g} + y_{p-1,t'}^{g} \le TC_p^{g,p-1}(t,t') \forall p = 2, \ldots, n, \forall g = 1, \ldots, m, \forall t, t' \in TM_g \qquad (19)$$

$$\sum_{t,t' \in TM_g} TC_p^{g,p-1}(t,t') = 1 \forall p = 1, \ldots, n, \forall p = 1, \ldots, n, \forall g = 1, \ldots, m \qquad (20)$$

$$\text{Constraints}(f_a) \qquad (21)$$

$$x_{d,p}^{t} \in \{0,1\} \qquad (22)$$

$$y_{p,t}^{g} \in \{0,1\} \qquad (23)$$

$$TC_p^{g,p-1}(t,t') \in \{0,1\} \qquad (24)$$

$$MC_p^{p-1}(g,g') \in \{0,1\} \qquad (25)$$

the preference inequalities (8). This model is general enough to capture all the relevant features of our problem instance.

3.2 Model for RMS

The operational framework employed in this section is adapted from [23,24], with specific enhancements to accommodate our unique research requirements. For an in-depth understanding of the constraints utilized, we refer the reader to the aforementioned authors. The variables and parameters employed are detailed in Table 2. The total production cost is denoted by f_c, as shown in equation (10). This function seeks to minimize the aggregate costs associated with machine changeovers, configuration adjustments, tool changeovers, processing, and additional fixed costs, represented as $AddCosts$. Similarly, the total production time, denoted by f_t and formulated in equation (11), focuses on minimizing the durations required for machine and tool setups, configuration changes, and actual processing time. Moreover, a comprehensive objective function f_a, defined in equation (12), incorporates multiple objectives ($a = 1, \ldots, A$) that could be essential for the optimization process such as environmental considerations like hazardous liquid waste and greenhouse gas emissions.. The constraint function associated with f_a, presented in equation (21), depends on various decision variables such as $x_{d,p}^t, y_{p,t}^g, MC_p^{p-1}(g,g'), TC_p^{g,p-1}(t,t')$, and Z_b. The parameter set Z_b includes additional factors that influence the optimization objectives.

3.3 Framework for Joint Model of PLOpt and RMS

We develop a joint model for PLOpt and RMS by integrating the cost function from RMS f_c (see equation (10)) into the profit function from PLOpt (see equation (3)). This integration requires us to account for the costs associated with various product attributes such as color, motorization, and others that are produced by RMS. Assuming each attribute can be managed by an RMS, we extend f_c to include all pertinent product attributes. The comprehensive costs for all attributes of a product j are represented as $Cost_j$, calculated by the equation:

$$Cost_j = \sum_{k=1}^{K} \sum_{l=1}^{L_k} f_c^{jkl} \cdot x_{jkl}, \quad (26)$$

where f_c^{jkl} indicates the production cost for level l of attribute k for product j, and $x_{jkl} = 1$ if product j includes attribute level l of attribute k. To incorporate these costs into the profit equation of PLOpt, we subtract $Cost_j$ from the selling price $Price_j$ of each product to determine its profit:

$$Profit_j = Price_j - Cost_j. \quad (27)$$

Considering the practical business environment, it is essential to acknowledge additional expenses like marketing and sales. However, in this model, we assume these are pre-determined and fixed, and thus they are not included in our optimization framework. Additionally, we compute the production time within the RMS framework as follows:

$$Time_j = \sum_{k=1}^{K} \sum_{l=1}^{L_k} f_t^{jkl} \cdot x_{jkl}, \tag{28}$$

where f_t^{jkl} represents the time taken to produce level l of attribute k for product j. This leads to a new multi-objective optimization model that combines PLOpt and RMS:

$$\max \sum_{j \in J} \sum_{i \in I} (Price_j - Cost_j) \cdot x_{ij} \tag{29}$$

$$\min \sum_{j \in J} Time_j \tag{30}$$

$$\min f_a \tag{31}$$

This model seeks to maximize the profit for a product line while managing the production costs and time, along with optimizing other critical factors.

4 Conclusions and Outlook

The integrated framework we have developed, which combines the PLOpt and the RMS, provides a comprehensive approach to managing and optimizing the production and profit of product-lines. By incorporating the cost function from RMS into the profit function of PLOpt, this model allows for a detailed consideration of production costs associated with various product attributes. This integration helps ensure that profit calculations reflect both the revenue and the underlying production expenses more accurately. The inclusion of $Cost_j$ within the profit equation provides a clear picture of the financial impact of each attribute on overall profitability, enabling better strategic decisions. Additionally, the framework calculates production times, which is crucial for optimizing operational efficiency and scheduling. The multi-objective optimization model not only maximizes profit by carefully managing costs but also minimizes production time, thereby enhancing overall productivity.

As a next step, we aim to tackle several instances of our novel optimization problem. Given the NP-hard nature of this problem within a multi-objective framework, it is essential to use algorithms that can effectively approximate the Pareto front. To this end, we plan to evaluate the performance of two genetic algorithms known as non-dominated sorting genetic algorithms, specifically NSGA-II [10] and NSGA-III [11], as well as an ant colony optimization method designed for multi-objective optimization [1]. Furthermore, we intend to conduct a thorough analysis of these algorithms' performances using simulation methods and statistical tests, similar to those described in [7].

References

1. Alaya, I., Solnon, C., and Ghedira, K.: Ant colony optimization for multi-objective optimization problems. In: 19th IEEE International Conference on Tools with Artificial Intelligence (ICTAI 2007), vol. 1, pp. 450–457. IEEE (2007)
2. Andersen, A.L., Lund, J.P., Thomsen, C.B.: Designing for change: a methodology for developing changeable and reconfigurable manufacturing systems. J. Manuf. Technol. Manag. (2024)
3. Andrisano, A.O., Leali, F., Pellicciari, M., Pini, F., Vergnano, A.: Hybrid reconfigurable system design and optimization through virtual prototyping and digital manufacturing tools. Int. J. Interact. Des. Manuf. **6**(1), 17–27 (2012)
4. Allenby, G.M., Arora, N., Ginter, J.L.: Incorporating prior knowledge into the analysis of conjoint studies. J. Mark. Res. **23**, 152–162 (1995)
5. Allenby, G.M., Ginter, J.L.: Using extremes to design products and segment markets. J. Mark. Res. **32**, 392–403 (1995)
6. Aydin, G., Ryan, J.K.: Product line selection and pricing under the multinomial logit choice model. In: Proceedings of the 2000 MSOM Conference, pp. 1–49 (2000)
7. Baier, D., Voekler, S.: One-stage product-line design heuristics: an empirical comparison. In: OR Spectrum, Springer, pp. 1–35 (2023)
8. Belloni, A., Freund, R., Selove, M., Simester, D.: Optimizing product line designs: efficient methods and comparisons. Manag. Sci. **54**(9), 1544–1552 (2008)
9. Bellos, I., Ferguson, M., Toktay, L.B.: The car sharing economy: interaction of business model choice and product line design. Manuf. Serv. Oper. Manag. **19**(2), 185–201 (2017)
10. Deb, K., Agrawal, S., Pratap, A., Meyarivan, T.: A fast elitist non-dominated sorting genetic algorithm for multi-objective optimization: NSGA-II. In: Schoenauer, M., et al. (eds.) PPSN 2000. LNCS, vol. 1917, pp. 849–858. Springer, Heidelberg (2000). https://doi.org/10.1007/3-540-45356-3_83
11. Deb, K., Jain, H.: An evolutionary many-objective optimization algorithm using reference-point-based nondominated sorting approach, part i: solving problems with box constraints. IEEE Trans. Evol. Comput. **18**(4), 577–601 (2013)
12. Debreu, G.: Review of R. D. Luce, individual choice behavior: a theoretical analysis. Am. Econ. Rev. **50**(1), 186–188 (1960)
13. ElMaraghy, H.A.: Changeable and Reconfigurable Manufacturing Systems. Springer, London (2008)
14. Foith-Förster, P., Winter, R., Roßmann, B., Breiter, H.: A planning approach for the structure of changeable and reconfigurable assembly systems. Assem. Autom. **35**(3), 233–247 (2015)
15. Galan, R., Racero, J., Eguia, I., Garcia, J.A.: A systematic approach for product families formation in reconfigurable manufacturing systems. Robot. Comput. Integr. Manuf. **23**(5), 489–502 (2007)
16. Gaul, W., Baier, D.: Simulations- und Optimierungsrechnungen auf Basis der Conjointanalyse. In: Conjointanalyse: Methoden – Anwendungen – Praxisbeispiele, pp. 163–182. Springer, Berlin, Heidelberg (2009)
17. Green, P.E., Rao, V.R.: Nonmetric approaches to multivariate analysis in marketing, Wharton School, University of Pennsylvania, Working Paper (1969)
18. Green, P.E., Krieger, A.M.: Models and heuristics for product line selection. Mark. Sci. **4**(1), 1–19 (1985)
19. Green, P.E., Krieger, A.M.: Choice rules and sensitivity analysis in conjoint simulators. J. Acad. Mark. Sci. **16**(1), 114–127 (1988)

20. Green, P.E., Krieger, A.M.: An application of a product positioning model to pharmaceutical products. Mark. Sci. **2**(1), 117–132 (1992)
21. Haddou Benderbal, H., Dahane, M., Benyoucef, L.: Modularity assessment in reconfigurable manufacturing system (RMS) design: an archived multi-objective simulated annealing-based approach. Int. J. Adv. Manuf. Technol. **94**, 729–749 (2018)
22. Hasan, F., Jain, P.K., Kumar, D.: Performance issues in reconfigurable manufacturing system. DAAAM Int. Sci. Book **24**, 295–310 (2014)
23. Khettabi, I., Benyoucef, L., Boutiche, M.A.: Sustainable reconfigurable manufacturing system design using adapted multi-objective evolutionary-based approaches. Int. J. Adv. Manuf. Technol. **115**(11), 3741–3759 (2021)
24. Khezri, A., Haddou Benderbal, H., Benyoucef, L.: Towards a sustainable reconfigurable manufacturing system (SRMS): multi-objective based approaches for process plan generation problem. Int. J. Prod. Res. **59**(15), 4533–4558 (2021)
25. Knieke, C., Rauch, E., Rottmann, L., Schmid, F.: Managed evolution of automotive software product line architectures: a systematic literature study. Electronics **11**(12), 1860 (2022). https://www.mdpi.com/2079-9292/11/12/1860
26. Kohli, R., Krishnamurti, R.: Optimal product design using conjoint analysis: computational complexity and algorithms. J. Oper. Res. **40**, 186–195 (1989)
27. Kohli, R., Krishnamurti, R.: A heuristic approach to product design. Manag. Sci. **33**, 1523–1533 (1987)
28. Kohli, R., Sukumar, R.: Heuristics for product-line design using conjoint analysis. Manag. Sci. **36**, 1464–1478 (1990)
29. Koren, Y.: The Global Manufacturing Revolution: Product-Process-Business Integration and Reconfigurable Systems. Wiley, Hoboken, NJ (2010)
30. Koren, Y.: The reconfigurable manufacturing system: the key to future manufacturing. J. Manuf. Syst. **58**, 239–252 (2020)
31. Koren, Y., et al.: Reconfigurable manufacturing systems. CIRP Ann. **48**(2), 527–540 (1999)
32. Koren, Y., Shpitalni, M.: Design of reconfigurable manufacturing systems. J. Manuf. Syst. **29**(4), 130–141 (2010). https://doi.org/10.1016/j.jmsy.2010.01.001
33. Koren, Y., Ulsoy, A.G.: Vision, principles and impact of reconfigurable manufacturing systems. Powertrain Int. **5**(3), 14–21 (2002)
34. Lenk, P.J., DeSarbo, W.S., Green, P.E., Young, M.R.: Hierarchical bayes conjoint analysis: recovery of part-worth heterogeneity from reduced experimental design. Mark. Sci. **15**(2), 173–191 (1996)
35. Luce, R.D., Tuckey, J.W.: Simultaneous conjoint measurement: a new type of fundamental measuring. J. Math. Psychol. **1**, 1–27 (1964)
36. Mussa, E., Rosen, S.: Monopoly and product quality. J. Econ. Theory **18**, 301–317 (1978)
37. Louviere, J.J., Woodworth, G.: Design and analysis of simulated consumer choice or allocation experiments: an approach based on aggregate data. J. Mark. Res. **20**(4), 350–367 (1983)
38. Moreno, A., Terwiesch, C.: The effects of product line breadth: evidence from the automotive industry. Mark. Sci. **36**(2), 254–271 (2017)
39. Park, J.S.: A pragmatic tool and new methodology to assess reconfiguration manufacturing system adaptability. J. Manuf. Syst. **45**, 280–297 (2017)
40. Petroodi, K., Shokouhyar, S., Madani, H.: A hybrid simulation-optimization approach for production planning and resource allocation in a reconfigurable manufacturing system. J. Intell. Manuf. **30**(2), 823–837 (2019)

41. Rösiö, C.: Assessing reconfigurability in manufacturing systems. J. Manuf. Syst. **53**, 174–186 (2019)
42. Selka, S., Baier, D.: Kommerzielle Anwendung Auswahlbasierter Verfahren der Conjointanalyse: Eine Empirische Untersuchung zur Validitätsentwicklung. Mark. ZFP **36**(1), 54–64 (2014)
43. Tseng, M.M., Jiao, J.: Mass customization. In: Handbook of Industrial Engineering: Technology and Operation Management (2001)
44. Tu, Q., Vonderembse, M.A., Ragu-Nathan, T.S.: Measuring modularity-based manufacturing practices and their impact on mass customization capability: a customer-driven perspective. Decis. Sci. **35**(2), 147–168 (2004)
45. Wägemann, T., Tavakoli Kolagari, R., Schmid, K.: Optimal product line architectures for the automotive industry. In: Modellierung 2018, pp. 119–134. Gesellschaft für Informatik e.V., Bonn (2018)
46. Wozniak, L., Clements, P.: How automotive engineering is taking product line engineering to the extreme. In: Proceedings of the 19th International Conference on Software Product Line, pp. 327–336 (2015)
47. Yang, J., Son, Y.H., Lee, D., Noh, S.D.: Digital twin-based integrated assessment of flexible and reconfigurable automotive part production lines. Machines **10**(2), 75 (2022)
48. Yelles-Chaouche, A.R., Hadjou, M., Khaled, B.M., Marir, F., Lahrech, M.: Reconfigurable manufacturing systems from an optimization perspective: a focused review of literature. Int. J. Prod. Res. **59**(21), 6400–6418 (2021)
49. Zufryden, F.S.: A conjoint measurement-based approach for optimal new product design and market segmentation. In: Shocker, A.D. (ed.) Analytic Approaches to Product and Market Planning, pp. 100–114. Marketing Science Institute, Cambridge, MA (1977)

Open Access This chapter is licensed under the terms of the Creative Commons Attribution 4.0 International License (http://creativecommons.org/licenses/by/4.0/), which permits use, sharing, adaptation, distribution and reproduction in any medium or format, as long as you give appropriate credit to the original author(s) and the source, provide a link to the Creative Commons license and indicate if changes were made.

The images or other third party material in this chapter are included in the chapter's Creative Commons license, unless indicated otherwise in a credit line to the material. If material is not included in the chapter's Creative Commons license and your intended use is not permitted by statutory regulation or exceeds the permitted use, you will need to obtain permission directly from the copyright holder.

Exploring Interactions in Autonomous Vehicles: A Comprehensive Evaluation of Various Interaction Methods for 2D and 3D Content

Zack Walker, Ansgar Gerlicher[✉], Axel Braun, Lea Pinnow, Daniel Heinemann, Simon Janik, and Elias Merzhäuser

Stuttgart Media University, Nobelstraße 10, 70569 Stuttgart, Germany
{walker,gerlicher}@hdm-stuttgart.de

Abstract. This study investigates the effectiveness of interaction modalities for working within 2-dimensional and 3-dimensional content in autonomous vehicles (AVs). In a Virtual Reality (VR) simulation, three interaction concepts - space mouse, gesture control, and touchpad - are evaluated independently and comparatively based on task completion, user feedback, User Experience Questionnaire (UEQ), and NASA Task Load Index (TLX) assessments. The results show that the space mouse concept is positively received for both 2D and 3D interactions, with high task completion rates and reduced task times. Gesture control is favorable when executed intuitively, while the touchpad performs well in 2D but struggles in 3D due to spatial disparities and technical limitations. UEQ and TLX results align with qualitative findings. Consequently, a hybrid system integrating gesture control and space mouse input is developed to address limitations and enhance user experience within autonomous vehicles.

Keywords: Autonomous Vehicles · Automotive UI · 3D Interaction Concepts · User Experience Design · Augmented Reality · Mixed Reality · Virtual Reality

1 Introduction

Assuming autonomous vehicles will eventually become an integral part of our everyday lives, it is plausible that everyday working life will also become significantly more mobile as a result.

Today, augmented reality (AR), virtual reality (VR) and mixed reality (MR) play an increasingly important role in the value chain of many companies - for example in the planning and construction of vehicles, buildings or machines. Some modern cars already offer basic AR capabilities in the form of head-up-displays. In this context other forms such as AR glasses are not commonly used today, but - as seen with the Apple Vision Pro [1] - could become more common. Technological advances in the development of AR glasses, which increase wearing comfort and make the glasses more suitable for everyday use, contribute to their increasing social spread [2]. With the "Audi activesphere concept" [3], a well-known car manufacturer is showing what a future could look like in which AR glasses are used in an overall concept for control, information display and processing and entertainment within an autonomous car. A meta-analysis of 150 papers between 2009 and 2020 found that scientific interest in AR in autonomous vehicles is increasing every year. Topics such as "safety and driver assistance", "UI design", "trust and acceptance" or "interaction modalities" were frequently dealt with. But papers that examine "interaction modalities", often use just AR head-up displays and no AR glasses [4].

Despite extensive and very differentiated research on interaction modalities in traditional driving, autonomous driving in general or augmented reality, there is only limited knowledge about how forms of interaction in autonomous vehicles have to be designed so that their users can work efficiently with AR glasses in a 2D or 3D environment.

The present study was part of the FlexCAR project (10/18 to 09/23), funded by the Federal Ministry of Research and Education, and compares three special forms of interaction within a VR simulation of an autonomous vehicle: AR/VR gesture control, 3D space mouse / joystick [5] and touchpad / smartphone.

All three forms of interaction have already been extensively studied in other environments [6–12], but not yet in the context of working in autonomous vehicles.

Thus, the aim of this study is to investigate the use of these forms of interaction in an autonomous vehicle. The underlying use case is working with AR glasses in a virtual 2D or 3D environment. A 2D environment in this context can be understood as a screen or a two-dimensional projection plane that can be interacted with, while a 3D environment adds depth to interactions.

2 Literature Review

The literature on how forms of interaction in autonomous vehicles have to be designed so that their users can work efficiently with AR glasses in a 2D or 3D environment is very limited.

Several studies have directed their attention towards investigating various interaction modalities, including gesture control, space mouse, and touchpad. However, the outcomes of these investigations have yielded conflicting results concerning the superior performance among these interaction modes [13–15]. As a result, we initiate our exploration by individually examining these three distinct forms of interaction:

2.1 Gesture Control

When looking at existing AR glasses such as the Holo Lens 2 or the Apple Vision Pro, it is noticeable that the gestures are quite similar. The ergonomics of these gestures were evaluated in many development tests before release and can therefore be assumed to be leading [16,17]. Selecting can be executed by tapping the thumb and forefinger together, to grab an object, the thumb and forefinger are held and the hand is moved while in this gesture to move the object. To zoom and rotate, hold thumb and forefinger first. The hands are then moved away from each other to zoom. When rotating, the hands are guided in a circular motion around the object (Fig. 1).

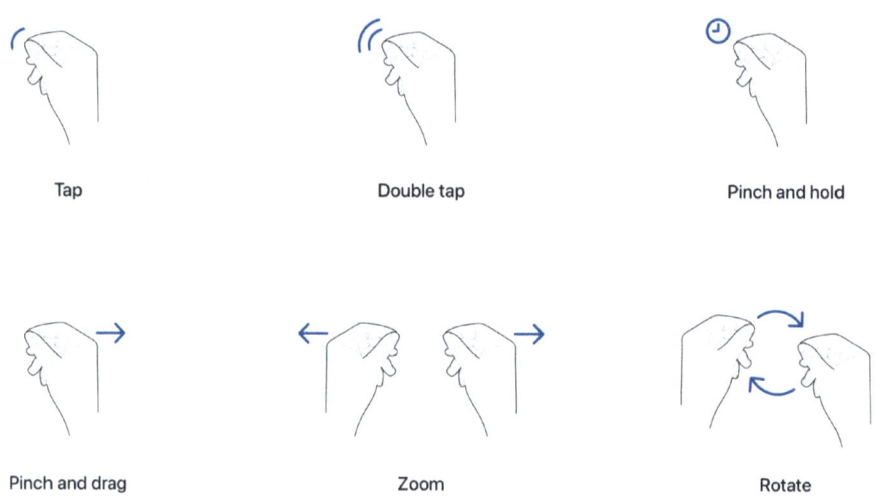

Fig. 1. Gestures of Apple Vision Pro [17]

Further studies confirm this set of gestures, with the exception of grabbing and rotating [6,7], where virtual 3D objects are gripped and rotated like real objects with the whole hand in most cases instead of the index finger-thumb combination.

2.2 Joystick / Space Mouse

Joysticks have always been used primarily in computer simulations, e.g. in flight simulation. Fine movements are required in order to move the controlled object precisely. Joystick-like hardware has also historically been used in cars like a rotary push button for infotainment navigation (see Fig. 2 left). For the creation of 3D CAD models, architects or engineers can also utilize specialized hardware such as a space mouse with six degrees of freedom (6-DoF), which makes it possible to move, rotate and scale 3D objects very precisely (Fig. 2 right).

Hardware-bound interaction offers the advantage of haptics - the user has something in their hand. 3D manipulation can be performed much more efficiently with a space mouse than with other forms of interaction, such as a normal mouse [13,14]. This study aims to further investigate the usage of a space mouse in the context of vehicles.

Fig. 2. Left: Rotary push button ("Dreh-Drücksteller"); Right: 6-DoF spacemouse

2.3 Touch Pad / Smartphone

Touchpads are widespread nowadays: Whether in smartphones, in cars or other electronic devices, people have become accustomed to touch surfaces. Manipulation of 2D content on a 2D surface is common and performs very well. However, the manipulation of true 3D objects poses a challenge. Technological concepts of 3D manipulation using a smartphone often involve rotating and moving the device in open space [18,19]. Other concepts propose the use of multiple fingers to solve this problem [20]. Another concept uses a depth sensor to determine the height of the finger above the display [21] so that the depth axis can be mapped. However, the literature on the manipulation of 3D objects using a standard smartphone on a flat surface is very manageable. Therefore this study examines 3D manipulation with a stationary smartphone and investigates how this performs in comparison to the other two forms of interaction.

3 Methodology

Based on the research conducted as outlined in the previous chapter, three unique interaction methods were selected relevant to addressing the research questions, namely a touchpad or touchscreen, a 3D space mouse and gesture control.

The touchpad was selected due to its widely recognized nature and highly intuitive interaction capabilities. Consequently, the integration of such a familiar device in autonomous vehicles is anticipated to be met with high enthusiasm, with it requiring little effort to master.

Given the incorporation of 3D holograms in the display concept, which demands a novel interaction approach, a space mouse with joystick-like controls emerged as the second chosen interaction modality. With its capacity for six mechanical degrees of freedom (6DoF), the space mouse promises easy maneuvering of content within three-dimensional space.

Gesture control was selected as the third interaction method. Centered around hand gestures, it obviates the necessity for physical touch or input devices, allowing hands-free interaction. Hand gestures hold the advantage of universal intuitiveness across diverse user backgrounds and ages. Beyond their intuitive nature, hand gestures present many possibilities for diverse interactions based on distinct gestures and their combinations.

To evaluate how well these three interaction methods work in an autonomous vehicle concept, a system was devised based around the the FlexCAR project. To accomplish this, the basic framework from the the pre-existing FlexCAR [22] project was utilized. This included the mixed reality simulator setup with the vehicle platform and car seats and the virtual shuttle asset. Additionally, the same VR hardware used for the FlexCAR project, namely the Varjo XR-3 headset, was used to simulate the vehicle and content for interaction. Due to the fact that the investigated use case was solely focused on interaction with content displayed in the vehicle itself, a stationary vehicle scenario was used whereby users would seat themselves in one of the car seats mounted on the vehicle platform and wear the VR goggles (Fig. 3).

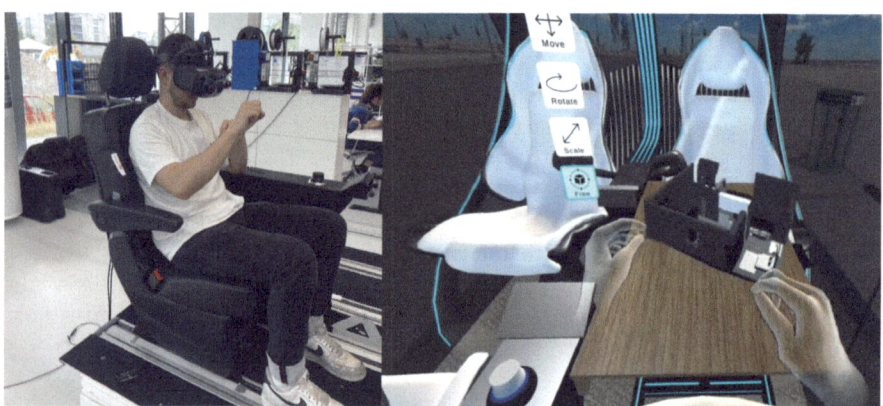

Fig. 3. A user sitting in the mixed reality simulator (left) and the corresponding view of the environment in the VR goggles (right)

Next to the car seat, a center console was situated with indentations to house the devices needed for interaction with the system.

3.1 Research Design

In order to evaluate the chosen interaction concepts, the study underwent two distinct test phases (Fig. 4). During the initial phase, a series of extensive pre-tests scrutinized each interaction method in isolation with the objectives of identifying individual strengths and weaknesses of these methods and gauging participant acceptance.

Then, building upon the insights gained from users' feedback and evaluations, a subsequent phase was implemented wherein the distinct interaction concepts were synthesized into a unified system, which was to capitalize on the collective advantages and strengths of each interaction method for an optimal user experience. This system was subsequently subjected to a second round of user tests, with the objective of discerning the specific scenarios in which different facets of the integrated system came into play and confirming its viability and user acceptance.

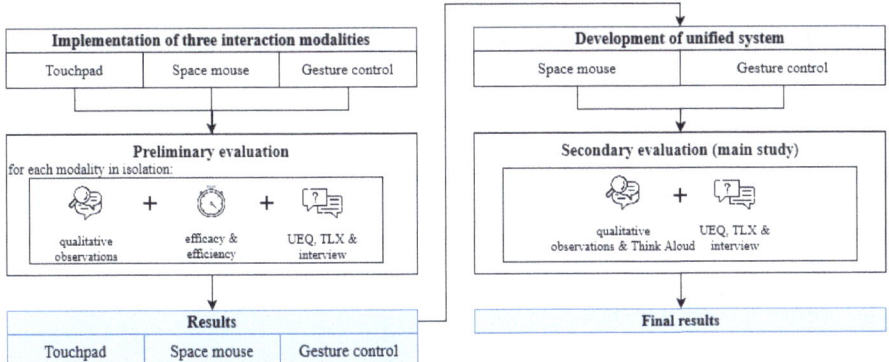

Fig. 4. Overview of the study design

During both testing phases a mixed-methods, within-subject design was adopted, combining quantitative and qualitative data collection. To avoid undesired sequence biases, a counterbalancing strategy was adopted for the presentation order of the interaction methods.

Before conducting the pre-tests, the following hypotheses were defined:

H1: For 2D interactions, conventional control concepts (touchpad and space mouse) are more efficient and preferred.
H2: For 3D interactions, gesture control is more efficient and preferred.
H3: Gesture control requires a high physical strain (compared to touchpad and space mouse).
H4: The space mouse performs worse in interactions with 3D projections because it is more difficult to learn and master compared to touchpad and gesture control.

3.2 Pre-test Design

The preliminary tests were executed with a total of nine participants. Participants were first given the time to freely explore the functionality of each modality, allowing to evaluate the inherent intuitiveness of the interaction concepts, while also offering the opportunity to identify potential improvements. In addition, the efficacy and efficiency of each modality was quantified by time measurement and binary classification of task completion (task completed = 1; task not completed = 0). Each participant was exposed to each interaction paradigm individually. To quantify user experience and perceived effort, the User Experience Questionnaire (UEQ-S) and NASA Task Load Index (TLX) were employed. Concluding each test, a brief interview was conducted to elicit participants' general sentiments toward each conceptual framework.

For the second iteration, the touchpad was omitted completely in favor of a combination of space mouse and gesture control to manipulate content. This system gave users the option to select their preferred mode of interaction based on the displayed content.

The functionality of the space mouse was largely unchanged while gesture control underwent a major overhaul. Gestures were changed to be more intuitive and multiple variations of gestures were included so that the system was more accurate. Furthermore, the content was updated to feature more variation and complexity in order to evaluate more realistic scenarios in the main study. This unified system then underwent a second phase of user tests in the main study.

3.3 Main Study Design

The main study was conducted with 20 participants, four of which had also taken part in the pre-tests. Similarly to the pre-tests, each subject first answered questionnaires about their experience level and perceived technical affinity before engaging with the system. Participants were again given time to explore the system in which they were able to use all aspects of each interaction concept simultaneously. In contrast to the pre-tests, however, participants were given an introduction to the operation of the system. This ensured that every participant had an identical understanding of the system and its possibilities before being given a task. The subsequent task involved the manipulation and navigation of content across both two-dimensional (2D) and three-dimensional (3D) domains. Due to the more qualitative nature of the main study compared to the pre-study, the tasks were not timed. Instead, we asked participants to engage in the Think Aloud method, in which users relayed their thoughts and feelings towards the system.

Upon task completion, participants once again completed the UEQ-S and NASA TLX questionnaires.

For the main study, four additional hypotheses were defined:

H5: The space mouse is preferred for tasks that require precision
H6: Since gesture control does not provide perfect reliability, users prefer to use the space mouse for most interactions
H7: The possibility of using whatever suits the user best leads to lower frustration overall
H8: Users decide on one form of interaction and do not change it afterwards

3.4 Limitations

Many of the participants were students or worked in a technical field. It can be assumed that the resulting high average technical affinity significantly contributed to the fact that test subjects were able to quickly and easily adapt to the technical system and its operation.

Furthermore, interactions were tested in a non-moving environment. In a driving car, external influences such as g-forces and vibrations that are caused by the vehicle making turns or road surface conditions can severely affect interaction effectivity, efficiency and precision. These external influences were not accounted for in our study.

The touchpad and space mouse were placed on the left side of the participants, due to the technical setup of the driving simulator. Thus the left-handed users could have an advantage compared to right handed users. This could slightly influence the overall performance of the users but it was mentioned neither negatively nor positively by the users in the study.

In regards to hardware, the Varjo XR-3 headset was used to simulate the environment. This specific device features two infrared cameras at the front of the headset that are used for detecting the user's hand positions and movements. For this to work reliably however, a users hands must be positioned in the cameras' field of view at all times. This technical limitation leads to the fact that users often cannot keep their elbows on the seat's armrests which in turn leads to higher physical strain, which may have negatively affected the acceptance and overall user experience.

4 Study Results

The preliminary study results show a highly positive reception for 2D and 3D interactions utilizing the space mouse, with a 100 percent task completion rate. Tasks were completed notably faster, especially with 3D content, compared to alternative methods. User feedback from Think Aloud protocols and questionnaires confirmed the system is widely regarded as "easy to use and intuitive." Contrary to expectations, users quickly grasped space mouse controls in both 2D and 3D contexts.

Gesture control was well received in both the 2D and 3D contexts, provided gestures were executed intuitively and reliably, without any noticeable delays in recognition. However, instances where certain users executed gestures with slight variations or where technical limitations led to improper recognition, substantial frustration with the system was observed. Regarding intuitive gestures, particularly in the 2D context, users attempted to interact with the display as if it were a touchscreen. Gestures such as tapping with a finger, grasping/pinching objects, and zooming with two fingers from one hand were performed without any guidance from the experimenters. Conversely, gestures like opening the left palm to initiate rotation were deemed highly unintuitive and consequently received negative evaluations. Moreover, in comparison to other modes of interaction, gesture control exhibited an advantage in terms of quicker albeit more coarse manipulation. Actions like scaling were perceived positively, as the execution of these gestures, once recognized, yielded rapid results.

The average time required for becoming familiar with gesture control was found to be longer in comparison to the learning curve associated with the space mouse. It is noteworthy that a considerable number of participants utilized both gesture control and the space mouse in their interactions during the main study. Precisely, for the 3D task, 14 out of 20 participants chose this hybrid approach, while 9 out of 20 participants did so for the 2D task. Interestingly, although participants generally perceived the space mouse as the more user-friendly option, all participants deliberately chose to incorporate gesture control during task execution. Upon closer examination, it became evident that participants often switched to using gestures when aiming for rapid task completion. Conversely, users prominently favoured the space mouse when precision in movement was required.

The touchpad yielded the most unfavorable results, particularly within the 3D context. As anticipated, interaction with the touchpad exhibited strengths within the 2D context. Given the high level of technical proficiency demonstrated by the test subjects, pre-existing mental models likely existed for operating a touchpad within a 2D environment. This assertion is underlined by the low task completion time observed in this context. However, within the context of the 3D use case, participants encountered substantial challenges in maneuvering the system, particularly concerning object rotation facilitated by a three-finger swipe gesture on the touchpad. This gesture was described as highly unintuitive and proved to be either difficult or inaccessible for the majority of participants. Due to their familiarity with everyday touch devices, participants gravitated toward established touch gestures, contributing to these difficulties.

The results from the UEQ mirror the qualitative findings outlined above. The interaction concept utilizing the space mouse exhibits excellent scores in pragmatic quality (Mean: 1,833) and above-average scores in hedonic quality (Mean: 1,056) compared to the benchmark dataset (Fig. 5). Gesture control fares poorly in pragmatic quality (Mean: 0,528) but attains an excellent rating (Mean: 1,806) in hedonic quality (Fig. 6). Conversely, the touchpad interaction receives low scores in both hedonic (Mean: 0,389) and pragmatic (Mean: -0,250) quality according to UEQ standards (Fig. 7).

The results from the NASA Task Load Index (TLX) predominantly affirm the assumption of elevated physical demands associated with gesture control (GC). This is evident in the relatively high scores observed for both physical demand and effort (Table 1). Comparable high scores are also apparent for touchpad (TP) input, despite only involving finger movement. Overall, the data highlights the perception of significant physical strain when utilizing gesture control. The space mouse (SM) consistently garners the most favorable ratings across almost all categories.

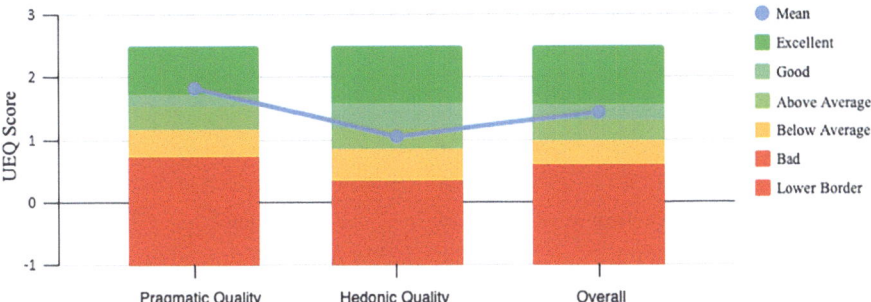

Fig. 5. Space mouse UEQ results (pre-study)

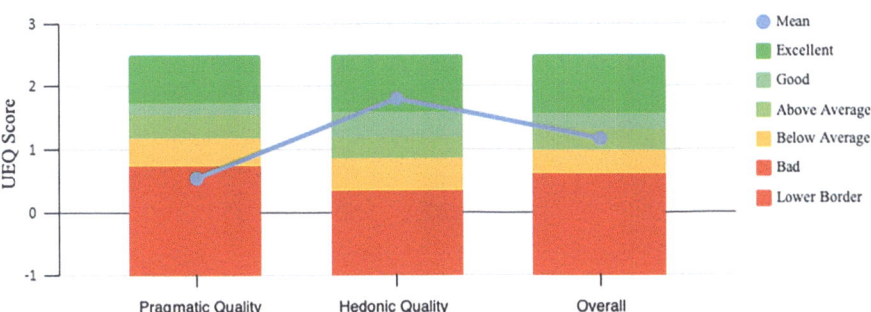

Fig. 6. Gesture control UEQ results (pre-study)

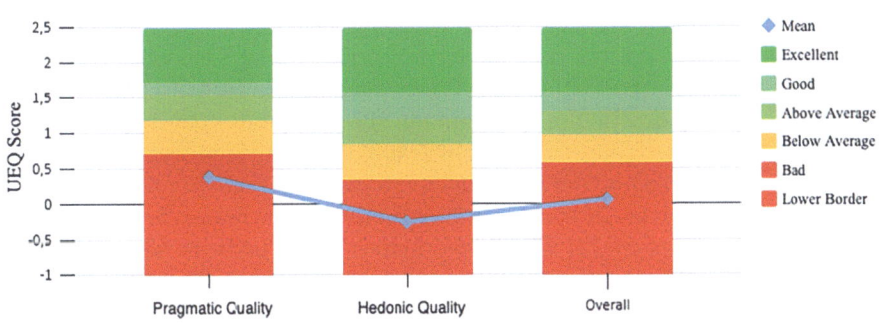

Fig. 7. Touchpad UEQ results (pre-study)

After combining space mouse and gesture control concepts for the main study, the user experience was gauged with the UEQ-S again (Fig. 8).

The evaluation results show an above-average pragmatic quality, indicating practical usefulness, as well as an excellent hedonic quality in terms of emotional appeal, essentially retaining the "best of both worlds" regarding user experience.

The assessment of workload via the unified system, as recorded using NASA TLX, produced the subsequent outcomes:

Table 1. NASA TLX results for each interaction modality (pre-study)

	Mental Demand	Physical Dem.	Tempolar Dem.
SM	40,56	22,78	32,78
GC	55,22	**60,56**	36,11
TP	48,89	36,11	31,11
	Performance	Effort	Frustration
SM	36,11	39,44	36,11
GC	51,11	**65,56**	**57,78**
TP	47,78	48,89	48,33

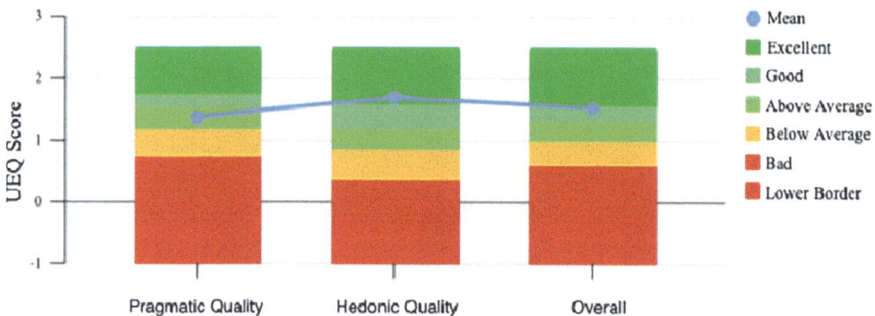

Fig. 8. Unified System UEQ-S Results (main study)

Regarding task load, the unified system displays improvements when compared to the pre-test results (Table 2). When compared to gesture control results from the pre-study, drastic improvements in the categories "physical demand", "performance", "effort" and "frustration" were observed (improvements of 45%, 44%, 40% and 63%, respectively). Compared to the space mouse results from the pre-study, physical demand is slightly increased (46% increase), however, other areas show improvements (21% in "performance", 40% in "frustration").

In addition, participants were surveyed regarding the efficacy of different interaction methods within the system for task completion. Findings from the 2D application evaluation revealed that gesture control had an average utility rating of 3.25 (SD = 1.12). In contrast, the space mouse received an average usefulness rating of 4.26 (SD = 1.15).

The results for the 3D application indicated a similar trend: the average usefulness of the gesture control was 3.5 (SD = 1.24). In comparison, the space mouse achieved an average usefulness rating of 4.1 (SD = 1.25).

With reference to the unified system, the participants were asked to share their perception of the ease of mastering the system in both 2D and 3D applications. The average score for ease of learning and using the system in the 2D application was 4.3 (SD = 0.98). Similarly, ease of use for the 3D application

Table 2. NASA TLX results for unified system (main study)

	Mental Demand	Physical Dem.	Temp. Dem.
Mean	42,25	33,25	27,75
SD	23,87	25,82	26,08
	Performance	Effort	Frustration
Mean	28,5	39,5	21,5
SD	17,78	21,33	23,40

was perceived to be high, with an average score of 4.25 for controlability (SD = 0.55).

The unified system's efficacy in both application contexts was also investigated. In the 2D application, the combined system achieved an average effectiveness rating of 4.5 (SD = 0.60). Similarly, the average effectiveness rating for the 3D application was 4.2 (SD = 0.77).

The sense of control over the unified system was assessed by the participants in the 2D application, with an average rating of 3.75 (SD = 0.79). Similarly, in the 3D application, the average rating for the sense of control was 4.1 (SD = 0.72).

Additionally, it appears that the two types of interaction employed in the user tests for the system have a harmonizing effect. The pragmatic value nearly matched the level attained when using the space mouse exclusively. The hedonic quality was nearly at its peak value when using gesture control alone.

Regarding the hypotheses that were defined before conducting the study, the following statements can be made after reviewing the study results:

H1: The hypothesis "2D interactions and conventional control concepts (touchpad and space mouse) are more efficient and preferred" can be partially confirmed. Users needed less time to complete the tasks in the 2D use case, especially for the space mouse but also for the touchpad. The space mouse also performed extremely well in the users' ranking.

H2: The second hypothesis, which states that gesture control would be preferred over 3D control, can be rejected. Observations show that there was no clear preference and that users often switched between the interaction types.

H3: The third hypothesis, which states that gesture control requires a high degree of physical effort compared to the touchpad and the space mouse, can be confirmed based on the measurement data from the NASA TLX.

H4: The fourth hypothesis, stating that the space mouse performs worse in interactions with 3D projections due to its challenging learning phase compared to the touchpad and gesture control cannot be confirmed. Both quantitative and qualitative findings show that it was instead easier and more convenient for the users to operate.

H5: Observations confirm hypothesis H5, indicating a preference for the space mouse in tasks requiring precision.

H6: The hypothesis stating that the space mouse is preferred for most interactions due to the imperfect reliability of gesture control, as stated in H6, can be disproven by the results of the main study.

H7: The results from the NASA TLX support hypothesis H7, which suggests that allowing users to freely choose between the two forms of interaction within the system results in reduced user frustration.

H8: The study's results contradict hypothesis H8, which suggests that users opt for a single form of interaction and stick with it. Instead, most users used a combination of both options, suiting their preferences.

5 Discussion

The pre-study results show that the space mouse clearly outperformed the gesture control in regard to efficiency, while the touchpad performed worst. Gesture control is less efficient and less preferred for 3D interactions than the space mouse. Gesture control in the pre-study included gestures that were deemed unintuitive by users, including opening the left hand in order to rotate objects. Additionally, when pinching, the middle-finger, ring-finger and small-finger had to be pointed outwards for the system to recognize the gesture. It can be assumed that these issues, paired with missing visual feedback for gestures led to the low pragmatic quality and high physical and mental demand as well as higher frustration. The majority of users attempted to grab the 3D object via pinch with both hands and manipulate scale and rotation simultaneously. Those gestures were identified as user preferences and subsequently, these gestures were implemented in the unified system.

The space mouse consistently performed better than both gesture control and touchpad in all categories, contradicting expectations. In the implementation the freedom of the space mouse was restricted, meaning that not all axes could be manipulated simultaneously. Specifically, rotational manipulation was confined solely to the y-axis. This simplification likely contributed to mitigating the risk of inexperienced users losing control over object interaction within the 3D environment, thereby enhancing overall user experience. However, no comparison was made to support this assumption.

The touchpad performed well in 2D contexts which was expected due to it's familiarity. In 3D contexts the touchpad performed worse than the space mouse and gesture control. The assumption is that this is due to the touchpad inherently being a 2D surface and missing familiar interaction concepts for manipulating 3D content. Thus, unintuitive touch-gestures such as swiping with three fingers had to be implemented which were rejected by users, leading the concept to be omitted in the second iteration. Touchpad control may still be a viable option, however, it first requires the development and acceptance of more intuitive gestures for 3D content.

Due to superior test results, the decision was made to include the space mouse in the second iteration. Despite the lack of reliability, not being easy to use and the resulting frustration, gesture control was chosen as an additional supporting method. This decision was made due to the high hedonic quality assessed with the UEQ-S. In addition, more intuitive gestures were identified during the pre-tests by observing participants attempts to guess the correct gestures. Implementing gestures that mirror real-life movements is also supported by the literature [23–25].

Even after implementing more intuitive gestures, gesture control does not provide perfect reliability, with users preferring to use the space mouse for most interactions. The reliability of gesture control is assumed to be a technical limitation however and not due to the nature of the interaction concept itself. As the second iteration of gestures were not tested in isolation like in the pre-study, an improvement can only be assumed. This assumption is supported by the statements of two participants that partook in both the pre-study and the main study, in which they mentioned significant improvements.

In the unified system, we observed that combined interaction methods lead to lower overall frustration levels. We assume that this is due to users not being forced to utilize a single option. Instead, users can utilize different methods based on what works best for them. In our observations users switched to gestures when they wanted to complete a task quickly. The higher sensitivity of the gesture control allows for quick movements of the 2D and 3D models. In contrast, for precise movements, some users opted to utilize the space mouse instead. There is also evidence that the gestures were sometimes not recognised correctly making precise navigation difficult. Furthermore, after the execution of an action using gesture control, there is a brief transitional period in which the system continues to register gestures despite the user not intending any. This is also a cause of sub optimal precision.

In both the pre-study and the main study, we observed a high perceived physical demand of the gesture control. This drawback of gesture control is also outlined in the literature [26]. It is important to note that the Varjo XR-3 headset used in this study only features front-facing cameras to detect a user's hands. With additional cameras facing downwards, like featured on other headsets like the Apple Vision Pro, the high physical strain may be reduced, as users would be able to rest their arms on a surface while performing gestures.

Even though most users reported that the space mouse was easier to use, all users nonetheless opted into utilizing gestures during the task. This observation may be based on the users curiosity towards the novel technology and they simply wanted to try it out. On the other hand it is possible that the strength of the interaction methods complement each other. This theory is supported by the users feedback and the outcome of the UEQ and NASA TLX questionnaire.

Subjects also expressed that using their hands was more fun than using the space mouse. This suggests that hedonic aspects might also be important for users interaction preferences. It is unclear in which contexts those hedonic aspects are relevant. Regarding the relatively high attractiveness of the gesture control it is also possible that this attractiveness decreases when the interaction form is no longer new to users anymore.

6 Conclusion

In conclusion, a unified system of hand-gesture control and a joystick-like hardware device like the 3D space mouse is viable and accepted by users in the realm of autonomous vehicles. The space mouse is especially advantageous when users need to perform tasks that require precision, like slowly zooming into a specific area or clicking a small button, while hand gestures provide benefits when dealing with tasks where precision is not a priority and speed is more important - for example when scrolling large lists or websites. A system, in which the user can choose an interaction modality is recommended, as each user has their own individual preference for which method works best for them in which situations. The interaction system should be left free in this regard and possibilities for interaction should not be restricted. Additionally, a haptic option should always be an option for users to fall back on, especially when considering accessibility and inclusion.

Looking forward, gesture control should be continuously improved, especially considering precision. In addition, visual feedback for hand gestures should be implemented to make the learning process easier and more intuitive. Looking back at the limitations of this study, a test setup that takes into account external influences e.g. g-forces and vehicle vibrations could potentially significantly alter the viability of the developed system.

In our main study, many users preferred using gesture control even after commenting that it was not as reliable as the space mouse. This may be due to the novelty of the interaction form. If this assumption is correct, it must be determined whether this novelty wears off after extended use. If users reject the usage of gesture control after getting used to it and there being a more reliable alternative, it must be evaluated whether gesture control is a viable interaction form at all.

References

1. HenryZhang99: Tesla cyber truck and apple vision pro (2024). https://www.youtube.com/shorts/NR9yM_S0Z1g
2. Tenzer, F.: Prognose Zum Absatz Von virtual-reality- und augmented-reality-Brillen Weltweit Von 2020 bis 2027, Technical report, Statista (2023)
3. Audi AG: The next sphere of future premium mobility, The Audi activesphere concept (2023). https://www.youtube.com/watch?v=QU552UYRi2I
4. Riegler, A., Riener, A., Holzmann, C.: A systematic review of virtual reality applications for automated driving: 2009–2020. Front. Hum. Dyn. **3**, 689856 (2021). https://doi.org/10.3389/FHUMD.2021.689856
5. 3DConnexion: Spacemouse compact (2023). https://3dconnexion.com/de/spacemouse/
6. Piumsomboon, T., Clark, A., Billinghurst, M., Cockburn, A.: User-defined gestures for augmented reality. In: Kotzé, P., Marsden, G., Lindgaard, G., Wesson, J., Winckler, M. (eds.) INTERACT 2013. LNCS, vol. 8118, pp. 282–299. Springer, Heidelberg (2013). https://doi.org/10.1007/978-3-642-40480-1_18

7. Buchmann, V., Violich, S., Billinghurst, M., Cockburn, A.: FingARtips: gesture based direct manipulation in augmented reality. In: Proceedings of the 2nd International Conference on Computer Graphics and Interactive Techniques in Australasia and South East Asia, pp. 212–221. Association for Computing Machinery (ACM) (2004). https://doi.org/10.1145/988834.988871
8. Billinghurst, M., Piumsomboon, T., Bai, H.: Hands in space: gesture interaction with augmented-reality interfaces. IEEE Comput. Graph. Appl. **34**(1), 77–81 (2014). https://doi.org/10.1109/MCG.2014.8
9. Satriadi, K.A., Ens, B., Cordeil, M., Jenny, B., Czauderna, T., Willett, W.: Augmented reality map navigation with freehand gestures. In: 2019 IEEE Conference on Virtual Reality and 3D User Interfaces (VR), pp. 593–603 (2019). https://doi.org/10.1109/VR.2019.8798340
10. Perelman, G., Serrano, M., Raynal, M., Picard, C., Derras, M., Dubois, E.: The roly-poly mouse: designing a rolling input device unifying 2D and 3D interaction. In: Conference on Human Factors in Computing Systems - Proceedings, pp. 327–336. Association for Computing Machinery (2015). https://doi.org/10.1145/2702123.2702244
11. Katzakis, N., Kiyokawa, K., Hori, M., Takemura, H.: Plane-casting: 3D cursor control with a smartphone. In: Proceedings of the ACM Conference on Human Factors in Computing Systems (2018). https://doi.org/10.1145/2212776.2212698. arXiv preprint arXiv:1801.05100
12. Mendes, D., Caputo, F.M., Giachetti, A., Ferreira, A., Jorge, J.: A survey on 3D virtual object manipulation: from the desktop to immersive virtual environments. Comput. Graph. Forum **38**(1), 21–45 (2019). https://doi.org/10.1111/cgf.13390
13. Jiang, H., Wachs, J.P., Pendergast, M., Duerstock, B.S.: 3D joystick for robotic arm control by individuals with high level spinal cord injuries. In: IEEE International Conference on Rehabilitation Robotics (2013). https://doi.org/10.1109/ICORR.2013.6650432
14. Balakrishnan, R., Baudel, T., Kurtenbach, G., Fitzmaurice, G.: The rockin' mouse: integral 3D manipulation on a plane. In: ACM SIGCHI (1997). https://doi.org/10.1145/258549.258778
15. Werkhoven, P.J., Groen, J.: Manipulation performance in interactive virtual environments. Hum. Factors **40**(3), 432–442 (1998). https://doi.org/10.1518/001872098779591322
16. Crash Wiki· Gestures & Hand Tracking in HoloLens 2 (2019). https://crasteroids.notion.site/Gestures-Hand-Tracking-in-HoloLens-2-b14713225a544affb348c2c3bc3f96af
17. Clover, J.: These Gestures Are How You Control Apple Vision Pro - MacRumors (2023). https://www.macrumors.com/2023/06/08/apple-vision-pro-gestures/
18. Tseng, P.H., Hung, S.H., Chiang, P.Y., Yao, C.Y., Chu, H.K.: EZ-manipulator: designing a mobile, fast, and ambiguity-free 3D manipulation interface using smartphones. Comput. Vis. Media **4**(2), 139–147 (2018). https://doi.org/10.1007/S41095-018-0105-0
19. Grandi, J.G., Berndt, I., Debarba, H.G., Nedel, L., Maciel, A.: Collaborative 3D manipulation using mobile phones. In: 2016 IEEE Symposium on 3D User Interfaces, 3DUI 2016 - Proceedings, pp. 279–280 (2016). https://doi.org/10.1109/3DUI.2016.7460079
20. Chase, T.F., Dooley, K.M., Kerner, R.F.: Method for using a two-dimensional touchpad to manipulate a three-dimensional image (2020). https://patents.justia.com/inventor/robert-f-kerner

21. Malik, S., Laszlo, J.: Visual touchpad: a two-handed gestural input device, pp. 289–296 (2004). https://doi.org/10.1145/1027933.1027980
22. Schneider, T., Hois, J., Rosenstein, A., Ghellal, S., Theofanou-Fülbier, D., Gerlicher, A.R.: ExplAIn yourself! Transparency for positive UX in autonomous driving (2021). https://doi.org/10.1145/3411764
23. Ultraleap: XR design principles - Ultraleap documentation (2023). https://docs.ultraleap.com/xr-guidelines/Getting%20started/design-principles.html
24. Meta: Designing accessible VR, Oculus developers (2023). https://developer.oculus.com/resources/design-accessible-vr/
25. Thakur, A., Rai, R.: User study of hand gestures for gesture based 3D CAD modeling. In: Proceedings of the ASME Design Engineering Technical Conference, 1B-2015 (2016). https://doi.org/10.1115/DETC2015-46086
26. Hansberger, J.T., et al.: Dispelling the Gorilla arm syndrome: the viability of prolonged gesture interactions. In: Lackey, S., Chen, J. (eds.) VAMR 2017. LNCS, vol. 10280, pp. 505–520. Springer, Cham (2017). https://doi.org/10.1007/978-3-319-57987-0_41

Open Access This chapter is licensed under the terms of the Creative Commons Attribution 4.0 International License (http://creativecommons.org/licenses/by/4.0/), which permits use, sharing, adaptation, distribution and reproduction in any medium or format, as long as you give appropriate credit to the original author(s) and the source, provide a link to the Creative Commons license and indicate if changes were made.

The images or other third party material in this chapter are included in the chapter's Creative Commons license, unless indicated otherwise in a credit line to the material. If material is not included in the chapter's Creative Commons license and your intended use is not permitted by statutory regulation or exceeds the permitted use, you will need to obtain permission directly from the copyright holder.

Enhancing an Autonomous Vehicle Simulation Through Holoride Technology Integration to Reduce Motion Sickness and Increase Immersion: A Proof of Concept and Empirical Evaluation

Zack Walker, Korbinian Kuhn, Ansgar Gerlicher[✉], and Axel Braun

Stuttgart Media University, Nobelstraße 10, 70569 Stuttgart, Germany
{walker,gerlicher}@hdm-stuttgart.de

Abstract. Looking towards a possible future in which fully autonomous vehicles are commonplace, it is imperative to already begin evaluating vehicle concepts. As of today level 5 autonomous vehicles are still under active research, mainly with a more technical perspective. Therefore evaluations often take place in virtual worlds, crafted for this specific use-case. However, stationary simulations fail to adequately represent the dynamic movements of a vehicle, which has inherent drawbacks, particularly a low sense of immersion and the development of motion sickness. To address this issue, this paper proposes the integration of Holoride technology, which leverages OpenStreetMap data and real-time sensor feedback from a vehicle to create a dynamic environment that closely mimics the real-world driving experience in Virtual Reality (VR). The hypothesis is a reduction in the risk of motion sickness and an increase in the perceived immersion. To test this hypothesis, a dynamic simulation is developed and evaluated in a study involving 18 participants, comparing it to a static simulation across various metrics including motion sickness, immersion and overall user experience. While the results did not show a noticeable reduction in motion sickness in the dynamic simulation, potentially attributable to technical inaccuracies of the holoride technology, there was a significant increase in perceived immersion and presence. Moreover, participants consistently rated the user experience higher in the dynamic simulation compared to the static counterpart.

Keywords: Autonomous Vehicles · Virtual Reality · Mixed Reality · Driving Simulation · User Experience Design · Motion Sickness

1 Introduction

Looking into the future, it seems inevitable that eventually, autonomous vehicles (AVs) will take over the roads and human drivers will no longer be allowed or only in special cases. By the end of 2019, there were about 31 million cars with some level of automation (SAE Level 2) in operation with this number expected to increase to at least 54 million by 2024 [1]. By 2025, it is projected, that about 63% of all vehicles sold worldwide will feature SAE Level 2 or higher automation systems [2] and by 2030, sales of vehicles with Level 3 autonomy will reach around 58 million units [3].

In order for automobile companies to adapt to this changing landscape and stay competitive, they must shift their focus towards this vision and already start designing and developing prototypes. Functional autonomous vehicle projects have been tested in designated areas for several years already [4–6]. However, these test projects focus mainly on the technical aspects of AVs. Improving the passenger experience, which is more personal and subjective, is crucial and requires rigorous user testing through User Centered Design. This iterative process benefits from Virtual Reality (VR) simulations, allowing quick adaptations to user feedback in various simulated environments. An example of such a VR setup is the FlexCAR[1] project, (10/18 to 09/23), funded by the Federal Ministry of Research and Education, of which this study is part of. Despite the advantages, such stationary VR simulations of vehicles have drawbacks. They lack the ability to simulate road-induced vibrations and gravitational forces experienced in real vehicles.

A study by Ahmad B.I. et al. found road vibrations increase touchscreen input errors by up to 70% [7], leading to uncertainty in VR user test results. Additionally, motion sickness is a significant issue, with 17% of U.S. consumers reporting it in VR experiences according to a 2022 survey by the National Research Group [8].

2 Related Work

2.1 Levels of Automation

The Society of Automation Engineers (SAE)[2] defines six levels of driving automation vehicles must pass through until they can be considered fully self-driving ranging from no autonomy or assistance systems (Level 0) to full drive autonomy (Level 5) [9]. The National Highway Traffic Safety Administration (NHTSA)[3] in the United States have introduced a similar six item classification. However, as the two definitions differ only in terminology and for conciseness, only the version by the SAE will be discussed here.

As of 2024, Level 2 classified assisted vehicles are commonplace on public roads and highways around the world, with many car manufacturers outfitting their vehicles with Level 2 classified ADS [10–13]. In July 2021, the German Federal Ministry for Digitalization and Transport passed a law permitting Level

---

[1] https://arena2036.de/de/flexcar.
[2] https://www.sae.org/.
[3] https://www.nhtsa.gov/.

3 automated systems on streets and motorways in Germany [14]. In December 2021, Mercedes Benz became the first automotive company in the world to meet the legal requirements for a Level 3 system enabling conditionally automated driving [11]. In the USA and China, self-driving car companies have been operating robotaxi services in select cities, which are considered Level 4 [6,15,16].

The SAE Level 5 classification explicitly allows for the possibility of a passenger of an AV to take over control of the vehicle, should they request it ("[The user] may become the driver after a requested disengagement" [9]). Considering the benefits – especially regarding safety – an extension of the six automation levels can be envisioned. A possible seventh level (Level 6) would distinguish itself from Level 5 by eliminating the option for passengers to assume control of the vehicle. At this point, the need for a steering wheel, pedals or gearbox would disappear, opening the vehicle interior to new possibilities regarding its design. According to User Centered Design principles, these new interior designs must be rigorously tested and evaluated in user studies. Given that fully autonomous vehicles without the option of assuming manual control are not broadly available as of 2024, many research institutes and companies rely on VR solutions to simulate vehicles [17–23]. These simulations can be static or dynamic. Static simulations are characterized by a non-moving platform on which parts of a vehicle are placed, such as seats or a steering wheel. The user sits on the platform and wears a VR headset, in which the vehicle, it's motion and the environment is displayed. One of these simulations can be found within the FlexCAR project (Fig. 1). Dynamic simulators on the other hand add an additional dimension by attempting to to replicate movements of the simulated vehicle to create a more realistic experience [18–20]. Despite the benefits of VR simulations, they are susceptible to the development of motion sickness in the user.

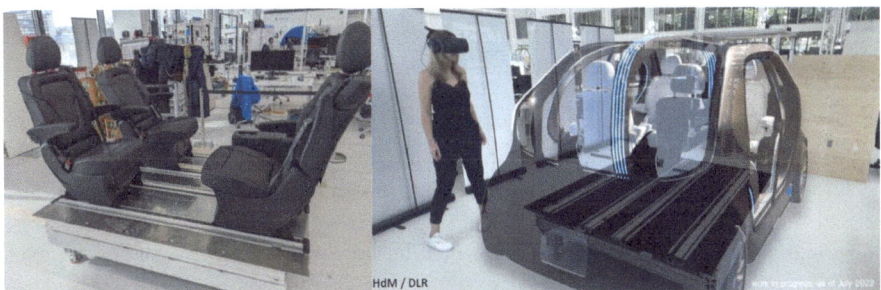

Fig. 1. The FlexCar simulator with its vehicle platform and four car seats (left). Augmented with the virtual shuttle (right)

2.2 Motion Sickness

Motion sickness, characterized by nausea, dizziness, and discomfort, affects up to 60% of the population and is common during travel in cars, flights, or boats [24,25]. About 46% of adults experience it as car passengers [26]. The generally accepted explanation as to how motion sickness occurs is the sensory conflict theory, which states that motion sickness happens when the movement an individual sees is different from what the inner ear senses [26,27].

A study conducted by Cheung and Nakashima found that lower frequencies of motion (<2 Hz) produced the highest amount of motion sickness incidence with its occurrence decreasing rapidly as the frequency increased [28]. Interestingly, increased frequency of motion does not lead to increased probability or severity of motion sickness. Instead, it only increases discomfort and injury (Fig. 2).

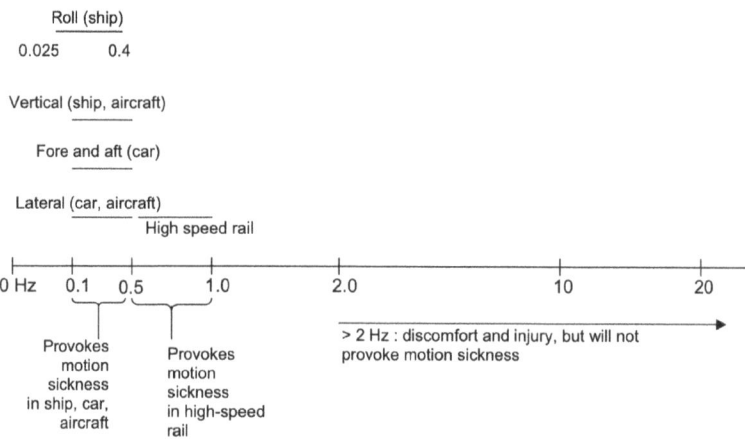

Fig. 2. Effects of frequency on motion sickness [28]

For motion sickness in cars it is notable that horizontal movements and linear acceleration are more relevant in producing motion sickness, in contrast to the vertical motion of ships or aircraft.

Motion sickness is not completely curable; however, multiple studies suggest that it is preventable to a certain extent. One prevention method is habituation. Constantly and gradually exposing individuals to the specific motion or stimuli that trigger their symptoms allows the sensory systems to acclimate and re-calibrate the sensory mismatch over time. Various investigations confirm this theory [29–33]. A more recent study claims it is the "most effective non-pharmacological method to reduce motion sickness" [34].

Motion sickness can also occur with the absence of physical movement. Here, the visual system drives the sensation of apparent self-motion, conflicting with the vestibular system, which does not feel the same motion sensation. Motion sickness in this context is commonly referred to as Visually Induced Motion

Sickness (VIMS) [34]. Emerging technologies and trends such as AVs and VR have made VIMS a topical issue [35]. Since the motion of AVs can no longer be controlled by a driver, the risk of motion sickness is significantly increased. Furthermore, many AV concepts feature backwards-facing seats, which can be an additional factor in producing motion sickness [34].

For VR, the occurrence and severity of motion sickness depends on multiple technical factors such as latency, visual fidelity, field of view and image refresh rate as well as personal factors like age, gender and history of motion sickness, but also environmental factors like illumination and general environment brightness [36]. Reported incident rates are as high as 60% [26,37].

While immersed in the FlexCAR simulation, participants are seated within a stationary module, wearing VR goggles that replace their actual surroundings with a virtual environment featuring the moving shuttle, creating the sensory conflict described above. The issue with static driving simulators in the context of VIMS versus dynamic ones is not unknown and multiple studies have attempted to adjust and improve upon static simulators with various solutions.

Multiple methods to reduce VIMS have been proposed, including adapting VR content to the movement of a vehicle by using gyroscopic sensors [38] or by employing a sophisticated hydraulic motion platform to dynamically replicate the motion of a vehicle in direct response to a driver's input [39]. The outcomes showed that symptoms of VIMS were able to be dramatically alleviated using these methods.

A further issue inherent to static simulations such as that found in the FlexCAR project, which must be tackled is immersion and perceived presence.

2.3 Immersion

There seem to be different and conflicting definitions of immersion [40–42]. A number of literature reviews have been conducted to bring together and summarize the vast number of different interpretations [42,43]. For one, Witmer and Singer describe immersion as "a feeling of being enveloped by, included in, and interacting with the virtual environment" [44]. Arsenault, McMahan, Ermi and Mäyrä and Adams and Rollings use similar descriptions for immersion while also extending the definition with different types of immersive experiences. They include definitions for immersion as a response to a narrative, such as the feeling of being absorbed in fictional stories and a fictional wold and it's characters [45–47] or "a state of intense and focused attention on the story world and the unfolding events and acceptance of these as real" [42]. Slater offers a more technical definition of immersion, saying that it is "a property of the technology mediating the experience" with the quality and fidelity of displays dictating the level of immersion [48].

Many other diverse definitions have been given by other researchers, linking immersion to a person's perception and imagination, a narrative, or the engagement in a world or system [42,43].

In this work immersion is defined as the feeling of being physically present in a virtual world with the feeling of being able to influence the environment and in turn the environment being able to influence the user impacting the perceived level of immersion. This combines established definitions mentioned above and contextualizes them within the study.

A feeling of presence in AV simulations can play a pivotal role in a testing environment. By immersing users in realistic virtual environments, individuals can experience a more authentic simulation of driving conditions [49]. It allows testing of a diverse range of scenarios, environments, weather and traffic conditions. Realism is generally considered a goal to strive for, as it significantly contributes to enhancing the overall user experience, as one study finds [50]. One well-known example that illustrates the importance of realistic testing conditions is that of touch-screens in cars: While generally proven to be effective on smartphones and in situations, in which the user is not moving [51,52], research has shown that the effectiveness of touchscreens can not be replicated in cars and the evaluation of their effect on user experience cannot be undertaken under solely static conditions [7,53–56].

Thus, it can be said that there is demand for realistic and immersive virtual simulations and experiences. More immersive VR experiences may also reduce the severity of experienced motion sickness. While there may not be sufficient research to back up this claim, it seems plausible due to immersion mitigating the extent of sensory conflict encountered.

The introduction of Holoride, which dynamically displays content based on real-time car rides, is expected to increase the realism of testing situations and further improve upon the approaches.

2.4 Holoride

Holoride, a spin-off from *Audi* in 2018, develops immersive in-car entertainment using VR and AR technologies, aiming to synchronize VR content with vehicle movements to reduce motion sickness and enhance travel. Studies show interleaving motion cues in VR can mitigate motion sickness [57–59], but Holoride acknowledges it doesn't eliminate it completely [60]. In a study by Audi and Holoride, participants with no motion sickness increased from 18% to 53% using Holoride, with severe and moderate symptoms reduced by 31% and 54%, respectively [61].

The technology works by connecting the holoride software running in a VR environment with real-time data points from the host vehicle, including acceleration, steering, traffic data, as well as travel route and time [62]. As the vehicle executes maneuvers like turns or navigates uneven terrain, the VR content dynamically adapts to convey corresponding sensations of motion, striving to maintain synchronization with the physical world.

There are two possibilities to use holoride. Starting in 2023, the holoride interface is integrated into some Audi models [63], allowing compatible VR headsets to connect directly with the vehicle. The second option is to utilize an external puck-like device called "holoride retrofit" to enable the functionality in any vehicle. In this study, the retrofit solution is utilized.

As the technology is focused on entertainment, with the company stating that their vision of the product is to "turn vehicles into theme parks" and to "add thrill to every ride" [64], it is noteworthy that, as of the present, there has been a conspicuous absence of public research applications employing Holoride technology. This void represents a substantial opportunity for exploration and experimentation.

3 Methodology

Utilizing Holoride's *Elastic SDK*, two variations of an AV simulation are developed: a virtual ride where participants sit in a static simulator and experience the ride through a VR headset (static simulation), and a dynamic virtual ride where participants experience the same VR ride while in a moving vehicle (dynamic simulation), in which the simulated vehicle reacts to the motion of the physical vehicle. Both simulations use the Elastic SDK to automatically generate the environment based on GPS location and OpenStreetMap data. The implemented simulation can be seen in Fig. 3.

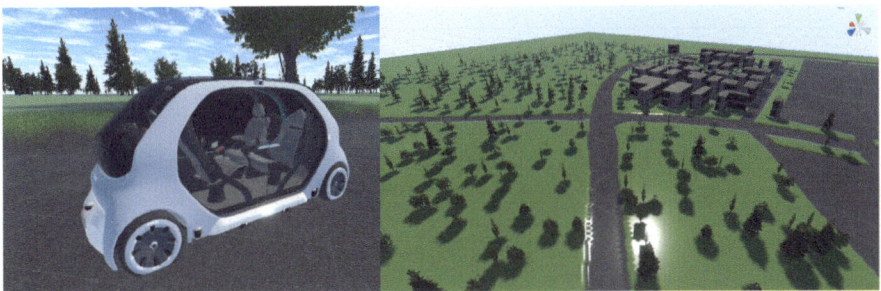

Fig. 3. The implemented simulation with the shuttle (left) and the procedurally generated environment (right)

The study aims to determine if significant differences exist between the two experiences in terms of motion sickness, immersion, and user experience. It employs a mixed-methods, within-subject design approach; an overview of which is shown in Fig. 4.

Participants first complete a preliminary questionnaire and then experience two simulation scenarios. One scenario uses a static simulator where a car ride is simulated via the holoride application, providing the sensation of movement without actual motion. The other scenario involves participants being driven in a physical vehicle while wearing a VR headset with the holoride application in dynamic mode, reacting to the car's real movements. During each simulation, participants use the Think Aloud method to share their observations and feelings. After each simulation scenario, test subjects were asked to rate their perceived level of motion sickness using the 7-point Motion Sickness Severity Scale (MSSS) [65], which ranges from no symptoms (0) to vomiting (6). The MSSS has the advantage of allowing a quick and easy assessment of whether any symptoms occurred during the simulation. Notably, it eliminates the need for distinguishing between symptoms accompanied by nausea and those devoid of it, as this distinguishment is not relevant for the study.

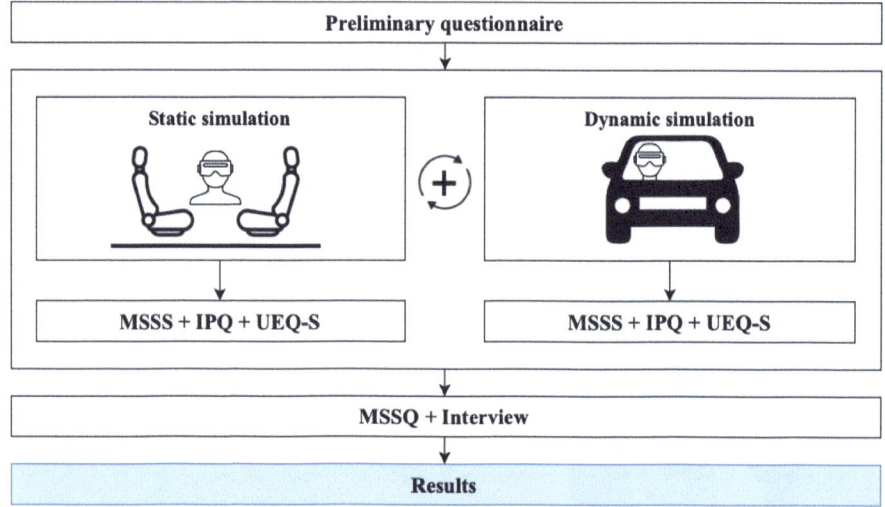

Fig. 4. Overview of the study design

To measure perceived immersion, the Igroup Presence Questionnaire (IPQ) was used [66]; a scale for measuring the sense of presence experienced in a virtual environment. This 14-item questionnaire is separated into three subscales: spatial presence (SP), involvement (INV) and experienced realism (REAL). An additional general item (G) assesses the "sense of being there".

For overall User Experience (UX), the 8-item Short User Experience Questionnaire (UEQ-S) is utilized [67]. The UEQ-S is an abbreviated version of the original 26-point User Experience Questionnaire, designed for studies in which the user experience of multiple variants of a single product must be evaluated. The UEQ-S groups the items into two scales: pragmatic quality (task-related UX aspects) and hedonic quality (non-task-related UX aspects). To evaluate the UX quality of a product, the UEQ-S offers a benchmark which classifies a product into five categories (Excellent, Good, Above average, Below average and Bad), based on a large data set [67]. To mitigate the potential confounding effects of the order of the simulation scenarios on the results, counterbalancing was utilized, meaning that half of the participants first experienced the static simulation with the ride replay while the other half experienced it the other way around.

After both scenarios, each participants general susceptibility to motion sickness was assessed using the Motion Sickness Susceptibility Questionnaire (MSSQ) [68]. This provides valuable context for interpreting individual differences in motion sickness responses observed during the study. Finally, the user test is concluded with a short qualitative interview in which they were able to express any further general remarks and go into detail about what they enjoyed and disliked about each simulation experience.

3.1 Hypotheses

Multiple hypotheses were established to serve as guiding principles for the empirical investigation:

H1. Among the entirety of participants, symptoms of motion sickness occur less frequently during the dynamic simulation compared to the virtual ride in the static simulator.
H2. If symptoms of motion sickness occur, they are milder during the dynamic simulation compared to the virtual ride in the static simulator.
H3. Participants who reported experiencing motion sickness more frequently in the past show a higher susceptibility to it, regardless of the simulation setting.
H4. Participants who report having more experience with VR are less likely to develop motion sickness, regardless of the simulation setting.
H5. The perceived immersion, measured by the IPQ, is higher during the dynamic virtual ride compared to the virtual ride in the static simulator.
H6. The user experience, measured by the UEQ-S, is rated more positively for the dynamic virtual ride compared to the virtual ride in the static simulator.
H7. Participants experiencing motion sickness prefer the simulation setting where symptoms were milder, regardless of how they rate the immersion.
H8. Participants who do not experience motion sickness prefer the simulation setting where they reported higher immersion.

3.2 Limitations

The study is limited by its reliance on convenience sampling for recruitment, which may restrict the generalizability of findings to the broader population. Additionally, while standardized measures like MSSS, IPQ, and UEQ-S were used, their inherent subjectivity as self-report methods introduces potential biases in capturing participants' experiences accurately. Technical limitations of the Holoride application, such as its dependency on stable internet and GPS connections, can disrupt the simulation's fidelity, impacting immersion and user experience. Moreover, conducting parts of the study in a real-world setting introduces uncontrollable environmental variables like weather and traffic conditions, which may influence participant responses and compromise the experiment's repeatability.

4 Results

The study was conducted with a total of 18 persons ($n = 18$; n being the number of participants) - twelve female and six male, aged 21 to 30 years. Most ($n = 13$) had some VR experience, but use it rarely, while others stated that they have either never used VR before ($n = 3$), occasionally use it ($n = 1$) or regularly use it ($n = 1$).

Regarding motion sickness susceptibility (measured with the MSSQ), most participants ($n = 15$) were in the 30th to 60th percentile. Two participants reported no motion sickness in the past ten years, placing them in the 0th percentile (the minimum value – meaning nobody being less susceptible), while one participant, frequently experiencing motion sickness, fell into the 95th percentile. About half ($n = 10$) reported rarely or sometimes feeling sick during car or bus rides, while the rest ($n = 8$) reported no symptoms.

Overall, twelve participants developed motion sickness in at least one of the simulation variants. Nine experienced symptoms during the static simulation, including stomach discomfort ($n = 6$) and mild nausea ($n = 3$). Eight participants had symptoms during the dynamic simulation, with stomach discomfort ($n = 5$), mild nausea ($n = 2$), and one severe nausea case. No significant gender difference could be observed, with 50% of both male and female participants experiencing symptoms in the static simulation, and 41.67% of females versus 50% of males reporting symptoms in the dynamic simulation.

In five cases, experienced symptoms were more severe in the dynamic simulation than in the static simulation. In six cases it was the other way around with symptoms being less severe in the dynamic simulation compared to its static counterpart. In the remaining seven cases, the reported symptoms were the same for both the static and the dynamic simulation.

Viewing the measured data, motion sickness severity seems not to have been affected by the simulation variant. A Wilcoxon signed-rank test showed that motion sickness severity did not differ significantly between the static simulation (median $Mdn = 0$) and the dynamic simulation ($Mdn = 0$), $W = 31$, $p = 0.856$.[4][5]

Regarding perceived presence, measured using the IPQ, the categories general immersion (G), spatial presence (SP) and involvement (INV) show only minor differences. A more clear difference can be observed in the realism category (REAL). A Wilcoxon signed-rank test showed that G, INV, as well as the overall perceived presence did not differ significantly between the static simulation and the dynamic simulation, whereas for SP and REAL a significant difference could be determined (Table 1).

---

[4] W: Wilcoxon signed-rank test statistic.
[5] p: p-value for the test.

Table 1. Results of the IPQ per category

Category	Static (Mdn)	Dynamic (Mdn)	W	p
G	5	5.5	23.5	0.216
SP	5.2	5.8	32.5	0.03
INV	5.125	4.25	45.5	0.141
REAL	2.75	4	15.5	0.007
Overall	4.42	4.83	46	0.085

Fig. 5 shows the UX benchmarks evaluated by the UEQ-S. For the static simulation, the pragmatic quality achieves a rating of "Above average" (mean $M = 1.389$, stadard deviation $SD = 0.739$, confidence interval $95\% CI : 1.048 - 1.73$) and the hedonic quality receives a rating of "Good" ($M = 1.458, SD = 0.796, 95\% CI : 1.09 - 1.826$). The overall UX is rated at "Good" ($M = 1.42, SD = 0.64, 95\% CI : 1.128 - 1.719$).

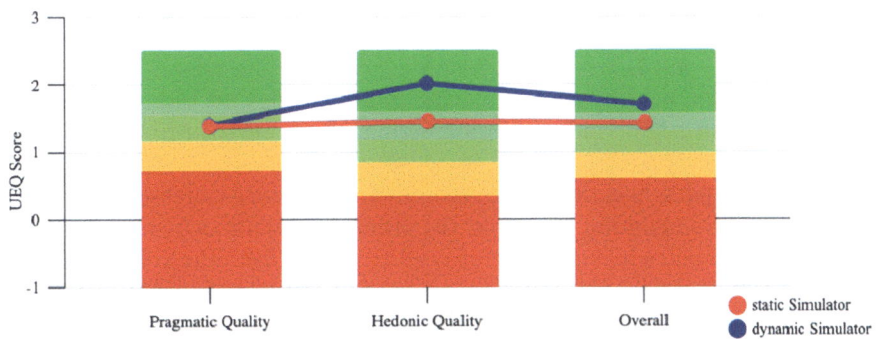

Fig. 5. UEQ-S benchmark results for the static and dynamic simulations

For the dynamic simulation, while the pragmatic quality achieves a similar value compared to the static simulation "Above average" ($M = 1.403, SD = 1.304, 95\% CI : 0.801 - 2.005$), the hedonic quality manages to score significantly higher, with it achieving the rating "Excellent" ($M = 2.014, SD = 0.957, 95\% CI : 1.572 - 2.456$). The overall UX is also measured at "Excellent" ($M = 1.708, SD = 1.009, 95\% CI : 1.242 - 2.175$).

In addition to the quantitative data presented in the preceding chapter, qualitative data was also raised during the study, through the Think Aloud method and a post-test interview, in which participants could select a preferred simulation variant and explain why they preferred it. Additionally, participants had the chance to describe what they felt was missing from both simulations.

Overall, the dynamic simulation was clearly favored with 14 participants stating that they preferred it versus only four participants preferring the static simulation. The reasoning behind the preferred simulation variants is summarized in Table 2.

Table 2. Reasoning behind preference towards a certain simulation variant

Preferred variant	Reasoning	Mentions
static	Less severe motion sickness	3
	Less distractions from real environment	1
dynamic	Felt more realistic	12
	More fun and exciting	4
	Less severe motion sickness	3
	More realistic environment sounds	2
	Novelty	1

Of the four individuals who preferred the static simulation, three cited less motion sickness compared to the dynamic variant as the reason. One person found external sounds like those from the environment and those omitted by the vehicle (motor sound, indicator) distracting, which also contributed to their preference.

Three participants mentioned reduced motion sickness as a reason for preferring the dynamic simulation. However, the primary reason was its realism (mentioned by twelve participants), as it allowed them to "feel" the vehicle's movements and included realistic sounds. Four participants also found the dynamic simulation more fun, aligning with the UEQ-S results on hedonic quality. Interestingly, one participant preferred the dynamic simulation despite experiencing more severe motion sickness, valuing realism over discomfort.

When asked about what they felt was missing from the simulations, participants primarily would have preferred a more real-looking environment. This includes adding traffic signs and traffic lights, integrating road unevenness with bumps and pot-holes at the real positions and adding other vehicles and pedestrians. Some participants complained that the virtual road was depicted as being completely even and flat, whereas in the real environment, the road was uneven and had multiple bumps and some pot-holes. Furthermore, in the real environment there were other vehicles and pedestrians influencing the motion of the vehicle during the dynamic simulation. This was not represented in the virtual environment, leading to some confusion as to why the vehicle sometimes stopped or waited for a longer period of time at an intersection, as participants were not able to see the real environment. Some participants mentioned that they would have liked to be able to anticipate the movement of the vehicle more. For one, this includes the previously mentioned point of being able to see other vehicles. Additionally, it was stated that it would be nice if it was possible to comprehend in which direction the car was going to turn in an upcoming intersection, before the vehicle actually moved in that direction. In the static simulation, there was no way of knowing if the vehicle was about to turn left, right or continue forward at a crossing. In the dynamic variant, participants could hear the sound of the vehicle's indicator, however they could not see which direction it was signaling.

Participants also noted that some kind of audio feedback - especially for the static simulation - would have been nice.

Even though it was intentionally omitted from the research goals of the study, participants stated that they would have liked to be able to interact with the vehicle or the environment more (e.g. listened to music during the ride or use an infotainment system.

5 Discussion

The fact that about 67% of participants said that they developed motion sickness symptoms in at least one of the simulation variants is higher than the literature suggests (about 50%) but generally is in the expected range, especially considering that the participants who stated that they developed motion sickness during the simulations being those who also reported developing symptoms during car or bus rides in the last ten years via the MSSQ.

However, the dynamic simulation did not seem to impact the development of motion sickness. It is not clear why no significant difference could be observed. A possible explanation for this may be the simulations inaccuracies. Participants noted issues like driving off-road, into buildings, and inaccuracy in turns, with the vehicle seeming to drift sideways. They were also surprised by bumps and potholes, which were not portrayed in the virtual world. Confusion about the vehicle's movements, such as brief pauses and uncertain path at crossroads, may have also contributed to motion sickness. The virtual simulation lacked crucial driving context, such as other traffic participants and signals, which may have exacerbated the problem. In the context of the presented literature, it makes sense that these factors played a significant role in the development of discomfort and nausea during the dynamic solution, as it exactly represents the sensory conflict between the visual and vestibular systems thought to be the leading cause for motion sickness.

In terms of immersion, both static and dynamic simulations scored similarly overall, reflecting their identical visual and aesthetic design. No Significant difference could be determined between the two simulation variants in the categories general immersion and involvement. This is most likely due to the non-interactive nature and general similarity of both simulations. However, the categories "Spatial presence" and "Realism" showed more significant differences, with the dynamic simulation scoring higher. Presumably, this is due to added sensations of vehicle motion, which participants praised.

For user experience measured with the UEQ-S, pragmatic quality (task-related aspects like efficiency and dependability) was similar for both simulations since participants were mere observers without interaction tasks. The hedonic quality (non-task-related aspects like stimulation and excitement) showed greater differences, with the dynamic simulation being rated as more fun, exciting, and novel.

5.1 Hypothesis Evaluation

Hypotheses **H1** and **H2** predicted that motion sickness symptoms would occur less frequently and be milder during the dynamic simulation compared to the static simulation among all participants. However, motion sickness was equally prevalent in both simulations, and the dynamic simulation did not mitigate symptoms when they occurred. Approximately half of the participants who experienced motion sickness reported worse symptoms during the dynamic simulation, including one case of severe nausea not seen in the static simulation. Therefore, both hypotheses were disproven by the study's findings. This also suggests that motion sickness susceptibility is governed by more than simply the motion of the vehicle.

In **H3**, it was hypothesized that individuals with a history of frequent motion sickness would show higher susceptibility, regardless of simulation type. However, no correlation was found between reported motion sickness susceptibility and occurrence, possibly due to the small sample size ($n = 18$). Therefore, this hypothesis remains inconclusive.

Hypothesis **H4** suggests that participants with more VR experience are less prone to motion sickness in both simulation types. However, with only two regular VR users showing minor symptoms in the static simulation, and several inexperienced participants reporting no symptoms, the small sample size (14 out of 18 participants with little to no VR experience) prevents a statistically significant conclusion on the correlation proposed in **H4**.

In contrast, hypotheses **H5** to **H8** could all be proven in this study.

H5 stated that the perceived immersion measured by the IPQ is higher during the dynamic simulation compared to the static simulation. The dynamic simulation did not out-score the static simulation in all categories of the IPQ, however, a more significant difference was measured in the category "Realism". Additionally, the qualitative results show that participants often perceived the dynamic simulation variant to be more realistic and immersive.

In **H6** it was expected that the measured user experience is generally rated more positively for the dynamic simulation than for the static simulation. The difference in the measured hedonic quality confirms this expectation. Due to the simulation not featuring any interactability, the results of the pragmatic quality measured by the UEQ-S do not hold significant relevance. Overall, despite considering **H6** proven, due the small study population caution is warranted in generalizing the findings to broader contexts or populations.

Hypothesis **H7** claimed that participants who experienced motion sickness would prefer the simulation setting in which symptoms were milder. In the vast majority of cases (17 of 18), this was proven to be true. In one case, the dynamic simulation was preferred despite minor discomfort, although the individual did not experience any motion sickness in the static simulation. This implies that users tend to accept minor symptoms of motion sickness, as long as they are limited to slight discomfort.

The final hypothesis, **H8**, stated that participants who did not experience motion sickness would prefer the simulation variant for which they reported higher perceived immersion. In total, six individuals did not experience any motion sickness, all of which preferred the dynamic simulation, for which they also reported experiencing higher immersion.

6 Conclusion

This work successfully developed a dynamic simulation using Holoride and its SDK, generating a virtual world from real-world data and responding to physical vehicle sensor input. However, practical application effectiveness hinges on factors like stable internet and GPS availability. The user study indicated that compared to stationary simulations, the dynamic simulation did not significantly reduce motion sickness, possibly due to insufficient fidelity in replicating real roads. Nonetheless, areas for improving the simulation to potentially reduce motion sickness were identified. Participants generally appreciated the immersive experience facilitated by Holoride integration, suggesting a possible link to reduced motion sickness, which warrants further investigation. This research provides valuable insights into Holoride's capabilities and limitations, serving as a foundation for future developments in addressing motion sickness and enhancing realism in autonomous vehicle simulations.

6.1 Future Research Directions

The study suggests foundational findings for further research. It found that motion sickness persists, likely due to sensory conflicts from factors like road unevenness absent in procedurally generated virtual environments and technical inaccuracies. The current Holoride SDK uses OpenStreetMap for terrain, but future improvements could integrate more vehicle data via LiDAR, cameras, and picture recognition algorithms. This data, accessible through a dedicated API for VR applications, could enable real-time virtual representations of actual traffic conditions, enhancing realism. Incorporating additional real vehicle data, such as indicators and vehicle state, into VR simulations may help reduce sensory conflicts and motion sickness. Further research is needed to validate these findings.

Additionally, Holoride technology shows promise for evaluating interaction concepts in realistic vehicle conditions, unlike static or simulated settings. This approach better reflects real-world usage scenarios, focusing on interaction modalities rather than precise vehicle motion accuracy. Regarding user experience, the task-related (pragmatic) quality of interaction strategies may also be more effectively evaluated in an advanced dynamic simulation. Overall, while promising, the potential of Holoride in mitigating motion sickness hinges on broader network coverage and deeper integration of vehicle and environmental data.

References

1. Placek, M.: Autonomous vehicles worldwide - statistics & facts. https://www.statista.com/topics/3573/autonomous-vehicle-technology
2. Statista: Level 2-4 autonomous vehicle sales as a share of total vehicle sales in 2025 and 2030, by automation level (2023). https://www.statista.com/statistics/1230101/
3. Statista: Number of autonomous vehicles globally in 2022, with a forecast through 2030. https://www.statista.com/statistics/1230664/projected-number-autonomous-cars-worldwide/
4. HOCHBAHN: Das projekt heat (2021). https://www.hochbahn.de/de/projekte/das-projekt-heat
5. The Knowledge Base on Connected and Automated Driving (CAD). Test sites (2023). https://www.connectedautomateddriving.eu/test-sites/
6. Bhuiyan, J.: San Francisco to get round-the-clock robo taxis after controversial vote (2023). https://www.theguardian.com/us-news/2023/aug/10/san-francisco-self-driving-car-autonomous-regulation-google-gm
7. Ahmad, B.I., Langdon, P.M., Godsill, S.J., Hardy, R., Skrypchuk, L., Donkor, R.: Touchscreen usability and input performance in vehicles under different road conditions: an evaluative study. In: Proceedings of the 7th International Conference on Automotive User Interfaces and Interactive Vehicular Applications, AutomotiveUI '15, pp. 47–54, New York, NY, USA, 2015. Association for Computing Machinery. https://doi.org/10.1145/2799250.2799284
8. Statista: Leading complaints about virtual reality (VR) headsets per consumers in the United States as of March 2022 (2023). https://www.statista.com/statistics/1337138/leading-complaints-vr-headset-us-consumers/
9. SAE International: Taxonomy and definitions for terms related to driving automation systems for on-road motor vehicles, p. 41 (2021). https://doi.org/10.4271/J3016_202104
10. Tesla, Inc.: Autopilot and full self-driving capability (2023). https://www.tesla.com/en_eu/support/autopilot
11. Mercedes Benz AG: The front runner in automated driving and safety technologies. https://group.mercedes-benz.com/innovation/case/autonomous/drive-pilot-2.html
12. Audi AG: Fahrerassistenzsysteme. https://www.audi-mediacenter.com/de/audi-technik-lexikon-7180/fahrerassistenzsysteme-7184
13. Volkswagen AG: Volkswagen focuses development for autonomous driving. https://www.volkswagen-newsroom.com/en/press-releases/volkswagen-focuses-development-for-autonomous-driving-15271
14. Bundesministerium für Digitales und Verkehr: Gesetz zum autonomen fahren tritt in kraft. https://bmdv.bund.de/SharedDocs/DE/Artikel/DG/gesetz-zum-autonomen-fahren.html
15. AutoX: Autox opens its fully driverless robotaxi service to the public (2021). https://www.autox.ai/blog/20210128.html
16. AutoX: Autox launches Shanghai's first fully driverless robotaxi operation. https://www.autox.ai/blog/20230322.html
17. Universität Stuttgart: Interior design engineering (2024). https://www.iktd.uni-stuttgart.de/forschung/ide/
18. Mercedes Benz AG: Fahrsimulator von Mercedes-Benz. (2024). https://group.mercedes-benz.com/innovation/produktinnovation/technologie/fahrsimulator.html

19. project:syntropy: BMW high-fidelity und high-dynamic Fahrsimulatoren (2024). https://www.project-syntropy.de/portfolio-item/bmw-high-fidelity-und-high-dynamic-fahrsimulatoren/
20. project:syntropy: DLR inst. transportation systems – dynamic driving simulator (2024). https://www.project-syntropy.de/en/portfolio-item/dlr-inst-transportation-systems-dynamic-driving-simulator/
21. Technische Universität Dresden: Driving simulator (2024). https://tu-dresden.de/bu/verkehr/ivs/vpsy/forschung/ausstattung?set_language=en
22. TUM School of Engineering and Design Technical University of Munich: Static driving simulator (2024). https://www.mec.ed.tum.de/en/lfe/research/labs/static-driving-simulator/
23. Forschungsinstitut für Kraftfahrwesen und Fahrzeugmotoren Stuttgart: The stuttgart driving simulator (2024). https://www.fkfs.de/en/test-facilities/driving-simulator/stuttgart-driving-simulator
24. Money, K.E.: Motion sickness. Physiol. Rev. **50**(1), 1–39 (1970). https://doi.org/10.1152/physrev.1970.50.1.1
25. Schmidt, E.A., Kuiper, O.X., Wolter, S., Diels, C., Bos, J.E.: An international survey on the incidence and modulating factors of carsickness. Transp. Res. Part F Traffic Psychol. Behav. **71**, 76–87 (2020). https://doi.org/10.1016/j.trf.2020.03.012
26. Keshavarz, B., Golding, J.F.: Motion sickness: current concepts and management. Curr. Opin. Neurol. **35**(1), 107–112 (2022)
27. Golding, J.F.: Motion sickness susceptibility. Auton. Neurosci. **129**(1–2), 67–76 (2006)
28. Cheung, B., Nakashima, A.: A review on the effects of frequency of oscillation on motion sickness, Defence R&D Canada Toronto Technical Report, pp. 10–11 (2006)
29. Glaser, E.M.: Prevention and treatment of motion sickness. Proc. Roy. Soc. Med. **52**(11), 965–972 (1959). https://doi.org/10.1177/003591575905201116
30. Shupak, A., Gordon, C.R.: Motion sickness: advances in pathogenesis, prediction, prevention, and treatment. Aviat. Space Environ. Med. **77**(12), 1213–1223 (2006)
31. Brainard, A., Gresham, C.: Prevention and treatment of motion sickness. Am. Fam. Physician **90**(1), 41–46 (2014)
32. Murdin, L., Golding, J., Bronstein, A.: Managing motion sickness. BMJ **343** (2011)
33. Leung, A.K.C., Hon, K.L.: Motion sickness: an overview. Drugs Context **8** (2019)
34. Keshavarz, B., Golding, J.F.: Motion sickness: current concepts and management. Curr. Opin. Neurol. **35**(1), 107–112 (2022)
35. Bertolini, G., Straumann, D.: Moving in a moving world: a review on vestibular motion sickness. Front. Neurol. **7** (2016). https://doi.org/10.3389/fneur.2016.00014
36. Chattha, U.A., Janjua, U.I., Anwar, F., Madni, T.M., Cheema, M.F., Janjua, S.I.: Motion sickness in virtual reality: an empirical evaluation. IEEE Access **8**, 130486–130499 (2020)
37. Chang, E., Kim, H.T., Yoo, B.: Virtual reality sickness: a review of causes and measurements. Int. J. Hum.-Comput. Interact. **36**(17), 1658–1682 (2020)
38. Cho, H.-J., Kim, G.J.: RoadVR: mitigating the effect of vection and sickness by distortion of pathways for in-car virtual reality. In: Proceedings of the 26th ACM Symposium on Virtual Reality Software and Technology, VRST '20, New York, NY, USA (2020). Association for Computing Machinery. https://doi.org/10.1145/3385956.3422115

39. Aykent, B., Merienne, F., Guillet, C., Paillot, D., Kemeny, A.: Motion sickness evaluation and comparison for a static driving simulator and a dynamic driving simulator. Proc. Inst. Mech. Eng. Part D J. Automobile Eng. **228**(7), 818–829 (2014)
40. Lee, J., Kim, M., Kim, J.: A study on immersion and VR sickness in walking interaction for immersive virtual reality applications. Symmetry **9**(5) (2017). https://doi.org/10.3390/sym9050078
41. Berkman, M.I., Akan, E.: Presence and immersion in virtual reality. In: Lee, N. (ed.) Encyclopedia of Computer Graphics and Games, pp. 1–10. Springer, Cham (2019). https://doi.org/10.1007/978-3-319-08234-9_162-1
42. Nilsson, N.C., Nordahl, R., Serafin, S.: Immersion revisited: a review of existing definitions of immersion and their relation to different theories of presence. Hum. Technol. **12**(2), 108–134 (2016)
43. Agrawal, S., Simon, A., Bech, S., Bærentsen, K., Forchhammer, S.: Defining immersion: literature review and implications for research on audiovisual experiences. J. Audio Eng. Soc. **68**(6), 404–417 (2020)
44. Witmer, B.G., Singer, M.J.: Measuring presence in virtual environments: a presence questionnaire. Presence **7**(3), 225–240 (1998)
45. Arsenault, D.: Dark waters: spotlight on immersion. In: GAMEON-NA International Conference, pp. 50–52. Eurosis (2005)
46. McMahan, R.P., Gorton, D., Gresock, J., McConnell, W., Bowman, D.A.: Separating the effects of level of immersion and 3D interaction techniques. In: Proceedings of the ACM Symposium on Virtual Reality Software and Technology, pp. 108–111 (2006)
47. Ermi, L., Mäyrä, F.: Fundamental components of the gameplay experience: analysing immersion. In: DiGRA Conference, pp. 7–8. Citeseer (2005)
48. Slater, M.: A note on presence terminology. Presence Connect **3**(3), 1–5 (2003)
49. Michael, D., Kleanthous, M., Savva, M., Christodoulou, S., Pampaka, M., Gregoriades, A.: Impact of immersion and realism in driving simulator studies. IJITN **6**(1), 10–25 (2014)
50. Newman, M., Gatersleben, B., Wyles, K.J., Ratcliffe, E.: The use of virtual reality in environment experiences and the importance of realism. J. Environ. Psychol. **79**, 101733 (2022)
51. Hoye, T., Kozak, J.: Touch screens: a pressing technology. In: Tenth Annual Freshman Engineering Sustainability in the New Millennium Conference April, vol. 10 (2010)
52. Jennings, A., Ryser, S., Drews, F.: Touch screen devices and the effectiveness of user interface methods. In: Proceedings of the Human Factors and Ergonomics Society Annual Meeting, vol. 57, pp. 1648–1652. SAGE Publications Sage CA, Los Angeles, CA (2013)
53. Eren, A., Burnett, G., Large, D.: Can in-vehicle touch screens be operated with zero visual demand? An exploratory driving simulator study (2015)
54. Ng, A., Brewster, S.A., Beruscha, F., Krautter, W.: An evaluation of input controls for in-car interactions. In: Proceedings of the 2017 CHI Conference on Human Factors in Computing Systems, pp. 2845–2852 (2017)
55. Pitts, M.J., Burnett, G.E., Williams, M.A., Wellings, T.: Does haptic feedback change the way we view touchscreens in cars? In: International Conference on Multimodal Interfaces and the Workshop on Machine Learning for Multimodal Interaction, pp. 1–4 (2010)

56. Angelini, L., et al.: A comparison of three interaction modalities in the car: gestures, voice and touch. In: Actes de la 28ième Conference Francophone sur l'Interaction Homme-Machine, pp. 188–196 (2016)
57. Li, J., Reda, A., Butz, A.: Queasy rider: how head movements influence motion sickness in passenger use of head-mounted displays. In: 13th International Conference on Automotive User Interfaces and Interactive Vehicular Applications, AutomotiveUI '21, pp. 28–38, New York, NY, USA (2021). Association for Computing Machinery. https://doi.org/10.1145/3409118.3475137
58. Qiu, Z., Mcgill, M., Pöhlmann, K.M.T., Brewster, S.A.: Manipulating the orientation of planar 2D content in VR as an implicit visual cue for mitigating passenger motion sickness. In: Proceedings of the 15th International Conference on Automotive User Interfaces and Interactive Vehicular Applications, AutomotiveUI '23, pp. 1–10, New York, NY, USA (2023). Association for Computing Machinery. https://doi.org/10.1145/3580585.3607157
59. Pöhlmann, K.M.T., Li, G., Mcgill, M., Markoff, R., Brewster, S.A.: You spin me right round, baby, right round: examining the impact of multi-sensory self-motion cues on motion sickness during a VR reading task. In: Proceedings of the 2023 CHI Conference on Human Factors in Computing Systems, CHI '23, New York, NY, USA (2023). Association for Computing Machinery. https://doi.org/10.1145/3544548.3580966
60. Holoride: FAQ, holoride (2023). https://www.holoride.com/en/faq
61. Audi AG: Holoride: virtual reality meets the real world. https://web.archive.org/web/20231129061609/https://www.audi.com/de/innovation/future-technology/virtual-reality/holoride-virtual-reality-meets-the-real-world.html
62. Holoride: First steps on a global stage (2019). https://www.holoride.com/en/newsroom/first-steps-on-a-global-stage
63. Audi AG: Virtual reality Im Fahrzeug Erleben. https://web.archive.org/web/20240413223726/https://www.audi.de/de/brand/de/service-zubehoer/audi-digital-service/audi-connect/holoride.html
64. Holoride: Revolutionizing in-car entertainment (2023). https://www.holoride.com/en/about-holoride
65. Polymeropoulos, V., et al.: Tradipitant in the treatment of motion sickness: a randomized, double-blind, placebo-controlled study. Front. Neurol. **11**, 09 (2020). https://doi.org/10.3389/fneur.2020.563373
66. Igroup: Igroup presence questionnaire (IPQ) overview. http://www.igroup.org/pq/ipq/index.php
67. Hinderks, A., Schrepp, M., Thomaschewski, J.: User experience questionnaire. https://www.ueq-online.org
68. University of Westminster: Motion sickness susceptibility questionnaire short-form (MSSQ-short). https://www.westminster.ac.uk/sites/default/public-files/general-documents/MSSQ-short-form.pdf

Open Access This chapter is licensed under the terms of the Creative Commons Attribution 4.0 International License (http://creativecommons.org/licenses/by/4.0/), which permits use, sharing, adaptation, distribution and reproduction in any medium or format, as long as you give appropriate credit to the original author(s) and the source, provide a link to the Creative Commons license and indicate if changes were made.

The images or other third party material in this chapter are included in the chapter's Creative Commons license, unless indicated otherwise in a credit line to the material. If material is not included in the chapter's Creative Commons license and your intended use is not permitted by statutory regulation or exceeds the permitted use, you will need to obtain permission directly from the copyright holder.

Qualitative Comparison of Tools for Handling Unstructured IIoT Data

Makki Ben Salem(✉), Philipp Niklas Rosenthal, and Abdelmajid Khelil

Institute for Data and Process Science, Landshut University of Applied Science, Am Lurzenhof 1, 84036 Landshut, Germany
{makki.ben-salem,philipp.niklas-rosenthal, khelil}@haw-landshut.de

Abstract. The Industrial Internet of Things (IIoT) generates vast amounts of data, often unstructured and heterogeneous. Processing and analyzing this data to gain insights and drive industrial automation requires sophisticated tools. This paper investigates the potential of various platforms for handling unstructured IIoT data, focusing on a qualitative comparative analysis of Node-RED, a visual programming tool, against established industrial solutions such as Apache NIFI, KNIME Analytics Platform, and Microsoft Azure Logic Apps. By evaluating these tools, we highlight their respective capabilities and limitations in managing unstructured data. Through this comparative analysis, we demonstrate the distinct features and performance of each tool and discuss their applicability in IIoT data management. The study shows that Node-RED is the most effective tool for handling unstructured IIoT data. Accordingly, we illustrate its use for a specific use case, i.e., value stream analysis.

Keywords: Industrial Internet of Things (IIoT) · Unstructured data · Node-RED · Data lineage · Data-driven technologies

1 Introduction

Industrial Internet of Things (IIoT) is a framework that employs IoT concepts to enhance automation systems for industrial processes. It encompasses the examination and integration of texts written in natural language through artificial intelligence and mathematical linguistics [1]. The ever-growing volume and complexity of data generated by the I IIoT pose significant challenges but also opportunities for organizations seeking to extract meaningful insights and automate processes.

Within the realm of Big Data and IoT, unstructured data poses a challenge as it is typically not easily searchable or computationally operable. This poses a hindrance to the analysis of data and the conversion of data into actionable information [2]. The objective of IIoT platforms is to consolidate production systems into a singular data model, generating both structured and unstructured data from a variety of IoT devices [3]. The conversion of unstructured data into structured data simplifies the processing of said data, enabling the visualization of data through images for more streamlined production management and operation [4].

This data, often unstructured and heterogeneous [5], requires specialized tools and techniques for effective handling. In this context, Node-RED [6] emerges as a promising contender, offering a visually intuitive and versatile platform for managing IIoT data. However, its suitability compared to other established tools in the IIoT data handling landscape remains an open question.

This paper aims to address this gap by conducting a comparative analysis of Node-RED against three prominent alternatives, i.e., Apache NiFi [7], KNIME Analytics Platform [8], and Microsoft Azure Logic Apps [9]. The comparison will focus on key aspects relevant to IIoT data management, including:

Supported data types: Assessing the ability of each tool to handle diverse IIoT data formats, such as sensor readings, log files, and images.

Supported communication protocols: Evaluating the compatibility of each tool with common IIoT communication protocols, such as Message Queuing Telemetry Transport (MQTT) [10] and Open Platform Communication Unified Architecture (OPC UA) [11].

Strengths and weaknesses: Identifying the unique advantages and limitations of each tool in handling IIoT data, considering factors like scalability, performance, and ease of use.

Use cases: Highlighting typical scenarios where each tool shines in the IIoT domain, enabling informed decision-making based on specific requirements.

Through this comparative analysis, we aim to provide valuable insights into the strengths and limitations of Node-RED as a tool for dealing with unstructured IIoT data. By understanding its position relative to other established options, users can make informed choices based on their specific data handling needs within the IIoT landscape.

The remainder of this paper is structured as follows. Section 2 discusses related work Sect. 3 presents the methodology, which contains selection of tools, criteria for comparison and the evaluation approach. Section 4 presents the comparative analysis. Section 5 describes the use case applied which is data collection and analysis for waste detection in production using the example of the Fischertechnik learning factory. Finally, the Conclusion summarizes the findings and suggests avenues for future research.

2 Related Work

Several academic studies have delved into testing and standardizing architectures for IoT platforms, focusing on various challenges and approaches. For instance, [12] compared Node-RED with kaa using three real-world scenarios, including a remote-controlled LED, a chat application, and a data transmitting sensor, finding that Node-RED performs better on a single server, while Kaa has better overall scalability and reliability with built-in security, device discovery, and support for clustering.

Another study [13] highlighted key challenges in testing open-source IoT software: virtualization limitations, expensive and often outdated testbeds, and complexities in interoperability testing due to potential incompatibilities across devices and platforms. The study suggests strategically combining different testing methods for a comprehensive evaluation.

While dedicated research on IoT scalability and performance testing remains limited, [14] offers valuable foundational knowledge on IoT concepts and definitions.

[15] emphasized the significance of Visual Programming Environments (VPE) in simplifying the development process for Artificial Intelligence (AI) and IoT-based projects. These environments enable Small Medium Enterprises (SMEs) to leverage advanced technologies without requiring deep expertise in algorithms and programming languages, with examples such as Node-RED, KNIME, MoD kit, and Open Roberta. However, this work lacks specific real-world case studies or practical implementations of the proposed software tools.

The Survey [16] focuses on AI methods to deal with heterogeneous data, however, does compare the state of the art tools to implement these AI methods.

3 Methodology

3.1 Selection of Tools

For the comparative analysis, three prominent alternatives to Node-RED were carefully chosen based on their widespread adoption and relevance to IIoT data management. Apache NiFi, KNIME Analytics Platform, and Microsoft Azure Logic Apps were selected given their established presence in the field and comprehensive features for handling diverse types of IIoT data. Apache NiFi is renowned for its data flow management capabilities and support for complex data routing and transformation tasks. KNIME Analytics Platform offers a visual workflow editor and a wide range of analytics and machine learning tools, making it suitable for advanced data processing requirements. Microsoft Azure Logic Apps provides a cloud-native approach to workflow automation, integration, and data orchestration, aligning with the growing trend of cloud adoption in IIoT deployments. These tools were deemed representative of different approaches to IIoT data management, ensuring a comprehensive comparison against Node-RED capabilities.

3.2 Criteria for Comparison

The criteria for comparison were carefully selected to encompass the essential aspects of IIoT data management, ensuring a comprehensive evaluation of Node-RED and its alternatives. Firstly, the supported data types criterion assesses the ability of each tool to handle the diverse and often unstructured nature of IIoT data, including sensor readings, log files, and multimedia content. Secondly, the communication protocols criterion evaluates the compatibility of each tool with prevalent IIoT protocols such as MQTT and OPC UA, crucial for seamless integration with IoT devices and systems. Scalability, the third criterion, examines the tools capacity to scale efficiently with increasing data volumes and processing demands, vital for accommodating the dynamic nature of IIoT deployments. Performance, the fourth criterion, focuses on evaluating the speed and efficiency of data processing and analysis, considering both real-time and batch processing scenarios. Lastly, the ease-of-use criterion assesses the user-friendliness of each tool interface and the availability of intuitive features for designing, deploying, and managing IIoT data workflows, aiming to facilitate adoption and streamline development efforts. These criteria collectively provide a holistic framework for comparing Node-RED with its counterparts, enabling a thorough assessment of their suitability for IIoT data management tasks.

3.3 Evaluation Approach

Given the scope of this study, the evaluation approach involves constructing a comparative table highlighting key features and capabilities of Node-RED and its identified alternatives: Apache NiFi, KNIME Analytics Platform, and Microsoft Azure Logic Apps. The comparison will be based on a thorough review of available documentation, official websites, user reviews, and technical specifications provided by the respective tool providers.

The comparative table will focus on essential aspects relevant to IIoT data management, including supported data types, communication protocols, scalability, performance, ease of use, and any other pertinent features or functionalities. Each tool will be assessed based on publicly available information, ensuring consistency and fairness in the evaluation process.

Additionally, to demonstrate the practical application of Node-RED, a use case scenario will be presented, showcasing how Node-RED can be utilized for IIoT data management and workflow automation. While direct testing of the other tools will not be conducted in this study, insights gathered from the comparative table will inform discussions on Node-RED strengths and potential areas for improvement relative to its alternatives.

4 Comparative Analysis

In this section, the comparative analysis is presented. The analysis is based on a review of key features and capabilities gathered from publicly available information, including official documentation, technical specifications, and user feedback [17-20].

4.1 Supported Data Types

Node-RED offers comprehensive support for a wide range of IIoT data types, including sensor readings, log files, images, and various other formats commonly encountered in industrial environments. Its modular and extensible architecture enables users to integrate additional data processing nodes and libraries as needed, providing flexibility in handling diverse data sources and formats.

Apache NiFi boasts robust support for handling diverse data types through its data flow management capabilities. With a graphical user interface for designing data pipelines and a rich set of processors for data transformation and enrichment, Apache NiFi facilitates seamless integration of disparate data sources and formats in IIoT environments.

KNIME Analytics Platform provides a visual workflow editor and a vast collection of nodes for data preprocessing, analysis, and visualization. While its primary focus is on data analytics and machine learning, KNIME's versatility allows users to process and analyze various data types, including IIoT data, with ease.

Microsoft Azure Logic Apps offers integration and workflow automation capabilities in a cloud-native environment. While its support for specific data types may be more tailored to cloud-based applications, Azure Logic Apps provides connectors and adapters for integrating with various data sources commonly used in IIoT deployments (Table 1).

Table 1. Comparison based on supported data types.

Feature	Node-RED	Apache Nifi	KNIME Analytics Platform	Microsoft Azure Logic Apps
Data types	Numbers, strings, objects, arrays, buffers, JSON, CSV, YAML, XML, HTML	AVRO, Parquet, ORC, JSON, XML, CSV	AVRO, Parquet, ORC, JSON, XML, CSV	JSON, XML, CSV, EDI, X12

4.2 Messaging Protocols

Node-RED supports a wide range of messaging protocols commonly used in IIoT environments, including MQTT, OPC UA, HTTP, DDS, Sparkplug and CoAP. Its modular design allows users to easily implement custom messaging protocols using available nodes or develop custom nodes as needed.

Apache NiFi provides extensive support for data ingestion from various sources, including IIoT devices, through its processors and connectors. While NiFi offers native support for common protocols like MQTT and HTTP, its extensible architecture allows integration with additional protocols through custom development or community-contributed extensions.

KNIME Analytics Platform focuses primarily on data analytics and processing tasks rather than messaging protocols. However, users can leverage its connectivity capabilities and available extensions to integrate with IIoT devices and systems using standard protocols supported by KNIME's nodes and connectors.

Microsoft Azure Logic Apps offers connectors for integrating with a wide range of services and systems, including IoT Hub, Event Hubs, and Azure IoT Central, facilitating seamless communication with IIoT devices and platforms. Its native support for MQTT and AMQP protocols simplifies integration with MQTT-enabled devices commonly used in IIoT deployments (Table 2).

Table 2. Comparison based on supported communication protocols.

Feature	Node-RED	Apache Nifi	KNIME Analytics Platform	Microsoft Azure Logic Apps
communication protocols	HTTP, MQTT, Web-Sockets, TCP, UDP, FTP, MODBUS, OPCUA, etc	HTTP, JMS, Kafka, etc	HTTP, FTP,	HTTP, TCP, UDP, FTP

4.3 Scalability

Node-RED provides scalability through its distributed deployment options and support for running on cloud platforms such as AWS and Azure. By leveraging containerization technologies like Docker and Kubernetes, Node-RED can scale horizontally to accommodate growing IIoT data volumes and processing demands.

Apache NiFi offers scalability through its clustered deployment architecture, allowing users to distribute data processing tasks across multiple nodes to handle large-scale data flows efficiently. Its built-in load balancing and failover mechanisms ensure high availability and reliability in IIoT deployments.

KNIME Analytics Platform scalability depends on the underlying infrastructure and resources available for data processing. While it may not offer native support for distributed computing out of the box, users can deploy KNIME workflows on scalable cloud platforms or distributed computing environments to achieve scalability as needed.

Microsoft Azure Logic Apps provides scalability through its cloud-native architecture, allowing users to scale resources dynamically based on workload demands. By leveraging Azure auto-scaling capabilities and serverless computing offerings, Azure Logic Apps can handle varying IIoT data processing loads efficiently (Table 3).

Table 3. Comparison based on scalability.

Feature	Node-RED	Apache Nifi	KNIME Analytics Platform	Microsoft Azure Logic Apps
Scalability	Can be scaled horizontally by adding more node	Highly scalable with distributed architecture	Can be scaled horizontally and vertically	Scales automatically based on demand

4.4 Performance

Node-RED is known for its lightweight and efficient runtime environment, making it well-suited for handling real-time and near-real-time IIoT data processing tasks. Its event-driven architecture ensures low latency and high throughput, enabling responsive data processing and analysis in IIoT deployments.

Apache NiFi offers robust performance for data ingestion, routing, and processing tasks through its distributed and parallel processing capabilities. By optimizing data flow execution and resource utilization across clustered nodes, Apache NiFi can achieve high throughput and low latency for IIoT data processing workflows.

KNIME Analytics Platform's performance depends on factors such as data volume, complexity of data processing tasks, and available computational resources. While it may not match the performance of stream processing frameworks for real-time data processing, KNIME capabilities for batch processing and offline analytics make it suitable for certain IIoT use cases.

Microsoft Azure Logic Apps provides scalable and reliable performance for IIoT data integration and workflow automation tasks in a cloud-native environment. By leveraging Azure's global infrastructure and managed services, Azure Logic Apps can deliver low-latency data processing and high throughput for IIoT applications (Table 4).

Table 4. Comparison based on performance.

Feature	Node-RED	Apache Nifi	KNIME Analytics Platform	Microsoft Azure Logic Apps
performance	Can be resource-intensive for complex flows	High performance with distributed processing	High performance with optimized execution engine	High performance with cloud-based infrastructure

4.5 Ease of Use

Node-RED is lauded for its intuitive visual programming interface, which allows users to design and deploy IIoT data workflows with minimal coding effort. Its extensive library of pre-built nodes and flow templates further simplifies development and accelerates time-to-deployment for IIoT applications.

Apache NiFi offers a user-friendly graphical interface for designing data flows and managing data processing tasks. Its drag-and-drop interface and extensive library of processors make it accessible to users with varying levels of technical expertise, facilitating rapid development and deployment of IIoT data pipelines.

KNIME Analytics Platform provides a visual workflow editor and a rich set of data processing nodes, making it easy for users to design and execute complex data analytics tasks. While its focus on data analytics may require additional configuration for IIoT-specific use cases, KNIME user-friendly interface and extensive documentation support ease of use for IIoT application development.

Microsoft Azure Logic Apps offers a low-code approach to workflow automation and integration, allowing users to design and deploy IIoT data workflows using visual designers and predefined connectors. Its integration with Azure services and enterprise-grade security features further enhances ease of use and accelerates development for IIoT applications (Table 5).

4.6 Use Cases

Node-RED is widely used for a variety of IIoT applications, including sensor data monitoring and visualization, predictive maintenance, and industrial automation. Its flexibility and extensibility make it suitable for implementing diverse IIoT use cases across different industries and domains.

Table 5. Comparison based on ease of use.

Feature	Node-RED	Apache Nifi	KNIME Analytics Platform	Microsoft Azure Logic Apps
Ease of use	Easy to learn and use with visual programming	Moderate learning curve, requires understanding of data flow concepts	Steeper learning curve, requires programming knowledge	Moderate learning curve, requires understanding of logic app concepts

Apache NiFi is commonly deployed for data ingestion, routing, and transformation tasks in IIoT environments, enabling real-time data processing, event-driven architectures, and data lake integration. Its support for complex data workflows and scalability features makes it ideal for handling large-scale IIoT data streams.

KNIME Analytics Platform is employed for data analytics, machine learning, and predictive modeling tasks in IIoT applications, such as anomaly detection, predictive maintenance, and quality control. Its comprehensive suite of analytics tools and integration capabilities enable advanced data analysis and decision-making in industrial settings.

Microsoft Azure Logic Apps is utilized for workflow automation, integration, and orchestration in IIoT deployments, facilitating seamless communication between IoT devices, cloud services, and enterprise systems. Its built-in connectors and adapters simplify integration with IIoT platforms and services, enabling efficient data exchange and business process automation (Table 6).

Table 6. Comparison based on use cases.

Feature	Node-RED	Apache Nifi	KNIME Analytics Platform	Microsoft Azure Logic Apps
Use cases	IoT data integration, rapid prototyping, simple automation tasks	Data ingestion, data integration, data streaming, data provenance	Data analysis, data visualization, machine learning	Enterprise application integration, business process automation, API integration

5 Use Case: Data Collection and Analysis for Waste Detection in Production Using the Example of the Fischertechnik Learning Factory

The use case was tested within the Fischertechnik Learning Factory [21] and implemented following the documentation of Node-RED.

5.1 Fischertechnik Learning Factory

The Fischertechnik Learning Factory is an educational tool designed to teach automation, mechatronics, and industrial manufacturing principles. The factory consists of five main components, the Vacuum Suction Gripper (VGR), the Automated high-Bay Warehouse (HBW), a Multi-Processing station with a kiln (MPO), a Sorting Line with color recognition (SLD), the environmental Station with Surveillance Camera (SSC) and a - and output station with color recognition and NFC reader (DPS). The SSC is not considered in this use case.

5.2 Data Collection

Figures 1, 2 and 3 show the Node-RED flow, which is used to capture and secure the data. The general structure is described as follows. In Fig. 1, auxiliary variables for the data points of the OPC UA server, the buffering of the values and for the Node-RED flow itself are initialized.

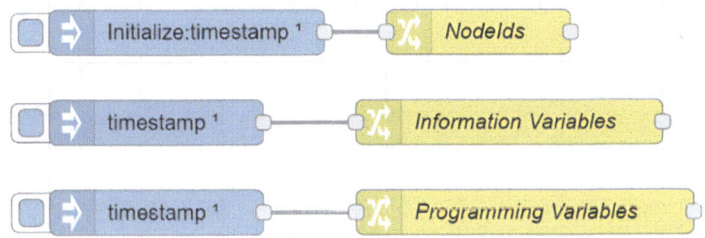

Fig. 1. Variable initialization

The Node-RED-contrib-iiot-opcua palette is used to collect data via OPC UA. This offers nodes for listening (listener node) on data points from OPC UA servers and other useful functions when dealing with OPC UA in Node-RED. Figure 2 refers to data collection. The flow starts with an OPC UA Inject Node, which creates a listener object for a predetermined data points and passes it on to the listener node.

The whole process works according to the publish/subscribe principle. The output is a msg object which contains the payloadobject, which contains the value of the associated data point, as well as an addressSpaceItems array, which contains information such as the name of the data point. Various function nodes (one for each machine or data point) are connected behind this listener node, which intercept this msg object and check which data point the information comes from. To do this, the nodeId in the addressSpaceItems array is checked and compared. If the information has been received by the correct function node, the addressSpaceItems array is now created with the remaining necessary data.

For example, in the case that the "nodeId" = "ns = 3"; s = "gtyp_Interface_Dashboard"."Subscribe"."State_VGR"."x_active", meaning the VGR has picked up the workpiece, the data points "x_target" (transport destination of the VGR) and "ldt_ts" (time stamp) are added to the addressSpaceItems array and then sent on to the "Read" node. The "Read" node can read and return multiple values on the OPC UA server at the same time. Next, the returned data is saved in a flow variable. A flow variable is a kind of global variable that only applies to the current flow. In addition, auxiliary variables are also set to allow you to recognize which process you are currently in. The final step occurs when the end of a running cycle is reached. There are four different ways in which a run ends.

1. Error-free delivery cycle (storage of the workpiece is the final signal)
2. Error in the workpiece during storage (detecting the error is the final signal)
3. Error-free ordering cycle (delivery of the workpiece is the final signal)
4. Error in the workpiece during the ordering cycle (detecting the error is the final signal)

After one of the signals is triggered, the cached data is saved in a MySQL database.

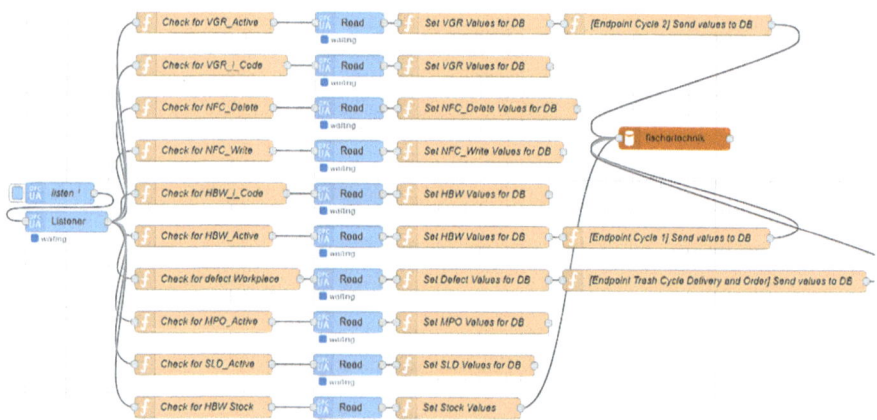

Fig. 2. Data collection through OPC UA

5.3 Data Analysis

The Node-RED flow depicted in Fig. 3 is responsible for data analysis, the input is a JSON file that is intended to represent a value stream without data in a standardized manner. The JSON files in this work are only examples and were created by an expert,

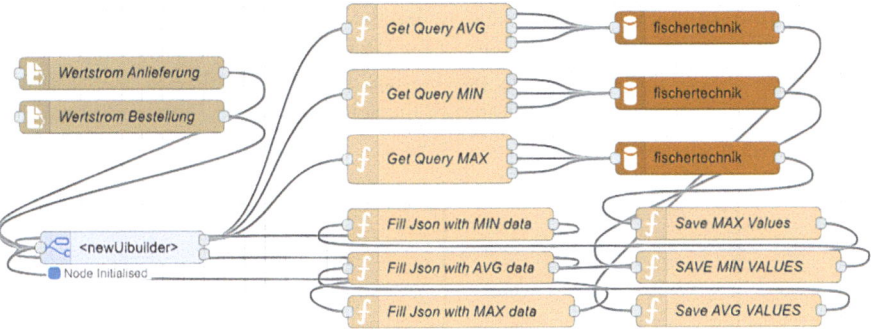

Fig. 3. The Node-RED flow for data analysis

as no standardization has been implemented up to this point. This JSON file is now displayed in a small web application that was created using the UI-Builder. The UI-Builder is part of the Node-RED-contrib-uibuilder palette and enables easy creation of data-driven front-end web applications in Node-RED. After the JSON file has been successfully transferred, the "empty" input file can now be filled with data which is retrieved from the database with the previously recorded data using a query based on the machine name and activity. This is then inserted into the correct fields of the JSON file. The web application allows you to select a start or end time, which indicates the time period for which the data should be displayed, down to the minute. It should also be mentioned that the average value for the processing times is always calculated for each process. It is possible to display the maximum or minimum value. The processing times of the activities within a process that has a defect are not included in the average, as this distorts the result. Thus, in a reject process, the activity times for the machines that occur after the defect is detected would all show 0, as these are no longer recorded.

The process can now be visualized in the form of a value stream (Fig. 4). The core of the value stream representation in the production process is a number of important key figures, which are determined from the database listed above. The key figures for each production process are shown in a data box (Table 7).

Table 7. Value stream metrics.

Identification	Description	Unit
Nfc, hbw, mpo, sld	Machine name	-
Delete, write, store	Activity	-
CT	Cycle time/Processing time	seconds
CP	Capacity	percent
dRate	Defect rate	percent

In Fig. 4, for example it can be seen here that the transport from the "write" activity of the NFC machine to the "store" activity of the HBWmachine takes an average of 23.6 s. It can also be seen that the delivery process has a defect rate of 0.08, which means that 8 percent of the waste was generated in the selected period. Finally, it should be mentioned that the average inventory is 2 items.

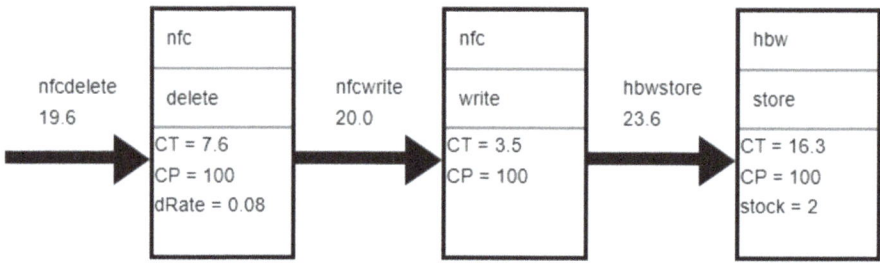

Fig. 4. Representation of the delivery process as a value stream with the average values

5.4 Results and Visualization of Waste

Based on the collected data, a waste detection system has been developed to identify and locate waste within the Fischertechnik Learning Factory. This system utilizes pre-existing waste model rules to analyze various parameters, it classifies each activity into three categories: value-added, non-value-added, and necessary non-value-added and provides a detailed breakdown of the percentage of each waste type—transport, defect, inventory, overprocessing, overproduction, waiting, and motion—present in specific activities. In this example, we will focus only on using transportation waste and defects. By analyzing transport times, the system can pinpoint potential areas of waste buildup along transportation activities. Similarly, by monitoring production data, the system can identify anomalies indicative of defects. Once the waste detection process is complete, the results are visualized in the Node-RED dashboard (Fig. 5).

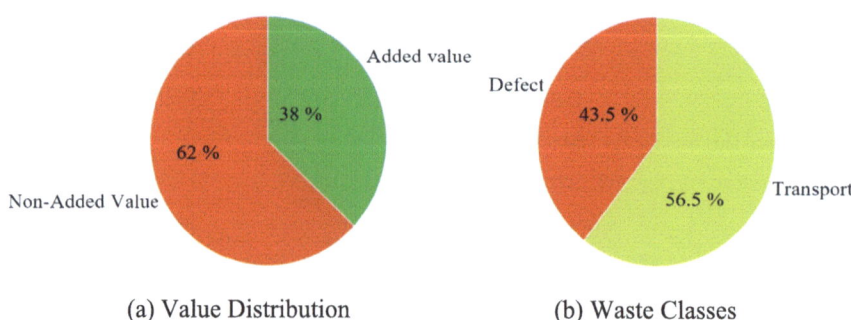

(a) Value Distribution (b) Waste Classes

Fig. 5. Waste visualization within the Node-RED dashboard

6 Conclusion

Node-RED emerges as a powerful tool for handling unstructured IIoT data, offering a versatile and user-friendly approach to data processing. Through our comparative analysis and its application in a waste detection use case, we have highlighted Node-RED strengths and showcased its potential to play a key role as an industrial data collection, processing and visualization tool. We plan to use Node-RED to implement AI driven Industry 4.0 applications to exploit its effectiveness and versatility.

References

1. Sergeeva, M.B., Voskobovich, V., Kukharenko, A.: Data processing in industrial internet of things (IIoT) applications: industrial agility. Wave Electronics and its Application in Information and Telecommunication Systems (WECONF) (2022)
2. Farrell, R., Yuan, X., Roy, K.: IoT to structured data (IoT2SD): a big data information extraction framework. 1st International Conference on AI in Cybersecurity (ICAIC) (2022)
3. Dhanaraj, R.K., Rajkumar, K., Hariharan, U.: Enterprise IoT modeling: supervised, unsupervised, and reinforcement learning. EAI/Springer Innovations in Communication and Computing (2020)
4. Chunfeng, W., et al.: An Industrial Mass Unstructured Data Processing Method and System (2019)
5. Sarikoz, S.K.: Examining Knowledge Extraction Processes from Heterogeneous Data Sources. Brilliant engineering (2023)
6. Node-RED Documentation. https://nodered.org/docs/. Accessed 15 June 2024
7. Apache NiFi Documentation. https://nifi.apache.org/docs.html. Accessed 15 June 2024
8. KNIME Documentation. https://docs.knime.com/. Accessed 15 June 2024
9. Azure Logic Apps Documentation. https://docs.microsoft.com/en-us/azure/logic-apps/. Accessed 15 June 2024
10. Cameron, N.: MQTT. In: ESP32 Formats and Communication. Maker Innovations Series. Apress, Berkeley (2023)
11. Rinaldi, J.S.: OPC UA - Unified Architecture: The Everyman's Guide to the Most Important Information Technology in Industrial Automation. 1st edition. CreateSpace Independent Publishing Platform (2016)
12. Scott, R., Ostberg, D.: A Comparative Study of Opensource IoT Middleware Platforms (2018)
13. Baccelli, E., Ortmann, L., Rosenkranz, P., Wahlisch, M.: A Distributed Test System Architecture for Open-source IoT Software. ACM. ISBN (2015)
14. Rotondi, D., Minerva, R., Biru, A.: Towards a Definition of the Internet of Things. http://iot.ieee.org/definition. Accessed 20 April 2024
15. Hauck, M., Machhamer, R., Czenkusch, L., Gollmer, K., Dartmann, G.: Node and block-based development tools for distributed systems with AI applications. IEEE Access (2019)
16. Kamm, S., Veekati, S.S., Müller, T., Jazdi, N., Weyrich, M.: A survey on machine learning based analysis of heterogeneous data in industrial automation. Computers in Industry, Bd. **149** (2023)
17. Node-RED Forum. https://discourse.nodered.org/. Accessed 16 Mar 2024
18. Apache NiFi Reviews on G2. https://www.g2.com/products/apache-nifi/reviews. Accessed 18 Mar 2024

19. KNIME Forum. https://forum.knime.com/. Accessed 18 Mar 2024
20. Microsoft Azure Feedback Forums. https://feedback.azure.com/forums/287593-logic-apps. Accessed 20 Mar 2024
21. Fischertechnik Lernfabrik 4.0 accompanying booklet version 6 (2023)

Open Access This chapter is licensed under the terms of the Creative Commons Attribution 4.0 International License (http://creativecommons.org/licenses/by/4.0/), which permits use, sharing, adaptation, distribution and reproduction in any medium or format, as long as you give appropriate credit to the original author(s) and the source, provide a link to the Creative Commons license and indicate if changes were made.

The images or other third party material in this chapter are included in the chapter's Creative Commons license, unless indicated otherwise in a credit line to the material. If material is not included in the chapter's Creative Commons license and your intended use is not permitted by statutory regulation or exceeds the permitted use, you will need to obtain permission directly from the copyright holder.

An Approach for a Human-Assisted Data Loop in Connected Manufacturing Systems

Matthias Weiß[1(✉)], Alexander Schön[2], Matthias Lück[3], Maximilian Schnierle[2], Stefan Carosella[2], Nasser Jazdi[1], Carmen Constantinescu[3], Peter Middendorf[2], and Michael Weyrich[1]

[1] University of Stuttgart (IAS), Pfaffenwaldring 47, 70569 Stuttgart, Germany
`matthias.weiss@ias.uni-stuttgart.de`
[2] University of Stuttgart (IFB), Pfaffenwaldring 31, 70569 Stuttgart, Germany
[3] Fraunhofer IAO, Nobelstraße 12, 70569 Stuttgart, Germany

Abstract. Today's production is characterized by shorter innovation cycles, leading to a highly dynamic shop floor environment. To tackle the need for fast and flexible adaptation, it is aimed for a complete digital integration of machines and a comprehensive data management. Benefiting from the vast amounts of heterogeneous data collected requires a thorough understanding of their dependencies and correlations as well as a human-assisted, yet automated analysis process. Still, up to this date, the realization of this vision proves to be very challenging in practice and requires a myriad of different new methods and technologies.

Therefore, the scope of this paper is to provide an outline of how such an end-to-end connectivity can be achieved along the real-world example of a lightweight construction process. As such, it is shown how its machines are upgraded and connected to the network in order to allow for collection of time series data using a central agent. A context provider is introduced, with which external data sources, such as wearables, can be integrated. After collecting and merging the data, its storage and sharing are managed using a Data Lakehouse. The data is analyzed by using a human-assisted feedback loop, effectively integrating expert knowledge into the learning process. Using the so-gained knowledge, optimization of the shop floor, intralogistics and IT infrastructure can be achieved.

Preliminary results from parts of this loop already in operation suggest significant potential benefits, paving the way for more comprehensive integration across production systems.

Keywords: Data Loop · Lightweight Construction · Context Integration · Correlation Analysis · Human Feedback · Optimization

1 Introduction

As reduced batch sizes and high time-to-market pressure become increasingly common in modern production environments, fast adaptation and introduction of automation in manufacturing systems is required. To enable this unprecedented degree of flexibility, an increasing amount of interconnected software applications are introduced to the shopfloor, transforming manufacturing processes to be software-defined by nature [1]. The seamless merging of the physical and digital worlds results in a high number of dependencies between hard- and software assets and leads to high volumes of data being produced. To cope with the increased complexity and to gain insights into the inner workings of the production environment and manufacturing process itself, connected manufacturing systems increasingly involve the use of artificial intelligence (AI) through machine learning (ML) at a fundamental level.

However, while machine learning models can process large amounts of data - usually in high-dimensional feature spaces - human intervention is often required to validate and interpret the results and make informed decisions based on the knowledge gained. This results in an expansion of knowledge through the analysis of the recorded process data and its evaluation using machine learning models [2]. In complex manufacturing processes, human intervention is invaluable, as adjustments can be made in real time based on experience and expertise. This ensures optimal process performance. Furthermore, human involvement is crucial when it comes to unexpected events such as anomalies or outliers in industrial processes [3, 4]. While automated systems can recognize known patterns, humans are capable of critical insight and problem-solving skills to creatively address novel problems that ensure both reliability and safety of operations. In addition, human expertise is essential for interpreting process data and understanding complex monitoring systems.

As such, the alignment of information flows through suitable interfaces to humans becomes essential. This human-centered approach is referred to as Industry 5.0 (I5.0) or the human-centered approach (HCA) [5, 6]. The prioritization of people can be achieved, among other things, by placing the process-related interests of people at the center of production monitoring and relying on technologies that help employees to initiate optimizations through the development of knowledge and skills.

The human-in-the-loop paradigm emphasizes the role of humans in monitoring and mediating (information) processes of increasingly automated and intelligent production systems. In this context, humans are the central point of contact for information flows. Future manufacturing systems will still need humans to mediate the supervisory level and information flows even in the most autonomous production scenarios. However, challenges such as providing technological support for humans and improving collaboration between humans and manufacturing equipment are still omnipresent.

To address these challenges, established tools for guiding information flows are combined with the idea of human-in-the-loop in the following [2]. The remainder of this paper is structured as follows: In Sect. 2, basics and potential use cases regarding HCA are introduced. In Sect. 3, the concept of this paper in the form of a human-assisted data loop is introduced. Section 4 highlights the the implementation results of the first data loop steps. Finally, Sect. 5 summarizes the paper and gives an outlook about future research directions.

2 Related Work

The integration of physical and digital systems in manufacturing under the umbrella of Industry 4.0 requires advanced techniques. One of the most important strategies in this area is predictive maintenance (PdM), which aims to optimize the condition and performance of industrial equipment by minimizing downtime and extending the useful life of components. Machine learning (ML) is crucial for PdM applications as it is able to process extensive operational data and effectively detect, diagnose and predict faults. The use of ML-powered PdM can significantly reduce maintenance costs, minimize machine downtime, increase operator safety and improve overall production efficiency. Current research emphasizes the importance of selecting appropriate ML algorithms, data types and data sizes to ensure effective PdM implementation while overcoming challenges such as data acquisition, model training and cross-validation. The combination of multiple ML models, the use of cloud-based implementations and the integration of PdM with new sensor technologies are proposed to improve prediction accuracy and data analysis capabilities [7].

Moving towards Industry 5.0, the focus shifts to worker well-being and sustainability. Industry 5.0 builds on the technological advancements of Industry 4.0 by incorporating resilience, sustainability, and human-centered approach (HCA). It integrates advanced technologies such as artificial intelligence, robotics, human-robot collaboration, and digitalization to create human-centric production systems. This new paradigm aims to combine human creativity and intelligence with machine efficiency to meet societal needs, highlighting the importance of human-robot collaboration, digital twins, the metaverse, blockchain, big data, cognitive computing, and augmented reality. Challenges for Industry 5.0 include social barriers, technology integration issues, and the need for lifelong learning for workers. Practical research and application scenarios are essential to fully exploit Industry 5.0's potential [5, 8].

The importance of adaptation and innovation in the manufacturing industry to achieve sustainable production using AI and ML technologies is increasingly recognized. AI and ML, as key technologies of Industry 4.0, are creating new industrial paradigms by improving sustainability through the optimization of energy resources, logistics, and supply chain management. However, there are significant research gaps, particularly in the integration of customer, environmental, and human-in-the-loop aspects in ML-PPC models. Challenges include the coupling of production planning and control (PPC) with logistics, product and process design, and the complexity of IoT data collection. The dynamic production conditions and the need for customized products at low cost and shortened time-to-market further complicate the landscape. There is a pressing need to improve data collection systems, integrate PPS with logistics and product design, and prioritize human interaction and environmental aspects for ethical production [9].

Current research highlights the significant potential of ML in PPC, which can help learn from historical and real-time data to respond to events. However, this requires substantial investment in data warehousing. To advance the field, it is suggested to focus on enhancing data collection systems, integrating PPC with logistics and product design, and emphasizing ethical production by prioritizing human interaction and environmental aspects [9, 10].

3 Concept for a Human-Assisted Data Loop

This chapter describes the concept of the human-assisted data loop that this paper proposes. For this, the first section lists requirements to the data loop based on the observed research trends and discussions with the involved stakeholders. Subsequently, the structure and features of the data loop are presented.

3.1 Requirements

In the rapidly evolving landscape of modern production, several key requirements must be met to maintain competitiveness and efficiency. The following requirements serve as a key motivation for implementing a human-assisted data loop within connected manufacturing systems.

1. **Data availability:** Today's production environments demand that data be accessible to a wide range of stakeholders, including operators, engineers, managers, and external partners. This accessibility ensures that all relevant parties can make informed decisions based on the latest information.
2. **Continuous changes in production processes:** The shift towards more agile and flexible manufacturing necessitates that production processes can quickly adapt to new specifications and demands. This often results in smaller production runs, requiring systems that can efficiently handle frequent changes.
3. **Continuous data analysis for anomaly detection:** To maintain high standards of quality and efficiency, it is essential to continuously analyze production data. This real-time analysis helps detect anomalies and deviations from desired behavior, enabling prompt corrective actions.
4. **Integration of contextual information:** Leveraging data allows for more precise control over production processes. By incorporating contextual data, it is possible to understand the causal relationships between various measurements, leading to more informed decision-making and process optimization.

3.2 Structure and Features

To fulfill the aforementioned requirements of modern production, the human-assisted data loop is characterized by several essential features. Firstly, it establishes comprehensive end-to-end connectivity between the shop floor, backend systems, and human operators. This connectivity ensures seamless data flow and communication, enabling real-time monitoring and control of production processes. Effective data management and access control are considered to be crucial components in this context. These mechanisms ensure that the vast amounts of data generated are handled efficiently and securely shared with multiple stakeholders, maintaining data integrity and confidentiality.

The loop also involves mining for dependencies and correlations within the measured time series data. This analysis is vital for understanding the intricate relationships within the production process and uncovering hidden patterns that can drive process improvements. Automation plays a significant role in this procedure, with automated data mining and diagnostic tools enhancing the efficiency of data analysis. Due to the dynamic and frequent changes in the shopfloor and data landscape caused by small batch sizes, establishing a strategy for a continuous adaptation of the analysis pipelines is paramount to ensure fast and reliable results. It has been suggested frequently that incorporating human expertise in a controlled process is beneficial for quick analysis adaptation [3]. In this context, the analysis tools provide insights and recommendations that can be reviewed by systems engineers. The so produced feedback can be used to improve future recommendations of the analysis tools, effectively establishing an automated self-learning loop.

Finally, the loop establishes a closed feedback system that incorporates decisions and newly discovered optimizations back into the production process. This iterative refinement and enhancement ensure continuous improvement, leading to sustained gains in efficiency and quality.

The basic structure of this data loop concept is highlighted in Fig. 1. Starting from left to right, Data of the shopfloor is collected and managed in a centralized Data Lakehouse. This Data Lakehouse provides different stakeholders (both internal and external) with access to the stored data according to their permissions and responsibilities. In this example, stakeholder 1 in particular is responsible for real-time data analysis, implementing the aforementioned self-learning loop between the systems engineer and the analysis procedures. Once the cause for a particular deviation is found, the systems engineer is involved into the decision-making process and can assist in the optimization of the shopfloor environment.

In order to enable such a human-assisted feedback loop and to understand the connections between different data time series, contextual information is required. Integrating context into data management is accomplished by utilizing a context provider, as can be seen in Fig. 2. For the purpose of this paper, two basic data sources are integrated: Machine data, i.e., time series data of the specific production process, and environmental data, e.g., temperature and humidity inside the shopfloor. The data is collected using a centralized data collector and subsequently sent to the Data Lakehouse. Finally, the context provider links the environmental data to the production data according to the rules and patterns learned from information provided, as of now, manually by engineers.

4 Example: Data Loop for a Lightweight Construction Process

For the purpose of this paper, parts of the described concept, i.e., data aggregation and context provisioning, have been implemented for a lightweight construction process. This chapter describes the resulting system architecture and shows the results of the integration process.

4.1 Connectivity of Machines (Lightweight Construction) and Environment Data

In cooperation with the manufacturer Tajima, an embroidery machine is upgraded so that in it makes its sensor and process data available to the shop floor network via OPC-UA. Additionally, the embroidery machine will also react to data from the shop floor. This data is also transferred to the machine via OPC UA.

A pump cart was upgraded to automatically record all data for the VARI process digitally. The data is transmitted to the data aggregator via WLAN using MQTT. The

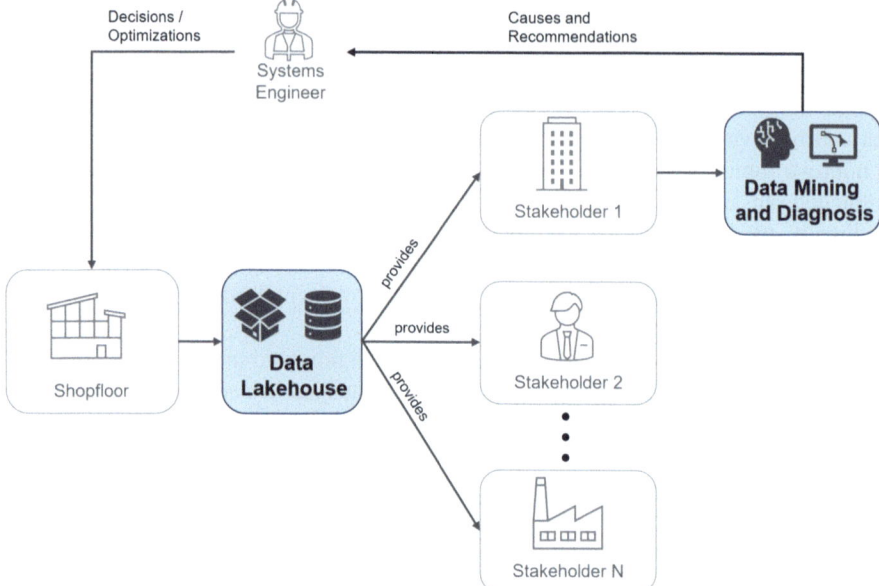

Fig. 1. Top-Level view of the human-assisted data loop. Shopfloor data is collected by the Data Lakehouse and distributed to stakeholders, which dan use the data for shopfloor adjustments.

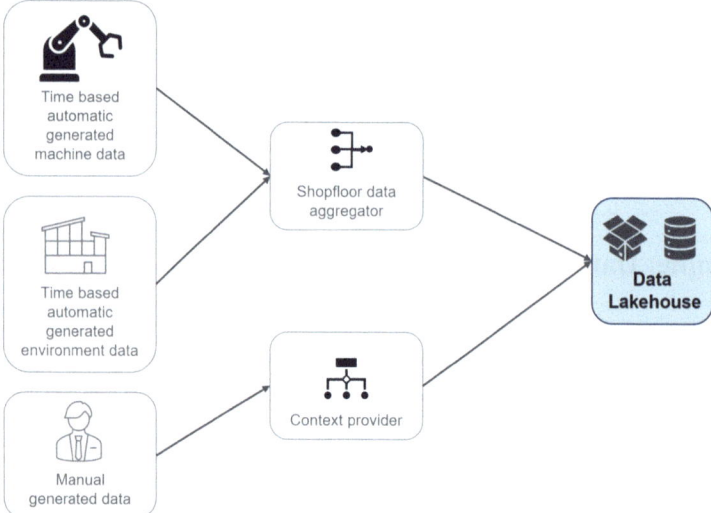

Fig. 2. Data aggregation and context provisioning of shopfloor data. As of now, context is provided manually by the systems engineer.

production hall was provided with sensors for environmental monitoring (temperature, humidity, air pressure) at several points. These sensors transmit their data to the data aggregator via MQTT.

4.2 Shopfloor Data Aggregator

The Connectware software from Cybus was selected as shopfloor data aggregator. Within this environment, Cybus Connectware acts as a technology-neutral layer between the IT and OT levels in the automation stack (see Fig. 2).

Cybus Connectware is a software ecosystem that bridges the gap between operation technology (OT) on the shop floor and information technology (IT) by building a universal data architecture. Additional Connectware is allowing internet-technology based applications to access industrial hardware in a secure and standardized way.

Connectware enables connectivity between various devices on the shop floor, such as sensors, signals, edge devices, and PLCs. At a controller level, Cybus Connectware can transmit data bidirectionally via the typical industrial communication protocols, such as OPC UA, MQTT, Modbus/TCP, Siemens S7, Ethernet/IP [11, 12].

This setup allows any-to-any communication and data flow for shop floor devices towards enterprise IT systems and vice versa. Using Cybus Connectware as a manufacturing data platform allows for a factory-wide data architecture that receives live data from the shop floor and manages access control as well as the distribution of data to users or applications in the cloud or on-premises. With DevOps capabilities, such as Ansible [13], Connectware enables the realization of complex use cases and data operations regarding the possibilities of Industry 4.0. A CI/CD for manufacturing is implemented with Gitlab.

Within Cybus Connectware, the hardware parameters are addressed via industry protocol schemes and are mapped into configurable hierarchical names that reflect individual accessible topics on an MQTT-based message broker [11].

Data access is secured by a fine grained (per data-endpoint) authentication and authorization system, i.e. Connectware controls who is allowed to access what parameter in which way. Finally, an entire set of industry protocols, parameter mappings, corresponding authentication and authorization schemes and optionally plugin-like applications for local pre- or postprocessing can be bundled in a so-called Service. Once created, Services can be deployed to the host running Cybus Connectware via configurable text-based Yet Another Markup Language (YAML) files, allowing for easy and highly automatable installation.

Connectware ships with an intuitive browser-based interface for administering and configuring the entire Cybus ecosystem [11]. The Cybus Connectware ecosystem enables rapid prototyping. This allows the system engineer to quickly and easily check whether the changes made lead to the desired result.

For the practical setup of this research work, a Cybus agent software collected the time-based series data in the machine net on the shop floor. The main Connectware instance then transported the data to the Data Lakehouse System shepard.

4.3 Data Lakehouse

The open-source software "shepard" [14] (see Fig. 3) is used as Data Lakehouse. Shepard consists of two different parts. The shepard backend is the heart of the system and manages data, metadata and users. While payloads are stored in specialized databases, internal objects are saved in Neo4j. The shepard frontend is provided as a web interface. Communication with the backend works exclusively via the documented REST API.

Additionally, the shepard Timeseries Collector (sTC) is responsible for data aggregation. If a lot of data is transferred individually to shepard, this overloads the system because the overhead of REST is too large compared to the user data. The sTC mitigates this problem by transferring the data to shepard as a batch. The sTC configuration is saved as YAML files. As such, the configuration of sTC can be included in a CI/CD pipeline in GitLab. Connectware prepares the data so that it can be accessed via sTC.

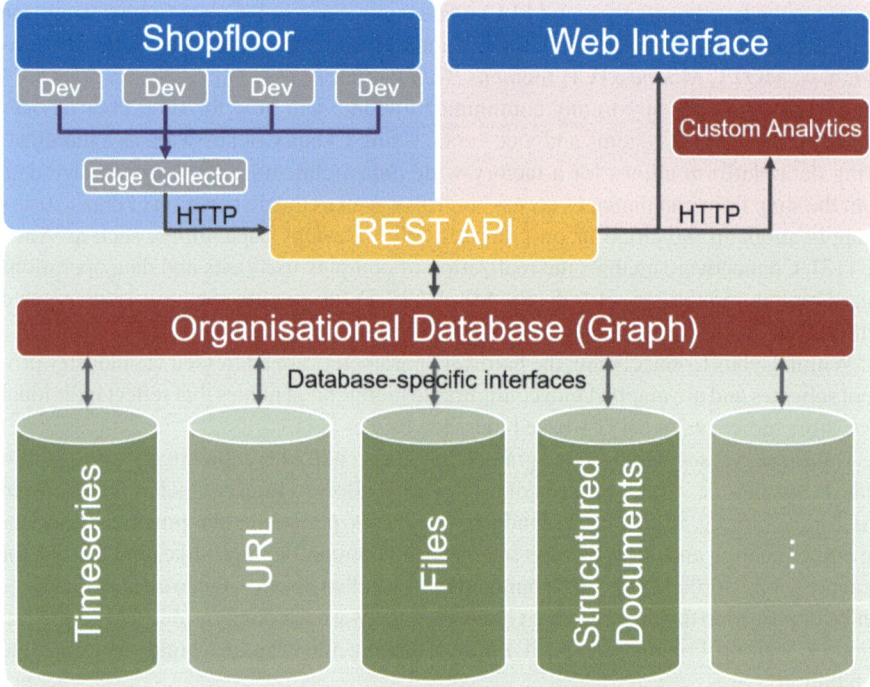

Fig. 3. Architectural overview of shepard, a multi-database storage system for highly heterogenous research data. Shopfloor data is collected and stored in specialized databases and can be viewed via web interface.

The data model consists of entities and references. These objects are used to create connections between payloads. In this way, metadata is stored together with payload data so that both can be easily found and used. Entities consists of Collections and Data Objects. Collections contain Data Objects. Each Data Object belongs to exactly one Collection. Data Objects can be related either as parent/child or as predecessor/successor.

While each Data Object can have at most one parent, multiple children, predecessors, and successors are allowed.

Every Data Object can contain an unlimited number of references. References are pointers to a specific data set. These data sets may or may not be stored in special databases. Data objects can contain the references to other files, timeseries, collections, data objects and URIs.

Containers allow users to store data and exist for different types of data, namely timeseries, structured data and files. Depending on the kind of data to be stored, a container can contain several data sets. Containers are seen as an enclosed box that contains data related to each other. References can point to one or more datasets within a single container.

4.4 Context Provider

The time-based series data collected from Connectware are stored in shepard in the corresponding containers. The context provider is responsible for combining all the different data in the various containers. It implements the data model designed by the engineer in shepard. This also includes all meta information.

As the context provider is a highly specialized application, it must be created explicitly for each process. For this paper, the context providers are created as websites. These web pages request all required information from the user and then create the collections, data objects and references in shepard. In addition, the website also offers the option of transferring manually created data to Shepard, e.g. a photo taken with a smartphone.

The respective context providers, in conjunction with the Data Lake, also configure the access rights of the individual users and groups to the data around privacy (ELSI). Using this method, data was made available to the respective stakeholders, such as data analysts and system engineers. The so-created human-assisted data loop offers the potential to make informed decisions about the collected data and its context and allows to perform modifications to the manufacturing and data collection processes. The latter is enabled by the availability of the context providers on GitLab, facilitating automated updates via the implemented CI/CD pipeline.

5 Conclusion

This paper presents a comprehensive approach for integrating a human-assisted data loop within connected manufacturing systems, emphasizing the fusion of human expertise and automated analysis to enhance the efficiency and adaptability of modern production environments. The outlined concept demonstrates how end-to-end connectivity can be achieved through the integration of machines, environmental sensors, and external data sources, managed by a central Data Lakehouse and context provider. By leveraging a human-assisted feedback loop, the system ensures that expert knowledge is incorporated into the data analysis process, leading to optimized shop floor operations, improved intralogistics, and enhanced IT infrastructure.

The implementation of the data loop in a lightweight construction process showcases the practical benefits of this approach. Preliminary results indicate significant potential for improving production processes through continuous data analysis and real-time anomaly detection. The use of Connectware and shepard for data aggregation and management highlights the importance of flexible, secure, and scalable data handling solutions in connected manufacturing systems.

Future research directions should focus on further refining the integration of human feedback within the construction process to enhance the adaptability and robustness of the data loop. This includes developing advanced machine learning models that can better interpret contextual information and anomalies, and improving the interfaces and tools that facilitate human interaction with the data. Additionally, exploring the scalability of this approach across different manufacturing domains and more complex production environments will be crucial in realizing the full potential of human-assisted data loops in Industry 5.0. Addressing challenges related to data privacy, security, and ethical considerations will also be vital to ensure the responsible deployment of these technologies in the manufacturing sector.

Acknowledgement. This contribution was funded by the Federal Ministry of Education and Research (BMBF) under grant 02P23Q843. The authors would like to thank all partners and coordinators of the Arena2036 research campus that shaped the project's objectives and are involved in its completion.

References

1. Weiß, M., Dettinger, F., Jazdi, N., Weyrich, M.: DevOps als enabler der kontinuierlichen funktionsverbesserung und automatisierten update-analyse in software-definierten systemen. In VDI Verlag eBooks, pp. 487–500 (2023). https://doi.org/10.51202/9783181024195-487
2. Lück, M., Hornung, T., Teklezgi, J.: Der mensch in der intelligenten fabrik. zeitschrift für wirtschaftlichen fabrikbetrieb/ZWF. Zeitschrift Für Wirtschaftlichen Fabrikbetrieb **119**(6), 456–459 (2024b). https://doi.org/10.1515/zwf-2024-1064
3. Weiß, M., Müller, M., Dettinger, F., Jazdi, N., Weyrich, M.: Continuous Analysis and Optimization of Vehicle Software Updates using the Intelligent Digital Twin (2023). https://doi.org/10.1109/etfa54631.2023.10275489
4. Weiß, M., Stümpfle, J., Dettinger, F., Jazdi, N., Weyrich, M.: Simulating Cloud Environments of Connected Vehicles for Anomaly Detection. arXiv (Cornell University) (2024). https://doi.org/10.48550/arxiv.2404.11740
5. Alves, J., Lima, T.M., Gaspar, P.D.: Is Industry 5.0 a human-centred approach? A systematic review. Processes **11**(1), 193 (2023). https://doi.org/10.3390/pr11010193
6. Nelles, J., Kuz, S., Mertens, A., Schlick, C.M.: Human-centered design of assistance systems for production planning and control: The role of the human in Industry 4.0 (2016). https://doi.org/10.1109/icit.2016.7475093
7. Çınar, Z.M., Nuhu, A.A., Zeeshan, Q., Korhan, O., Asmael, M., Safaei, B.: Machine learning in predictive maintenance towards sustainable smart manufacturing in industry 4.0. Sustainability **12**(19), 8211 (2020). https://doi.org/10.3390/su12198211
8. Leng, J., et al.: Industry 5.0: prospect and retrospect. J. Manuf. Syst. **65**, 279–295 (2022). https://doi.org/10.1016/j.jmsy.2022.09.017
9. Usuga Cadavid, J.P., Lamouri, S., Grabot, B., Pellerin, R., Fortin, A.: Machine learning applied in production planning and control: a state-of-the-art in the era of industry 4.0. J. Intell. Manuf. **31**(6), 1531–1558 (2020). https://doi.org/10.1007/s10845-019-01531-7

10. Kaasinen, E., Anttila, A., Heikkilä, P., Laarni, J., Koskinen, H., Väätänen, A.: Smooth and resilient Human–Machine Teamwork as an Industry 5.0 design challenge. Sustainability **14**(5), 2773 (2022b). https://doi.org/10.3390/su14052773
11. Cybus. Overview (2024). https://docs.cybus.io/latest/user/overview.html. Accessed 26 June 2024
12. Cybus. Connectivity Portfolio (2024). https://www.cybus.io/connectivity-portfolio/. Accessed 26 June 2024
13. Erichsen, J.: Connectware Orchestration Using Ansible (2024). https://www.cybus.io/learn/connectware-orchestration-using-ansible/. Accessed 26 June 2024
14. Haase, T., Glück, R., Kaufmann, P., Willmeroth, M.: Shepard - Storage for Heterogeneous Product and Research Data (2021). https://doi.org/10.5281/ZENODO.5091604, https://zenodo.org/record/5091604

Open Access This chapter is licensed under the terms of the Creative Commons Attribution 4.0 International License (http://creativecommons.org/licenses/by/4.0/), which permits use, sharing, adaptation, distribution and reproduction in any medium or format, as long as you give appropriate credit to the original author(s) and the source, provide a link to the Creative Commons license and indicate if changes were made.

The images or other third party material in this chapter are included in the chapter's Creative Commons license, unless indicated otherwise in a credit line to the material. If material is not included in the chapter's Creative Commons license and your intended use is not permitted by statutory regulation or exceeds the permitted use, you will need to obtain permission directly from the copyright holder.

Cycle Time Measurement Using AI-Based Object Detection and Tracking in Industrial Processes

Tim Staudenrausch and Bernd Lüdemann-Ravit[✉]

University of Applied Sciences Kempten, Institute for Production and Informatics, Mittagstraße 28A, 87527 Sonthofen, Germany
{tim.staudenrausch,bernd.luedemann-ravit}@hs-kempten.de

Abstract. This paper presents an AI-based system for improving cycle time measurement in industrial environments, leveraging YOLOv8 for object detection and ByteTrack for tracking. Our non-invasive approach analyzes video from an Azure Kinect camera to calculate cycle times by detecting objects and monitoring their state changes. Tested at the University of Applied Sciences Kempten's demo plant, the system showcased high accuracy against ground truth data, highlighting its potential to enhance production line monitoring and efficiency significantly. This work contributes to industrial automation by offering a real-time, accurate method for cycle time analysis, promising substantial advancements in manufacturing process optimization.

Keywords: Industrial automation · object tracking · cycle time measurement · computer vision · artificial intelligence · process monitoring

1 Introduction

The measurement of cycle times in industrial settings is a cornerstone for optimizing efficiency, reducing costs, and enhancing the overall productivity of manufacturing processes. Traditionally, this task has relied on manual measurements using stopwatches, manual video, video evaluation, or simple automated systems based on traditional algorithms, which often fall short in accuracy and flexibility.

This paper introduces an advanced system that leverages artificial intelligence (AI) for object detection and tracking, specifically utilizing state-of-the-art technologies like YOLOv8 [1] and ByteTrack [2], to measure cycle times of objects' processes in industrial environments. By utilizing sophisticated AI algorithms in conjunction with camera technology, we propose a non-invasive visual monitoring solution that measures cycle times with high accuracy and offers robustness against the dynamic and often unpredictable nature of industrial settings.

The significance of this system lies in its potential to improve how industries monitor and optimize their production lines, thereby contributing to the broader field of industrial automation and efficiency. By overcoming the limitations of traditional methods, our AI-based approach provides a flexible, scalable, and precise tool for real-time cycle time measurement, essential for informed decision-making and process optimization in modern manufacturing.

2 State of the Art

Cycle time measurement is critical in industrial production for optimizing efficiency and productivity. Traditional methods for cycle time measurement include manual timing with stopwatches, use of light barriers, photocells, and mechanical counters. These methods, while effective in controlled environments, often lack flexibility and require physical alterations to the production line, which can be invasive and disruptive.

Light barriers and photocells are commonly used for detecting the presence or absence of objects at specific points in the production line. These sensors can provide precise timing data but are limited to fixed locations and specific types of movements. They require installation and maintenance, and their accuracy can be affected by environmental factors such as dust, vibration, and lighting conditions [3].

In recent years, computer vision technologies have been explored for cycle time measurement and monitoring in various domains. Techniques such as object detection and activity recognition, using state-of-the-art algorithms like YOLOv8, have enabled real-time monitoring of construction progress by identifying and tracking equipment on construction sites [4]. In manufacturing, people tracking systems have been implemented to monitor worker movements and optimize workflows, with some systems using deep learning techniques to overcome challenges like occlusions and dynamic backgrounds [5].

Advancements in AI-based object detection and tracking have opened new possibilities for non-invasive and flexible cycle time measurement. Convolutional Neural Networks (CNNs) and deep learning methods have been applied to industrial applications for object detection and classification. One notable work is by Wozniak et al. [6], which focuses on the robustness assessment of AI-based 2D object detection systems in industrial environments. Their method includes guidelines for evaluating the robustness of object detection models, addressing practical challenges encountered during the integration of these models into industrial systems.

Another significant study is by Buongiorno et al. [7], which compares different training strategies for AI-based depalletization systems in industrial applications. They found that deep learning methods, particularly convolutional neural networks (CNNs), significantly improve the detection and handling of various objects under unstructured scenarios. This study emphasizes the importance of training data quality and volume, and explores methods to optimize the model's performance with minimal data.

Furthermore, Choi et al. [8] present a real-time object detection and tracking system designed for local dynamic map generation, using a modified YOLOv4 and a DeepSORT-based tracker. This system, implemented on a Qualcomm SoC, highlights the feasibility of deploying AI-based object detection and tracking on embedded edge devices, showcasing the adaptability of such systems in various industrial contexts.

These studies highlight the increasing focus and active research efforts in applying AI-driven object detection and tracking technologies to enhance industrial processes. Building on these advancements, our work aims to provide a robust, real-time cycle time measurement system, utilizing YOLOv8 and Byte-Track, for use in industrial production environments.

2.1 Comparison with Traditional Methods

In the initial stages of this study, traditional methods for cycle time measurement were explored, utilizing standard OpenCV implemented trackers such as the Boosting, KCF, and MOSSE trackers. These trackers rely on an initial bounding box manually placed around the object to begin tracking, distinguishing them as object trackers rather than detectors. To identify moving parts, traditional methods such as frame differencing and background subtraction were explored. Although effective at detecting movement, these techniques indiscriminately identified all changes in the frame, necessitating the use of thresholds, such as bounding box size, to filter out irrelevant motion.

Despite their real-time performance, the results obtained with traditional tracking methods were not satisfactory for applications in industrial settings. While these methods can track objects under stable conditions, their performance deteriorates in dynamic environments with rapid changes, such as those found in industrial plants. Tracking often fails when the visual characteristics within the bounding box change abruptly due to factors like the sudden appearance of other objects or shifts in lighting conditions. These limitations were evident during our tests on the demo plant.

The inconsistency and lack of robustness in traditional tracking technologies led to the exploration of an AI-based approach, as detailed in this paper. The AI method's superior ability to handle dynamic scene influences and maintain consistent tracking accuracy under various operational conditions represents a significant improvement over the conventional techniques previously employed.

Our system's use of advanced AI models allows for more accurate and reliable cycle time measurements, addressing the shortcomings of traditional methods. The non-invasive implementation, coupled with the ability to monitor multiple objects simultaneously, provides a significant advantage in industrial applications where flexibility and scalability are essential.

3 Methodology and Implementation

3.1 System Overview

Our system offers a modern approach to measuring cycle times in industrial settings, utilizing advancements in artificial intelligence and camera technology. At its foundation, the system utilizes a Microsoft Azure Kinect camera, selected for its high-quality color camera capabilities, to capture detailed 2D images and videos of the manufacturing process. While the Azure Kinect camera serves as an exemplary model due to its high fidelity imaging, the architecture of our system is designed to be compatible with a broad range of camera technologies in mind.

The object detection was done using a YOLOv8 model fine tuned on a custom dataset comprised of annotated images from an exemplary demo industrial plant at the University of Applied Sciences Kempten (see Fig. 1). This demo plant, albeit small, provided a comprehensive array of moving parts typical of industrial settings, serving as an ideal training ground for the model. The training process allowed the YOLOv8 model to achieve high precision in detecting objects relevant to the cycle time measurement task.

Fig. 1. Industrial demo plant at the Institute for Production and Informatics, University of Applied Sciences Kempten, used for our experiments in testing our cycle time measurement system.

Video feeds from the cameras are used by our system, with the frames being directly fed into the AI processing unit. Here, the YOLOv8n model performs real-time object detection, identifying and classifying objects as they appear in the camera's field of view. Subsequently, the detected objects are tracked through their movement across the plant using the ByteTrack multiple object tracking algorithm. The tracking data of each object is then automatically processed by our system to determine cycle times based on changes in the object's moving state. Utilizing the 2D trajectory data from the object tracker, we calculate the object's speed for each frame to determine changes from standstill to moving state, thus indicating a change in process.

Our system's architecture is designed for flexibility and adaptability, initially demonstrated using a single-camera setup. This foundational configuration, centered around an Azure Kinect camera, serves to capture detailed images and videos from a strategic vantage point within the manufacturing process. A prospective multi-camera setup is intended to extend coverage across various angles and sections of the manufacturing process, aiming for a more comprehensive monitoring capability. Although the practical application of a multi-camera system remains untested and is earmarked for future exploration, the potential for such an expansion underscores our system's capability to evolve and adapt to more complex industrial environments, thereby enhancing its utility in providing precise cycle time measurements across different production stages.

3.2 AI Implementation

The implementation of AI technologies within our system is centered around two key components: YOLOv8n for object detection and ByteTrack for object tracking. These technologies were selected for their state-of-the-art performance in real-time processing and their ability to accurately identify and follow objects through complex industrial environments.

Central to our system's functionality is the AI object detection. Our system utilizes the YOLOv8n (nano) model for object detection, a variant of the YOLOv8 family optimized for efficiency and speed without significantly compromising detection accuracy. Pre-trained YOLOv8 models are available in various sizes, each offering a trade-off between speed, accuracy, and computational resource requirements [9]. The nano model is designed for applications requiring high speed and low computational overhead, while the larger models offer progressively higher accuracy at the cost of increased processing time and computational resources (see Fig. 2 for performance graphs).

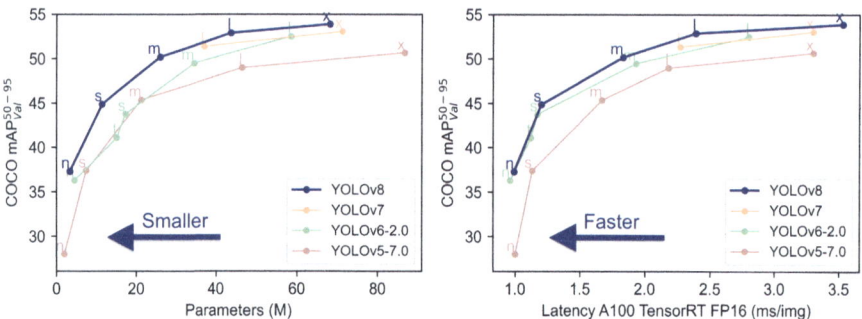

Fig. 2. Comparison of different YOLO models based on the Mean Average Precision (mAP) metric on the COCO dataset, relative to parameter count and latency [9].

For our system, YOLOv8n was configured to identify some of the moving components within the exemplary demo industrial plant at the University of

Applied Sciences Kempten. For this we fine-tuned the YOLOv8n model using a specialized dataset of annotated images from the demo plant, ensuring high precision in object detection tailored to the specific nuances of our application environment.

ByteTrack complements YOLOv8n's detection capabilities by offering robust tracking of the detected objects across video frames. This tracking algorithm is designed to maintain the identity of objects even in scenarios of partial occlusion or when objects move out of the camera's view temporarily. ByteTrack's integration into our system ensures that once an object is detected by our custom YOLOv8n model, its movement can be tracked continuously throughout the production process.

In the integrated system, video frames from the Azure Kinect camera are first processed by our fine-tuned YOLOv8n model for object detection, which generates bounding boxes around each detected object. These bounding boxes, which outline the object's location and size in the 2D image, are then fed into the ByteTrack algorithm. ByteTrack assigns unique identifiers to each object and tracks them across frames. This streamlined process combines the precision of YOLOv8n's detection with ByteTrack's effective tracking, enabling the system's process analysis and cycle time measurement.

To enable real-time processing demands relevant in the industrial environment, both YOLOv8n and ByteTrack utilized the GPU processing power of a Nvidia RTX 3070 graphics card. Additionally the camera's images were downscaled to 640 × 480 pixels resolution for all processing steps. This setup enabled rapid processing, ensuring that object detection and tracking could be conducted in real time.

3.3 Training and Data Processing

The training and data processing phase of our system's development was critical in ensuring the AI models, particularly YOLOv8n, performed optimally under the specific conditions of our industrial environment. This phase was characterized by two main processes: the creation of a tailored dataset through annotation and the strategic augmentation of this dataset to enhance the model's robustness and accuracy.

Initially, we compiled a concise yet representative dataset consisting of approximately 50 images captured from Azure Kinect video footage of the demo industrial plant at the University of Applied Sciences Kempten. These images were selected to cover a wide range of scenarios and important object types that the system would need to detect in a real-world industrial setting, in this case the moving parts of the plant. To annotate these images, we used Roboflow [10], which streamlined the process of creating the dataset. With Roboflow's annotation tools, we manually drew bounding boxes around each object of interest. This meticulous process ensured high-quality and precise annotations, accurately defining the target objects for detection.

Given the relatively small size of our initial dataset, image augmentation became a pivotal strategy to artificially expand our data pool. Through augmentation techniques such as rotation, scaling, cropping, and lighting adjustments, we increased the diversity of our dataset without the need for additional real-world images. This process resulted in a final dataset size of approximately 150–200 images of 640 × 480 pixels resolution, significantly enhancing the variety of training examples available for the model. Augmentation is widely acknowledged as an effective method to combat overfitting, particularly in scenarios where limited data is available. By simulating various real-world conditions, augmentation enables the model to learn a broader range of features, ultimately improving performance on unseen data [11,12].

This approach is especially valid in industrial applications, where collecting a large and diverse dataset can be difficult or costly. Studies have demonstrated that augmentation can effectively generate a more robust dataset by introducing variations that mimic real-world conditions, such as lighting changes or different perspectives [11,13]. Augmentation has been shown to be effective across various fields, including industrial image recognition, by enhancing the model's ability to generalize from limited training data [12].

The augmented dataset was then used to fine-tune the YOLOv8n model. The training process involved experimenting with different configurations for the number of epochs, batch and image sizes. Through the careful annotation of images, strategic dataset augmentation, and meticulous model training, our system was equipped with a highly capable AI model, poised to transform the efficiency and accuracy of cycle time measurement in industrial settings.

3.4 Automatic Process Analysis Based on Object Tracking Data

The core of our system's capability to measure cycle times revolves around the automated analysis of object tracking data generated by the ByteTrack object tracker. Our approach leverages 2D trajectory data extracted from the object tracker, which records the object's position across frames alongside corresponding timestamps. This data is then used to determine cycle times by identifying transitions in the object's state of motion, specifically changes from a standstill to a moving state, and vice versa. These transitions are indicative of changes in the monitored process.

The initial step involves extracting relevant tracking data of the objects, including their bounding box center pixel coordinates (x, y), timestamps (t), and frame indices. To mitigate the effects of noise in object detection and tracking, and to ensure the accuracy of subsequent calculations, we apply a smoothing technique to the x and y coordinates. This smoothing is achieved through a moving average, utilizing a rolling window of a specified size (N), which averages the points within the window to produce a smoother trajectory.

$$x'_t = \frac{1}{N} \sum_{i=t-(N-1)/2}^{t+(N-1)/2} x_i \tag{1}$$

$$y'_t = \frac{1}{N} \sum_{i=t-(N-1)/2}^{t+(N-1)/2} y_i \tag{2}$$

where:

- x'_t and y'_t represent the smoothed values of x and y at time t, respectively.
- N is the smoothing window size.
- x_i and y_i represent the original values of x and y at time i, respectively.

Following smoothing, we calculate the object's speed by first determining the distance between successive center points and the time elapsed between these points. Speed is then calculated as the distance divided by the time interval. Given that raw speed calculations may still exhibit fluctuations, we apply the same rolling smoothing technique to these speed values to achieve a more consistent speed profile.

$$d_t = \sqrt{(x'_{t+1} - x'_t)^2 + (y'_{t+1} - y'_t)^2} \tag{3}$$

$$\Delta t = t_{t+1} - t_t \tag{4}$$

$$v_t = \frac{d_t}{\Delta t} \tag{5}$$

$$v'_t = \frac{1}{N} \sum_{i=t-(N-1)/2}^{t+(N-1)/2} v_i \tag{6}$$

where:

- d_t is the distance between the center points at time t.
- x'_t and y'_t are the smoothed coordinates at time t.
- Δt is the time difference between consecutive timestamps.
- v_t is the speed of the object at time t.
- N is the smoothing window size.
- v'_t is the smoothed speed at time t.

To detect changes from standstill to motion, we examine the smoothed speed data for transitions where the speed crosses a predefined threshold v_{thresh} (e.g., $v_{\text{thresh}} = 15$). This threshold is critical for distinguishing between motion and immobility. Transitions are captured when the speed exceeds the threshold, indicating motion, or drops below the threshold, indicating a return to immobility. The exact moments of these transitions are captured, and their corresponding frame indices are identified using the following condition:

$$\text{transition} \quad \text{if} \quad v'_t > v_{\text{thresh}} \quad \text{or} \quad v'_t < v_{\text{thresh}} \tag{7}$$

Mathematically, this can be expressed as:

$$t_{\text{trans}} = \{t \mid \Delta(\text{v}'_t > v_{\text{thresh}}) \neq 0\} \tag{8}$$

where:

- t_{trans} represents the times of transitions.
- $\Delta(\text{v}'_t > v_{\text{thresh}})$ represents the change in the condition where the speed exceeds the threshold.

However, not all detected transitions are significant; some may represent minor fluctuations rather than true changes in process state. To address this, we implement a filtering step that excludes transitions occurring too close to each other, based on a minimum duration criterion Δt_{min}. This ensures that only meaningful transitions, indicative of process changes, are considered in the cycle time analysis.

$$\Delta t_{\text{trans}} > \Delta t_{\text{min}} \tag{9}$$

where:

- Δt_{trans} is the time difference between consecutive transitions.
- Δt_{min} is the minimum duration criterion.

With the significant transitions identified, we proceed to calculate the duration between these transitions, representing the cycle times of interest. Additionally, we compute the average speed between transitions to determine the object's state of motion. If the average speed within a cycle exceeds the motion threshold v_{thresh}, the object is considered to be moving during that cycle.

$$\overline{v}_i = \frac{1}{t_{i+1} - t_i} \sum_{t=t_i}^{t_{i+1}} v_t \tag{10}$$

$$\text{moving if } \overline{v}_i > v_{\text{thresh}} \tag{11}$$

where:

- t_i and t_{i+1} are the times of the significant transitions.
- $\overline{v}_i$ is the average speed during the i-th cycle.
- v_t is the speed at time t.
- v_{thresh} is the motion threshold.

3.5 Multi-Camera-Setup

While the initial implementation of our system utilized a single-camera setup to validate the concept and efficiency of AI-based cycle time measurement, a multi-camera approach is planned for future development to further enhance the system's capabilities. This envisioned setup aims to extend the system's tracking capabilities across a broader area, covering multiple viewpoints within an industrial environment to ensure no object goes untracked, even in complex manufacturing landscapes.

The multi-camera setup involves strategically positioning cameras to cover various segments of the manufacturing process, ensuring comprehensive visibility. This arrangement would require a meticulous planning phase to determine the optimal locations and angles for each camera, ensuring overlap areas are minimized and the entire area of interest is covered effectively.

Synchronization of multiple cameras is paramount to achieve seamless object tracking across different viewports. This involves two primary challenges: temporal synchronization and spatial alignment. Temporal synchronization ensures that the video feeds from all cameras are aligned in time, enabling the system to accurately track objects as they move from one camera's view to another. This could be achieved through network time protocol (NTP) servers or by embedding synchronization signals directly into the video streams.

Spatial alignment, on the other hand, involves calibrating the cameras in such a way that the system can understand the spatial relationship between different camera views. This requires sophisticated calibration techniques to establish a cohesive 3D model of the environment, allowing the system to accurately map the position of an object from one camera's coordinate system to another. Advanced algorithms can then be employed to stitch together these different views, creating a unified tracking environment that accurately represents the object's trajectory through the entire manufacturing process.

The integration of multiple cameras into the system would necessitate the development of robust algorithms capable of identifying and tracking an object across different camera feeds. This includes the challenge of re-identifying objects that may have been temporarily occluded or have moved out of one camera's view and into another's. Techniques such as feature matching and machine learning models trained on the specific geometry of the environment could be utilized to address this challenge.

While the practical application of a multi-camera setup remains a future goal, the theoretical framework laid out here provides a solid foundation for its development. Implementing such a system would significantly enhance the capability of industrial environments to monitor and measure cycle times more comprehensively, providing a more detailed insight into the efficiency and productivity of the manufacturing process.

4 Results

The results of our system's implementation are based on a prototype using a single Microsoft Azure Kinect camera at 30 FPS pointed at the demo industrial plant at the University of Applied Sciences Kempten. The setup of the camera in the evaluation phase matches the one of our training dataset for the fine-tuning of the Yolov8n object detection model. Ground truth data for the duration of each object's processes was created manually by analyzing the video footage for the start and end frame index of each process. A process is defined as a change of an object's state from standstill to moving, and vice versa. Our system calculates this automatically based on object tracking data, as described in Sect. 3.4.

The analysis highlighted in Table 1 demonstrates that our system closely approximates ground truth values with a high degree of accuracy. The differences between the system-calculated durations and the manually observed ground truth durations are minimal, suggesting that our approach is effective in accurately detecting the start and stop points of object movements within a real-world industrial production environment.

Table 1. Calculated processes for object 3 using the proposed method in this paper compared to manually determined ground truth values. See Fig. 5 for corresponding camera images and Fig. 3 for graphs. Idx stands for index. Gt stands for ground truth.

process id	moving	start_idx	end_idx	duration in s	start_idx (gt)	end_idx (gt)	duration in s (gt)
1	False	1	287	9.80	1	284	9.43
2	True	287	327	1.20	284	321	1.23
3	False	327	373	1.63	321	370	1.30
4	True	373	423	1.53	370	420	2.00
5	False	423	463	1.47	420	457	1.23
6	True	463	539	2.20	457	536	2.63
7	False	539	814	9.63	536	810	9.13
8	True	814	836	0.37	810	844	1.13

4.1 System Performance

The average error margin between the system-calculated cycle times and the ground truth was consistently under 5%. This margin is considered acceptable in many industrial applications, where minor variances are expected due to inherent process fluctuations, such as machine performance deviations, material inconsistencies, or environmental factors [14].

However, the acceptability of this margin depends on the specific requirements of the application. In high-precision sectors, such as semiconductor manufacturing, stricter tolerances may be required to maintain product quality [15]. In contrast, in industries like general manufacturing or bulk production, a 5% margin is often more than sufficient, given the acceptable levels of variability within these processes [14]. Therefore, for many real-world industrial applications, a 5% error margin strikes a suitable balance between accuracy and robustness, ensuring reliable performance under typical operating conditions.

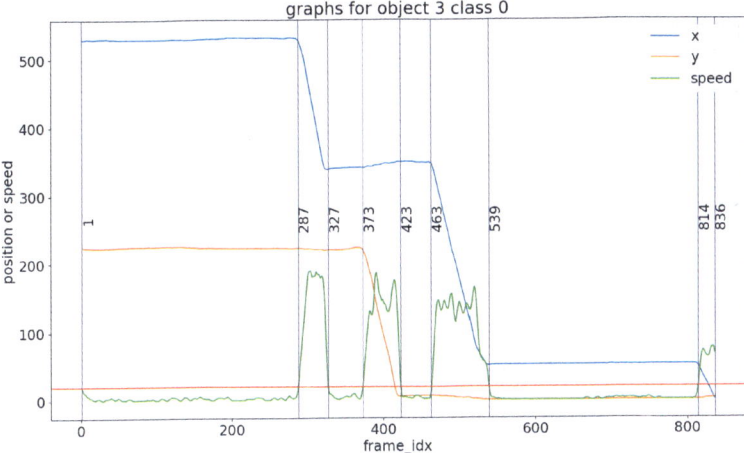

Fig. 3. Graph describing the tracked X and Y pixel values of object 3 using the proposed method in this paper. The calculated speed of the object was used in conjunction with a speed threshold to determine process start and end, indicated by the vertical blue lines.

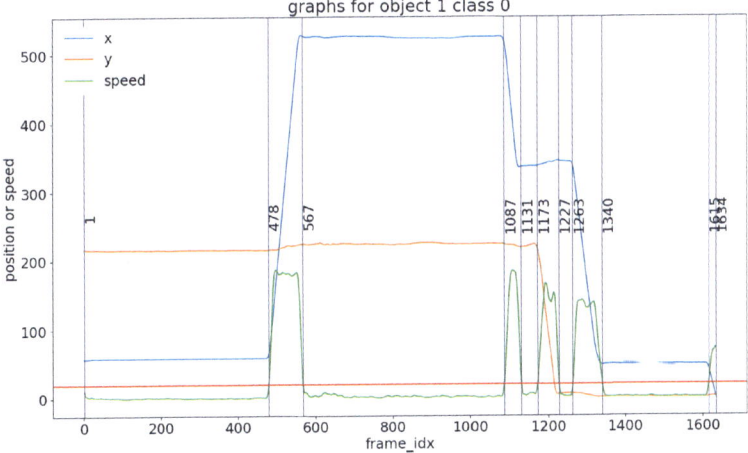

Fig. 4. Graph describing the detected X and Y pixel values of object 1 using the proposed method in this paper. The calculated speed of the object was used in conjunction with a speed threshold to determine process start and end, indicated by the vertical blue lines.

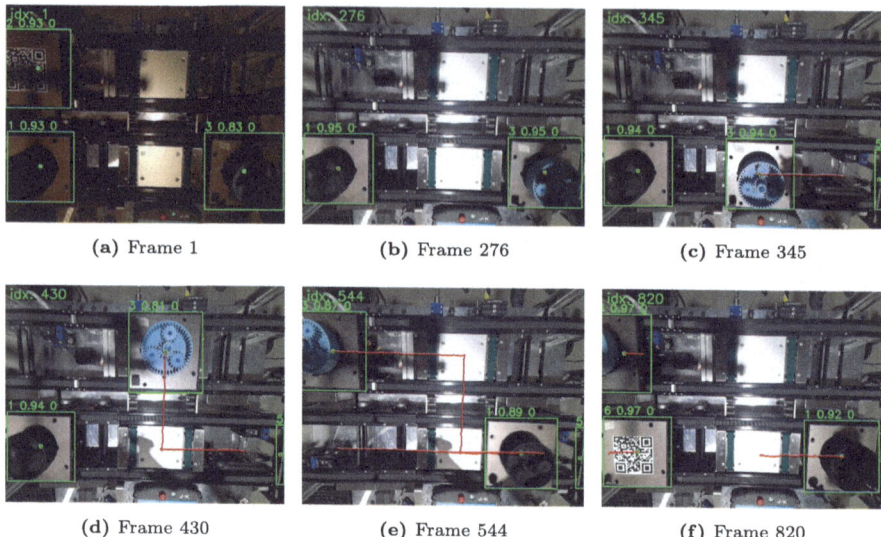

Fig. 5. Tracking of objects on industrial demo plant. Each tracked object's detection box is labeled with a tracking/object id, confidence value and object class. The red line shows the trajectory. See result graphs for object 1 (see Fig. 4) and object 3 (see Fig. 3) for comparison.

4.2 Error Analysis

The sources of errors in our system can be attributed to several factors:

1. **Latency in Detection and Tracking:** The system's frame rate (30 FPS) and processing time introduce delays, which can cause slight shifts in the identified start and end points of object movements.
2. **Object Localization Inaccuracies:** While YOLOv8 and ByteTrack provide high accuracy, minor inaccuracies in object localization can affect speed calculations, potentially leading to slight deviations in detecting state changes.
3. **Threshold Sensitivity:** The speed threshold $v_{\text{thresh}} = 15$ used for detecting transitions between movement and standstill influences when state changes are registered. Altering this threshold can lead to earlier or later detections, which may affect the output.
4. **Smoothing Window Size:** The smoothing window N impacts the system's responsiveness to speed changes. A larger window smooths data more effectively but may delay transition detection, while a smaller window may introduce noise into the speed data.

These errors are systematic and can be minimized through calibration and optimization of the system parameters. Compared to traditional methods, which may have similar or larger error margins due to mechanical limitations and environmental factors, our AI-based approach offers improved accuracy and robustness. The non-invasive nature of the system eliminates errors introduced by physical sensor installations, and the adaptability of the AI models allows for continuous improvement through retraining with additional data.

5 Conclusions

This paper presents an AI-based system for cycle time measurement that significantly advances the capabilities of traditional methods. By integrating YOLOv8 for object detection and ByteTrack for object tracking, our system offers a robust, accurate, and non-invasive solution for monitoring and optimizing industrial production processes.

The system demonstrated high accuracy in cycle time measurement, closely aligning with manually annotated ground truth data. Its robustness was evident in our tests with the demo plant at the University of Applied Sciences Kempten, where traditional methods would fail. This AI-driven approach provides real-time, reliable data, enhancing operational efficiency and reducing the need for manual monitoring.

In an industrial context, the system could improve production scheduling, reduce downtime, and increase overall productivity. Its non-invasive nature ensures minimal disruption to existing workflows, making it a flexible and efficient tool for a wide range of manufacturing applications.

Our AI-based system represents an important step forward in industrial automation. Future research could further enhance its capabilities through multi-camera integration and expanded AI model training, driving continued innovation in intelligent manufacturing processes.

References

1. Jocher, G., Chaurasia, A., Qiu, J.: Ultralytics YOLOv8, Version 8.0.0, License: AGPL-3.0 (2023). https://github.com/ultralytics/ultralytics
2. Zhang, Y., et al.: Bytetrack: multi-object tracking by associating every detection box. In: European Conference on Computer Vision, pp. 1–21. Springer (2022)
3. Chunjiao, Z., Du, Z.: The application and development of photoelectric sensors. In: Intelligence Computation and Evolutionary Computation, vol. 180, pp. 671-677. Springer, Berlin, Heidelberg (2020). https://doi.org/10.1007/978-3-642-31656-2_91
4. Yang, J., Wilde, A., Menzel, K., Sheikh, M.Z., Kuznetsov, B.: Computer vision for construction progress monitoring: a real-time object detection approach. arXiv preprint arXiv:2305.15097 (2023)
5. Jiang, F., Liu, L., Xie, H., Wang, Z.: Computer vision-based deep learning for supervising excavator operations and measuring real-time earthwork productivity. J. Supercomput. **78**, 11601–11620 (2022). https://doi.org/10.1007/s11227-022-04803-x

6. Wozniak, A.-L., Segura, S., Mazo, R.: Robustness assessment of AI-based 2D object detection systems: a method and lessons learned from two industrial cases. Electronics **13**(7), 1368 (2024)
7. Buongiorno, D., et al.: Object detection for industrial applications: training strategies for AI-based Depalletizer. Appl. Sci. **12**(22), 11581 (2022)
8. Choi, K., Moon, J., Jung, H.G., Suhr, J.K.: Real-time object detection and tracking based on embedded edge devices for local dynamic map generation. Electronics **13**(5), 811 (2024). https://doi.org/10.3390/electronics13050811
9. Ultralytics: YOLOv8 model documentation (2023). https://docs.ultralytics.com/de/models/yolov8/#supported-tasks-and-modes
10. Roboflow: Roboflow Annotate (2023). https://roboflow.com/annotate
11. Zhou, P., Jin, F., Wang, X., Huang, H.: Review of image augmentation used in deep learning-based material microscopic image segmentation. Appl. Sci. **13**(11), 6478 (2023). https://doi.org/10.3390/app13116478
12. Nanni, L., Paci, M., Brahnam, S., Lumini, A.: Comparison of different image data augmentation approaches. J. Imag. **7**(12), 254 (2021). https://doi.org/10.3390/jimaging7120254
13. Shorten, C., Khoshgoftaar, T.M.: A survey on image data augmentation for deep learning. J. Big Data **6**, 60 (2019). https://doi.org/10.1186/s40537-019-0197-0
14. Cao, H., Ji, X.: Prediction of garment production cycle time based on a neural network. Fibres Text. East. Eur. **29**(1), 8–12 (2021). http://fibtex.lodz.pl/article2263.html
15. Zhang, Y., et al.: iRazor: current-based error detection and correction scheme for PVT variation in 40-nm ARM Cortex-R4 processor. IEEE J. Solid-State Cir. **53**(2), 619–631 (2018). https://doi.org/10.1109/JSSC.2017.2749423

Open Access This chapter is licensed under the terms of the Creative Commons Attribution 4.0 International License (http://creativecommons.org/licenses/by/4.0/), which permits use, sharing, adaptation, distribution and reproduction in any medium or format, as long as you give appropriate credit to the original author(s) and the source, provide a link to the Creative Commons license and indicate if changes were made.

The images or other third party material in this chapter are included in the chapter's Creative Commons license, unless indicated otherwise in a credit line to the material. If material is not included in the chapter's Creative Commons license and your intended use is not permitted by statutory regulation or exceeds the permitted use, you will need to obtain permission directly from the copyright holder.

A Data-centric Evaluation of Leading Multi-class Object Detection Algorithms Using Synthetic Industrial Data

J. Moises Araya-Martinez[1(✉)], Sarvenaz Sardari[1], Mats Lambert[1], J. Alexander Zak[1], Florian Töper[1], Jörg Krüger[2], and Jens Lambrecht[3]

[1] Mercedes-Benz AG, Future Automotive Manufacturing, Benz-Straße Bau 40, Sindelfingen, Germany
jose_moises.araya_martinez@mercedes-benz.com
[2] TU Berlin, Industrial Automation Technology, Pascalstraße 8-9, Berlin, Germany
[3] TU Berlin, Industry Grade Networks and Clouds,, Straße des 17. Juni 135, Berlin, Germany

Abstract. Object detection is crucial in many industrial computer vision applications. However, the reliance on manually annotated data prevents a cost-effective deployment of supervised models in domain-specific industrial tasks.

This research work presents an evaluation of the real-world performance of leading algorithms in the context of industrial multi-class object detection while being trained solely on synthetic data. We train widely-used models on an order-of-magnitude less synthetic industrial data than the current state-of-the-art and demonstrate a mAP@50-95 of 75% under high-variability environments. Our work also offers an ablation study to narrow the sim-to-real domain gap based on context-aware identification of synthetic features that contribute the most to closing the sim-to-real gap. We show that by employing guided domain randomization based on low-level and on semantic contextual features, it is possible to reduce the amount of required synthetic images by a factor of three while affecting mAP@50-95 on real data by only 2%.

Keywords: Object detection · photorealistic rendering · sim-to-real domain gap · data-centric AI

1 Introduction

In recent years, the domain of machine learning has attracted considerable attention. Notable milestones have been reached within the field of deep learning [1,2] and particularly in computer vision with the introduction of AlexNet [3], U-Net [4], and Fast R-CNN [5], to mention a few examples. Subsequently, deep learning models have surpassed human-level performance across various domains, such as image recognition [6] and object detection [7]. Moreover, multiple evaluation metrics have emerged to quantify such advances, with Mean Average Precision (mAP) being one of the most widely-used in the context of multi-class object detection [8–10].

Inconveniently, the supervised nature of these models requires a large amount of annotated data during the training process [11]. Thus, multiple publicly available datasets of common objects have supported the development of supervised object detection algorithms [9,10]. Nonetheless, in industrial environments, data availability remains one of the main limiting factors preventing full-scale adoption of machine learning [12–14]. This annotation burden prompts efforts to reduce image acquisition and annotation demands through synthetic data generation [15]. Data frequently exist in the form of Computer-Aided Design (CAD) files in many industrial settings, presenting an opportunity to leverage annotation to automated synthetic data generation techniques [14,16,17]. Even though fully-automated synthetic data generation does not require human supervision, it still may be computationally expensive concerning valuable time and energy. Moreover, previous research shows that redundant features do not contribute significantly to a model's performance but potentially make the training process inefficient [18]. Thus, in order to optimize the training process, we select relevant images from a synthetic dataset based on domain-specific features prior to training our object-detection models by using both low-level and high-level features. We identify relevant features by using a set of contextualized real images to provide domain-specific information in the form of low-level and high-level features. This approach has the benefit of only requiring unlabeled images for contextualization.

Low-level features contain pixel information such as contrast, blur, amount of lighting, and color. High-level features contain more complex semantic information beyond just the pixel data. Semantic information in this context could be a set of objects suggesting a factory setting, a meeting room, or an open space. We rely on a simple image hashing [19] approach instead of learning-based methods to filter synthetic images by their semantic content. Low-level features are also extracted with common computer vision algorithms contained in the OpenCV [20] library. The matching process is done by clustering images with the highest similitude on features like lighting, blur, contrast, and color.

The scientific contributions of our data-driven object detection work can be summarized as follows:

- **Sim-to-real gap bridging on multi-class industrial object detection in uncontrolled scenarios.** We demonstrate that bridging the sim-to-real gap is possible even with one order of magnitude fewer synthetic images than the current state of the art in the context of multi-class object detection of industrial objects while achieving mAP@50-95 = 75%. Our data-centric experiments are demonstrated across multiple leading algorithms and a variety of scenarios, as shown in Fig. 1.
- **Automated image generation framework with domain randomization capabilities** We developed an automated synthetic data generation framework based on Blender [21], capable of rendering and annotating randomized synthetic scenes on specific variable ranges to help bridging the sim-to-real gap by domain randomization. We demonstrate promising results based on data generated with the real-time, non-physically accurate EEVEE [21] rendering engine, which requires about 30 times less rendering time than its physically-accurate counterpart, Cycles [22].

– **Ablation study of synthetic features for guided domain randomization.** Our extensive ablation study demonstrates that context-aware feature extraction can help reducing the amount of synthetic images by a factor of three while still achieving mAP@50-95 > 70% on our industrial multi-class scenario. Our findings may prompt future work on more efficient data generation and training of out-of-the-box object detection algorithms based on context-aware guided domain randomization based on real, unlabeled data.

The following section discusses available scientific literature related to datasets, techniques and models that inspired our work. We also mention some of their current limitations that justify our further research on the topic. In Sect. 3, we explain the methodology and experimental setup in which our experiments took place. Our quantitative and qualitative results are shown in Sect. 4. Finally, in Sect. 5, we briefly conclude our findings and propose future work.

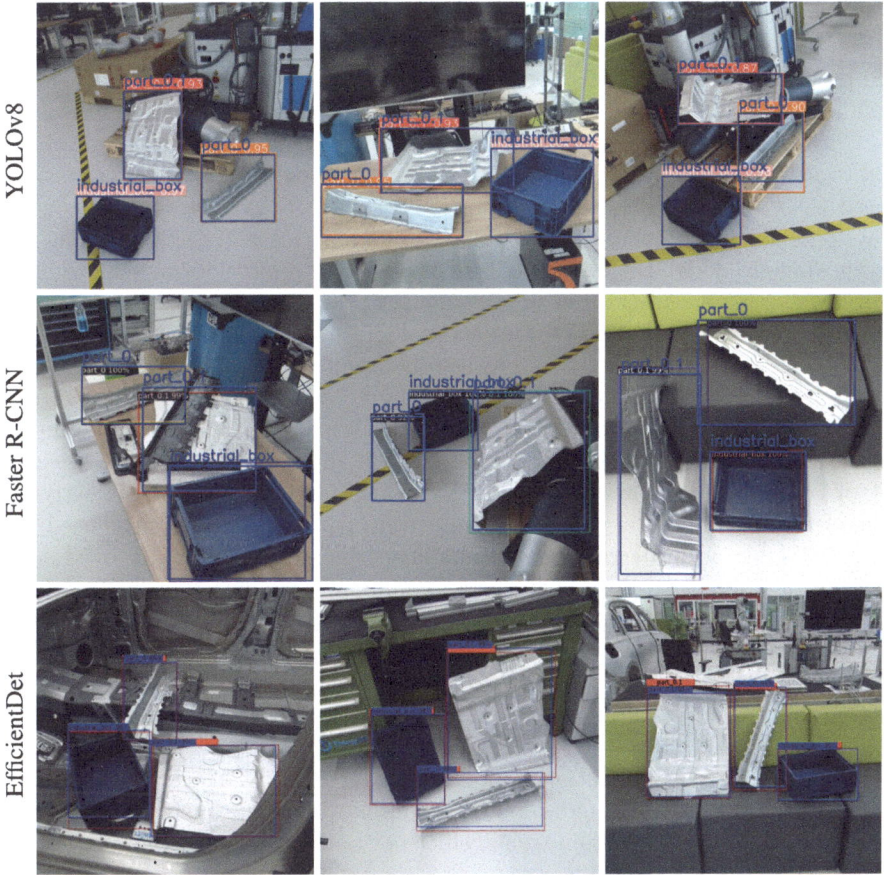

Fig. 1. Sample inferences of YOLOv8 (top), Faster R-CNN (middle) and EfficientDet (bottom) with an mAP@50-95 = 100%. All models were trained on our multi-class synthetic industrial dataset and tested on real images.

2 Related Works

In the effort to replicate the human ability to recognize and distinguish objects nearly instantaneously, convolutional neural networks (CNNs) are widely applied in the field of object detection to mimic the visual cortex [23]. This research on improved models to increase accuracy, avoid overfitting, and train faster can be denoted as a model-based approach [24].

Conversely, supervised machine learning models derived from the model-based approach typically require a substantial amount of annotated data for effective training. This prompts efforts to reduce or avoid time-consuming annotations by generating synthetic data with automated annotations [14,25,26], augment existing data [27] or to employ transfer learning to re-use previously learned features and require less domain-specific data [11]. Research in this domain can be described as data-centric approaches [28].

Model-based Approaches

Within this domain, we consider three state-of-the-art algorithms to offer empirical proof of the generalization of our data generation and selection across multiple model architectures.

Cutting-edge object detection algorithms can be divided into single-stage and two-stage architectures. Two-stage architectures like Faster R-CNN (Region-based CNN) divide object detection into a first feature extraction step to generate regions of interest and a subsequent bounding-box regression and classification stage to derive the desired output labels. This approach can achieve high object detection accuracies but requires longer processing times [29].

One-stage approaches like EfficientDet [30] and the You Only Look Once (YOLO) class of object detectors, on the other hand, utilize a one-step detection strategy [29]. This is achieved by transforming object detection into a direct regression task of bounding-box positions and class probabilities from an input image [29]. With a single forward pass through the network using an efficient set of computational parameters, the YOLO variants are capable of deriving high classification accuracy in real-time settings [7]. This addresses the need for fast and highly accurate computer vision algorithms that can be deployed onto computationally constrained edge devices commonly found in industrial settings [23].

In a comparative study by Sportelli et al. [31], YOLO v8 and EfficientDet were evaluated for their performance in a single-shot learning application. The findings indicate that EfficientDet is inferior in terms of mAP@50 and mAP@50-95.

All three of the aforementioned models - Faster R-CNN, EfficientDet and the YOLO variants - are used in a wide range of applications due to their high object detection accuracy and computational efficiency and are amongst the top-ranked object detection methods [32].

Data-centric Approaches

Within this set of approaches, it is typical to focus on data quality, data augmentation, and domain knowledge [14,26]. For instance, a subfield of artificial intelligence considers transfer learning [24,33] for novel yet related learning tasks. A common approach is to use a pre-trained deep learning model as a feature extractor by freezing the weights outside the last layers. This utilizes the pre-trained model to identify edges or shapes and uses the last layers to learn the novelty. In [34], the authors utilized this approach by training on synthetic data and predicting real-world data. This is especially prominent within the domain of few-shot learning, which addresses scenarios where only a limited number of labeled examples are available for the learning task. On a different approach, recent research suggests that allowing all layers to re-train during the transfer learning stage improves the achieved accuracy if trained with synthetic data [26].

Different methods have been researched to generate data for real-world applications synthetically. Authors in [35,36] argue that the sim-to-real gap can be bridged by using synthetic data alone. In addition, other research highlights the efficacy of employing physics-aware rendering engines and photorealistic data generation [17,37].

In [38], authors advocated for domain randomization, which incorporates non-physical scenarios. The authors demonstrated that this methodology enhances model performance when trained in conjunction with real data. Using synthetically generated data mitigates the annotation overhead. Additionally, research introduces distractor objects in synthetic scenes to create occlusions and clutter as part of the augmentation process [34].

Eversberg et al. [17] undertake a comparative analysis of physics-based engines and domain randomization within the context of industrial applications. The authors achieve good results with 5000 synthetic images and suggest using YOLO as future work. In their approach, features such as textures are varied but no reference contextual information is used to explore Guided Domain Randomization (GDR) [14].

In [14] the authors introduce a methodology that synthetically generates images for industrial environments based on exploring effects of object relations with their surroundings, texture and material attributes and varied illumination scenarios. Even though they achieve good results with synthetic training sets ranging from 2500 to 5700 images, their findings still need to be challenged in a multi-class object detection scenario. Adopting a similar methodology, Vanherle et al. in [26] find that a synthetic replica of a real scenario is not better than light and pose randomization. Additionally, they describe that transfer learning has a larger impact than synthetic data augmentation alone.

Guided Domain Randomization (GDR) is an improvement of domain randomization on which reference features can be essential in creating a training set with high informativeness [14,25]. In this context, perceptual hashing, which extracts semantic features in a compact, fingerprint-like representation, can play an important role if reference images for contextualization are available [39].

3 Methodology and Experimental Setup

Our research focuses on industrial multi-class object detection, addressing the challenges of limited real-data availability and substantial annotation overhead. We employ a data-centric Guided Domain Randomization (GDR) approach on synthetic data to bridge the sim-to-real gap.

We assume Computer-Aided Design models are available for all parts to be detected, a common scenario in industries like automotive manufacturing. Additionally, we assume access to at least one image of the deployment environment as a domain reference. This facilitates context-aware generation and filtering of synthetic data, optimizing rendering and training times by minimizing irrelevant features.

Fig. 2, shows our synthetic data generation pipeline. One component of the pipeline involves the acquisition of real images from a context-related scene. The other component entails the generation of synthetic data via a rendering pipeline. Various algorithms contextually match the real and synthetic data to construct the training dataset.

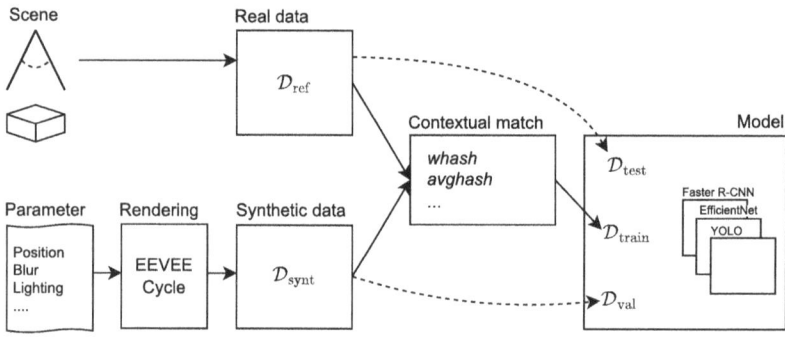

Fig. 2. Synthetic data generation pipeline. It comprises a domain randomizer framework for photorealistic image generation and a context-aware image selection based on high-level (semantic) and low-level (pixel information) features.

3.1 Real and Synthetic Datasets

As shown in Fig. 2, real images are required for synthetic contextualization. Thus, we acquire a total of 75 real images from our experimental setup with a Panasonic DMC-FZ100 camera with a resolution of up to 4320×3240 pixels and with a Leica DC Vario-Elmarit 4.5–108mm f/2.8-5.2 ASPH objective lens. Furthermore, we establish an evaluation protocol by splitting the collected real data into a reference dataset $\mathcal{D}_{\text{ref}}$, with 25 unlabeled images and a test dataset $\mathcal{D}_{\text{test}}$, with 50 annotated frames [40]. Additionally, during the acquisition process of $\mathcal{D}_{\text{ref}}$ and $\mathcal{D}_{\text{test}}$, we ensure variation of lighting conditions, object poses, backgrounds and optical focus to increase data diversity.

The synthetic training dataset $\mathcal{D}_{\text{train}}$, comprising 1000 images, is generated with either the Cycles [22] or EEVEE [21] rendering engines to evaluate the importance of physically-based rendering, as shown in Sect. 4.1. Also, as discussed deeper in Sect. 3.2, synthetic images have been rendered with different spatial positioning, backgrounds, occlusions, lighting conditions, and blur.

Then, by using $\mathcal{D}_{\text{ref}}$, different hashing and low-level filtering methods are used for context-aware filtering such that, for instance, the average hashing generates a dataset $\mathcal{D}_{\text{avg}} \subset \mathcal{D}_{\text{train}}$. The number of data points is in $\{5, 10, 15, 30, 75, 100, 150, 300, 750\}$. Finally, the model is trained on $\mathcal{D}_{\text{train}}$ and validated on $\mathcal{D}_{\text{val}}$, which comprises 10% of $\mathcal{D}_{\text{train}}$. The trained model is evaluated on $\mathcal{D}_{\text{test}}$ to assess the performance with mAP@50-95. The first row of Fig. 3 illustrates samples of all three classes in our multi-class $\mathcal{D}_{\text{train}}$ dataset. We chose this selection of target objects to have minimal sample of various industry-relevant materials, textures, colors and symmetries. The second row of Fig. 3 shows exemplary frames from the test, reference and train datasets.

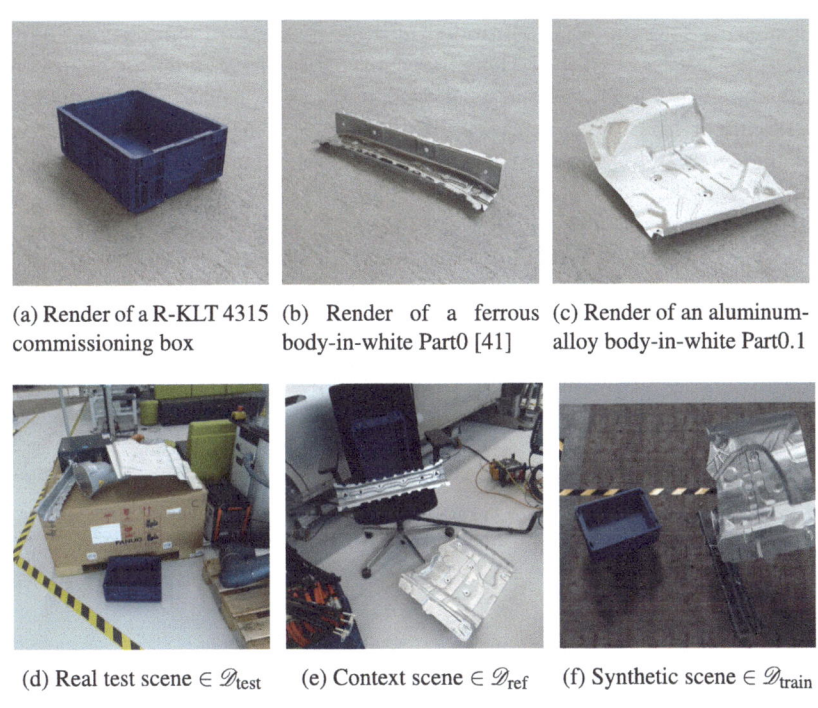

(a) Render of a R-KLT 4315 commissioning box (b) Render of a ferrous body-in-white Part0 [41] (c) Render of an aluminum-alloy body-in-white Part0.1

(d) Real test scene $\in \mathscr{D}_{\text{test}}$ (e) Context scene $\in \mathscr{D}_{\text{ref}}$ (f) Synthetic scene $\in \mathscr{D}_{\text{train}}$

Fig. 3. Top row: renderings of all three target objects present in our industrial dataset. Bottom row: (d) real photo belonging to the test set, (e) unlabeled reference frame for contextualization and (f) a rendered frame $\in \mathcal{D}_{\text{train}}$ with elements from (a), (b) and (c).

3.2 Photorealistic Synthetic Rendering of Multi-class Industrial Objects

We aim to avoid the annotation overhead of real images. Similarly to some approaches mentioned in Sect. 2, we take advantage of the availability of Computer-Aided Design models in the industry to create a suitable framework for synthetic dataset generation.

The developed framework aims at bridging the sim-to-real gap by creating data with features that can, in the inference stage, be extrapolated to real-world features. Several approaches exist to achieve this goal, one of the most prominent in the field is called domain randomization [14,16,26].

Domain randomization consists of randomizing a part of the data generation process such that later real-world variations are interpreted as just one more randomized element in the input data. Controversy, creating the right amount of randomized variables is a non-trivial task. Experiments hint that randomization of camera and object positions outperforms results compared to carrying out training on static scenes [26].

Considering the analysis mentioned above, we create a simulation environment based on version 4.0 of the open-source Blender rendering engine [21]. We constructed a versatile scene as shown in Fig. 4, consisting of the following elements: one or multiple cameras, multiple sources of light comprising environmental and area light sources with variable emission intensity, movable background planes with varying textures and colors, objects of interest with variable six-degrees-of-freedom-pose and distractor objects with multiple shapes and textures with variable position in the same volume as the objects of interest.

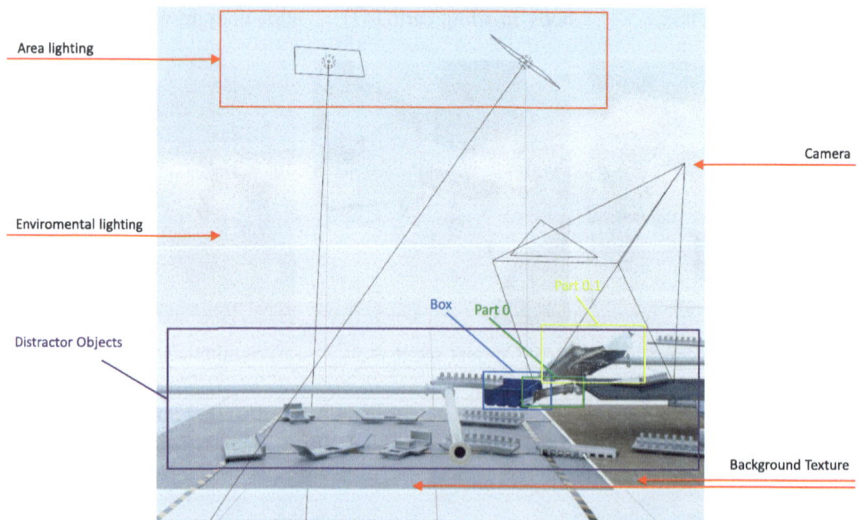

Fig. 4. Elements present in a typical scene of our synthetic data generation framework.

As shown in Fig. 4, we have multi-class target objects along with distractor objects with variable six-degrees-of-freedom pose. This introduces the possibility of collisions between target objects and distractor objects. Thus, it is important to note that we explore an approach in which no simulation of physical interactions between objects is computed before the rendering step takes place. This reduces scene realism in the event of object collisions but simplifies the rendering pipeline [42].

Similar to Hinterstoisser et al. [34], while generating synthetic frames, we are interested specifically in the relative object-to-camera pose to randomize the object's rotations and translations with the camera as the reference. Thus, before the camera projection takes place, we define the object's 6D pose in the 3D space based on a transformation matrix $\mathbf{T_{obj2cam}}$. This transformation matrix belongs to the Special Euclidean Group $\mathbf{SE(3)}$ [43] and is comprised by a rigid rotation $\mathbf{R}$ and a translation matrix $\mathbf{t}$. It can be expressed as:

$$\mathbf{T_{obj2cam}} = [\ \mathbf{R}\ |\ \mathbf{t}\ ] \in \mathbf{SE(3)} : \mathbb{R}^3 \to \mathbb{R}^3. \tag{1}$$

Additionally, we employ the pinhole camera model to simulate how points in 3D space are projected as pixels onto the 2D camera plane [43]. As shown in Table 1, our automated rendering framework accepts object's rotation in Euler angles $\mathbf{R} = f(\alpha, \beta, \gamma)$ and translation variables $\mathbf{t} = f(t_x, t_y, t_z)$ as input to randomize the object's position within the scene with six degrees of freedom. The permutation of these variables across their respective ranges offers a source of variation to strive for domain randomization and bridging the sim-to-real gap.

Table 1. List of accepted input variables in our rendering engine. The given ranges are given as an example and correspond to the specific values used in this work. The user can freely set new values if required.

Variable	Range
Object-to-camera translation $f(t_x)$	$[0\,\text{m}, 1.6\,\text{m}]$
Object-to-camera translation $f(t_y)$	$[0\,\text{m}, 1.2\,\text{m}]$
Object-to-camera translation $f(t_z)$	$[0.75\,\text{m}, 4.15\,\text{m}]$
Object-to-camera rotation $f(\alpha, \beta, \gamma)$	$\{(\alpha, \beta, \gamma)\ |\ [-180°, 180°] \in \mathbb{R}^3\}$
In-frame object percentage	$[0\%, 100\%]$
Amount of distractors	$[0, 10] \in \mathbb{N}$ obstacles per frame
Environment lighting	$[0, 100]$ W/m^2
Area lighting	$[0, 300]$ W
Background textures	$[0, 10] \in \mathbb{N}$ moving planes
Blur / depth of field	F-Stop variation $[1.0, 10.0]$
Amount of images	1000

3.3 Context-Aware Image Selection

Domain generalization alone is a non-trivial task, as finding a general solution to generate enough synthetic variability to enclose real-world variations while avoiding unnecessary scene complexity remains an ongoing research topic [42]. Therefore, instead of adopting a pure domain generalization approach, we investigate its combination with a context-aware synthetic data selection of the randomized frames [38]. This approach has been called Structured Domain Randomization (SDR) [25] or GDR [14], as exposed in Sect. 2. For this data selection process, we apply GDR on two primary feature's domains as follows:

- **Semantic features:** Image Hashing offers a simple yet effective method to match images based on their semantic content while being resilient to changes in specific pixel values [19]. During our implementation, we use the ImageHash library [44] to calculate image hashes and measure semantic differences in terms of Hamming distance. We offer comparisons for the following image hashing algorithms: average hashing, color hashing, difference hashing, perceptual hashing, and wavelet hashing [45].
- **Low-level features:** We also filter synthetic images based on their pixel contents, specifically we consider clustering images by their brightness, sharpness and blur values. During our experiments, we use the Encord Active [46] open-source library to compare the feature domains of synthetic datasets with reference real images. We filtered the images based on their fitness within the real dataset domain.

For instance, brightness is calculated as the average normalized pixel value across each image. Let $\mathbf{I}$ be an image with M rows (height) and N columns (width). The pixel value at position (i, j) is represented by $\mathbf{I}_{i,j}$, and $\mathbf{I}_{\max}$ denotes the maximum possible pixel value (e.g., 255 for an 8-bit image). The brightness $\mathbf{B}$ of the image is given by:

$$\mathbf{B}_{\text{brightness}} = \frac{1}{M \cdot N} \sum_{i=1}^{M} \sum_{j=1}^{N} \frac{\mathbf{I}_{i,j}}{\mathbf{I}_{\max}} \qquad (2)$$

Blurriness and sharpness are determined by applying a Laplacian filter to each image and computing the variance of the output, which effectively measures the "amount of edges" in each image.

First, the Laplacian filter, denoted by $\mathcal{L}$, is applied to the image $\mathbf{I}$. The Laplacian kernel $\mathbf{L}$ is defined as:

$$\mathbf{L} = \begin{bmatrix} 0 & -1 & 0 \\ -1 & 4 & -1 \\ 0 & -1 & 0 \end{bmatrix}$$

The result of applying the Laplacian filter to the image $\mathbf{I}$ is given by:

$$\mathcal{L}(\mathbf{I}) = \sum_{m=-1}^{1} \sum_{n=-1}^{1} \mathbf{L}_{m,n} \mathbf{I}_{i+m, j+n} \qquad (3)$$

where $\mathbf{I}_{i+m, j+n}$ represents the pixel value at position $(i+m, j+n)$ in the image.

Next, the sharpness (or inversely, blurriness) score **S** is computed by calculating the variance of the Laplacian-filtered image:

$$\mathbf{S} = \text{Var}(\mathcal{L}(\mathbf{I})) \tag{4}$$

Here, Var(·) denotes the variance operation.

3.4 Models' Parameters

Table 2 summarizes relevant training hyperparameters we used for training the exposed models. It should serve as a guidance to replicate our results.

Table 2. Training hyperparameters for YOLOv8, Faster R-CNN and EfficientDet on pure synthetic data

Model	Batch Size	Learn. Rate	Epochs	Patience	Optimizer	LR - Scheduler
YOLOv8	102	0.001	200	50	AdamW	Cosine
Faster R-CNN	8	0.001	100	15	AdamW	Cosine
EfficientDet	32	0.001	110	30	AdamW	Reduce On Plateau

4 Results

As discussed in Subsect. 3.2, we randomize variables present in Table 1 within the given ranges to generate enough variation. With this approach, we aim at bridging the sim-to-real gap with domain randomization. Additionally, as shown in image 2, we use domain-specific unlabeled data to filter our synthetic images based on both; semantic information and low-level features. This context-aware training step is not mandatory but allows to potentially reduce the amount of input images required to achieve similar performance if compared to random image picking.

Unless otherwise explicitly stated, along this work, we evaluate inference performance with the Mean Average Precision mAP@50-95 metric [10]. Also, we train our models exclusively on synthetic data, ranging from 5 to 900 images and evaluate them on our real dataset $\mathcal{D}_{\text{test}}$ which comprises 50 images with all three target objects, as depicted in Fig. 3.

4.1 Sim-To-Real Gap with Domain Randomization

Domain randomization has proven to be a better approach than realism to bridge the sim-to-real gap [17, 26]. Thus, we first evaluate if a rendering backbone with higher photorealistic capabilities (Cycles) presents any advantage against a simpler but faster rendering engine (EEVEE).

Comparison of Rendering Engines: EEVEE vs. Cycles

EEVEE [21] is a real-time render engine that enables fast render times using rasterized rendering techniques. Its advantages include fast render times, immediate real-time previews, and low resource requirements. Cycles [22], on the other hand, is a physically-based render engine that uses ray tracing techniques to produce realistic light and shadow simulations. It is particularly suitable for photorealistic renderings and demanding visualizations due to its high image quality and physical accuracy. The dataset mentioned in Table 1 was rendered using both render engines. Rendering-specific settings and required computing times are shown in Table 3.

Table 3. Comparison of rendering times of 1000 images between Cycles and EEVEE in a RTX 1080 Ti GPU. Note that the samples variable impacts the quality of each image.

Render engine	Number of images	Samples per image	Resolution	Total time
Cycles	1000	250	512 × 512	350 min
EEVEE	1000	64	512 × 512	13 min

Figure 5 offers a visual comparison based on features present on an identical frame rendered with Cycles and EEVEE. For easier comparison, notable shadow differences are enclosed in blue and reflexes in red bounding boxes.

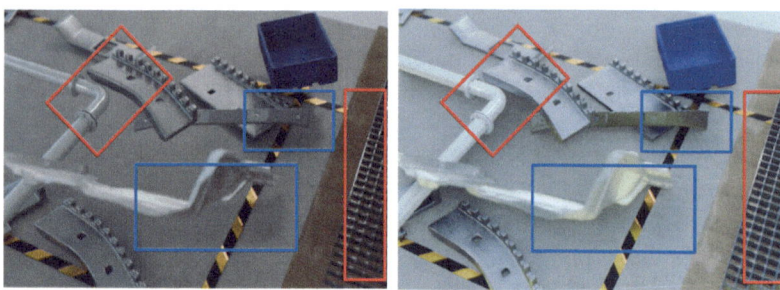

(a) Synthetic frame generated with Cycles. (b) Synthetic frame generated with Eevee.

Fig. 5. Visual comparison of identical frames generated with Cycles and EEVEE. Differences in shadows are marked with blue and reflexes with red.

Figure 6 shows mAP@50-95 results after training on $\mathcal{D}_{\text{train}}$ and evaluating on $\mathcal{D}_{\text{test}}$. Our experiment confirms previous state-of-the-art findings suggesting that higher photorealism is not strictly related to further bridging of the sim-to-real gap [17, 26].

Taking into consideration these results and also considering Table 3, where it is explicit that EEVEE is almost 30 times faster than Cycles, in the following analysis, we exclusively use synthetic data generated with EEVEE.

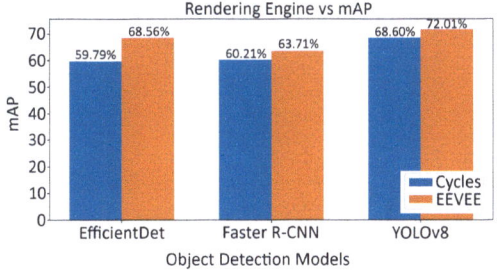

Fig. 6. Comparison of mAP50-95 results between EfficientDet, YOLOv8, and Faster R-CNN on our $\mathcal{D}_{\text{train}}$ i.e. 900 synthetic images rendered with Cycles and EEVEE.

4.2 Ablation Study on Context-Aware Synthetic Features

As discussed in Sect. 3 and shown in Fig. 2, after generating a randomized synthetic dataset, we use context-aware images to narrow down the scope of randomized features and give priority to the most relevant ones to our application's domain.

Figure 7 shows the relation of mAP against context-aware image selection based on semantic content and low-level features. Additionally, we offer a comparison against random image picking, which we consider to be the baseline to determine the efficacy of other image selection methods. The graphs show also the mAP value at 900 images, which corresponds to training the algorithms on the whole training dataset $\mathcal{D}_{\text{train}}$. Note that graphs in Fig. 7 use a logarithmic scale for better visualization of mAP values with the amount of images lower than 100.

Considering the results shown in Fig. 7, we select 300 images to be the benchmark at which we evaluate the efficacy of image selection of the multiple evaluated methods. It is noteworthy to mention that taking 300 synthetic images for training object detectors represents only a fraction of the number of images used in similar previous works aiming at closing the sim-to-real gap in industrial environments [14,17,26].

Figure 8 shows a sorted graph from lowest to highest mAP at 300 images for the different context-aware image selection methods across all evaluated modes. The red curve in Fig. 8 shows the average mAP across YOLOv8, EfficientDet, and Faster R-CNN for a given method of feature extraction.

By taking the random image picking method as a reference, we can observe from Fig. 8 that feature extraction based on both high-level and low-level context-aware features can perform better than random image picking. Notably, in our experiments image matching based on difference hashing offers in average 5.7% higher mAP@50-90 across all models if compared to random image picking, at no cost in the training time and without requiring annotated real images.

As depicted in Fig. 8, synthetic image filtering using perceptual hashing and with the low-level feature of brightness also offer superior results than random

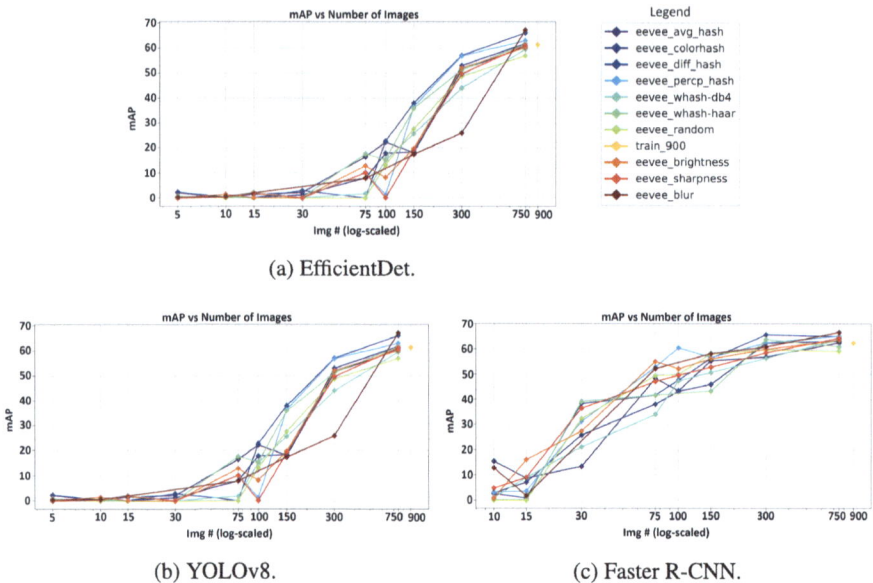

Fig. 7. mAP@50-95 vs. amount of selected images by context-aware semantic content (multiple hashing algorithms) or low-level features (brightness, sharpness and blur). We evaluate on EfficientDet (a), YOLOv8 (b), and Faster R-CNN (c).

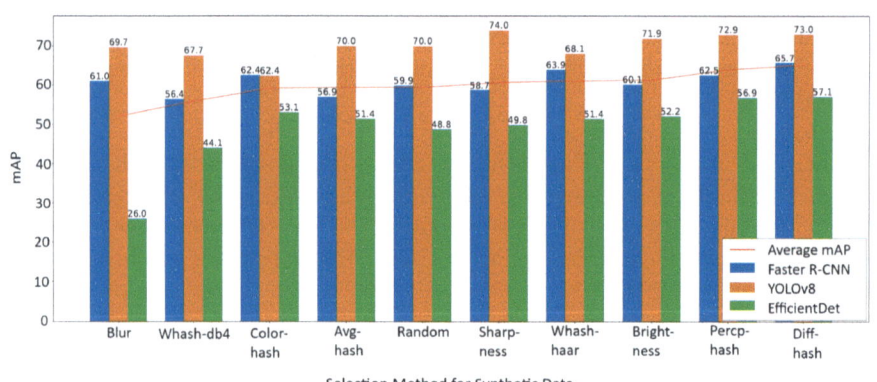

Fig. 8. Comparison of mAP@50-95 results between EfficientDet, YOLOv8, and Faster R-CNN on 300 synthetic images rendered with EEVEE and selected by different methods.

image selection. Additionally, we explored the effect of varying the bit-length of multiple image hashing algorithms but our experiments did not show any direct correlation to mAP results as shown in Fig. 9. Also if the reader is interested on per-class results of all three models explored, please refer to Fig 10 in appendix B where we offer the resulting confusion matrices for all three algorithms trained on 900 synthetic images.

5 Future Work and Conclusion

Taking into consideration our results and the state-of-the-art aiming at the reduction of the sim-to-real gap in industrial scenarios, we think that there is a high potential to use real-time, non-physically accurate rendering engines in combination with context-aware unlabeled images to improve feature selection in synthetic image generation. Therefore we propose as a follow-up work to this research to delve into the following aspects: i) explore the extraction of low-level and high-level features employing AI-based feature extraction models. ii) create a feature-based feedback mechanism towards the rendering engine to generate on-demand contextualized images ii) explore combining generative AI models and rendering engines to create more diverse synthetic data.

Our work explores the applicability of pure synthetic training on industrial multi-class object detection tasks. To this end, we benchmark three widely-used object detection models i.e. YOLO, Faster R-CNN and EfficientDet. We train the models on $\mathcal{D}_{\text{train}} \in \{5, 10, 15, 30, 75, 100, 150, 300, 750\}$ synthetic images and evaluate them on 50 real images containing industry-relevant objects in an uncontrolled environment. Under these conditions, YOLO offers mAP@50-95 values of 75% for 900 images whereas EfficientDet and Faster R-CNN fall behind by a sort margin. Additionally, our ablation study sheds light on the applicability of low-level and semantic feature filtering based on brightness, sharpness, blur, and image hashing for context-aware guided domain randomization. Our results show that context-aware filtering can improve results if compared to random image selection. Notably, if using difference hashing as guided randomization method, YOLO achieves a mAP@50-95 of 73% if trained with only 300 synthetic images, i.e. only 2% lower results than training it on three times more images.

Acknowledgment. This research work has been funded by the German Federal Ministry for Economic Affairs and Climate Action based on a resolution of the German Bundestag, financed by the European Union. We are also grateful to Gautham Mohan and Freddy Fernandes for their technical discussions and implementation support during the experimental section of this work.

A Effect of Hash Size in mAP

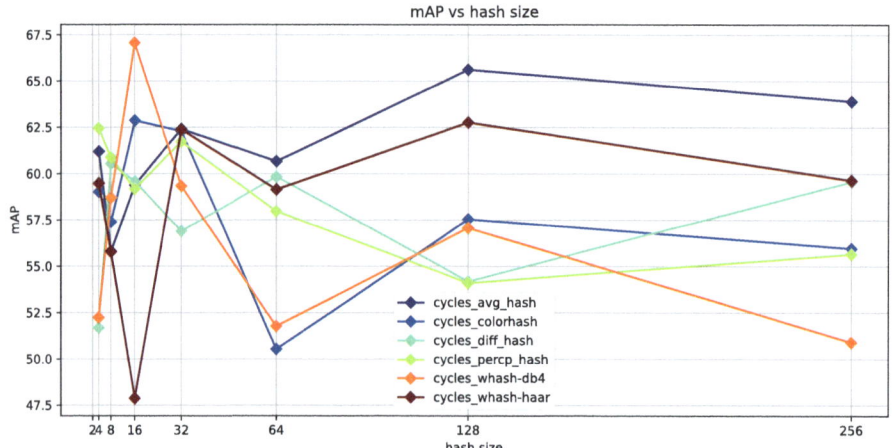

Fig. 9. Effect of hash size variation on mAP@50-95 in the context-aware selection of synthetic images based on their semantic similitude.

B Confusion Matrices of Evaluated Models

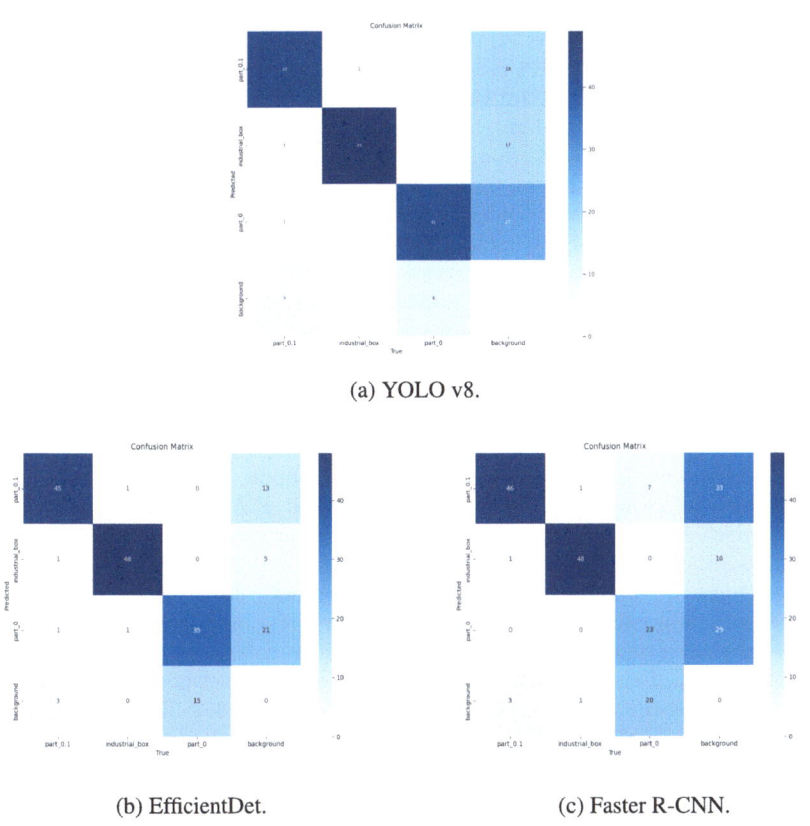

Fig. 10. Confusion matrices of YOLO v8, Faster R-CNN, and EfficientDet after being trained on 900 synthetic images and tested on our multi-class real $\mathcal{D}_{\text{test}}$ set.

References

1. LeCun, Y., Bengio, Y., Hinton, G.E.: Deep learning. Nature **521**(7553), 436–444 (2015). https://doi.org/10.1038/nature14539
2. Goodfellow, I., Bengio, Y., Courville, A.: Deep learning, Adaptive Computation and Machine Learning Series. The MIT Press (2016). https://mitpress.mit.edu/9780262035613/deep-learning/

3. Krizhevsky, A., Sutskever, I., Hinton, G.E.: Imagenet classification with deep convolutional neural networks. In: Pereira, F., Burges, C., Bottou, L., Weinberger, K. (eds.) Advances in Neural Information Processing Systems, vol. 25. Curran Associates, Inc., Lake Tahoe, USA (2012). https://papers.nips.cc/paper_files/paper/2012/file/c399862d3b9d6b76c8436e924a68c45b-Paper.pdf
4. Ronneberger, O., Fischer, P., Brox, T.: U-Net: convolutional networks for biomedical image segmentation. In: Navab, N., Hornegger, J., Wells, W.M., Frangi, A.F. (eds.) MICCAI 2015. LNCS, vol. 9351, pp. 234–241. Springer, Cham (2015). https://doi.org/10.1007/978-3-319-24574-4_28
5. Girshick, R.: Fast r-CNN. In: Proceedings of the IEEE International Conference on Computer Vision, pp. 1440–1448 (2015)
6. He, K., Zhang, X., Ren, S., Sun, J.: Deep residual learning for image recognition. In: 2016 IEEE Conference on Computer Vision and Pattern Recognition (CVPR), pp. 770–778 (2016). https://doi.org/10.1109/CVPR.2016.90
7. Redmon, J., Divvala, S., Girshick, R., Farhadi, A.: You only look once: unified, real-time object detection. arXiv preprint arXiv:1506.02640 (2016)
8. Chen, X., et al.: Microsoft coco captions: data collection and evaluation server. arXiv preprint arXiv:1504.00325 (2015)
9. Lin, T.-Y., et al.: Microsoft COCO: common objects in context. In: Fleet, D., Pajdla, T., Schiele, B., Tuytelaars, T. (eds.) ECCV 2014. LNCS, vol. 8693, pp. 740–755. Springer, Cham (2014). https://doi.org/10.1007/978-3-319-10602-1_48
10. Everingham, M., Van Gool, L., Williams, C.K., Winn, J., Zisserman, A.: The Pascal visual object classes (VOC) challenge. Int. J. Comput. Vis. **88**, 303–338 (2010)
11. Nowruzi, F.E., Kapoor, P., Kolhatkar, D., Hassanat, F.A., Laganiere, R., Rebut, J.: How much real data do we actually need: analyzing object detection performance using synthetic and real data. arXiv preprint arXiv:1907.07061 (2019)
12. Demlehner, Q., Laumer, S.: Shall we use it or not? Explaining the adoption of artificial intelligence for car manufacturing purposes. In: European Conference on Information Systems (2020). https://api.semanticscholar.org/CorpusID:219416061
13. Hansen, E.B., Bøgh, S.: Artificial intelligence and internet of things in small and medium-sized enterprises: a survey. J. Manuf. Syst. **58**, 362–372 (2021)
14. Mayershofer, C., Ge, T., Fottner, J.: Towards fully-synthetic training for industrial applications. In: LISS 2020, pp. 765–782. Springer, Singapore (2021)
15. De Roovere, P., Moonen, S., Michiels, N., Wyffels, F.: Sim-to-real dataset of industrial metal objects. Machines **12**(2), 99 (2024)
16. De Roovere, P., Moonen, S., Michiels, N., Wyffels, F.: Dataset of industrial metal objects. arXiv preprint arXiv:2208.04052 (2022)
17. Eversverg, L., Lambrecht, J.: Generating images with physics-based rendering for an industrial object detection task: realism versus domain randomization. Sensors **21**(23), 7901 (2021). https://doi.org/10.3390/s21237901
18. Wang, C.Y., Liao, H.Y.M., Wu, Y.H., Chen, P.Y., Hsieh, J.W., Yeh, I.H.: CSPNet: a new backbone that can enhance learning capability of CNN. In: Proceedings of the IEEE/CVF Conference on Computer Vision and Pattern Recognition Workshops, pp. 390–391 (2020)
19. Mıhçak, M.K., Venkatesan, R.: New iterative geometric methods for robust perceptual image hashing. In: ACM Workshop on Digital Rights Management, pp. 13–21. Springer (2001)
20. Bradski, G.: OpenCV library 4.5 (2000). https://github.com/opencv/opencv
21. The Blender Foundation: Blender 4.0 (2023). https://projects.blender.org/blender/blender.git

22. The Blender Foundation: The cycles render engine (2023). https://projects.blender.org/blender/cycles.git
23. Hussain, M.: YOLO-v1 to YOLO-v8, the rise of YOLO and its complementary nature toward digital manufacturing and industrial defect detection. Mach. Tooling **11**, 677 (2023). https://doi.org/10.3390/machines11070677
24. Kornblith, S., Shlens, J., Le, Q.V.: Do better imagenet models transfer better? In: Proceedings of the IEEE/CVF Conference on Computer Vision and Pattern Recognition, pp. 2661–2671 (2019)
25. Prakash, A., et al.: Structured domain randomization: bridging the reality gap by context-aware synthetic data. In: 2019 International Conference on Robotics and Automation (ICRA), pp. 7249–7255. IEEE (2019)
26. Vanherle, B., Moonen, S., Van Reeth, F., Michiels, N.: Analysis of training object detection models with synthetic data. arXiv preprint arXiv:2211.16066 (2022)
27. Dwibedi, D., Misra, I., Hebert, M.: Cut, paste and learn: surprisingly easy synthesis for instance detection. In: Proceedings of the IEEE International Conference on Computer Vision, pp. 1301–1310 (2017)
28. Motamedi, M., Sakharnykh, N., Kaldewey, T.: A data-centric approach for training deep neural networks with less data. arXiv preprint arXiv:2110.03613 (2021)
29. Soviany, P., Ionescu, R.T.: Optimizing the trade-off between single-stage and two-stage deep object detectors using image difficulty prediction, pp. 209–214 (2018)
30. Tan, M., Pang, R., Le, Q.V.: EfficientDet: scalable and efficient object detection. arXiv preprint arXiv:1911.09070 (2019)
31. Sportelli, M., et al.: Evaluation of YOLO object detectors for weed detection in different turfgrass scenarios. Appl. Sci. **13**(14), 7901 (2023). https://doi.org/10.3390/app13148502
32. Papers with Code: Object detection models (2024). https://paperswithcode.com/methods/category/object-detection-models
33. Russell, S., Norvig, P.: Artificial Intelligence, Global Edition, 4 edn. Pearson (2021). https://elibrary.pearson.de/book/99.150005/9781292401171
34. Hinterstoisser, S., Lepetit, V., Wohlhart, P., Konolige, K.: On pre-trained image features and synthetic images for deep learning. arXiv preprint arXiv:1710.10710 (2017)
35. Zhu, X., Bilal, T., Mårtensson, P., Hanson, L., Björkman, M., Maki, A.: Towards sim-to-real industrial parts classification with synthetic dataset. In: 2023 IEEE/CVF Conference on Computer Vision and Pattern Recognition Workshops (CVPRW), pp. 4454–4463 (2023). https://doi.org/10.1109/CVPRW59228.2023.00468
36. Baumgart, N., Lange-Hegermann, M., Mücke, M.: Investigation of the impact of synthetic training data in the industrial application of terminal strip object detection (2024)
37. Hodaň, T., et al.: Photorealistic image synthesis for object instance detection. In: 2019 IEEE International Conference on Image Processing (ICIP), pp. 66–70 (2019). https://doi.org/10.1109/ICIP.2019.8803821
38. Tremblay, J., et al.: Training deep networks with synthetic data: Bridging the reality gap by domain randomization (2018)
39. Monga, V., Banerjee, A., Evans, B.: A clustering based approach to perceptual image hashing. IEEE Trans. Inf. Forensics Secur. **1**(1), 68–79 (2006). https://doi.org/10.1109/TIFS.2005.863502
40. Chollet, F.: Deep Learning with Python. Manning Publications (2018)

41. Druskinis, V., Araya-Martinez, J.M., Lambrecht, J., Bøgh, S., de Figueiredo, R.P.: A hybrid approach for accurate 6d pose estimation of textureless objects from monocular images. In: 2023 IEEE 28th International Conference on Emerging Technologies and Factory Automation (ETFA), pp. 1–8. IEEE (2023)
42. Hodaň, T., et al.: Photorealistic image synthesis for object instance detection. In: 2019 IEEE International Conference on Image Processing (ICIP), pp. 66–70. IEEE (2019)
43. Blanco-Claraco, J.L.: A tutorial on $\mathbf{SE}(3)$ transformation parameterizations and on-manifold optimization (2022)
44. Buchner, J.: Imagehash v4.3.1 (2024). https://github.com/JohannesBuchner/imagehash
45. Farid, H.: An overview of perceptual hashing. J. Online Trust Saf. **1**(1) (2021)
46. Encord: Encord active (2024). https://github.com/encord-team/encord-active

Open Access

Open Access This chapter is licensed under the terms of the Creative Commons Attribution 4.0 International License (http://creativecommons.org/licenses/by/4.0/), which permits use, sharing, adaptation, distribution and reproduction in any medium or format, as long as you give appropriate credit to the original author(s) and the source, provide a link to the Creative Commons license and indicate if changes were made.

The images or other third party material in this chapter are included in the chapter's Creative Commons license, unless indicated otherwise in a credit line to the material. If material is not included in the chapter's Creative Commons license and your intended use is not permitted by statutory regulation or exceeds the permitted use, you will need to obtain permission directly from the copyright holder.

Comparison of Active Learning and Self-Training as Adaptation Strategies for Robust Classification in a Dynamic Production Environment

Andreas Seitz[1(✉)], Florian Liebgott[1], Dominik Track[1], Daniel Kessler[1], and Hans-Peter Beise[2]

[1] Balluff GmbH, Schurwaldstr. 9, 73765 Neuhausen a.d.F., Germany
andreas.seitz@balluff.de
[2] Department of Computer Science, Trier University of Applied Sciences, 54293 Trier, Schneidershof, Germany

Abstract. Machines operating in a production environment inevitably undergo changes throughout their usage life span, especially due to environmental influences, modifications or wear and tear. This introduces complexities in leveraging machine learning to monitor their condition for maintenance purposes. Depending on the significance of the changes, the performance of the condition monitoring system could degrade drastically. Especially, if changes occur frequently, retraining the system from scratch is not feasible.

Therefore, we investigated active learning and self-training as adaptation strategies to enhance the robustness of the condition monitoring process. We evaluated both strategies using a demonstrator consisting of several electric motors and a single vibration sensor. The classification task is to identify, which motors are running based on their superimposed vibration data. We emulated different changes occurring in typical production environments and rated them as small, medium and large scale changes.

For a comparative analysis of both strategies, we trained a benchmark model based on a convolutional neural network and evaluated the performance of active learning and self-training for the different changes in relation to this reference model.

The results indicate that employing both active learning and self-training in a production environment to adapt the model can enhance its robustness. Self-training is the preferred option for small changes, as it adapts the model without the need for user interaction. For medium and large scale changes, on the other hand, self-training can fail, while active learning is a sensible strategy, despite the operator having to label some of the data.

Keywords: condition monitoring · vibration sensing · active learning · self-training · convolutional neural network

1 Introduction

In modern production environments, the durability and reliability of machines are crucial for an efficient production process. Therefore, knowledge about the condition of the machines is essential to plan maintenance and downtimes beforehand. To monitor the condition of machines, vibration sensors can be used and based on the vibrational patterns the condition of the machines can be derived [1]. However, machines are dynamic systems and are subject to environmental influences – for instance, a gradual alteration resulting from wear and tear, a modification due to the transition of production to a different product, or the substitution of faulty hardware. Machines are also continuously adapted to the conditions on site and the process for products is optimized. Furthermore for new products, new machines are procured and integrated into existing processes and machine landscapes. Machines and their environment can therefore be considered a complex, interconnected system, in which all parts influence each other.

The article by Hakami [2] discusses a method for improving the condition monitoring of machines using neural networks, particularly in situations with a lack of labeled data. The author uses a combination of a generative adversarial network (GAN) and a long short-term memory (LSTM) to draw conclusions about potentially underrepresented errors. However, Hakami also points out that this process is computationally expensive and the GAN's performance remains contingent on the quality of the training data. Other studies have utilized active learning with random forests to enhance fault diagnostics for predictive maintenance [3,4]. Despite their success in training models suitable for predictive maintenance, none of these studies have taken into account the variations of different magnitudes induced by environmental changes.

In our previous work [5], we investigated the classification of motor states based on superimposed vibration signals with a single vibration sensor. We showed that the classification of motor states is possible with a convolutional neural network (CNN) and a discrete Fourier transform (DFT) as preprocessing step when working with raw unprocessed vibration data. However, we did not investigate the robustness of the model in a dynamic production environment and the possibility of failure due to changes in the environment. For manufacturers and users of intelligent sensors, it is essential to know how their products behave in terms of durability and reliability in a machine environment with dynamic processes and environmental influences. Another important aspect for manufacturers is the possibility for customers to be able to adjust the models in the intelligent sensors themselves to ensure a separation between product and service. The present work examines the classification performance and adaptability of a CNN in a dynamically changing environment.

Therefore, in this work, we augment the experiment used in [5] to emulate a dynamic production environment. We introduce changes with different magnitudes to the setup and evaluate the performance of the model before starting an adaptation process. We use supervised learning to train a CNN model as benchmark and measure its performance with the different changes in the environment. To enhance the model's accuracy after a change, we compare the performance of active learning and self-training as adaptation strategies. We chose active learning due to the promising results in other domains, see e.g. [6–8]. In [9,10], the authors showed that self-training can achieve good results adjusting the classification accuracy of a pre-trained model as well. The advantage of self-training is that it can be used without the oracle needed by active learning. As described in [11], both methods have overlaps in their approach and therefore offer a natural comparability. Either is based on a model that is pre-trained with labeled data to obtain a starting prediction on unlabeled data. Furthermore, both methods meet the requirement of enabling autonomous operation of the product without the manufacturer's intervention. The author also refers to self-training as the complementary technology to active learning with an uncertainty sampling strategy.

We show that the use of active learning and self-training can enhance the robustness of the model in a dynamic production environment and therefore potentially enables the separation of product and service.

2 Theoretical Background and Methods

2.1 Self-training

Self-training is a discipline of semi-supervised learning and is classified as a wrapper algorithm, as the approach is based on a supervised learning process [12]. It is based on the assumption that the models have high confidence in the correctness of their predictions.

Our implementation follows the process described by Yarowsky [13]. In the first step, a model is trained with labeled data using a supervised learning method. Subsequently, this model is used to predict the labels of the unseen, unlabeled data. The samples that reach a prediction probability above a predefined threshold are included as labeled data in an extended training data set, which is then used to create a new model. This iterative process of data extraction and model training can be repeated until either the targeted classification accuracy is reached or no unlabeled data remains. Further information regarding the structure and implementation of the self-training process is presented in Sect. 3.2.

2.2 Active Learning

The idea behind active learning is that a supervised learning algorithm delivers better results if it is allowed to select the data which is used for training [11]. Figure 1 shows an example of a pool-based active learning process. At the beginning, there is a – usually very small – labeled training data set $\mathcal{L}$ with which an initial model is trained. Subsequently, new data, which are considered helpful for prediction, are selected from the unlabeled data pool $\mathcal{U}$ using a query strategy. The queried data gets labeled by an oracle and is added to the labeled training data set.

Lewis and Gale [14] recommend pool-based sampling when there is a scarcity of labeled data in comparison to a larger pool of unlabeled data. According to Settles [11], this method is well-suited for practical problems, given the typical abundance of unlabeled data. For this reason, the experiments in this paper are conducted with pool-based active learning. With this methodology, the unlabeled data in the pool is examined for their information value during the query step, labeled by an oracle and merged into the training data set, as illustrated in Fig. 1.

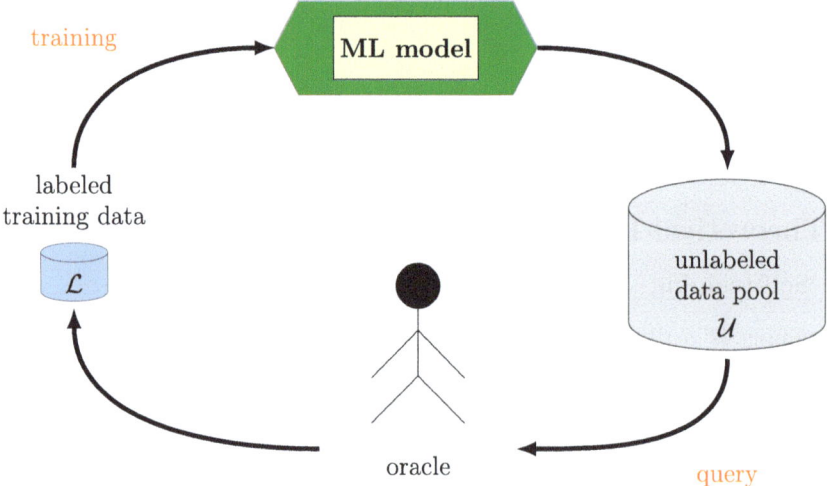

Fig. 1. Process of pool-based active learning (based on: [11])

As query strategies, the probability-based sampling approaches uncertainty sampling[1], margin sampling and entropy sampling are examined. To be able to contextualize the results, random sampling is used as a baseline. For each sample $\underline{x}$, each query strategy calculates an informativeness measure $\phi(\underline{x})$, which models its value for the training of the model.

The uncertainty sampling approach used in this paper was first introduced in [14] and was adapted to a multiclass problem [11]. Its informativeness measure $\phi_U(\underline{x})$ is determined from the probability $P(\hat{y}|\underline{x})$, where $\hat{y}$ is the class label with the highest posterior probability for a sample $\underline{x}$, and is defined as

$$\phi_U(\underline{x}) = 1 - P(\hat{y}|\underline{x}). \quad (1)$$

The distribution of the class probabilities is not considered by this strategy. To correct for this limitation, the margin sampling approach was introduced in [15]. The authors employ the proximity of class probabilities as a measure of informativeness. The classification of a sample becomes more challenging as the

---
[1] In some publications, this approach is referred to as least confident sampling.

distance between the class probability of the most probable class $\hat{y}_1$ and the second most probable class $\hat{y}_2$ decreases. The informativeness measure $\phi_M(\underline{x})$ for margin sampling is thus given as

$$\phi_M(\underline{x}) = P(\hat{y}_2|\underline{x}) - P(\hat{y}_1|\underline{x}). \tag{2}$$

The third probability-based sampling approach used in this work is entropy sampling, which is based on Shannon entropy introduced in [16] as a measure of information content. Its informativeness measure $\phi_H(\underline{x})$ is defined as

$$\phi_H(\underline{x}) = -\sum_i P(y_i|\underline{x}) \log P(y_i|\underline{x}), \tag{3}$$

which is maximum if the probability of all classes y_i is distributed equally.

3 Experimental Setup

The experiment is conducted in multiple stages and is divided in four scenarios. At first, an initial model A is trained with labeled data (scenario 0). Subsequently, changes are made to the setup and the model is evaluated with the respective new data (scenario 1–3). After that the model gets adjusted by labeling some new data with self-training and active learning, respectively.

In all scenarios, the state of four motors is monitored simultaneously. For each motor combination, the same number of recordings was created. To emulate a machine with different vibrational sources, we use the same demonstrator as in our previous work [5] but with two servo motors (servo motor 1, servo motor 2) and two DC motors (DC motor 1, DC motor 2), which differ in size. Every motor performs its specific movement, independent of the state of the other motors. The movement patterns are loosely based on real tasks of machines in production lines, e.g. a drilling or lifting process. Figure 2 shows the placement of the motors on the demonstrator.

Since the experiment is performed under laboratory conditions, all recordings are available with labels. This enabled us to juxtapose the classification accuracy attained via the adjustment process with the accuracy reached through supervised training, if all available data was labeled. As the classes are equally distributed, the metric used to evaluate our experiments is the test accuracy determined by a 5-fold cross-validation.

3.1 Scenarios

Scenario 0 – Initial Setup and Training: The initial scenario 0 (S0) is based on a delivery process of a new machine in which it gets set up by a service technician on site. A part of the process includes installing a vibration sensor to monitor the health status of the machine. Therefore, the technician has to record sample data and label it accordingly to train a supervised model which determines the health status. This scenario is used as a benchmark for the classification accuracies of the changes later on.

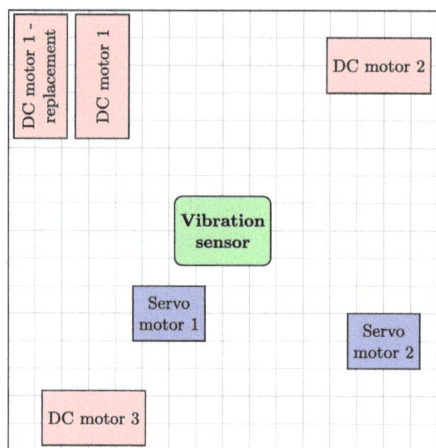

Fig. 2. Positions of the motors and the vibration sensor on the demonstrator

Scenario 1 – Small Changes: Machines are influenced by environmental conditions, modifications or wear and tear. This can, for example, be the wear of a tool, such as a drill, or temperature fluctuations in the operating environment. Both cases may call for an adjustment or optimization of the production process. Depending on the data which was used to train the initial model, such comparably small changes can already have an influence on the classification accuracy. For scenario 1 (S1), we emulate this kind of behavior by changing the characteristic curve of DC motor 1.

Scenario 2 – Medium Changes: If machines have more than one component that wears out in the process, it can happen that their influence on the classification accuracy simultaneously reaches a critical state. Building on scenario 1, we changed the characteristic curve of servo motor 1 in addition to the one of DC motor 1 for scenario 2 (S2). The DC motor emits significantly stronger vibrations than the servo motor and the motors' patterns differ significantly.

Scenario 3 – Large Changes: When machines are part of a production park, nearby vibrations are automatically part of the initial data used for the training of the model. But changes in the area around the machines can cause problems in the classification accuracy as well. This can, for example, originate from new machines being set up subsequently. For scenario 3 (S3), we use DC motor 3 as an external disruptive factor and in this way emulate that a new machine is set up nearby. All the other motors use the characteristic curve from scenario 0.

3.2 Model Adjustment Process

The model adjustment process with self-training and active learning, respectively, to increase the classification accuracy is shown in Fig. 3. For this process, only unlabeled data is used. The first step is preprocessing the data, for which we followed the procedure described in [5]. Initially, the data undergoes windowing, followed by a Fourier transform. The Fourier transform was computed over a window length of 4096 with an overlap of 400. These parameters were optimized before running the experiments.

At the beginning, an initial test accuracy is calculated with model A which was trained with the data from scenario 0. Next, the class probabilities are calculated to query the unlabeled data. The query process for self-training and active learning is carried out similarly. In iteration 0, the prediction probability is calculated with model A. Starting with iteration 1, model A is replaced with model B which is created as part of the adjustment process. The query strategy for self-training is based on a threshold. For our experiments, we set it to the probabilities 75 % and 99 %, respectively. To use the data, it has to be split in training and validation data next. Following the split, data gets merged with the initial data from scenario 0 and the labeled data from the iterations which took place before. Thus, the training data set grows in each iteration. With this new data set, model B gets trained and evaluated afterwards.

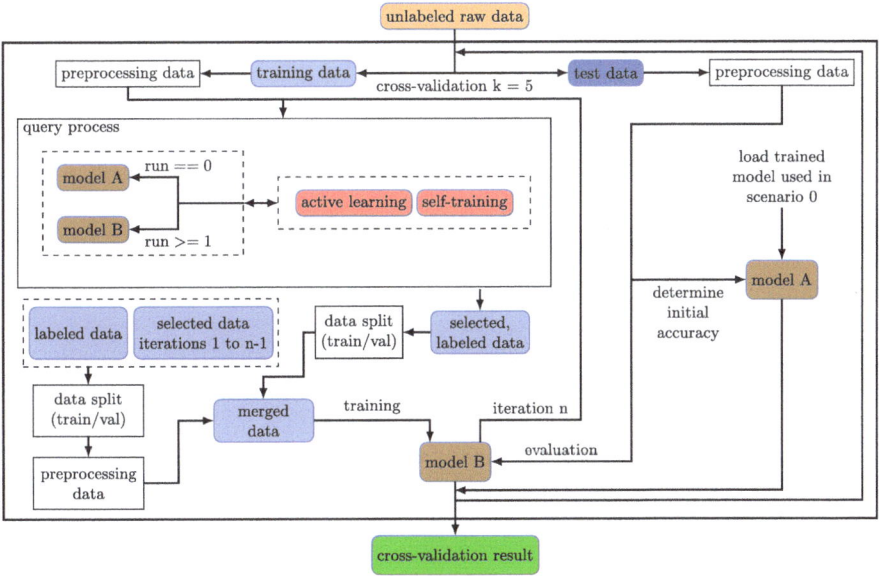

Fig. 3. Self-training and active learning process setup

Depending on the number of iterations n, the process of calculating probabilities, query unlabeled data, merge with labeled data and training/evaluation is repeated accordingly. Since the quantity of the queried data per iteration cannot be predicted beforehand when using self-training, we defined a minimum of two data points queried or a maximum number of five iterations as stopping criterion. For active learning, the number of queried data is limited by the defined sample size and hence the number of iterations serves as stopping criterion. To evaluate the influence of the availability of the labeled data from scenario 0 for active learning, we further conducted adjustment process experiments, utilizing 5% and 100%, of this data, respectively, for the training of the revised model.

To avoid mixing up training and validation data, preprocessing ideally should take place before the split because adjacent windows are likely to yield similar results post-Fourier transformation. This cannot be performed here because the data needs to be preprocessed to calculate the class probabilities. It is then no longer possible to draw conclusions between the raw and the preprocessed data. A possible mixing of the training and validation data, and thus a potential impairment of the training process for its optimization, is explicitly accepted. The test data is not affected by this and the accuracy is therefore not biased.

4 Results

4.1 Evaluation

Table 1 shows the evaluation accuracies μ_{eval} for scenario 0 and the adjusted scenarios 1–3. Also, the performance potentials[2], when all available data is labeled for training is shown for scenarios 1 to 3. It can clearly be seen that the classification accuracy μ_{eval} consistently decreases from scenario 0 onwards. From one scenario to the next, the decline in accuracy roughly doubles. Upon retraining with all data labeled, all three scenarios reach a performance potential over 90% with a standard deviation σ_{rt} of around 1%.

Table 1. Comparison of the evaluation results μ_{eval} for the initial scenario 0 and the adjusted scenarios 1–3. Furthermore, the performance potential based on the mean classification accuracy μ_{rt} and standard deviation σ_{rt} are given.

Scenario	μ_{eval}	μ_{rt}	σ_{rt}
0	97.52%	-	-
1	90.36%	97.97%	0.14%
2	74.69%	93.37%	1.09%
3	33.08%	91.51%	1.12%

4.2 Self-Training

Figure 4 illustrates the progression of the model adjustment with self-training for scenario 1, 2 and 3. It shows the classification accuracy $\mu_{\text{st-test}}$ on the test data, the accuracy $\mu_{\text{st-query}}$ on the queried data and the number of newly labeled

---

[2] In this paper, the classification accuracy μ_{rt} and standard deviation σ_{rt} is determined by cross-validation during the retraining process and is referred to as performance potential or reference value.

samples n_{new} over five query iterations. All quantities are displayed for both the thresholds $\theta = 75\,\%$ and $\theta = 99\,\%$.

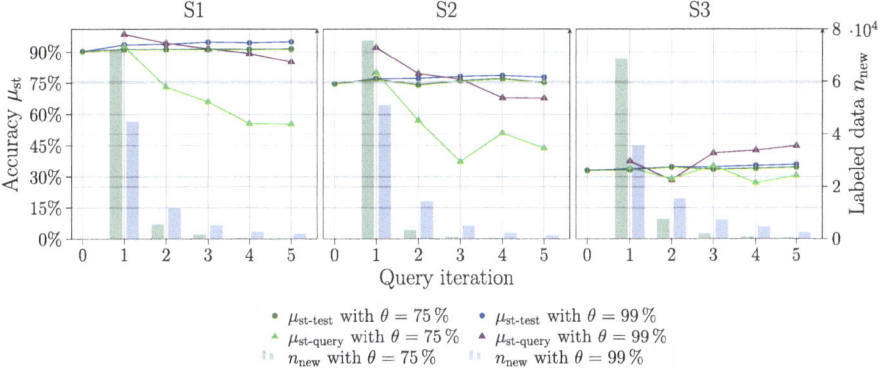

Fig. 4. Comparison of the progression of model adjustments with self-training for scenario 1, 2 and 3

The subfigure on the left in Fig. 4 illustrates that for scenario 1, self-training can produce a steady improvement with a threshold of 99 %. It reaches a classification accuracy of over 95 % which equates to an increase of more than 4.5 percentage points from the starting accuracy of 90.36 %. With a threshold of 75 %, an increase of under 1.5 percentage points can be reached. For both thresholds, the classification accuracy per query is continuously decreasing over 5 iterations. With a threshold of 99 %, the query accuracy starts with a correct classification rate of over 98 % and drops to about 85 % after iteration 5. The query accuracy with a 75 % threshold starts about 5 points lower with 93 % and ends with a query rate of about 55 %. The number of labeled data per iteration is decreasing continuously. In iteration 1 with a 75 % threshold, over 70,000 data points can be labeled, while a 99 % threshold results in over 45,000 data points. After that, a steep drop-off in the amount of data labeled is visible. After 4 iterations, there is almost no data left to label with self-training and a 75 % threshold. With a 99 % threshold, there are about 2,000 unlabeled data points remaining after 5 iterations.

The self-training adjustment for scenario 2 is displayed in the middle subfigure of Fig. 4. For both thresholds, a slight improvement is visible over 5 iterations although the final classification rate with 99 % is only about 3 points higher than the starting rate of 74.6 %. With a 75 % threshold, self-training yields an even lower classification rate. The labeling amount per iteration is similar to scenario 1. Given that the classification accuracy only improves marginally, despite using all available data, we do not anticipate significantly better results, if more data was collected.

Scenario 3 in the subfigure on the right confirms that self-training is not a suitable adjustment strategy for models with a high drop-off in accuracy due to large changes. Although almost all the new data gets labeled, there is essentially no increase in classification accuracy. Unlike scenarios 1 and 2, the query accuracy partially improves after the second iteration for both thresholds, but it never reaches a comparable level. Since there is no clear increase for the classification accuracy on the test data this is neglectable.

Overall, self-training is a suitable strategy for small changes in the data. For medium and large changes, the strategy is not suitable because the classification accuracy drop-off with the data after the changes is too significant to be compensated. In all three scenarios, we observed an enhancement in classification accuracy. However, only in scenario 1 a model potentially apt for classification could be achieved. While the amount of labeling per iteration was decreasing with each iteration, it cannot serve as an indicator of successful model adaptation.

4.3 Active Learning

The accuracy progression μ_{al} of the model adjustments with different query strategies for active learning is displayed in Fig. 5. All scenarios were run for 10 iterations with 100 samples being queried in each iteration. For scenario 1, the highest level of accuracy with 96% after 10 iterations can be reached with margin sampling and 100% of the initially labeled data in the training data set. This is an increase of around 6 percentage points compared to the model without adjustment. With 5% of the labeled data contained in the training set, the classification accuracy is 4 points lower with a maximum of 92%. The second-highest accuracy of 94% is reached with uncertainty sampling and 100% of the labeled data from scenario 0. Its 5% counterpart yields a classification rate of about 90%. With the results of the strategies random sampling and entropy sampling (both with 5% of the data from scenario 0 in the training set), this represents the lowest accuracy scored over 10 iterations. Apparently, no strategy can achieve a steady increase of classification accuracy over 10 iterations. The accuracy declines after iteration 1 with all strategies but entropy sampling and 100% labeled data. The highest drop-off in accuracy to about 78% occurs with entropy sampling and 5% of the labeled data from scenario 0.

The accuracy fluctuations of scenario 1 can also be seen in scenario 2 with all sampling strategies. As in scenario 1, the highest classification accuracy of 86% is reached with margin sampling as query strategy and 100% of the initial data of scenario 0. This corresponds to an increase in classification accuracy of approximately 11 percentage points and unlike the other strategies, there is no classification rate drop-off after iteration 1. The second-highest score of 82% is accomplished with random sampling with 100% labeled data. Entropy sampling with 5% labeled data from scenario 0 yields the lowest accuracy with about 61% after iteration 2. After 10 iterations, it ends up with a classification rate of about 73% which is 1 point lower than its maximum with 74% after iteration 7. With this result, it represents the lowest score for all optimization strategies in scenario 2. It is worth noting that in this case – in contrast to all other query strategies – for all iterations, the accuracy lies below the accuracy achieved without any model adaptation.

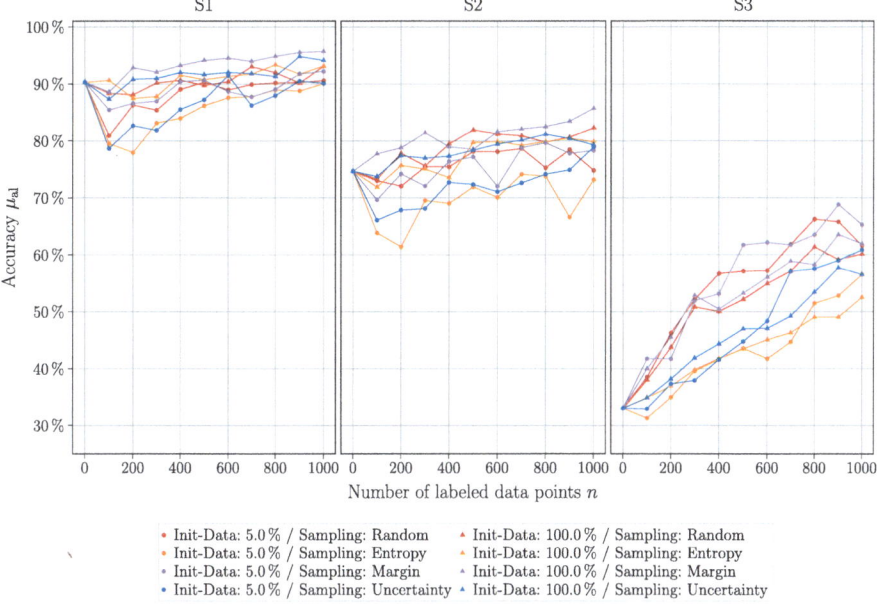

Fig. 5. Comparison of the accuracy progression μ_{al} of model adjustments with active learning for scenario 1, 2 and 3

A slight fluctuation in accuracy is visible for all query strategies in the model adjustment process of scenario 3 as well, but in contrast to the other scenarios, no significant drop-off is apparent over 10 iterations. After 9 iterations, the biggest increase in accuracy of 35 points is achieved with the margin sampling strategy, which reaches about 69 % with 5 % initially labeled data, although a drop-off to 65 % can be seen after 10 iterations. With the same amount of labeled data, random sampling reaches a classification rate of 66 % after 8 iterations, but also shows a drop-off to about 61 % after iteration 10. The query strategies margin sampling and random sampling end up with accuracies of over 60 % with 100 % labeled data, which is comparable to uncertainty sampling which uses 5 % of the labeled data from scenario 0. The lowest score after 10 iterations is reached by entropy sampling and 100 % labeled data.

It is apparent that the accuracy at the beginning of the adjustment process has a lower impact on the learning process when using active learning in comparison to self-training. Given the possible accuracy development potential for all three scenarios, active learning seems to be a good option regardless of the selected strategy.

To test the accuracy development over a longer timespan, we ran an adjustment process with margin sampling for all three scenarios over 50 iterations with a sample size of 100. As seen in Fig. 5, it seems to be advantageous for the model adjustment process when it can access more previously labeled data, if the classification accuracy is high at the beginning. This can be seen when we juxtapose scenario 1 with scenario 3. While in scenario 1, the active learning process with 100 % labeled data from scenario 0 achieved faster and higher accuracy results, the roles were reversed for scenario 3 with the model with 5 % initially labeled data performing better during the adjustment process. Hence, we decided to utilize the entire initially labeled data of scenario 0 for scenarios 1 and 2, while for scenario 3, we used just 5 % of it.

The accuracy results μ_{al} of the active learning adaption process, executed over 50 iterations for model refinement, is illustrated in Fig. 6. In scenario 1, the highest classification accuracy of 97.4 % can be reached after 50 queries. This represents a 7% point increase from the initial value and is about 0.57 percentage points lower than a training with the complete data set. After 32 iterations, the CNN reaches an interim peak value of over 97 % which corresponds to about 93 % of the total increase. Also, it is visible that the classification accuracy decreases by about 2 percentage points after iteration 1. With 200 labeled data points, the classification accuracy increases by about 3 percentage points and exceeds the initial value of 90.36 %.

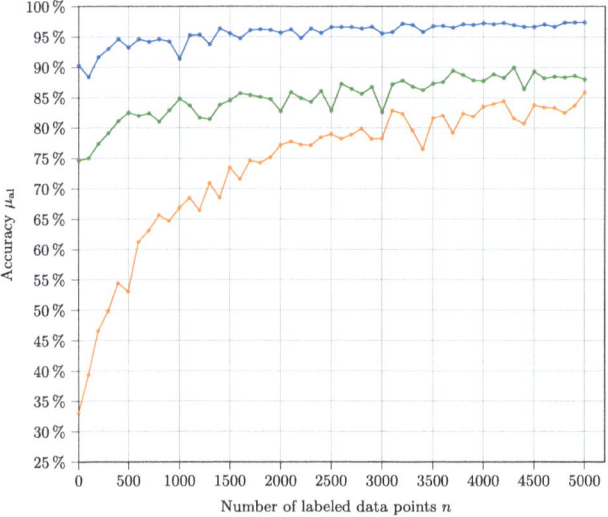

Fig. 6. Comparison of the classification accuracy μ_{al} with margin sampling as active learning query strategy over 50 queries

The highest accuracy rate in scenario 2 of about 90 % is reached after 43 iterations which is an increase of over 15 points. This result is 3 points lower than the accuracy rate that was reached with the retraining process using all available data. After 20 % of the iterations, the classification accuracy score is 85 %, which corresponds to two thirds of the total accuracy increase.

For scenario 3, a total accuracy increase of almost 53 points to 86 % can be achieved after iteration 50. This result is 5.5 points lower than the retraining score. After iteration 31, a temporary peak of 83 % is reached with a steep drop-off to 76.5 % only 3 iterations later. This peak point is about 3 points lower than the high score after 50 iterations, which requires labels for 2900 additional data points.

These results show that an adjustment of the models with active learning is possible regardless of the magnitude of the change. In all scenarios, accuracies close to the values of the retraining with the complete data set being labeled could be achieved. The highest value in scenario 1 even exceeds the classification accuracy of the model adjusted with self-training by 2.4 points. However, the fluctuation during the learning process is also clearly visible in the graph and can cause issues when assessing the performance of the adaptation.

5 Discussion

The findings in this work show that both methods of model adaptation, aimed at enhancing robustness, hold promise, each offering its own unique advantages. With active learning in all three scenarios, a classification accuracy improvement can be achieved irrespective of the starting accuracy and therefore the magnitude of the change. For significant changes, active learning is thus recommended due to its consistent accuracy improvement regardless of the initial accuracy. Nonetheless, it must be noted that this method necessitates the use of an oracle to label parts of the data, which entails additional effort and resources. Additionally, the risk of the oracle misclassifying also needs to be taken into account [17,18]. The learning process's inherent fluctuations also complicate the identification of the optimal point for the model's readiness for application.

Self-training, on the other hand, is apt for small alterations in the data, but it becomes unsuitable for medium and large changes. In these cases, the information available for learning is obviously limited and without additional insight into the change, the model cannot adapt to the data distribution shift. While small changes in the data distribution can be compensated by self-training, bigger changes cannot be handled by this method. This is due to the fact that self-training relies on the classification of the new, shifted data with the previous model, which is only feasible up to a certain point. Self-training is thus suitable for applications that require automated adjustment without external interactions, provided that the changes are minimal. For instance, in a consistent process of change, such as the wear of workpieces, it can be a feasible solution for continuous adjustment of the model.

But regardless of the method, a successful adjustment relies heavily on evaluating the performance correctly. In our case, the evaluation of the reduced classification accuracy is possible because of our laboratory environment and the availability of correctly labeled data. In practice, this conclusion can usually only be drawn by creating labeled test data sets or by subjective assessments. When a change in the process causes the classification performance to drop below a certain level a switch from self-training to active learning is vital. When the optimal point is accurately determined, a high classification rate can be maintained while utilizing the oracle as sparingly as possible. A continuous evaluation of the process is therefore an essential part for a successful adaptation process.

6 Conclusion

In our study, we examined the feasibility of self-training and active learning for the adaptation of models in a dynamic production environment and compared both methods. The results show that a change in the vibration through adjustments in the process (S1, S2) or an external disturbance (S3) can have serious effects on the classification accuracy. For small changes, the classification accuracy decreases by only a few percentage points, while larger changes can result in unusable models. Both self-training and active learning are able to improve the models in all scenarios considered within this study. However, self-training should be used with caution, as it is only suitable for use cases where the changes are minimal, while hardly increasing the classification accuracy for the scenarios with medium and large changes. With active learning, on the other hand, CNNs can achieve high classification results with few data points regardless of the magnitude of the change. An adaptation independent of the process change is thus possible at any time. In conclusion, self-training is a practicable solution to adapt a model in a dynamic production environment, if only small changes are expected, while active learning is suitable for all kinds of changes, but comes with the cost of having to label a small part of the data. Future research should explore the potential of integrating active learning and self-training to enhance the model adaptation process, leveraging the advantages of both methods. The strategic decision of when to employ each method will be a pivotal aspect.

Acknowledgment. This publication resulted in part from the research project KI-MUSIK4.0 funded by the German Federal Ministry of Education and Research (grant number 16ME0070).

References

1. Aburakhia, S., Tayeh, T., Myers, R., Shami, A.: Similarity-based predictive maintenance framework for rotating machinery. arXiv preprint arXiv:2212.14550 (2022)
2. Hakami, A.: Strategies for overcoming data scarcity, imbalance, and feature selection challenges in machine learning models for predictive maintenance. Sci. Rep. **14**(1) (2024). https://doi.org/10.1038/s41598-024-59958-9

3. Chen, J., Zhou, D., Guo, Z., Lin, J., Lyu, C., Lu, C.: An active learning method based on uncertainty and complexity for gearbox fault diagnosis. IEEE Access **7**, 9022–9031 (2019). https://doi.org/10.1109/access.2019.2890979
4. Andriani, A.Z., Kurniati, N., Santosa, B.: Enabling predictive maintenance using machine learning in industrial machines with sensor data. In: Proceedings of the International Conference on Industrial Engineering and Operations Management, vol. 2019, p. 2366 (2021)
5. Seitz, A., Liebgott, F., Rotter, D., Kessler, D., Beise, H.P.: Enabling single-sensor simultaneous condition monitoring of several vibration-emitting machine parts using neural networks, pp. 179–189. Springer (2023). https://doi.org/10.1007/978-3-031-27933-1_17
6. Liebgott, F., Yang, B.: Active learning with cross-dataset validation in event-based non-intrusive load monitoring. In: 2017 25th European Signal Processing Conference (EUSIPCO). IEEE (2017). https://doi.org/10.23919/eusipco.2017.8081216
7. Sener, O., Savarese, S.: Active learning for convolutional neural networks: a core-set approach. arXiv preprint arXiv:1708.00489 (2018)
8. Cohn, D.A., Ghahramani, Z., Jordan, M.I.: Active learning with statistical models. J. Artif. Intell. Res. **4**, 129–145 (1996). https://doi.org/10.1613/jair.295
9. Baek, S., Yoon, H.S., Kim, D.Y.: Abnormal vibration detection in the bearing-shaft system via semi-supervised classification of accelerometer signal patterns. Procedia Manuf. **51**, 316–323 (2020). https://doi.org/10.1016/j.promfg.2020.10.045
10. Karisani, P.: Neural networks against (and for) self-training: classification with small labeled and large unlabeled sets. arXiv preprint arXiv:2401.00575 (2023)
11. Settles, B.: Active learning literature survey, Computer Sciences Technical Report 1648, University of Wisconsin–Madison (2009)
12. Chapelle, O., Scholkopf, B., Zien, A. (eds.): Semi-Supervised Learning. The MIT Press (2006). https://doi.org/10.7551/mitpress/9780262033589.001.0001
13. Yarowsky, D.: Unsupervised word sense disambiguation rivaling supervised methods. In: Proceedings of the 33rd Annual Meeting on Association for Computational Linguistics. Association for Computational Linguistics (1995). https://doi.org/10.3115/981658.981684
14. Lewis, D.D., Gale, W.A.: A sequential algorithm for training text classifiers (1994). https://arxiv.org/abs/cmp-lg/9407020
15. Scheffer, T., Decomain, C., Wrobel, S.: Active hidden Markov models for information extraction. In: Hoffmann, F., Hand, D.J., Adams, N., Fisher, D., Guimaraes, G. (eds.) IDA 2001. LNCS, vol. 2189, pp. 309–318. Springer, Heidelberg (2001). https://doi.org/10.1007/3-540-44816-0_31
16. Shannon, C.E.: A mathematical theory of communication. Bell Syst. Tech. J. **27**(3), 379–423 (1948). https://doi.org/10.1002/j.1538-7305.1948.tb01338.x
17. Du, J., Ling, C.X.: Active learning with human-like noisy oracle. In: 2010 IEEE International Conference on Data Mining. IEEE (2010). https://doi.org/10.1109/icdm.2010.114
18. Chakraborty, S.: Asking the right questions to the right users: active learning with imperfect oracles. Proc. AAAI Conf. Artif. Intell. **34**(04), 3365–3372 (2020). https://doi.org/10.1609/aaai.v34i04.5738

Open Access This chapter is licensed under the terms of the Creative Commons Attribution 4.0 International License (http://creativecommons.org/licenses/by/4.0/), which permits use, sharing, adaptation, distribution and reproduction in any medium or format, as long as you give appropriate credit to the original author(s) and the source, provide a link to the Creative Commons license and indicate if changes were made.

The images or other third party material in this chapter are included in the chapter's Creative Commons license, unless indicated otherwise in a credit line to the material. If material is not included in the chapter's Creative Commons license and your intended use is not permitted by statutory regulation or exceeds the permitted use, you will need to obtain permission directly from the copyright holder.

Robotic Wiring Harness Bin Picking Solution Using a Deep-Learning-Based Spline Prediction and a Multi-stereo Camera Setup

Manuel Zürn[1(✉)], Carsten Schmerbeck[2], Andreas Kernbach[3], Mara I. Kläb[3], Alper Yaman[3], Daniel Bragmann[4], Michael Heizmann[2], Marco Huber[3], Werner Kraus[4], Armin Lechler[1], and Alexander Verl[1]

[1] ISW, University of Stuttgart, Seidenstr. 36, Stuttgart 70174, Germany
manuel.zuern@isw.uni-stuttgart.de
[2] IIIT, Karlsruhe Institute of Technology, Hertzstraße 16, Karlsruhe 76187, Germany
[3] IFF, University of Stuttgart, Allmandring 35, Stuttgart 70569, Germany
[4] IPA, Fraunhofer Institute for Manufacturing Engineering and Automation, Nobelstr. 12, Stuttgart 70569, Germany

Abstract. The automation of wire harness handling and installation in the automotive industry presents a challenge due to the inherent flexibility of cables, the high variance in wire harnesses and plug combinations, and the intricate spatial configurations required for accurate installation. Addressing this challenge requires the integration of sensors for accurate pose estimation with high-dexterity robotic systems. This work introduces a novel approach to automate the process of grasping of wiring harnesses for autonomous installation using a robotic arm. The methodology encompasses several stages. Initially, a multi-stereo camera setup creates a high-accuracy representation of the working area. Next, a deep learning model predicts a spline representing the segment with the biggest connector attached to it for 6D grasp pose estimation. The final stage uses a skill-based robot program to perform the grasping of the wiring harness, which is evaluated using 50 random configurations inside a bin. As a result, the proposed solution achieves an accuracy of 82% of successful wiring harness bin picking grasps, where success is defined when the result is that a specific connector on the wiring harness is in a predefined spot after grasping. Future work will use another robot to grasp the connector from the first robot to install it in an automotive demo door using reinforcement learning.

Keywords: Wire harness · Correspondence matching · Skill-based robotic control · Machine Vision · Deep Learning

---

M. Zürn and C. Schmerbeck—contributed equally to this work.

1 Introduction

The automotive wiring harness market is poised for continuous growth up to 2033, as indicated by market research [1]. This growth is driven by various factors, e.g., electrification and smart assistant systems. Considering the wiring harness value chain, costs are significant and encompass material, assembly, and installation. A bottleneck in the value chain is the final installation process, which consumes 70% of total production time and relies heavily on manual labor, with over 90% of tasks performed manually [2]. The manual-intensive process requires over-specification due to uncontrolled manipulation forces, which also may pose safety risks. Automation solutions emerge to address these challenges. By implementing automation, the industry aims to achieve several objectives: (i) Reduction of over-specification: Automation technologies can decrease dimensions and monitor manipulation forces, leading to a decrease in over-specification and minimizing material wastage. (ii) Decrease in manual installation time: Even partial automation solutions, e.g., human-robot collaboration, in wiring harness manipulation reduce stress on the worker due to the heavy automotive wiring harness. (iii) Enhanced Safety: Automation enables advanced monitoring systems that ensure precise and controlled installation, thereby improving safety standards and reducing the risk of errors or accidents.

Despite the benefits of automation, challenges persist in developing automation solutions, particularly in assembly tasks. Research from [3] underscores the complexities involved in wiring harness perception and dynamic robot programming, hindering the adoption of automation in this domain. Wiring harness manipulation requires different domain experts in 3D vision, advanced correspondence estimation for (i) wiring harness topology estimation and (ii) 6D component pose estimation and lastly advanced sensitive robotic control. This paper addresses a wiring harness bin-picking solution for manipulating medium-sized wiring harnesses by combining advanced multi-stereo 3D vision and wiring harness correspondence estimation with skill-based robotic control. The wiring harness is the size of a door wiring harness. We contribute a modular low-cost stackable camera system with high resolution for accurate 3D reconstruction, applying a recent implementation of generic intrinsic camera calibration [4], applying a modified iterative-closest-point algorithm [5] for extrinsic calibration in a multi-stereo approach, novel deep-learning architectures for estimating the topology of wiring harnesses predicting cubic splines and an evaluation of grasping a wiring harness in almost random positions in a bin. Furthermore, we contribute the dataset, model weights and the tensorboard log including quantitative and qualitative results in Supplementary information.

1.1 Related Works

Research in wiring harness perception mostly focuses on using 2D or 3D camera systems for perception with different goals. Applications range from visual inspection [6–8] to grasp point estimation on multiple wiring harnesses in a bin

[9] to robotized assembly [10,11] of wiring harnesses or artificial dataset augmentation [7,12]. Estimation of wiring harness states, e.g. done in [9,10], are done using probabilistic model estimation. Deep learning solutions have been applied for estimation segmentation masks [6,7] or directly valid grasp points in a bin as in [9]. However, solutions for segmentation masks usually have no information about the topology of the wiring harness required for robotic installation as in [10], as well as direct grasp point estimation as in [9].

There exists analytical topology estimation as in [13], which is limited to a restricted initial configuration of the wiring harness and therefore not directly suitable for bin picking. Solutions, as proposed in [11], which attach clamps with visual markers on the wiring harness, seem unpractical, as they introduce extra weight to the wiring harness. As concluded in [14], there exist vision-based solutions with deep learning offering a promising tool to solve these, it remains to be shown how accurate and how reliable the vision-based solutions perform. This study proposes and evaluates a deep learning approach for partially predicting the topology by predicting a specific segment of a wiring harness inside a bin, providing a reliable solution that tries to impose minimal constraints on the initial configuration of the wiring harness.

2 Methodology

2.1 Process Overview

The process setup is shown in Fig. 1. It consists of starting with a wiring harness in a random configuration in a bin. The bin is recorded from a multi-camera setup in different orientations and positions in a robotic cell. A deep learning model predicts a spline using 4 interpolated keypoints on each image. In the end, a grasp point on the wiring harness used for robotic wiring harness bin picking is estimated.

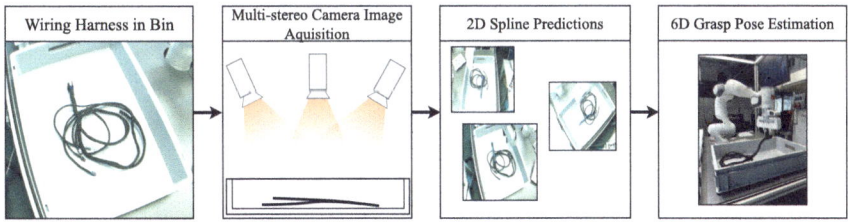

Fig. 1. Proposed wiring harness bin picking process. It consists of first, estimating three 2D splines representing a wiring harness segment and second estimating a 6D grasping pose.

2.2 Multi-stereo Camera System

A multi-stereo camera system is used to achieve accurate pose estimation of the wire harness. A passive vision system with multiple viewpoints was chosen because it does not require special illumination (such as structured light or laser scanners) [15] and the cameras can be directed or placed near the regions of interest in the scene, see Fig. 2. Furthermore, there is the possibility that the working area of the vision system is obstructed by, for example, the robot arm.

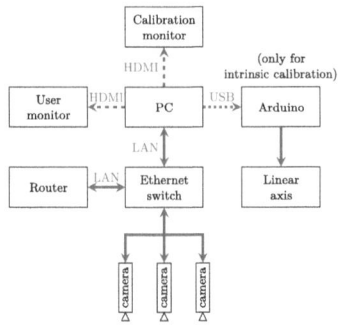
(a) Structure of the vision system

(b) Camera module with swivel mount

Fig. 2. Software and hardware setup of the camera modules

To achieve this, a modular camera system that allows the usage of a easily expandable number of cameras has been created, as illustrated in Fig. 2a. The mounting system enables the cameras to be securely mounted within a frame surrounding the robot cell, while still allowing for easy placement around the cell's perimeter. Each camera as shown in Fig. 2b utilizes a Raspberry Pi HQ camera with a C/CS lens mount, offering a resolution of 4056 × 3040 pixels and a 6 mm lens [16]. Additionally, the square design of the camera case permits stacking in any direction, facilitating stereo or light-field configurations. The experimental setup includes three camera modules aimed at the pick-up area where the cable is retrieved from a bin.

Generic Intrinsic Camera Calibration. The precise geometric calibration of camera systems is important for computer vision applications. Typically, these systems are modeled by using a perspective projection with a single projection center and are often characterized by low-dimensional, parametric models with minimal intrinsic parameters, such as the pinhole model [17, Chapter 2.3.1][4, Chapter 5.1]. The drawback of low-dimensional models lies in their limited descriptive capacity, which is insufficient to perfectly characterize every pixel among the vast number present in modern cameras. As a result, novel camera models have recently been developed. These generic models can be characterized as versatile imaging systems, independent of the used camera type, thereby enabling precise calibration [4].

The generic camera model was first introduced by Grossberg and Nayar [18]. In this model, any imaging setup is represented as a non-parametric discrete entity, like a black box, containing photosensitive components. Each individual

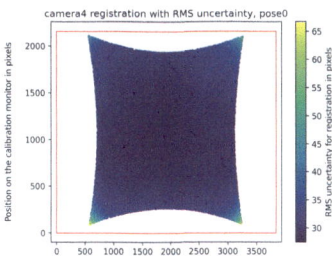

(a) Intrinsic calibration setup with sinusoidal pattern displayed on the monitor.

(b) The monitor is depicted by the red contour. The color scale shows the root mean square (RMS) error for the registration in monitor pixels.

Fig. 3. Calibration setup and example registration between camera and monitor.

pixel within this system captures a ray of light, referred to as raxel, comprising geometric ray coordinates alongside radiometric parameters. The entirety of these raxels represents the imaging model of the camera. In this work, intrinsic camera calibration is performed in two steps for each camera depicted in Fig. 3a.

(i) Initially, sequential images of sinusoidal patterns with phase-shift coding shown on a monitor are captured (Fig. 3a), and the registration between the camera pixels and monitor coordinates is determined using probabilistic phase unwrapping (Fig. 3b). A modified version of Uhlig's implementation [4, Chapter 4] is used for this purpose. This approach is relatively robust against external factors such as lighting. However, its robustness is limited in cases of strong reflections or glare on the calibration monitor. Therefore, the calibration setup is enclosed in a box to block unwanted light from the outside. The recording of the monitor is done from 10 equally spaced distances to allow for the accurate estimation of the raxels. For ease of use, this process is automated by placing the camera in the calibration setup on a linear axis, as seen in Fig. 3a. The display of the 96 calibration images (48 per direction, 4 frequencies [1,4,16,64] and 12 phase angles), the recording, and the movement of the camera (10 poses) are automatically controlled [19]. The number of poses and configuration of images is identical to the one use in [4].

(ii) Thereafter, the intrinsic calibration is found by an alternating minimization-based process of both estimating the set of rays $\mathcal{L}$ intersecting the registered points on the different monitor poses and estimating the set of monitor poses (translation $\mathcal{T}$ and rotation $\mathcal{R}$). This can be formulated as a least-squares minimization problem:

$$f(\mathcal{R}, \mathcal{T}, \mathcal{L}) = \sum_{k,i} d\left(\mathbf{p}_{ik}, \mathbf{l}_i\right)^2 . \tag{1}$$

Here, index i represents the individual rays and index k the reference coordinate system. $d(\cdot)$ is a ray-to-point distance measure between points on the calibration monitor $\mathbf{p}_{ik}$ and $\mathbf{l}_i$ being a individual ray. $\mathbf{R}_k \in SO(3), \mathbf{t}_k \in \mathbb{R}^3$ are the corresponding transformations to the camera coordinate system [4, Chapter 5].

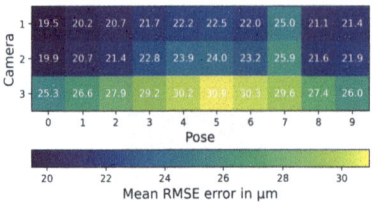
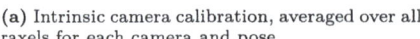
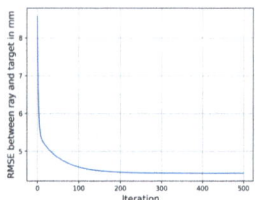

(a) Intrinsic camera calibration, averaged over all raxels for each camera and pose.

(b) Example of RMSE distances for extrinsic calibration between calibration rays and calibration targets over 500 iterations in mm.

Fig. 4. Quantitative results of the intrinsic and extrinsic camera calibration

The intrinsic calibration process described by Uhlig [4] returns an RMSE for each ray at each monitor pose representing the distances between the registration on the monitor and the intersecting ray. To display them in Fig. 4a the individual rays were averaged per pose and per camera. The results between roughly 20 μm and 30 μm are significantly smaller than the pixel pitch of the monitor used for calibration of 185 μm (BenQ PD2705U) proving the subpixel accuracy capabilities stated by Uhlig [4]. In his work, an accuracy of 32 μm was achieved with different cameras and monitors [4].

Generic Extrinsic Camera Calibration. After the intrinsic calibration of each camera, an extrinsic calibration is performed. This is done by attaching a calibration target with ArUco markers to the robot end effector. The markers are found by an ArUco detector implementation in OpenCV [20]. The robot moves the calibration target to known positions while the camera system takes the corresponding pictures, resulting in known calibration points P in the 3D space. To compute extrinsic calibration, an initial guess for each camera pose relative to the known calibration points $\mathcal{P}$ is calculated using the SolvePnP method implemented in OpenCV [20]. This method employs a standard pinhole camera model where the camera parameters are derived exclusively from the camera and lens data sheet [16] and assume an ideal camera. This yields an initial solution for the subsequent iterative process.

The exact pose of the cameras is estimated using a modified point-to-line iterative closest point (ICP) algorithm. The generalized approach of an ICP algorithm was described by Besl&McKay [5]. The set $\mathcal{P}$ describes a set of known calibration points in the scene. Here they are the corners of the ArUco markers on calibration targets. Only the rays intersecting the known calibration points $\mathcal{P}$ are selected from the intrinsic calibration model and called calibration rays $\mathcal{X}$. Initial values are denoted with a subscript 0.

ICP is an iterative process, consisting of four steps:

1. **Compute the closest points** in the model shape $\mathcal{X}_k$ to their corresponding data points in $\mathcal{P}$. $\mathcal{Y}_k = c(\mathcal{P}, \mathcal{X}_k)$ denotes the set of closest points. Besl&McKay

describe their algorithm for a limited set of geometries [5], whereas in our implementation, the closest point on a ray is calculated [21, Chapter 10.2.1].

2. **Compute the registration.** The distances d_k between the closest points are calculated $(\mathbf{Q}_k, d_k) = \mathcal{Q}(\mathcal{P}_0, \mathcal{Y}_k)$ where $\mathbf{Q}_k$ denotes the registration. A implementation of Murugan [22] is used here.
3. **Apply the registration.** Besl&McKay apply the registration to the points that should be fitted to the model. $\mathcal{P}_{k+1} = \mathbf{Q}_k(\mathcal{P}_0)$ In our case, the model should be translated and rotated accordingly, to fit to the known points. $\mathcal{X}_{k+1} = \mathbf{q}_k(\mathcal{X}_0)$
4. **Repeat until termination.** Here, the termination condition is either a maximum number of steps performed or an RMSE below a threshold τ.

As illustrated in Fig. 4b, the proposed generic ICP method can substantially lower the RMSE distances compared to the initial solution provided by SolvePnP [20]. Preliminary results from a static scene with manually positioned and measured calibration targets (refer to Fig. 5a) indicate that this technique can iteratively decrease the RMSE reprojection error for 468 visible calibration points $\mathcal{P}$ to as low as 2.85 mm.

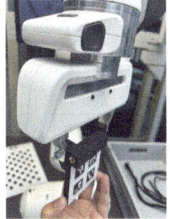

(a) Example of a static scene with ArUco markers used for extrinsic calibration.

(b) Used calibration target with ArUco markers. The target can be screwed into the robotic gripper to have a fixed transformation $^{EE}\mathbf{T}_{Target}$

Fig. 5. Markers for extrinsic calibration

Table 1. RMSE and maximum distances between the calibration points $\mathcal{P}$ and the calibration rays $\mathcal{X}$ after the extrinsic calibration.

Camera	RMSE in mm	Max in mm
1	7.80	17.65
2	6.54	12.25
3	4.41	10.59

As seen in Table 1, the results for the extrinsic calibration using the robot (see Fig. 5b) are slightly worse than using a static scene. This discrepancy could be caused by inaccuracies in the robot kinematics and requires further research to enable more accurate extrinsic camera calibrations. Nevertheless, the RMSE errors during the keypoint estimation are significantly lower than the values achieved during the extrinsic calibration (see Subsect. 3.1). Therefore one could assume that the inaccuracies in the robot kinematics may be mean-free and average themselves out by using multiple poses of the robot during the calibration.

2.3 2D Spline Prediction

Deep Learning Architecture. A deep learning model is trained to predict a spline on the first segment of the wiring harness. As the deep learning model predicts a specific segment on the wire harness, the process is referred to as partial direct topology prediction. The deep learning model consists of a backbone that extracts features of an RGB input image, a neck that extracts features from different feature map resolutions of the backbone and a head that predicts the interpolating cubic spline points of the spline. A reference image of the architecture is shown in Fig. 6. We used a tiny Swin Transformer from [23] as backbone. The backbone extracts features in 4 depth layers. By using a feature pyramid network [24] as neck, we ensure the capture of multi-scale feature representations which should improve model performance. Next, the different feature maps are concatenated and upsampled according to a U-Net architecture [25], which ensures faster learnability with skip connections.

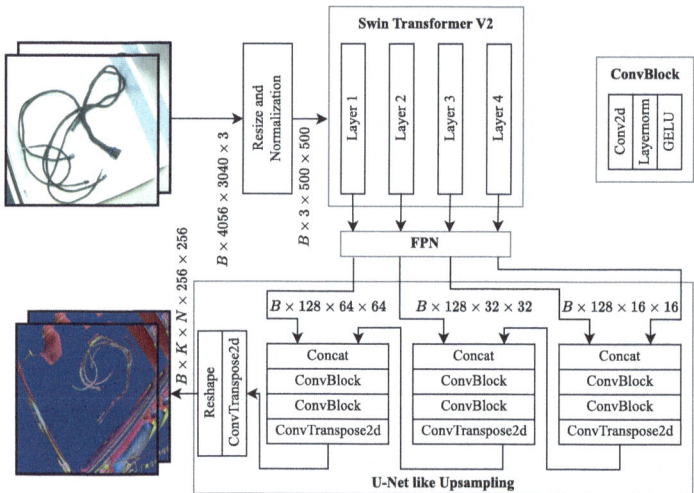

Fig. 6. Deep learning architecture using a Swin Transformer backbone, Feature Pyramid Network (FPN) Neck with a U-Net-like Convolutional Upsampling Head to predict natural cubic spline segments for wiring harnesses. This can be used to extract a partial directed graph for direct topology prediction as proposed in [13].

The output of the model is a heatmap with $B \times C \times H \times W$ features, where B corresponds to batch size, C to keypoint dimension times number of segments, H

to the height of the heatmap and W to the width of the heatmap. By reshaping C, we can learn different segments, such that the final dimension is $B \times K \times N \times H \times W$, where K corresponds to the number of segments and N to the number of keypoints on each segment. In this paper, we chose to learn one segment (K=1) modeled with 4 keypoints (N=4) to represent the cubic spline.

Loss Function and Training. Annotation takes place by approximating the shape of the segment using a natural cubic spline. The annotation tool was written in Python using NiceGui[1] and stores the segments and nodes in .json format. Standard deep learning praxis in e.g., facial landmark detection, use heatmaps to estimate keypoints or directly predict the keypoints by using regression and L1/L2 loss. Heatmaps tend to converge faster, while they usually can't be used for direct coordinate regression due to loss of gradient in non-differentiable keypoint extraction. Therefore, we use a way of estimating the keypoints without losing the gradient proposed in [26]. We apply L1 loss on the keypoint estimation and conducted a hyperparameter study with various variations of the loss functions and the hyperparameters. The hyperparameter study was conducted using

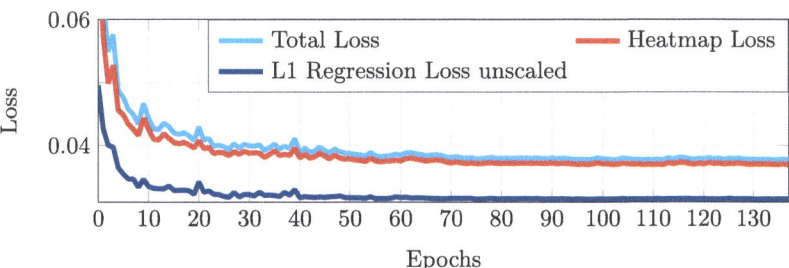

Fig. 7. Validation loss over a dataset with 584 annotated images. Hyperparameters: Batch size 5, learning rate 1.1e-4 with 139 epochs. Regression loss is scaled with 0.5 and added to the heatmap loss to get the total loss, which achieves the best results. Training/validation data split is 0.8/0.2.

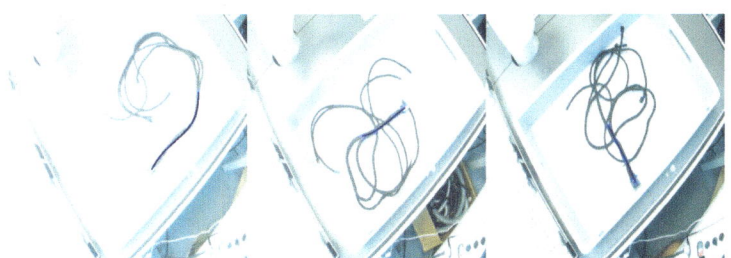

Fig. 8. Validation dataset samples of the natural cubic spline prediction (blue) depicted as overlay in the camera images. (Color figure online)

---

[1] *doi.org/10.5281/zenodo.13623828*.

optuna [27] with a hyperband pruner and a TPESampler for sampling in the learning rate between 10^{-5} and 10^{-3}, batch size between 5 and 20 and epochs between 100 and 300. The best trial is shown in Fig. 7. Samples of the prediction of the validation dataset are shown in Fig. 8.

2.4 3D Point Reconstruction and 6D Grasp Pose Estimation

For 3D point reconstruction we need the correspondence in each image represented by keypoints. The keypoints are given in pixel coordinates (s, t) as seen by the camera. For each keypoint, the corresponding ray from each camera is selected and stored together as a set of rays that should theoretically intersect at the sought-after keypoint. Finding a solution for the closest point to n (generally) skewed lines is done by a Nelder-Mead minimization of the sum of squared distances between the point and the set of lines [28]. Each 3D point reconstruction thus returns an RMSE value representing the uncertainty of the estimate. For robotic grasping, we need a 6D grasp pose consisting of an orientation $\mathbf{R} \in SO(3)$ and translation $\mathbf{t} \in \mathbb{R}^3$. Two 3D points and the $\mathbf{xy}$ plane are used to estimate the 6D pose. First, the connector point of the spline is used, as it is usually the one with the highest estimation certainty. Second, the first keypoint of the spline is used apart from the connector. Both are transformed into two 3D points using the intersection point of the rays as previously described. The orientation is restricted to an orientation about the $\mathbf{z}$ axis of the robot base, as the box $\mathbf{xy}$ plane is parallel to the one of the robot base coordinate system. $\mathbf{R}$ can then be calculated by subtracting one point from the other and projecting and normalizing the vector, resulting in the $\mathbf{x}$ axis of $\mathbf{R}$. As $\mathbf{z}$ is fixed, $\mathbf{y}$ can be calculated using the cross product of $\mathbf{z}$ and $\mathbf{y}$. The translational part is calculated by going $\frac{2}{3}$ of the distance between the estimated connector point and the estimated first keypoint in the keypoint direction, which is just an arbitrary position on the segment and has no particular meaning.

2.5 Skill-Based Robotic Architecture

The proposed wiring harness bin picking is evaluated by executing the states according to Fig. 9. Starting from a home position, a valid grasp point is estimated. The robot moves then and tries to grasp the valid position.

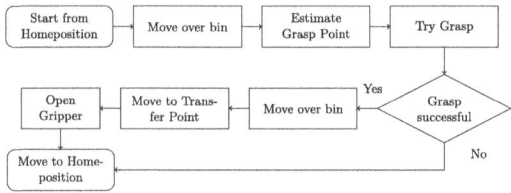

Fig. 9. Flow chart of the robot validation program. The different states in the flow chart are composed of parameterizable robot skills. The transfer point indicates a station where in future work a second robot grasps the connector of the wiring harness for plugging purposes.

Failure might occur due to the gripper colliding with the box, the robot moving into singularities due to limited rotation around its hand joint or due to collision of the gripper with the wiring harness resulting from incorrect grasp point estimations. The results are shown in Sect. 3.

3 Experiments

3.1 Validation of Keypoint Measurement Accuracy

When determining the positions of keypoints in space, it is of interest to achieve this goal as accurately as possible. Therefore, a scene with calibration targets randomly placed in the pickup box in 13 poses was prepared and recorded, see Fig. 10. The recordings were then evaluated for two different metrics: (a) The accuracy with which the three rays of the cameras intersect at the reference points and (b) the accuracy of length measurements in the scene. For both, the corners of the ArUco markers were used as reference targets.

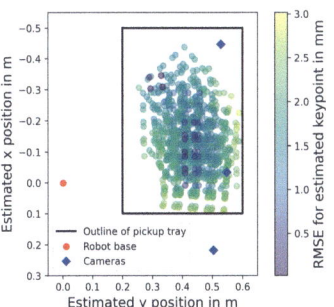

(a) Scatterplot of estimated keypoints on a calibration target with 64 reference points in 14 poses, viewed from above. The color denotes the RMSE distance between the rays for the estimate and the resulting point.

(b) Histogram of the errors in mm while measuring distances of 40 mm on the calibration target with 64 reference points in 14 poses and an example of four length measurements.

Fig. 10. Validation of the 3D point estimation

(a) After estimating the position of the reference target, the distances between the rays and the corresponding estimated target positions were calculated and displayed as shown in Fig. 10a. (b) For this evaluation, the distances between the corners of the ArUco markers were calculated along their four sides, as depicted in Fig. 10. Then the actual length of the marker sides of 40 mm was subtracted from the measurements, resulting in an absolute error. One can see, that the average length measurement is longer than the actual value. This could be caused by the cameras being slightly out of focus, possibly resulting in OpenCV estimating the targets to be larger than they actually are. These results show that estimating keypoints and lengths in the scene is possible, but further research in this area is needed to investigate possible sources for errors and improve the quality of the measurements.

3.2 Wiring Harness Bin Picking

The experiment validates if the proposed uncertainties in the form of intrinsic and extrinsic calibration uncertainties and the uncertainties in the 2D spline prediction with 3D grasp point evaluation are low enough for autonomous robotic wiring harness bin picking applications. 50 consecutive trials according to the flowchart in Fig. 9 were recorded and evaluated using a Franka Emika Panda 7 Axis Cobot. Each time, the wiring harness is almost randomly positioned inside the bin. Three simplifications were used for placing the wiring harness inside the bin. (i) The pickable segment can't face in the direction of the robot base to avoid robotic kinematic hand singularities. (ii) The pickable segment can't be too close to the box's border to avoid collisions between the robotic hand and the box and (iii) the pickable segment is always on top of the wiring harness. Figure 11 shows 6 example trials.

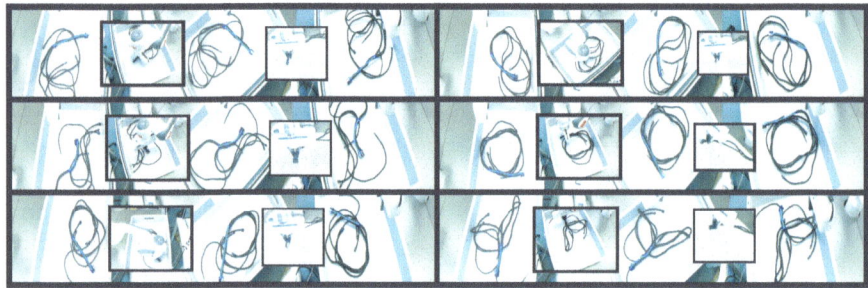

Fig. 11. Six exemplary images of the proposed robotic wiring harness bin picking application. The resulting grasps and the transfer points' final positions are shown in one trial.

Inside one trial there are the three spline predictions from the three cameras, an image right after the grasping and an image showing the connector in a fictive transfer point for further manipulation, which is not part of this paper. The quantitative results of the experiment are shown in Table 2.

Table 2. Experiment results out of 50 trials. If robotic errors aren't accounted for, the overall success rate is 91%.

Outcome	Count	Robot Kinematic Singularities	Collision with Box	Wrong Estimation Wrong Grasp
Successful	41 (82%)	-	-	-
Failed	9 (18%)	1 (2%)	5 (10%)	3 (6%)

The most interesting parts are the failures, explained in the following. Robotic singularities and collisions with the box are not directly addressed in this paper and therefore not part of any further investigation. The interesting parts concern the wrong estimation and wrong grasp section. Three different failure scenarios occurred. Once, the grasp was unsuccessful as the estimated grasp point was too close to the connector, which can be detected using the resulting gripper width, and ended up in a failed attempt. In another trial, the estimated depth was too large, such that the robot grasped an underlying cable as well. The third failure occurred due to a failed spline prediction in one image, detected by a low certainty score of an average of 1.23 for the spline prediction, compared to an average of around 5–6 for the successful spline predictions.

4 Conclusion

This paper presents a robotic bin-picking solution for unstructured wiring harness with minimal initial configuration constraints. It consists of a multi-stereo camera solution and a deep-learning based spline segment estimation of the pickable wiring harness segment. The overall success rate over 50 trials with various configurations is 82%. However, if trials are neglected which have failed due to robotic singularities or collision with the box, the success rate increases to 91%. This paper presents a method for generic intrinsic and extrinsic camera calibration using modified and novel methods. Furthermore, a novel deep-learning architecture with novel spline loss is presented. The proposed system was able to deal with highly overlapped wires. A transfer point was chosen as a final state for the robotic handling, which requires a second robot in the future to grasp the connector and install it into a car door.

The camera system generates extensive, accurate sensor data, of which only a small portion is used for point estimation. A multi-stage low- to high-resolution approach could enhance accuracy. 3D reconstruction algorithms like space carving [29] and shape-from-silhouette [30] could further improve results. For the robot, advanced control with singularity and collision avoidance, and strategies for selecting viable grasping points can be added. Currently, only a single segment of the harness is considered. Future work should address full topology estimation and develop solutions that generalize across different wiring harnesses.

Supplementary information Dataset with camera calibration and validation: *doi.org/10.35097/1n5nqty3c1rxekhk*.
Dataset with annotation, model weights logs: *doi.org/10.18419/darus-4236*.

Acknowledgment. The authors would like to thank the Ministry of Science, Research and Arts of the Federal State of Baden-Württemberg for the financial support of the projects within the InnovationsCampus Future Mobility (ICM).

References

1. Automotive wiring harness market (2024). https://www.precedenceresearch.com/automotive-wiring-harness-market
2. Nguyen, H.G., Kuhn, M., Franke, J.: Manufacturing automation for automotive wiring harnesses. Procedia CIRP **97**, 379–384 (2021). https://doi.org/10.1016/j.procir.2020.05.254
3. Salunkhe, O., et al.: Review of current status and future directions for collaborative and semi-automated automotive wire harnesses assembly. Procedia CIRP (2023). https://doi.org/10.1016/j.procir.2023.09.061
4. Uhlig, D.: Light Field Imaging for Deflectometry, PhD thesis, Karlsruher Institut für Technologie (KIT) (2023)
5. Besl, P.J., McKay, N.D.: A method for registration of 3-D shapes. IEEE Trans. Pattern Anal. Mach. Intell. **14**(2), 239–256 (1992). https://doi.org/10.1109/34.121791
6. Nguyen, T.P., Yoon, J.: A novel vision-based method for 3D profile extraction of wire harness in robotized assembly process. J. Manuf. Syst. **61**, 365–374 (2021). https://doi.org/10.1016/j.jmsy.2021.10.003
7. Kicki, P., et al.: Tell me, what do you see?-interpretable classification of wiring harness branches with deep neural networks. Sensors **21**(13), 4327 (2021). https://doi.org/10.3390/s21134327
8. Guo, J., Zhang, J., Gai, Y., Wu, D., Chen, K.: Visual recognition method for deformable wires in aircrafts assembly based on sequential segmentation and probabilisitic estimation. In: 2022 IEEE 6th Information Technology and Mechatronics Engineering Conference (ITOEC), vol. 6, pp. 598–603. IEEE (2022)
9. Zhang, X., Domae, Y., Wan, W., Harada, K.: Learning efficient policies for picking entangled wire harnesses: an approach to industrial bin picking. IEEE Rob. Autom. Lett. **8**(1), 73–80 (2023). https://doi.org/10.1109/lra.2022.3222995
10. Wnuk, M., Zürn, M., Paukner, M., Ulbrich, S., Lechler, A., Verl, A.: Case study on localization for robotic wire harness installation. In: Kiefl, N., Wulle, F., Ackermann, C., Holder, D. (eds,) Advances in Automotive Production Technology – Towards Software-Defined Manufacturing and Resilient Supply Chains, pp. 333–343. Springer, Cham (2023)
11. Jiang, X., Koo, K., Kikuchi, K., Konno, A., Uchiyama, M.: Robotized assembly of a wire harness in car production line. In: 2010 IEEE/RSJ International Conference on Intelligent Robots and Systems, pp. 490–495 (2010). https://doi.org/10.1109/IROS.2010.5653133
12. Žagar, B.L., et al.: Copy and paste augmentation for deformable wiring harness bags segmentation. In: 2023 IEEE/ASME International Conference on Advanced Intelligent Mechatronics (AIM), pp. 721–726 (2023). https://doi.org/10.1109/AIM46323.2023.10196168
13. Zürn, M., Wnuk, M., Lechler, A., Verl, A.: Topology matching of branched deformable linear objects. In: 2023 IEEE International Conference on Robotics and Automation (ICRA), pp. 7097–7103 (2023). https://doi.org/10.1109/ICRA48891.2023.10161483
14. Wang, H., et al.: Overview of computer vision techniques in robotized wire harness assembly: current state and future opportunities. Procedia CIRP **120**, 1071–1076 (2023)
15. Moons, T.: 3D reconstruction from multiple images part 1: principles. Found. Trends Comput. Graph. Vis. **4**(4), 287–404 (2010). https://doi.org/10.1561/0600000007

16. Raspberry Pi: Raspberry Pi Documentation - Camera. https://www.raspberrypi.com/documentation/accessories/camera.html
17. Zhang, Y.-J.: 3D Computer Vision, 1st edn. Springer, Singapore (2024). https://doi.org/10.1007/978-981-19-7603-2
18. Grossberg, M.D., Nayar, S.K.: A general imaging model and a method for finding its parameters. In: Proceedings Eighth IEEE International Conference on Computer Vision. ICCV 2001, vol. 2, pp. 108–115 (2001). https://doi.org/10.1109/ICCV.2001.937611
19. Hezel, D.: Design and programming of a setup for automated camera calibration, Bachelor's thesis, Karlsruhe Institute of Technology (KIT), Karlsruhe, Germany (2024)
20. Bradski, G.: The OpenCV Library. Dr. Dobb's J. Softw. Tools (2000)
21. Schneider, P.J., Eberly, D.H.: Geometric Tools for Computer Graphics. The Morgan Kaufmann Series in Computer Graphics. Elsevier Science (2003)
22. Murugan, A.: Iterative closest point (ICP) algorithm in python (2020). https://github.com/iitaakash/icp_python/tree/master
23. Liu, Z., et al.: Swin transformer v2: scaling up capacity and resolution. In: International Conference on Computer Vision and Pattern Recognition (CVPR) (2022)
24. Lin, T.Y., Dollár, P., Girshick, R., He, K., Hariharan, B., Belongie, S.: Feature pyramid networks for object detection. In: 2017 IEEE Conference on Computer Vision and Pattern Recognition (CVPR), pp. 936–944 (2017). https://doi.org/10.1109/CVPR.2017.106
25. Ronneberger, O., Fischer, P., Brox, T.: U-Net: convolutional networks for biomedical image segmentation. In: Navab, N., Hornegger, J., Wells, W.M., Frangi, A.F. (eds.) MICCAI 2015. LNCS, vol. 9351, pp. 234–241. Springer, Cham (2015). https://doi.org/10.1007/978-3-319-24574-4_28
26. Nibali, A., He, Z., Morgan, S., Prendergast, L.A.: Numerical coordinate regression with convolutional neural networks. arXiv preprint arXiv:1801.07372 (2018)
27. Akiba, T., Sano, S., Yanase, T., Ohta, T., Koyama, M.: Optuna: a next-generation hyperparameter optimization framework (2019)
28. Virtanen, P., et al.: Fundamental algorithms for scientific computing in Python. SciPy 1.0. Nat. Methods **17**, 261–272 (2020). https://doi.org/10.1038/s41592-019-0686-2
29. Kutulakos, K.N., Seitz, S.M.: A theory of shape by space carving. Int. J. Comput. Vis. **38**, 199–218 (2000)
30. Szeliski, R.: Rapid octree construction from image sequences. CVGIP Image Underst. **58**(1), 23–32 (1993)

Open Access This chapter is licensed under the terms of the Creative Commons Attribution 4.0 International License (http://creativecommons.org/licenses/by/4.0/), which permits use, sharing, adaptation, distribution and reproduction in any medium or format, as long as you give appropriate credit to the original author(s) and the source, provide a link to the Creative Commons license and indicate if changes were made.

The images or other third party material in this chapter are included in the chapter's Creative Commons license, unless indicated otherwise in a credit line to the material. If material is not included in the chapter's Creative Commons license and your intended use is not permitted by statutory regulation or exceeds the permitted use, you will need to obtain permission directly from the copyright holder.

Reinforcement Learning to Improve Finite Element Simulations for Shaft and Hub Connections

Muhammad Saeed[1,2(✉)], Hassaan Muhammad[3], Narmeen Sabah[3], Jan Falter[1], Markus Wagner[1], Boris Eisenbart[2], and Matthias Kreimeyer[1]

[1] University of Stuttgart, Pfaffenwaldring 9, 70569 Stuttgart, Germany
msaeed@swin.edu.au
[2] Swinburne University of Science and Technology, Hawthorn, VIC 3122, Australia
[3] National University of Science and Technology, 12, Islamabad 44000, Pakistan

Abstract. Advancements in technology and numerical methods have shifted from slow, resource-intensive software to faster predictive solutions powered by artificial intelligence (AI). An exemplary case is the analysis of interference fit connections between a cylindrical shaft and hub, which has the potential to redefine optimal design, minimizing stress and maximizing torque transmission. Traditional experimental analysis using Finite Element Method (FEM) simulations is undeniably time-consuming, inefficient, and complex, thus necessitating the deployment of AI as a pivotal tool in industrial applications. This paper unequivocally introduces a cutting-edge technique that harnesses two powerful AI approaches: Supervised Learning and Reinforcement Learning. The Reinforcement Learning approach expounded in this paper impeccably predicts the shaft-hub geometry set, eliminating the need for iterative simulations and drastically streamlining the optimization process. In order to address this challenge, a Supervised Learning model is rigorously trained using limited data obtained from experimental structural analysis. Subsequently, the predictions from this model serve as the environment for the Reinforcement Learning (RL) algorithm. The customized environment in Reinforcement Learning ingeniously employs the model to refine predictions by adjusting the input parameters for different geometric sets through respective actions on the environment.

Keywords: Interference Fit Connections · Supervised learning · Reinforcement learning · Custom environment

1 Introduction

The shaft-hub connections are essential to machine components that transfer power within the drive systems [1]. Within the drive system, the frictional forces at the shaft-hub connection induce elastic deformation, causing stress, torque, and resulting rotation [2]. Cylindrical interference fits are widely used because of their considerable torque transfer, low manufacturing costs, high shock resistance, and excessive stresses, making them more efficient and durable [2, 3]. For these interference fits, changes are made to the geometrical features of the connection to optimize the stresses (first principal stress, joint pressure, notch effect) [3]. Regardless of the type of shaft-hub connection, determining an optimum geometry of the connection is necessary to ensure maximum power transfer and improved efficiency of the dynamic system [4]. Traditionally, the quantization of geometry changes is based on the complex experimental stress analysis of the connection, which requires expert knowledge alongside expensive and time-consuming simulation tools to perform finite element analysis (FEA) [2, 5]. The interference fit design has been rigorously optimized through the automated Finite Element Analysis process using a robust Artificial Intelligence approach, resulting in significantly increased time efficiency, as demonstrated in [5]. However, these approaches still need to be improved for general application due to their focus on approximate iterative solutions, which leads to a lack of precision and a retained dependency on time-consuming finite element simulations [2]. Hence, this paper introduces the optimization technique of shaft-hub connections independent of the finite element (FE) simulation, made possible through a robust implementation of an Artificial Neural Network (ANN) Machine Learning Model (ML) that is trained on existing simulation results, thereby reducing time inefficiencies [6]. Reinforcement learning is a subdivision of machine learning, and its methodology involves an agent that interacts with the environment by exploring and engaging with the surroundings for knowledge acquisition, with the ultimate objective of acquiring the most effective approach to achieve a specific goal [7], which in our case is geometry optimization for most efficient resulting stress [2]. As mentioned earlier, the Artificial Neural Network Machine Learning model is coupled with a policy optimization Reinforcement Learning Algorithm and generates accurate predictions for optimum geometry and stress of the shaft-hub connection [5]. Hence, the research explores how optimization of shaft-hub connections can be achieved through minimal and pre-existing finite element analysis iterations. The format of the research paper is outlined as follows. The use of Artificial Intelligence in Shaft and Hub connections is discussed in Sect. 2 and 0. Section 4 encompasses the initial evaluation of Reinforcement Learning to support Finite Element-based Simulations. In Sect. 6, the summary and conclusion of the paper, along with the prospective future development is, discussed.

2 Use of Artificial Intelligence in Shaft and Hub Connections

The concept of Artificial Intelligence (AI) revolves around endowing computer systems with the ability to replicate human thought processes and actions and comprehend intricate relationships within data [7, 8]. In the provided context, Fig. 1 classifies terms associated with Artificial Intelligence and Machine Learning. This categorization showcases the potential impact of AI on the realm of Finite Element simulations. Finite Element simulations, commonly utilized in engineering and scientific domains, often necessitate extensive computational resources and specialized expertise, resulting in time-consuming and expensive processes [2]. Integrating AI within the FEA Algorithm presents an opportunity to mitigate these challenges by reducing the computational resources and time required for these simulations [4]. One promising approach involves utilizing AI techniques, specifically neural networks, to train on historical simulation data, thereby replacing the necessity to conduct Finite Element simulations from the ground up [6]. Additionally, Reinforcement Learning algorithms enable real-time adaptation of system components in response to evolving operating conditions or environmental factors [3]. This enables the dynamic adjustment of simulation parameters or control strategies to optimize performance and bolster resilience against uncertainties or disruptions through ongoing monitoring of system dynamics and feedback signals [9].

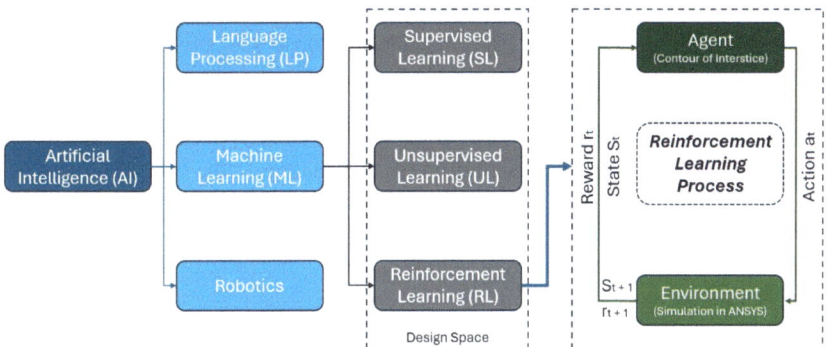

Fig. 1. Schematic Illustration of AI and RL

3 Use of Machine Learning in Shaft and Hub Connections

In AI, machine learning is divided into stress or geometry optimization subfields, as shown in Fig. 1 (bottom-left). Machine learning encompasses unsupervised, supervised, and reinforcement learning approaches [6, 8]. It includes unsupervised, supervised, and reinforcement learning approaches. Unsupervised learning finds hidden relationships in data, while supervised learning uses labelled data for accurate predictions. Reinforcement learning aims to maximize the value of the reward function for optimal outcomes [6, 10]. Implementing reinforcement learning alongside polynomial regression-based machine learning in finite element simulations helps automate and enhance the simulation process [9]. Reinforcement learning (RL) is a branch of AI that enables an agent to learn how to make decisions to achieve a specific goal by interacting with an environment [10]. RL algorithms optimize decision-making by maximizing the value or strategy function [7, 8]. RL can automatically adjust simulation parameters to optimize structural integrity, performance, and reliability for shaft and hub connections in FE simulations [2] based on predefined performance objectives and constraints.

3.1 Use of Supervised Leaning in Shaft and Hub Connections

Supervised learning is employed to predict the optimized stress values derived from the changing geometry of the shaft-hub-connections. There is an observed relation between the geometry set of the connections and real-time pressure values obtained from simulations. A linear fitting relation is derived from the design parameters of shaft and hub connections, which can be interpreted and applied by the machine learning model [8].

Consequently, the data becomes amenable to linear regression analysis, which assesses the relationship between two variables by predicting a dependent variable based on one or more independent variables. Linear Regression is a widely utilized statistical modelling technique, as highlighted by Sutton [11]. The supervised learning regression model is trained on the dataset obtained from previous iterative simulations and can then predict the Shaft and Hub Connections pressure values with input parameters being geometry set and the step size [2, 3]. The model was then evaluated on a test set, resulting in the Evaluation Metrics given in Table 1. This research harnesses the reinforcement learning framework to ensure precision and optimal performance. It employs Mean Absolute Error (MAE) and Root Mean Squared Error (RMSE) to evaluate the models in reinforcement learning (RL) rigorously. These metrics are utilized to gauge the accuracy of predictions, assess the performance of components such as value function approximators and policy models, and scrutinize state-action value estimations. The adaptability in choosing between MAE and RMSE empowers us to select the most suitable metric for the framework based on the specific requirements of the task. In this research, MAE is deemed more appropriate than RMSE in ensuring robust performance, especially in noisy data or outliers.

Table 1. Evaluation Metric between MAE and RMSE

Evaluation Metrics	
MAE	17.844
RMSE	22.025

3.2 Use of Reinforcement Leaning in Shaft and Hub Connections

Reinforcement Learning is a machine learning technique that involves an agent interacting with the environment using the Markov Decision Process [12]. At each time step "t" belongs to the number of iteration "N", the agent's subsequent Action (A_t) on the environment is 'reinforced' by the State (S_t) and produces a Reward ($R_t + 1$) of the environment. The goal of the reinforcement learning algorithm is to maximize the Reward (R_t). This is shown in Fig. 1 (right) [6]. The reinforcement learning framework iteratively employs a trained regression model to predict pressure values. It is the reinforcement learning agent that utilizes feedback mechanisms based on rewards to guide specific actions. Limited simulation data is efficiently used for training the machine learning model, reducing time constraints. The reinforcement learning agent plays a crucial role in ensuring that the input to the machine learning model represents the optimal geometric set value for the next iteration, aiming to optimize the shaft and hub connection. The Reward (R_t) for training the model is contingent upon the alignment of predicted pressure values with the training pressure data obtained from experimental shaft and hub connections analysis.

3.3 Choosing the Reinforcement Leaning Framework

Reinforcement Learning (RL) algorithms play a crucial role in guiding the decision-making of an RL agent within its environment [5]. These algorithms fall into two main categories: 'on-policy' algorithms, such as Proximal Policy Optimization (PPO) and A2C, which learn from current states and actions, and 'off-policy' algorithms, such as Q-Learning and Deep Q-Learning, which learn from random states and actions. The Q-Learning algorithm, as described in [1], has proven effective in accelerating Finite Element Simulations. However, it is highly time-consuming when used to eliminate simulations, as depicted in Table 2 entirely. To determine the most suitable algorithm, a comprehensive comparative analysis of the A2C, PPO, and DQN (Deep Q-Learning Network) algorithms was conducted.

Table 2. Computational Time vs. AI framework

	Reinforcement Learning	Q-Learning
Number of iterations	1999	14
Time taken	9.6 s	7.78 h

As demonstrated in Table 3, DQN was eliminated due to its time inefficiency for optimization. An on-policy RL algorithm is deemed the best fit for optimizing shaft and hub connections. This is because changes in the shaft geometry follow an iterative pattern to achieve optimization. Proximal Policy Optimization, which supports continuous action spaces [5, 13], was selected as the most suitable RL algorithm for shaft and hub optimization.

Table 3. Different RL framework comparison

A2C	PPO	DQN
• High Variance – varies drastically with minor changes in the parameters • Slower that PPO w.r.t increase in complexity	• Takes the least amount of time as the complexity of the environment increases • Maximum reward in minimum steps • Low variance	• Simple algorithm • Low convergence rate – will take a lot of time to converge to an optimum value of pressure

The continuity in the action space directly impacts the pressure/stress values, as our Reinforcement Learning Model does not require discrete changes in these values. The selection of PPO is further influenced by a quantitative comparison of 'on-policy' algorithms, as depicted in Table 4. The "proximal agent's clipping" technique curtails the RL agent's policy changes with each update, ensuring stability in its learning rate and maintaining a balanced sensitivity within the algorithm.

Table 4. Different RL framework comparison

	Loss	Value Loss	No. of updates	Total reward
A2C	-	0.0667	1999	0.995
PPO	15.3	69.8	40	0.741
SAC	8.46	-	9811	0.664
TD3	14	-	9558	0.659
DDPG	22.6	-	9558	0.688

4 Initial Evaluations of Machine Learning to Support Finite Element Based Simulations

The supervised learning regression model is trained using real-time Finite Element Analysis data as shown in Fig. 2. This model serves as an environment for the reinforcement learning PPO algorithm and is initially provided with a geometry set. The linear regression model predicts the pressure array for the given geometry set, which the reinforcement learning model then optimizes. The state of the reinforcement learning model is the predicted pressure array, and it acts by adjusting the step size and geometry set for the next prediction of the supervised learning regression model. It then compares the resulting pressure array with the previous array and the boundary condition to determine a reward that will influence the following action. The resulting visualizations of pressure optimization are displayed below.

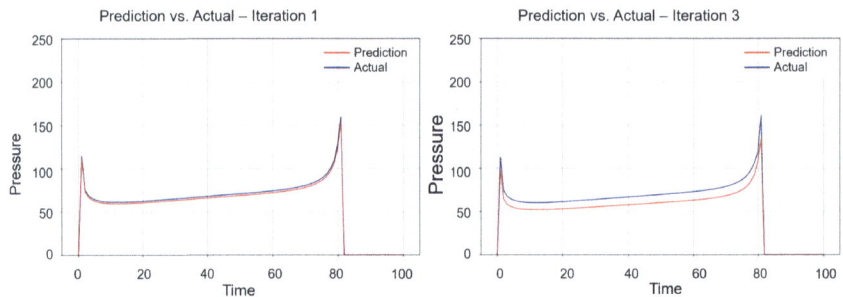

Fig. 2. Application of RL model on FE based data

5 Further Experimentation and Room for Improvement

Finite element analysis data is optimized using a reinforcement learning model, where the process completes as it meets standard boundary conditions. Key challenges identified for improvement include the number of iterations required and the accuracy of the predicted pressure array. By employing iterative optimization cycles through the RL model, progressively fine-tuned results are achieved, leading to more stable convergence in the number of iterations and more precise pressure predictions, as demonstrated in Table 5.

Table 5. Results from fine-tuning framework

Fine-Tuning	Iterations
First Fine-tuning	10
Second Fine-tuning	10
Third Fine-tuning	11

The efficiency of this optimization process is strongly influenced by the supervised learning model's performance and the extent of its training. As a result, the model diverges due to underfitting when exposed to excessively large training datasets. This leads to decreased accuracy in the predicted pressure array with increased iterations in the RL environment, highlighting the need for more balanced model training to improve the accuracy and efficiency of the system, as shown in Fig. 3.

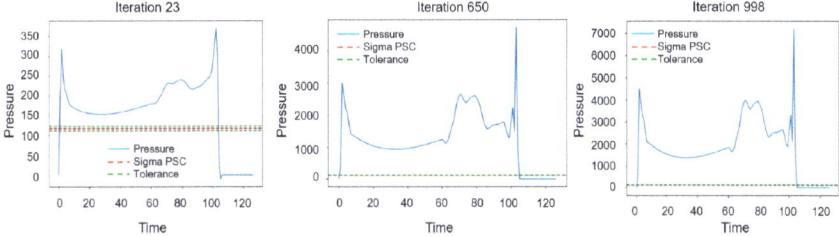

Fig. 3. The series of graphs show the optimization process which fails to converge

There remains substantial potential for improvement in the fine-tuning process of the optimization framework. The current approach, utilizing a simple linear regression model, could be significantly enhanced by integrating more sophisticated techniques such as Support Vector Machines (SVM), Random Forest, or Gradient Boosting, which are expected to provide improved predictive accuracy and convergence. Additionally, ensuring that boundary conditions are more closely aligned with the characteristics of the dataset would further enhance the effectiveness of Reinforcement Learning (RL) environment. Implementing dynamic step-size strategies within the RL environment, driven by agent feedback, could lead to more efficient optimization outcomes and improved convergence performance.

6 Conclusion and Future Outlook

This research presents a novel method for optimizing interference fit connections between cylindrical shafts and hubs by leveraging artificial intelligence, specifically supervised and reinforcement learning. The approach is designed to diminish the requirement for time-consuming finite element simulations by training a supervised learning model using a limited dataset derived from experimental analyses. Subsequently, this model is incorporated into a reinforcement learning framework to refine the shaft and hub geometry iteratively, aiming to achieve optimal stress distribution and torque transmission. The outcomes illustrate that integrating these AI techniques can efficiently and effectively enhance the design of shaft and hub connections, resulting in precise predictions and improved performance. This approach highlights the potential of AI to revolutionize traditional engineering practices, rendering them more adaptive and less resource intensive. Future research endeavors should explore the application of this AI-driven method to other mechanical systems, refining reinforcement learning algorithms to enhance scalability and expanding datasets to improve model accuracy.

Acknowledgement. The authors acknowledge the support of the Global Innovation Linkage (GIL) grant awarded by the Australian Federal Government and the Institute for Engineering design and industrial design (IKTD) at Stuttgart University, the Swinburne University of Technology and Students from National University of Science and Technology.

References

1. VDI-Gesellschaft Produkt- und Prozessgestaltung, 9. VDI-Fachtagung Welle-Nabe-Verbindungen 2022: Dimensionierung - Fertigung - Anwendungen und Trends : Leinfelden-Echterdingen bei Stuttgart, 23. und 24. November 2022. Düsseldorf: VDI Verlag GmbH (2022)
2. Saeed, M.S., Falter, J., Dausch, V., Wagner, M., Kreimeyer, M., Eisenbart, B.: Artificial intelligence techniques for improving cylindrical shrink-fit shaft-hub couplings. Proc. Des. Soc. **3**, 645–656 (2023). https://doi.org/10.1017/pds.2023.65
3. Dausch, V., Kröger, J., Kreimeyer, M.: An ai-based approach to optimize stress in shrink fits. Proc. Des. Soc. **2**, 1549–1558 (2022). https://doi.org/10.1017/pds.2022.157
4. Falter, J., Binz, H., Kreimeyer, M.: Investigations on design limits and improved material utilization of press-fit connections using elastic-plastic design. 2666–4968, **13**, p. 100124 (2023). https://doi.org/10.1016/j.apples.2022.100124
5. Schulman, J., Wolski, F., Dhariwal, P., Radford, A., Klimov, O.: Proximal policy optimization algorithms. OpenAI (2017). http://arxiv.org/pdf/1707.06347
6. Kidwai-Khan, F., Wang, R., Skanderson, M., Brandt, C.A., Fodeh, S., Womack, J.A.: A roadmap to artificial intelligence (AI): methods for designing and building AI ready data to promote fairness. J. Biomed. Inform. **154**, 104654 (2024). https://doi.org/10.1016/j.jbi.2024.104654
7. Liu, C., Liu, D.: Deep reinforcement learning algorithm based on multi-agent parallelism and its application in game environment. Entertainment Computing **50**, 100670 (2024). https://doi.org/10.1016/j.entcom.2024.100670
8. Skansi, S.: Introduction to Deep Learning: From Logical Calculus to Artificial Intelligence. Springer, Cham (2018)

9. Charniak, E.: Introduction to artificial intelligence. Undergraduate Topics in Computer Science: Springer (1985)
10. Netto, S.M.B., Paiva, A.C., Neto, A.A., Silva, A.C., Leite, V.R.C.: Application on reinforcement learning for diagnosis based on medical image. Erscheinungsort nicht ermittelbar: INTECH Open Access Publisher (2008). https://directory.doabooks.org/handle/20.500.12854/64664
11. Sutton, R.S., Barto, A.G.: Reinforcement Learning: An Introduction (Adaptive computation and machine learning): An introduction. MIT Press, Cambridge, Massachusetts (1998). https://ieeexplore.ieee.org/book/6267343
12. Xiang, D.: Reinforcement learning in autonomous driving. ACE **48**(1), 17–23 (2024). https://doi.org/10.54254/2755-2721/48/20241072
13. Kulkarni, P.: Reinforcement and systemic machine learning for decision making. Wiley-Blackwell, Chichester, UK (2012). https://onlinelibrary.wiley.com/doi/book/10.1002/9781118266502

Open Access This chapter is licensed under the terms of the Creative Commons Attribution 4.0 International License (http://creativecommons.org/licenses/by/4.0/), which permits use, sharing, adaptation, distribution and reproduction in any medium or format, as long as you give appropriate credit to the original author(s) and the source, provide a link to the Creative Commons license and indicate if changes were made.

The images or other third party material in this chapter are included in the chapter's Creative Commons license, unless indicated otherwise in a credit line to the material. If material is not included in the chapter's Creative Commons license and your intended use is not permitted by statutory regulation or exceeds the permitted use, you will need to obtain permission directly from the copyright holder.

Design Automation of Fibre Composite Parts via Graph-Based Design Languages

Jonas Braiger[✉], Johannes Baur, Jakob Gugliuzza, Stephan Rudolph, Stefan Carosella, and Peter Middendorf

Institute for Aircraft Design (IFB), University of Stuttgart, Pfaffenwaldring 31, 70569 Stuttgart, Germany
`braiger@ifb.uni-stuttgart.de`

Abstract. The design of fibre composite parts spans across various design phases, starting with requirements collection and decomposition, going on with geometry definition, mechanical calculations, optimization of fibre arrangements, production planning and so on and ending with manufacturing. Typically, changes in one design phase may affect all subsequent phases or even affect the entire design process. Graph-based design languages (GBDLs) offer a unified and consistent digital data model capable to store and propagate relationships of design objects via linked graph nodes, thereby expressing the different mutual dependencies for design modifications within the model. By defining the contents of the knowledge domains in a vocabulary, rules and a rule sequence, the entire design process can be described and executed automatically. This paper shows how GBDLs can be applied to the design process of fibre composite parts by using them to automatically generate a FEM simulation for a given geometry and then using the result to model and plan the production sequence for the created composite part.

Keywords: graph-based design languages · fibre composite parts · design process · digital engineering

1 Introduction

When designing fibre composite parts, a wide variety of requirements must be considered and coordinated. As is typical for every kind of design process, not all requirements and dependencies of the various components are known right from the beginning, but only emerge during the various development phases and require revisions of the initial design proposal. This may be due to the fact that it's initially unclear which initial variants will deliver promising designs later on, and which will not need to be pursued further. In addition, domain knowledge is rarely concentrated in one place, so that designs that receive a positive evaluation in an earlier design phase might be discarded later, because the detailed requirements or limitations from a specific domain were not known to the team producing the initial design. The same problems occur if requirements are changed or revised during the course of the project. With a purely linear approach, where every design step is carried out strictly after each other, such revisions may often lead to long delays and additional costs, especially if problems only become apparent during a late stage in the design process.

2 Overview

2.1 Graph Based Design Languages

Graph based design languages (GBDLs) offer a framework to model complex engineering problems in a coherent and self-consistent form. To represent the components of a given engineering domain via a GBDL, the relevant concepts are expressed as classes within vocabularies, based on the concept of a visual modeling language like the UML (Unified Modeling Language) [1]. These concepts in the vocabulary usually stem from different ontologies underlying the various engineering domains that are relevant to the specific model. As an example, typical concepts in the geometry domain are *points*, *lines*, *surfaces* and *bodies*, which might be used to construct a certain geometry when creating a CAD model. In order to be independent of a vendor-dependent DSL (Domain Specific Language) as an input format to a dedicated CAD-kernel, corresponding classes of *points*, *lines*, *surfaces* and *bodies* are available in an ontology for this domain, in this case a so-called abstract geometry, as shown in Fig. 1.

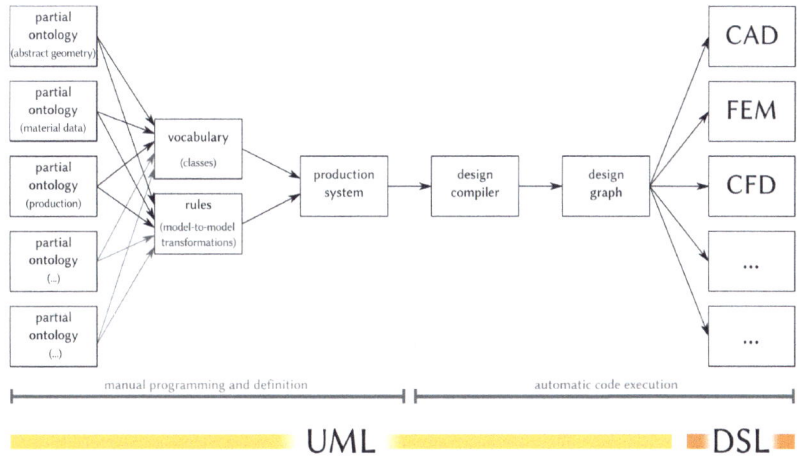

Fig. 1. General schematic of graph based design languages (adapted from [3])

Each class may have additional attributes, parameters and links to other classes to describe the behaviour of the concept it represents. Likewise, rules are used to describe the behaviour of those classes, as well as how they may modify or interact with each other. The order in which these rules are applied is defined in a production system. A design language compiler is necessary to construct the design graph from a given design rule sequence. In this work, the software package *Design Compiler 43* (for short: *DC43*) by IILS mbH [2] is used.

In Fig. 1, the instances of the class objects are stored in a design graph, with each instance represented as a node, and each association between two different instances as a link between the two nodes. By continuously modifying and expanding the design graph via the execution of appropriate rules, the overall design can be represented within the design graph. This resulting design graph can then be used as a central model for further mappings into the different domain-specific languages. Using this approach, the design graph and therefore the designs generated from it do not only differ parametrically but also topologically, meaning that different designs explicitly contain different classes in different arrangements. The resulting design may then be mapped and exported via interfaces to the software tools used in each domain involved into the design.

2.2 Fibre Reinforced Composite Manufacturing Process

There are many different manufacturing methods for fibre reinforced composite parts that highly influence the properties of the produced parts [4,5]. A common method to produce high quality, low-cost components in small and medium quantities is via vacuum assisted resin infusion (*VARI*). For this process, layers are cut from large sheets of fibre material and draped into a mould. The entire assembly is then sealed and infused with resin using vacuum. After the resin has hardened, either at room temperature or with the help of a heat source, the finished part is removed from the mould, trimmed and ready for further processing. However, when using different fibre types for instance the described process sequence has to be varied or extended. The general procedure is shown in Fig. 2.

To represent the manufacturing process of fibre reinforced composite parts in the digital framework, it is broken down into discrete steps. During each step multiple process parameters and sub-steps are considered. As an example, the layer draping process step can be further broken down into sub-steps like the stacking of the layers into the mould, the application of various seals, foils and other components and the testing of the vacuum setup. In the case of natural fibres, the preparation additionally includes the drying of the fibres in the oven for a certain time at a certain temperature.

This manufacturing process is well suited for modelling in GBDLs. The aim of the model is to make predictions about key process parameters such as production costs and process durations. This enables different parts, production types or components to be compared with each other in order to quickly and reliably select the best possible process, even with many possible variants. While methods to automate various aspects of the process chain for fiber manufacturing have been presented [6–8], the use of graph-based design languages offers a unique way to model the entire design process in a single consistent and comprehensible data model and have been used for applications such as ply placement and assembly modeling [9] and estimating production costs and durations for specific composite part fabrication methods [10,11].

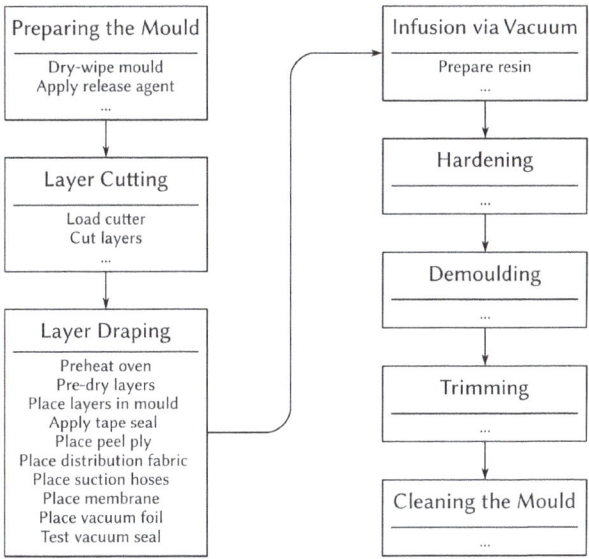

Fig. 2. Flowchart of the VARI manufacturing process, including selected sub-steps

3 Implementation

The process chain starts with a finite-element optimization of a given part in order to adjust layer sizes and strengthen the geometry with additional material where necessary, while reducing thickness of the part in less strained areas. The results of the optimization are then used to analyze the production process and calculate key process parameters. Two GBDLs are used for this purpose, which can either be run separately or in combination, as shown in Fig. 3.

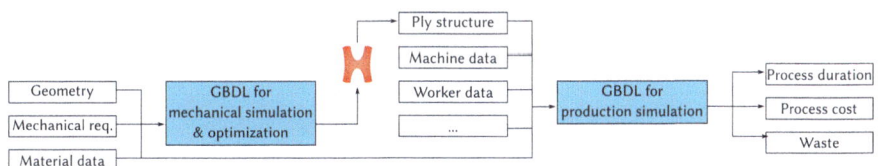

Fig. 3. Entire process chain with two GBDLs and integration of a FE-solver

As presented in Fig. 1, it is first required to model the key entities from the specific domains into classes of the design language. For the production process of fibre composite parts, these ontologies contain material data, mechanical requirements for the optimization, the geometry of the part to be analyzed, as well as information about machines, workers, work stations and other objects directly involved in the manufacturing process.

Initially, a geometry of the component to be modeled must be provided, which can either be designed directly in DC43 using the integrated geometry operators or may alternatively be imported as an .stl file, should it come from an external source. The boundary conditions, on the basis of which the FEM simulation is to design the component, must also be known. These include mechanical targets such as applied loads, moments or bearing and fixing points, but also specifications regarding the material, including mechanical characteristic values, as well as specifications for the layer structure and orientation of the fibre sheets [12].

The relevant classes from the chosen simulation program and their behaviour are modeled in the first GBDL. This way, the entire optimization problem is set up within DC43, where the geometry is first defined or imported, after which load and material parameters and finally the exact type of optimization are specified. For this project, Altair Hypermesh and Optistruct are used to simulate and optimize the part. To transfer the resulting design graph (as depicted in Fig. 4) to the FEM solver, a .tcl-script is created from the generated DC43 objects, which contains a description of the entire optimization problem. The script is automatically passed to Hypermesh via the included interface, where it is used to set up the simulation and start the optimization.

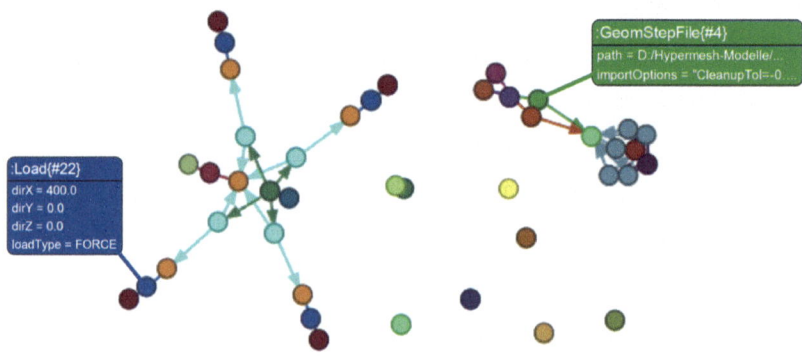

Fig. 4. Example design graph for an optimization problem, with two objects expanded to show their parameters

The optimization provides a proposal for the layer structure of the composite component. Based on the defined loads, a optimal material distribution is calculated, where areas that are structurally relevant are reinforced, while other regions that don't experience heavy loads are thinned, in order to save material and weight. The result represents the plies of composite material that need to be stacked on top of each other during the manufacturing process in order to produce a structurally sound component.

After the optimization, the ply information is used to simulate the production sequence via a second GBDL. While both design languages can be linked to form a continuous simulation chain with a singular, consistent data model, splitting the optimization from the production section ensures that both parts can also be operated fully independently, should the need arise. The GBDL responsible for the production sequence uses the ply information from the optimization process to estimate key parameters based on the chosen production process. These include process durations, waste and costs incurred, as well as machinery needed to carry out the selected manufacturing process.

To model a specific manufacturing process, the entire workflow is separated into logical steps like shown in Fig. 2. Each of those steps may be further subdivided into sub-steps that represent the actions that need to be carried out to complete the task. Every sub-step is associated with the machinery, workers, work stations and tools necessary for its completion. Other manufacturing processes can be modeled by switching out the rules for different production steps. This approach allows for a modular approach where predefined rules can be strung together in different ways for different processes, leading to a high degree of code re-usability.

The process time required for each sub-step is based on an individual equation for each action. Some tasks have a fixed duration, while others depend on the geometry of the part produced, such as draping plies into the mould, or applying a seal around the perimeter of the part. These exact process times are estimated based on the experience of production specialists, as well documented data for each task often doesn't exist. Based on the calculated process durations, process costs can also be estimated, using the energy consumption of the machine in use or the wage of a worker who carries out the task. Other key manufacturing parameters like accumulated waste can also be estimated via their own equations. Once the process parameters for all actions are calculated and all manufacturing steps and sub-steps are written to the design graph, the entire model can be used as a base for further analyses with regards to the planning of manufacturing resources.

4 Application

The approach shown here has been implemented and tested for a practical use case. Within the CYCLOMETRIC-project, a center console for a car was to be designed from the ground up [13]. A 3D-CAD model of the console was provided for the simulation. Natural flax fibre composite was chosen as a material, and the load requirements are based on real life data from industry experts.

The simulation begins by importing the console geometry into the first GBDL. The material and the load requirements are modeled in corresponding classes, as well as information about fibre orientation, optimization constraints and various other constraints needed for the mechanical optimization. After running the GBDL and creating the design graph, simulation and optimization are started directly from inside DC43.

After the optimization run has finished, the results are evaluated. The key parameters of the plies such as ply area or circumference are calculated. Together with a database of available workers and machines, process steps and sub-steps are modeled and joined together. As each sub-step has its own calculations for duration, cost and weight, the respective information about the entire production process is derived as an aggregation from process steps and sub-steps. The results can be used to evaluate different designs (that have been derived by modifying the design graph) in terms of their mechanical, economic or ecological performance.

It has been shown that an optimized design of the center console with a fine tuned material distribution not only has obvious weight advantages compared to a model with a uniform thickness, but could also lead to shorter production times due to the multiple smaller plies being easier to drape into the mould than single large plies that span the entire part. It has to be noted however that draping and positioning durations are highly dependant on the shape, size and complexity of the plies, the skills and experience of the worker and the flexibility of the material. The relationship of these different factors is complex, hard to quantify and not yet fully understood. Therefore, the results may serve as a rough indicator for this specific process step, but should not be taken as exact values. Other process steps can be analyzed more accurately, such as cutting the plies from large sheets or heating and curing the entire part in the oven. In those cases, the speed or heating rates of the respective machines are known and can be used to generate reliable estimations for the process durations.

5 Discussion

GBDLs have been shown to be a useful tool in assisting the design process of fibre reinforced composite parts and enable a high degree of flexibility and precision in the design phase. By combining the aspects of structural dimensioning and production planning in GBDLs, a big part of the larger general design process has been consolidated into a single continuous process chain ready for a repeated, machine-executed design process chain. In consequence, GBDLs facilitate the exploration of various design configurations and optimization of fibre orientations and placements to achieve desired mechanical properties in accordance to proposed requirements, opening this way the perspective for a systematic *" computer-based exploration of the design and manufacturing space"*.

A key point for further studies and improvements is the collection of more reliable estimations for specific process durations, as those heavily influence not only the time each task takes, but also production cost and the overall evaluation of a generated design. As process durations are highly dependent on a multitude of factors, such as part geometry or worker experience, this poses a major challenge in order to improve the process chain in pursuit of more accurate results.

Acknowledgments. The project CYCLOMETRIC was funded by the German Federal Ministry for Education and Research (BMBF), grant number 02J21E032. We would also like to extend our thanks to all other project partners for valuable input.

References

1. Schmidt, J., Rudolph, S.: Graph-based design languages: a lingua franca for product design including abstract geometry. IEEE Comput. Graph. Appl. (2016). https://doi.org/10.1109/MCG.2016.89
2. IILS mbH: Design Compiler 43. Design Compiler 43 is a trademark of IILS mbH. https://www.iils.de. Accessed 13 June 2024. IILS Ingenieurgesellschaft für intelligente Lösungen und Systeme mbH, Trochtelfingen, Germany
3. Rudolph, S.: On design process modelling aspects in complex systems. In: 13th NASA-ESA Workshop on Product Data Exchange (PDE), Cypress, 11-12 May 2011 (2011)
4. Grisin, B., Carosella, S., Middendorf, P.: Dry fibre placement: the influence of process parameters on mechanical laminate properties and infusion behaviour. Polymers (2021). https://doi.org/10.3390/polym13213853.
5. Heudorfer, K., et al.: Method of manufacturing structural, optically transparent glass fiber-reinforced polymers (tGFRP) using infusion techniques with epoxy resin systems and e-glass fabrics. In: Polymers (2023). https://doi.org/10.3390/polym15092183.
6. Swery, E.E., et al.: Complete simulation process chain for the manufacturing of braided composite parts. Compos. Part A: Appl. Sci. Manuf. **102** (2017). https://doi.org/10.1016/j.compositesa.2017.08.011.
7. Deden, D., Frommel, C., Glück, R., Larsen, L., Malecha, M., Schuster, A.: Towards a fully automated process chain for the lay-up of large carbon dry-fibre cut pieces using cooperating robots. In: SAMPE Europe Conference 2019, Nantes, France, 17-19 September 2019 (2019)
8. Jayasekara, D., Lai, N.Y.G., Wong, K., Pawar, K., Zhu, Y.: Level of automation (LOA) in aerospace composite manufacturing: Present status and future directions towards industry 4.0. J. Manuf. Syst. **62** (2022). https://doi.org/10.1016/j.jmsy.2021.10.015.
9. Kormeier, T., Rudolph, S.: Rule-based material topology modelling of composite structures. In: 16th ICED, Paris, France, 28-31 July 2007 (2007)
10. Holland, M., Paul, N., Linder, C., Elsafty, H.F.M., Ernis, G., Geinitz, S.: Model-driven approach for integrated design and process planning of fiber composite aerostructures. In: SAMPE, Madrid, Spain, 3-5 October 2023 (2023)
11. Holland, M., Schuster, J.: Model-based design for manufacturing of automated fiber placement (AFP) composite structures. In: SAMPE Europe Conference 2024, Belfast, Northern Ireland, 24-26 September 2024 (2024)
12. Ramsaier, M.: Integration der Topologie- und Formoptimierung in den automatisierten digitalen Entwurf von Fachwerkstrukturen. Dissertation, Bergische Universität Wuppertal (2020)

13. CYCLOMETRIC: Lebenszyklusorientierte Entwicklung von Fahrzeugkom-po-nen-ten. https://www.iao.fraunhofer.de/de/presse-und-medien/aktuelles/lebenszyklusorientierte-entwicklung-von-fahrzeug-komponenten.html. Access 13 June 2024. Fraunhofer-Institut für Arbeitswirtschaft und Organisation IAO, Stuttgart

Open Access This chapter is licensed under the terms of the Creative Commons Attribution 4.0 International License (http://creativecommons.org/licenses/by/4.0/), which permits use, sharing, adaptation, distribution and reproduction in any medium or format, as long as you give appropriate credit to the original author(s) and the source, provide a link to the Creative Commons license and indicate if changes were made.

The images or other third party material in this chapter are included in the chapter's Creative Commons license, unless indicated otherwise in a credit line to the material. If material is not included in the chapter's Creative Commons license and your intended use is not permitted by statutory regulation or exceeds the permitted use, you will need to obtain permission directly from the copyright holder.

Modeling and Optimization of Sustainability Criteria Along the Product Engineering Process of Handling Systems

Johannes Scholz[✉], Florian Koessler, and Jürgen Fleischer

Wbk Institute of Production Science, Karlsruhe Institute of Technology (KIT), Kaiserstrasse 12, 76131 Karlsruhe, Germany
johannes.scholz@kit.edu

Abstract. The importance of sustainability continues to grow. Considering the entire life cycle of a product is of great importance in evaluating and optimizing sustainability criteria. The greatest impact on sustainability is in the early phase of product development, but the information about the developed system is low. However, resolving the trade-offs between economic, environmental, and technical criteria requires information about the product over its entire life cycle. Many options need to be evaluated, especially in the early stages of product development. It is time-consuming to create detailed simulation models for each option and each stage of the life cycle. Therefore, the knowledge about the product in the phases of the product development process was analyzed, and measurable sustainability indicators for the phases were identified. Furthermore, the necessary models for the evaluation of the sustainability indicators are selected. Based on these findings, the paper presents a framework for the development of sustainable handling systems, which are part of nearly every automated production system, across all life phases. Starting from the requirements of the handling system, this framework helps to identify suitable solutions, evaluate them, and optimize them with respect to their sustainability score.

Keywords: sustainable manufacturing · product engineering process · production equipment

1 Introduction

Sustainability is one of today's biggest challenges. With its high energy and resource demand, the manufacturing industry is a main factor in improving sustainability [1]. Establishing sustainable production is also part of the Sustainable Develoment Goals of the UN. Legal regulations are increasing the pressure on companies to measure and optimize sustainability indicators. [2] The Corporate Sustainability Reporting Directive (CSRD) requires large companies to submit a report on various environmental factors based on international standards. Even if small and medium-sized enterprises are currently still exempt from the reporting obligation, the pressure regarding environmental sustainability along the supply chain is being increased. In addition, the industry is under high cost pressure, which is why ecological requirements must be brought into line with costs. The triple bottom-line approach defines three pillars of sustainability the economic sustainability, ecological sustainability and social sustainability. [3] The focus of this paper is on the economic and ecological aspects. Over 80% of ecological impact and costs are defined during product development [4]. In [5], the need to integrate ecological indicators in the product development phase was presented. In order to avoid optimizations in the use phase being overcompensated by increased use of resources in production, a cross-life-phase approach is necessary. A common methodology for evaluating the ecological impact of products is the life cycle assessment (LCA). Guidance for LCA is provided by [6], but the data acquisition is very time-consuming. In addition, the amount of data available is very small, especially in the early stages of the development of a product. The same challenge arises with regard to costs. Although there are several approaches to the use lifecycle costing, as in [7]. Due to the data situation, the existing approaches are only used to a limited extent. Overcoming the data availability, model-based systems engineering (MBSE) in combination with digitalization shows great possibilities.

Since handling systems are part of nearly every automated production system and engineered in an engineer-to-order process with high time pressure, the paper presents an economic and ecological evaluation model for handling systems. According to [8], a handling system consists of a handling device such as an industrial robot, a gripper system, and further peripherals. Based on the evaluation model a framework based on MBSE is presented for the efficient development of economically and ecologically sustainable handling systems.

The paper is structured in 5 sections. After the introduction Sect. 2 presents related research in the field of product engineering and approaches to design handling systems. Sect. 3 presents the developed evaluation model and the required data. In Sect. 4 the resulting framework to develop a handling system with regard to the evaluation model is presented. The Sect. 5 closes the paper with a summary and an outlook.

2 Related Research

2.1 Existing Approaches for Product Engineering

Several process models for product development exist in the literature. A very common one is the VDI 2221 [9], which presents a procedure for problem-solving during product development and defines required activities during problem-solving. Although the guideline refers to the solution evaluation step, it does not provide a specific procedure or possible solutions. Furthermore, the integration of the product life cycle in the development process is not considered.

The VDI 2206 [10] mentions that complex products like cyber-physical systems require a more detailed development process and presents the V-model. The focus of the approach is to generate a system architecture to decompose the system in its domains of software, mechanics, and electricity. After the systems are developed separately they are integrated, validated and verified to the system requirements.

Although the guideline recommends going through the process several times, there is no indication of the level of detail of the individual development cycles. In addition, the interaction between product, material and production is not considered with regard to sustainability criteria. Furthermore, an allocation of methods for finding solutions or optimizing them is not given.

However, the importance of modeling the product from different perspectives over the whole development cycle is mentioned. In [11], even a diamond as development model is recommended instead a "V". The idea is to have the development "V", like in [10], for the physical development. In parallel along an inverted "V" the associated model is developed.

As a combination of [9–11], a model with a focus on lightweight design was developed. The model demonstrates the importance of integrating materials, manufacturing, and joining throughout the product development process, and frames the process through modeling and optimization. The product is described in accordance with the RFLP approach. The approach describes the product by requirements "R", functions "F", logical elements "L" and physical design "P". The requirements and functions describe the problem space. The logical elements describe possible solutions without any geometry information. To overcome the big gap between "P" and the "L" the technical view "T" is introduced in this model. The description is used in three levels of detail. The system level considering the whole system, the subsystem level considers subsystems fulfilling subfunctions, and the component level considers individual components for function fulfillment. [12] A possible digital toolchain with a focus on the integration of the production was described in [13]. The focus here was on incorporating production and development by means of technology chain planning. This approach makes it possible to optimize individual components in terms of cost, weight, and CO2 but lacks a holistic evaluation model based on economic and ecological criteria as well as suitable modeling and optimization of design variables across all phases of the product life cycle in accordance to economic and ecological sustainability.

2.2 Existing Approaches for Optimizing Handling Systems

In particular, the optimization and selection of industrial robots using various evaluation criteria has been an intensive research topic in recent years. For industrial robot selection, many times pre-defined evaluation criteria are used, which are evaluated with a Multi-Criteria Decision Making method based on information provided by technical sheets. [14] The methods also lack in integrating models in the selection process. The integration of sustainability criteria like material consumption and emissions during manufacturing are not mentioned due to missing information. In [15], an approach was provided dealing with uncertainties in decision-making during the selection and design of components for handling systems. However, the approach does not consider how the development chain and modeling approaches need to be designed for an optimized design along the product engineering process. Furthermore, the approaches do not consider sustainability criteria during the requirements and function definition. The used requirements like operating space and degree of freedom are selected based on expert knowledge. No systematic method is used for minimizing the required degree of freedom or the required length of the axis.

To evaluate the energy consumption and life cycle costs of assembly systems, Damrath et al. [16] used physical simulation models. Several energy optimization measures were provided to the simulation model and evaluated regarding the life cycle costs. Stuhlenmiller et al. [17] used a physical simulation model for industrial robots to analyze the effect of optimized path planning and payload on the life cycle cost and greenhouse gas emissions over the entire life cycle of a handling system. Although these two approaches demonstrate the potential of using physical simulation in the early development stages such as requirements engineering, functional design and logical design information regarding mass or inertia is not available already.

Ramirez [18] analyses first the required degree of freedom based on the required robot path. This makes it possible to select the kinematics with the fewest axes and to optimize based on the inverse kinematics with regard to selectable criteria. Baumgaertner et al. [19] formulate the robot optimization problem based on kinematic models as a trajectory planning problem by modeling the length of each axis with a linear axis. This leads to faster calculation and the possibility to include the positioning of the industrial robot in its environment.

In summary, the approaches presented are mostly methods that are presented as standalone solutions for optimizing industrial robots. There is no approach that supports the development continuously based on an evaluation model for sustainability criteria and the application of a digital toolchain for information aggregation.

First of all, this requires an evaluation model which covers ecological, economic as well as technical evaluation criteria over the entire life cycle. Based on the evaluation model required information for evaluating needs to be defined. The information required results in development requirements that call for an end-to-end methodology with an associated digital toolchain.

3 Evaluation Model

An evaluation model that considers the entire life cycle is needed to develop economically and ecologically sustainable handling systems. Figure 1 visualizes the lifecycle with actors and processes along the lifecycle of a handling system. The system integrator is responsible for the design of the handling systems. Based on the customer requirements for the system the system integrator selects components like grippers from component manufacturers and designs as well as manufactures its own parts. With the decisions during the design process, the system integrator has a great impact on energy consumption and behavior during usage and end-of-use. Furthermore, the resource consumption and emissions in the earlier stages of the product lifecycle are also defined by the decisions made during product development. To do this, decisions must be made based on an evaluation model that considers the most important aspects of economic and ecological sustainability over the entire life cycle, in order to avoid overcompensating for optimizations at certain stages of the life cycle. Furthermore, it is important to consider the whole system to identify interdependencies between subsystems and components. For evaluating the life cycle of a product system LCA is a common methodology. As first step of the LCA, the system boundaries need to be defined. [6] Due to the requirement to consider the entire life cycle, the boundary on the left side is the raw material extraction, and on the right side is the end-of-use. For evaluating the product system, the input and output flows, as well as the intermediate product flows, need to be analysed and evaluated. The aim of the evaluation model is to evaluate the economic and ecological sustainability of the product. For evaluating ecological sustainability, the Product Carbon Footprint (PCF) is used, which is the sum and removal of greenhouse gas emissions in a product system expressed as CO2 equivalents and is based on the life cycle assessment with climate change as single impact category. [20] Executing an LCA on Fig. 1 the intermediate product flow starts with the raw material extraction. The inputs to the processes along the life cycle are energy and auxiliary materials. During usage also, maintenance as well as spare parts are required and need to be considered. The outputs are waste as well as emissions.

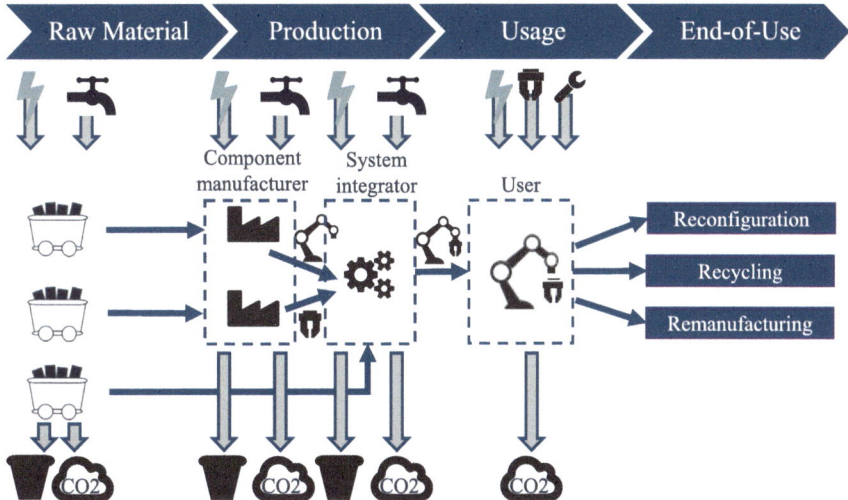

Fig. 1. Life cycle of a handling system with the in- and output flows of the actors

In Table 1 the structure of the PCF regarding the life cycle stages and the correlations with product properties are presented. It becomes clear that properties such as the type and quantity of material influence several life phases. In addition, the energy mix used to convert energy consumption into the PCF is a significant influencing factor.

For evaluating the product system from a cost perspective, the Life Cycle Costing (LCC) can be applied. For calculating the manufacturing cost the material flow costing in accordance with [21] is used. This procedure makes it possible to allocate the costs to the product and to the resulting material loss, which increases transparency. This leads to a similar structure as the PCF from a cost perspective. Especially in the production phase, labor costs and machine costs must be considered. The factors influencing the costs and the PCF are very similar, but a separate analysis is still necessary to be able to map the influence of various optimization measures. For example, the use of green power can minimize PCF manufacturing energy, but it can lead to increased costs. For evaluating the PCF and costs resulting from spare parts, the mean time between failures (MTBF) can be used to calculate the probability of default [22]. Combined with the costs and PCF of the spare parts itself the indicators can be calculated.

Table 1. Interdependencies between product system properties and the PCF

	Product Carbon Footprint (PCF)				
	PCF manufacturing		PCF usage	PCF transport	PCF end-of-use
Variables	PCF material extraction	PCF manufacturing energy			
Material type	x	x			x
Material amount	x	x			x
Waste	x				
Manufacturing time		x			
Manufacturing process		x			
Energy mix	x	x	x		x
Energy consumption			x		
Maintenance			x		
Transport distance				x	
Transport type				x	
End-of-use strategy					x

The technical requirements are another important aspect of the evaluation model. Although these must be fulfilled, the degree of fulfillment is also a good indicator of optimization potential. The description of the technical indicators depends on both the level of detail and the requirements defined for the developed system. The description of the technical indicators at system level includes, for example, the permissible velocities and the load capacity for the overall system. Subsystems such as the gripping system can be described via the force that can be applied. A possible indicator could be the ratio between necessary and possible force of a gripper. If this ratio is smaller than one, improvement is possible. At the component level, the technical indicators refer to the utilization of the components in terms of load. It can be used the same ratio between permissible load and permitted load.

In Fig. 2, the bottom-up structure of the system is shown. The three levels describe the level of detail. On the system level, the whole system has the function of moving an object between two positions. This can be separated into the functions of moving and holding. This results in the two subsystems, the industrial robot and the gripping system. These systems require several functions which are fulfilled with components. The economic and ecological indicator, which describes every system, subsystem, and component, is calculated bottom-up in the system. For identifying optimization potentials, the values of PCF and costs can be used. But if a comparison between several system solutions is required the three indicators need to be normalized and summed up to one sustainability score. This adds the possibilities of giving different weights to the indicators.

For a detailed analysis on the component level, each component needs to be described with the required information for the calculation. However, this results in two major challenges. The first challenge is the missing knowledge regarding subsystems and components in the beginning of product development. In this phase, the impact is highest, but a detailed evaluation is not possible due to missing knowledge. The second challenge is the data acquisition over the life cycle for detailed evaluation in later phases of the product development. For this reason, suitable models are needed for the individual life cycle phases to enable the evaluation and optimization of a solution.

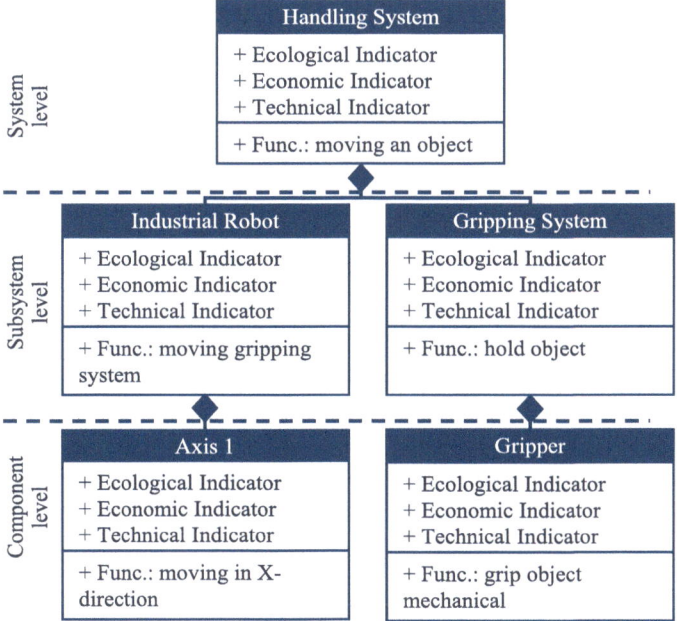

Fig. 2. Structure of the evaluation model and the functions of the systems and components with an UML class diagram

4 Methodology and Digital Toolchain

This section presents the basic methodology and its implementation in a digital toolchain to develop ecologically and economically sustainable handling systems based on the presented evaluation model. To overcome the challenge of missing information at the beginning of product development, the evaluation model needs to be broken down into measurable indicators without detailed component information. For this purpose, several design variables, which can be evaluated based on kinematic models, are collected from literature and evaluated regarding their influence on the overall evaluation model on the life cycle stages of the system. A kinematic model does not take physical effects into account compared to physical models, but the modeling and simulation effort is less due to the reduced information content. For this reason, they are very useful in the early stage of development because during this stage the knowledge of the system is low and a high amount of solution options need to be evaluated.

Table 2 presents the identified design variables and their optimization direction for generating a positive impact in the life cycle stages affected by the design variables.

Table 2. Summary of design variables based on kinematic models and their impact on the life cycle stages (↓ minimize, ↑ maximize, ? undefined)

Design Variables	Raw Material	Production	Usage	End-of-use
Number of Axis	↓	↓	↓	?
Length of Axis	↓	↓	↓	?
Velocity			↓	↓
Moving Distance			↓	↓
Acceleration			↓	↓
Operation Time			↓	
Stiffness			↑	
Manipulability			↑	

Reducing the number of axes reduces the number of parts of a kinematic. This has a direct impact on the raw material phase and the manufacturing phase. Through the reduced number of axes fewer motors, gears, and links are required. In addition to the reduced raw material amount, fewer parts need to be manufactured which reduces costs and emissions during production.

Reducing the length of the axis of the kinematic leads to a reduction of material consumption and reduces the effort during manufacturing of the mechanical parts.

Both of the mentioned design variables have a positive impact on the usage. Through fewer axes and smaller links, the carried masses are reduced. This leads to smaller torques in the joints which lead to less energy consumption and longer lifetime of the joints. Furthermore, the reduced torque requirement can lead to the selection of smaller motors with higher efficiency and less cost.

In contrast to the number of axes and the length of axes the velocity, moving distance, and acceleration do not describe the mechanical structure. These design variables describe the usage of the kinematic which is affecting the mechanical structure as well as the economic and ecological sustainability during usage and end-of-use.

The velocity, acceleration, and moving distance should be decreased to have a positive impact on energy consumption and maintenance during usage. Furthermore, the components were not so heavily loaded, which have a positive impact on the end-of-use strategy. The reduction of the velocity and acceleration can lead to an increase in the operation time. First of all, the operation time is a boundary condition that needs to be fulfilled. Furthermore, decreasing the operation time reduces energy consumption. For this reason, the right trade-off between the design variables needs to be found. Additionally, the length of the axes and the number of axes can affect the required velocities and accelerations to reach the required operation time. Increasing the stiffness and manipulability has a positive impact on the technical indicator during the usage phase.

Based on the evaluable design variables with kinematic models and the evaluation model, a model-driven development process and the corresponding digital tools were developed. The development process and the associated toolchain are presented in Fig. 3. First of all, the boundary conditions need to be defined. This consists of the planned robot path, the orientations of the handling object along the path, acceleration, and velocity boundaries as well as the cycle time. Additionally, the weight of the handling object is needed, which affects the mechanical structure as well as the selection of the gripper.

As previously mentioned reducing the number of axes has a high impact on several phases of the life cycle. The number of axes is dependent on the required degrees of freedom along a path. More degrees of freedom require more axes. Based on the required path the minimum number of required degrees of freedom can be identified. For this purpose, a tool was developed to calculate the required degrees of freedom based on the approach of [18]. The necessary degrees of freedom are described in a binary vector with six entries. The first three entries represent the three translational degrees of freedom and the last three entries describe the three orientational degrees of freedom. If a degree of freedom is not required the entry is zero. If a movement is required, the entry is one. [18] Furthermore, the tool is able to identify if a linear axis for moving the kinematics is required. For this purpose, the distances between the maximum positions of X-, Y- and Z-coordinates are checked. If the distance in one direction is greater than 3 m, a linear axis along this direction is required as the first axis of the kinematic.

The resulting required motion vector in combination with the additional linear axis sets the boundary conditions for further development. Using the path analyzing tool reduces the effort for analyzing the requirements and reduces the impact of expert knowledge. This can lead to new solutions.

To define the optimal axis structure, a database with kinematics with different numbers of axes, different axes types, and different axes orientations is used. The kinematics are described with Unified Robot Description Files (URDF) in accordance with the description from [19]. In this approach each link between two joints of the kinematic is described with a linear axis. In the database each link is described with two linear axes. If it is a linear axis the degree of freedom of the two linear axes, which describe the design variables for axis length, are orthogonal to the degree of freedom of the linear axis. If it is a revolute joint, one linear axis is orthogonal to the rotation axis of the revolute joint. The other linear axis is parallel to the direction of the rotation axis of the revolute joint. The optimization goal needs to be defined individually based on the design variables described in Table 2. If the focus is on short axes and a low number of axes, it needs to be considered that the length of a linear axis depends on its length of movement. For this purpose, it is necessary to integrate the maximum moving distance in the objective function. Furthermore, it should be considered that the structure of a linear axis is more complex than a rotational one.

The selected optimal structure, as well as the length of the axes, are the requirements for the model generation step. The CAD-model generation with its main focus on component level can be supported by a database, which consists of supplier components and components from previous developments. Based on the accelerations and velocities from the kinematic model the required gripping forces can be described. This leads to a better understanding of the process and reduces over-engineering. The resulting effect of reduced requirements leads to a reduced mass of grippers, which has a positive effect on the usage phase, and reduced material consumption, which has a positive effect on raw material consumption and production.

Furthermore, by taking similar components from previous developments the generation of CAD-models is faster. As the development is rarely a completely new development, reference systems already exist in most cases. However, a standardized description of the solutions is required to enable the identification of similarities between the handling system to be developed and previous development projects.

To execute a more detailed evaluation and optimization of the system now, a physical simulation model needs to be set up. The components are described based on the geometry information from the CAD-model. Especially the weight, volume, center of gravity, and moment of inertia are important. To determine the energy consumption of electric motors, their efficiency factor or a detailed electric model of the motor is required to transform the necessary torques in energy consumption. Additionally, physical models enable the consideration of friction parameters which increases the level of detail of the model. Based on the calculated trajectory by the kinematic optimizer the movement of the kinematic during usage can be simulated. The outputs of physical simulation models are torques and forces on the components.

The torque can be translated through the electrical model of the electrical motor into an energy consumption. Furthermore, the lifetime of components can be calculated based on this information which enables the consideration of maintenance during product development. For this purpose, the information is included in the MTBF calculation.

The resulting forces and torques are input parameters to the Finite Element Analysis (FEA) calculations, too. The combination of the physical simulation and the FEA simulation makes it possible to identify positive effects on other parts through weight reduction. Based on the results of the FEA calculation, the technical indicator is evaluated and helps to identify over-engineered parts. To evaluate the technical indicator, a material database with necessary material properties like yield strength is required.

For the raw material phase and the manufacturing phase, the possible manufacturing processes of the self-developed components need to be analyzed. As presented in [13] the component is analyzed with a technology chain planning approach. Based on the geometry information of the CAD model and the material properties, as well as further information like tolerances and surface quality, possible manufacturing chains can be generated. For this purpose, a manufacturing database with capabilities of manufacturing processes, material consumption, energy consumption, and auxiliary material is required. Furthermore, it is important to model the manufacturing time. If the manufacturing process is already defined, Computer Aided Manufacturing tools can be used for

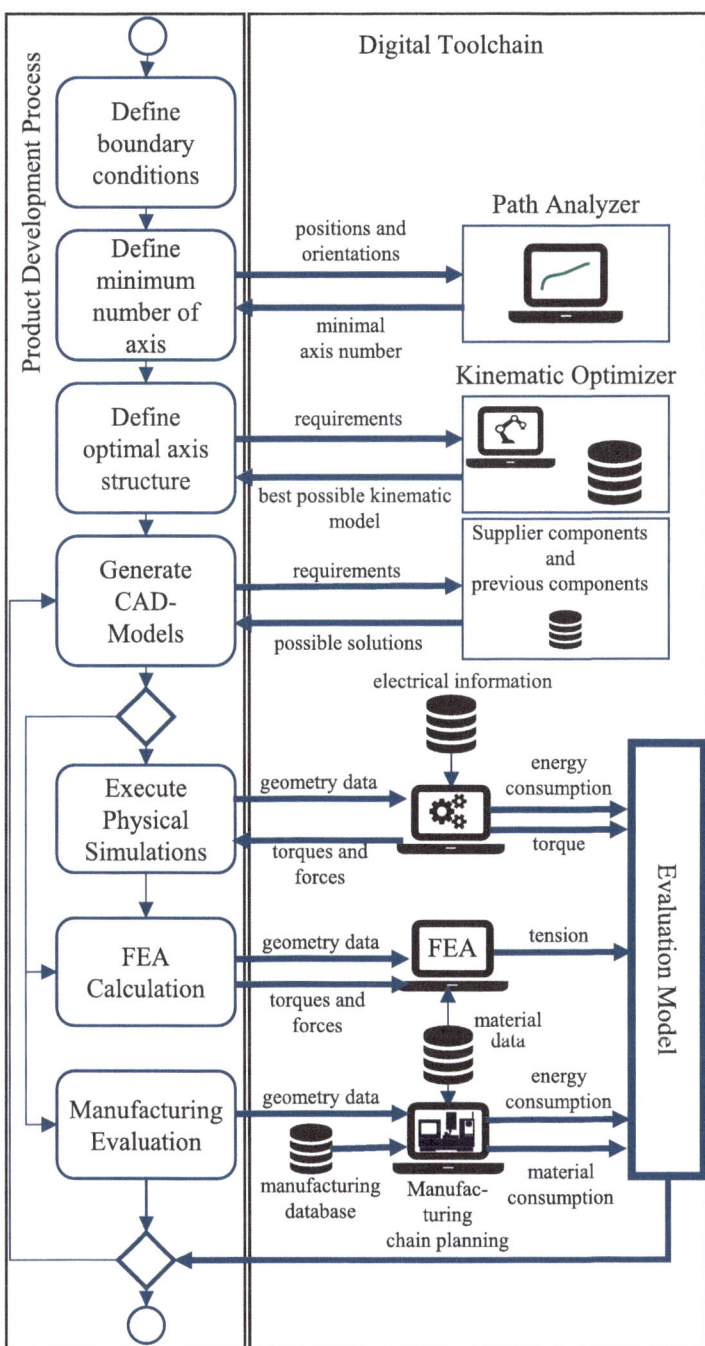

Fig. 3. Visualization of the specific process model and the corresponding toolchain

generating manufacturing time. In the case of a component that is recycled at the end of the life cycle, the costs and PCF are also calculated and integrated into the evaluation model.

For purchased components, such as grippers, the raw material and manufacturing phase are difficult to evaluate. In the best case, the supplier provides detailed information. If this is not the case, the evaluation criteria need to be estimated based on the given information regarding geometry and used material. Through evaluating the handling system from the raw material, production, usage, and end-of-use perspective during the whole development a local optimum that is over-compensated by other life cycle phases is avoided. In addition, the end-to-end methodology with the digital tools avoids time-consuming manual evaluation.

For a broad application of the methodology, the interfaces between the tools along the process must be further specified and implemented.

To compare different variants of the systems, the evaluation can be stored and the next variant can be evaluated with the methodology. If one design variant is selected, optimization potentials can be identified, and further optimization can be made based on the modeling tools and the evaluation models. For example, the trajectory can be optimized based on the physical model regarding energy consumption. Through the combination with the FEA calculation reduced forces to several components can be identified. This enables the engineer to use less material or change the material. The effect of this measure can be evaluated through the manufacturing evaluation. Furthermore, it is possible to execute optimization only from a manufacturing perspective through process optimization for example.

5 Summary and Outlook

This paper presented an approach to integrate and optimize the sustainability of handling systems over their entire life cycle in the product engineering process. For this purpose, an evaluation model, a development process, and the associated toolchain for evaluation and optimization were presented. In the state of the art, several development processes were analyzed, and the missing sustainability evaluation models over the entire life cycle were identified. Furthermore, approaches for selecting components for handling systems as well as optimizing industrial robots were presented. These approaches either consider the pure selection of components without including the overall system or are based on specific phases in the product development process. To overcome the missing evaluation model, the paper presents an evaluation model considering the entire life cycle from the raw material extraction until the end-of-use. It is separated into three indicators: economic indicator, which is calculated based on LCC; ecological indicator, which is described by the PCF; and technical indicator, which evaluates the fulfillment of functions and requirements. The analysis of the input variables to evaluate the PCF clearly showed the challenges regarding the required information. For this reason, a methodology was developed which enables to optimize and evaluate the handling system with less knowledge at the beginning of the development process and reduces the modeling effort. Furthermore, the process describes the transition from the system and subsystem level to the component level and describes the procedure to generate the required data for evaluation based on a digital toolchain. Design variables were identified that can be evaluated based on kinematic models. Based on the analysis of the influence of the variables on the life phases, methods were selected from the state of the art and further developed.

With the presented methodology and associated digital toolchain, an efficient development under consideration of sustainability criteria can be enabled. The application of the method to a specific use case is part of further research. It supports the engineer from the translation of boundary conditions into the required functions up to the detailed evaluation of the system. Furthermore, it is possible to evaluate optimization measures regarding their impact on the ecological and economic sustainability of handling systems. Possible measures may include approaches to increase circularity, topology optimization or improvements to manufacturing processes, and more. Additionally, for a better understanding of the impacts of the design variables for kinematic models on the sustainability of the system, tests with the path analyzer and the kinematic optimizer are required. The digital toolchain itself can be improved by automating workflow through the development of standardized interfaces between the software tools.

Acknowledgement. The authors thank the "Technologietransfer-Programm Leichtbau (TTP LB)" of the Federal Ministry for Economic Affairs and Climate Action (BMWK) for supporting their research within their project "SyProLei" (03LB2007H). Furthermore, the authors thank the Federal Ministry of Education and Research (BMBF) for supporting their research within the project InMicroBatt (03XP0542C) and the Projektträger Jülich for organizational support.

References

1. Bundesregierung Ziele nachhaltiger Entwicklung: Nachhaltig produzieren und konsumieren. https://www.bundesregierung.de/breg-de/themen/nachhaltigkeitspolitik/produzieren-konsumieren-181666. Accessed 10 May 2024
2. United Nations Sustainable Development Goals. https://sdgs.un.org/. Accessed 10 May 2024
3. Elkington, J.: The triple bottom line. In: Russo, M.V. (ed.) Environmental management: Readings and cases, 2. ed., **2**. SAGE, Los Angeles, Calif., pp. 49–66 (2008)
4. Scholz, U., Pastoors, S.: Nachhaltige Produktentwicklung. In: Scholz, U., Pastoors, S., Becker, J.H., et al. (eds.) Praxishandbuch Nachhaltige Produktentwicklung, 2–2009. Springer Berlin Heidelberg, Berlin, Heidelberg, pp. 63–77 (2018). https://doi.org/10.1007/978-3-662-573 20-4_7
5. Ruschitzka, C., Reimann, S., Kuhn, M., et al.: Structuring ecological indicators in machinery and plant engineering for product development. In: Bauernhansl, T., Verl, A., Liewald, M., et al. (eds.) Production at the Leading Edge of Technology, vol 44. Springer Nature Switzerland, Cham, pp. 316–325 (2024). https://doi.org/10.1007/978-3-031-47394-4_31
6. DIN Deutsches Institut für Normung e. V. DIN EN ISO 14040:2021-02: Umweltmanagement - Ökobilanz - Grundsätze und Rahmenbedingungen (ISO 14040:2006 + Amd 1:2020); Deutsche Fassung EN ISO 14040:2006 + A1:2020. https://doi.org/10.31030/3179655
7. Landscheidt, S., Kans, M.: Method for assessing the total cost of ownership of industrial robots. Procedia CIRP **57**, 746–751 (2016). https://doi.org/10.1016/j.procir.2016.11.129
8. Schmalz, J., Reinhart, G.: Automated selection and dimensioning of gripper systems. Procedia CIRP **23**, 212–216 (2014). https://doi.org/10.1016/j.procir.2014.10.080

9. Verein Deutscher Ingenieure. VDI-Richtlinie 2221 Blatt 1: Entwicklung technischer Produkte und Systeme: Modell der Produktentwicklung (2019). https://www.vdi.de/richtlinien/details/vdi-2221-blatt-1-entwicklung-technischer-produkte-und-systeme-modell-der-produktentwicklung
10. Verein Deutscher Ingenieure. VDI-Guidline 2206: Development of mechatronic and cyber-physical systems (2021). https://www.vdi.de/en/home/vdi-standards/details/vdivde-2206-development-of-mechatronic-and-cyber-physical-systems
11. Seal, D.: The System Engineering 'V' - is it still relevant in the digital age? (2018). https://gpdisonline.com/wp-content/uploads/2018/09/Boeing-DanielSeal-The_System_-Engineering_V_Is_It_Still_Relevant_In_the_Digital_Age-MBSE-Open.pdf?pdf=Boeing-DanielSeal-The_System_-Engineering_V_Is_It_Still_Relevant_In_the_Digital_Age-MBSE-Open. Accessed 24 May 2024
12. Scholz, J., Kaspar, J., Quirin, S., et al.: Konzept eines systemischen Entwicklungsprozesses zur Hebung von Leichtbaupotenzialen. Zeitschrift für wirtschaftlichen Fabrikbetrieb **116**, 797–800 (2021). https://doi.org/10.1515/zwf-2021-0182
13. Scholz, J., Zeidler, S., Koessler, F., et al.: Systematic lightweight design of production equipment with a digital toolchain. In: Bauernhansl, T., Verl, A., Liewald, M., et al. (eds.) Production at the Leading Edge of Technology, vol 116. Springer Nature Switzerland, Cham, pp. 24–33 (2024). https://doi.org/10.1007/978-3-031-47394-4_3
14. Prakash Garg, C., Görçün, Ö.F., Kundu, P., et al.: An integrated fuzzy MCDM approach based on Bonferroni functions for selection and evaluation of industrial robots for the automobile manufacturing industry. Expert Syst. Appl. **213**, 118863 (2023). https://doi.org/10.1016/j.eswa.2022.118863
15. Scholz, J., Dilger, L.J., Friedmann, M., et al.: A methodology for sustainability assessment and decision support for sustainable handling systems. Procedia CIRP **116**, 47–52 (2023). https://doi.org/10.1016/j.procir.2023.02.009
16. Damrath, F., Strahilov, A., Bär, T., et al.: Method for energy-efficient assembly system design within physics-based virtual engineering in the automotive industry. Procedia CIRP **41**, 307–312 (2016). https://doi.org/10.1016/j.procir.2015.10.004
17. Stuhlenmiller, F., Weyand, S., Jungblut, J., et al.: Impact of cycle time and payload of an industrial robot on resource efficiency. Robotics **10**, 33 (2021). https://doi.org/10.3390/robotics10010033
18. Ramirez, D., Kotlarski, J., Ortmaier, T.: Combined structural and dimensional synthesis of serial robot manipulators. In: Parenti-Castelli, V., Schiehlen, W. (eds.) ROMANSY 21 - Robot Design, Dynamics and Control, **569**. Springer International Publishing, Cham, pp. 207–216 (2016). https://doi.org/10.1007/978-3-319-33714-2_23
19. Baumgärtner, J., Kanagalingam, G., Puchta, A., et al.: One Problem, One Solution: Unifying Robot and Environment Design Optimization. arXiv (2023)
20. DIN Deutsches Institut für Normung e. V. DIN EN ISO 14067:2019–02: Treibhausgase - Carbon Footprint von Produkten - Anforderungen an und Leitlinien für Quantifizierung (ISO 14067:2018); Deutsche und Englische Fassung EN ISO 14067:2018. https://doi.org/10.31030/2851769
21. DIN Deutsches Institut für Normung e. V. DIN EN ISO 14051:2011–12, Umweltmanagement - Materialflusskostenrechnung - Allgemeine Rahmenbedingungen (ISO 14051:2011); Deutsche und Englische Fassung EN ISO 14051:2011. https://doi.org/10.31030/1803739
22. Zuverlässigkeit im Fahrzeug- und Maschinenbau. VDI-Buch. Springer-Verlag, Berlin/Heidelberg (2004)

Open Access This chapter is licensed under the terms of the Creative Commons Attribution 4.0 International License (http://creativecommons.org/licenses/by/4.0/), which permits use, sharing, adaptation, distribution and reproduction in any medium or format, as long as you give appropriate credit to the original author(s) and the source, provide a link to the Creative Commons license and indicate if changes were made.

The images or other third party material in this chapter are included in the chapter's Creative Commons license, unless indicated otherwise in a credit line to the material. If material is not included in the chapter's Creative Commons license and your intended use is not permitted by statutory regulation or exceeds the permitted use, you will need to obtain permission directly from the copyright holder.

The Application of LCA Data Uncertainty Analysis in the Sustainable Development Process

Ruiyang Deng[✉] and Sebastian Kilchert

Department of Sustainable Systems Engineering (INATECH),
Albert-Ludwigs-Universität Freiburg, Emmy-Noether-Straße 2,
79110 Freiburg, Germany
ruiyang.deng@inatech.uni-freiburg.de

Abstract. Environmental performance is increasingly being emphasized during the design phase. Although the lack of quantity and quality of data leads to inaccurate environmental results in the early stages of design, the ability of life cycle assessment (LCA) to influent environmental performance of the product is significant during this phase. This article aims to evaluate the data uncertainty of LCA during the design phase, integrate the data quality assessment results into the sustainability-oriented development process (NEP), and provide more accurate environmental influence information in the early stages of the design process to avoid major changes later. To that purpose, the life cycle inventory (LCI) data of hydrogen tank production based on pedigree matrix method is used to analyse the data quality of existing data in the early design stages. Methodologies are investigated which link these pedigree matrix methods to data of contribution analysis and probability distributions in order to predict data uncertainty in life cycle impact assessment (LCIA) results. A framework will then be proposed which allows to decide if further data updating or process refinement is needed based the calculated uncertainty for the predicted LCIA result. Through data quality analysis the reliability of LCA results can be improved at early design stage, and according to combination with the sustainable development proc ess LCA can identify the environmental influence during design phase, to avoid the large-scale change due to that in the late stages of design.

Keywords: Life cycle assessment · Data uncertainty analysis · Pedigree matrix · Sustainable development process

1 Introduction

As ecological requirements play an important role in product development, LCA is widely used to provide environmental information during the design phase. However, LCA is typically conducted by the end of product development, when making significant changes to the product is not feasible. This limitation is primarily due to the low availability and quality of data in the early design phase, stemming from the limited knowledge about the product at this stage. Consequently, environmental requirements are often only implemented by iterating through the product development cycles multiple times and then performing assessments

As shown in Fig. 1, data availability and quality increase along the stages of the design phase, while the capability to influence environmental performance decreases. Therefore, for product development, conducting environmental assessments at an early stage and providing environmental impact information for key decisions are crucial for the sustainable development of products. The sustainable development process aims to enable a fully digital and sustainability-oriented product design by considering environmental indicators as additional optimization targets in the early stages of concept development. The objective of the sustainable development process is to optimize existing process chains for vehicle and component development and validation, while simultaneously achieving significant ecological and economic benefits at the various design stages.

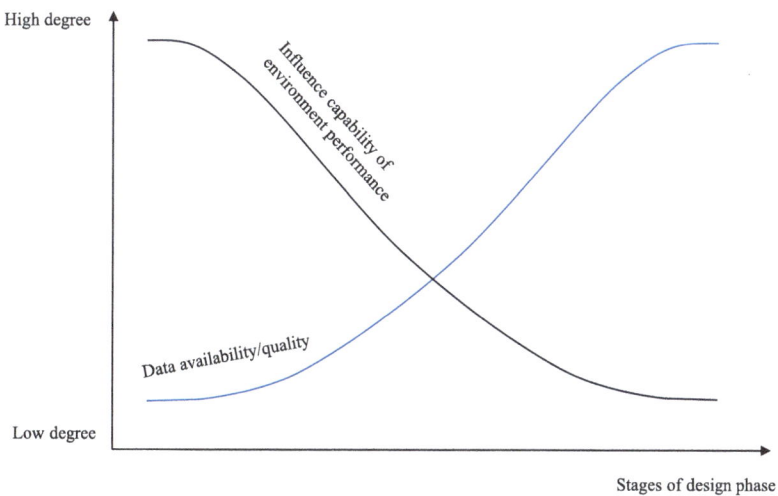

Fig. 1. The ability to influence environmental performance and the availability of data during design stages [1]

The use of a pedigree matrix to quantify data quality for LCA was first introduced by Weidema and Wesnaes [2], inspired by the Numeral Unit Spread Assessment Pedigree (NUSAP) proposed by Funtowicz and Ravetz [3]. This approach has been utilized in the Ecoinvent database since 2005 [4] and has seen widespread use since then. The method employs five data quality indicators (DQIs) to quantify data uncertainty: reliability, completeness, temporal correlation, geographical correlation, and further technological correlation. Each process is scored by expert judgement on a scale from 1 to 5 (with 1 being the best). These scores determine the uncertainty characteristics of the data, which are then used to model the probability density function (PDF).

In the Ecoinvent database, the lognormal distribution is used by default to express the PDF, as it captures many of the key characteristics commonly observed in real-world data. It also accommodates the positivity constraint of most real-life parameters and remains scale-independent in terms of standard deviation [5]. The lognormal distributione is expressed by the geometric standard deviation (GSD) with the 95% confidence interval and geometric mean value [6]. The data uncertainty (expressed by PDF) obtained from the pedigree matrix can then be used to quantitatively manage uncertainty in LCA results.

The pedigree matrix allows for the quantification of data uncertainty through expert judgement without measured variability information in the early design phase. Data quality typically improves as the design phase progresses, but escalating LCA model complexity doesn't always lead to a proportional decrease in overall errors compared to simpler models [7]. Thus, it's crucial to evaluate the data quality and environmental influence of each process at the early stages of product development to balance complexity and accuracy. This study aims to identify data uncertainty at the early stage, reducing the number of iterations required for the environmental model in the product development process and simplifying the model while ensuring accurate results.

2 Methodology

The methodology presented in this paper aims to identify uncertainty hotspots in the early design phase. It consists of three steps, illustrated in Fig. 2. Initially, the environmental impacts are calculated based on the product input data, and a contribution analysis is performed to ascertain the impact share of each individual process. At this stage, uncertainty information is not included, which will be addressed in the next step, where the uncertainty within the input data is assessed. The pedigree matrix is used to realize the semi-quantify approach of the data uncertainty and to determine PDF of each process. Next, Monte Carlo simulations are employed to estimate uncertainty. The Monte Carlo simulation allows for the evaluationof statistical information from LCA predictions through random sampling, based on the previously identified data uncertainty. This enables the analysis of the coefficient of variation and process contributions to identify inventory hotspots and reduce overall model uncertainty. Ultimately, the goal is to improve designers' comprehension of environmental impact assessment and streamline the process of conducting LCA during the design phase.

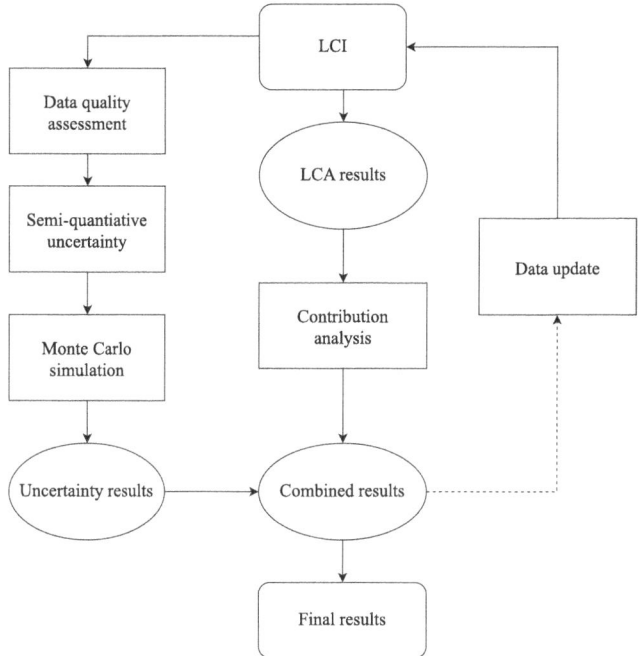

Fig. 2. Flowchart of methodology

2.1 Contribution Analysis

Contribution analysis is used to evaluate the significance of individual processes in influencing final outcomes. By examining the contributions of each process, it becomes possible to discern hotspots, thereby pinpointing areas of particular importance or concern. Especially during the early design phase, when lack of knowledge is common and data uncertainty tends to be heightened, the integration of contribution and uncertainty analyses becomes particularly crucial. This contribution-uncertainty combined approach enables a more in-depth identification of key processes within the model, allowing designers to make better improvements in accuracy and reliability compared to using either method alone.

2.2 Semi-quantitative Approach

The quantitative approach to evaluating data uncertainty utilizes five DQIs, derived from expert judgements using the pedigree matrix approach (Table 1). DQIs are used to represent the data quality at the unit process level, indicating the degree to which each input is adapted to the process.

Individual DQIs are assumed to be equivalent, to calculate the GSD from pedigree matrix, the conversion factors in Table 2 are used, which is also

Table 1. Pedigree matrix based on Ecoinvent 3.0 [5]

DQI	1	2	3	4	5
R	Verified data from measurements	Partially verified data or non-verified data from measurements	Partially based on qualified estimates	Qualified estimate	Non-qualified estimate
C	Representative data from all relevant market sites, sufficient time to balance fluctuations	Representative data from >50% of relevant market sites, sufficient time to balance fluctuations	Data from some relevant market sites (<<50%) or >50% sites but shorter periods	Data from only one relevant market site or some sites but shorter periods	Data representativeness unknown or from few sites and shorter periods
TiC	<3 years difference to dataset period	<6 years difference to dataset period	<10 years difference to dataset period	<15 years difference to dataset period	Age of data unknown or >15 years difference
GC	Data from area under study	Average data from larger area including study area	Data from area with similar conditions	Data from area with slightly similar production conditions	Data from unknown or distinctly different area
TeC	Data from studied enterprises	Data of same technology but different enterprises	Data from studied processes but different technology	Data from related processes	Data from laboratory scale or different technology

R - Reliability, C - Completeness, TiC - Temporal correlation, G - Geographical correlation, TeC - further technological correlation

applied by Ecoinvent 3.9.1. The GSD is calculated using Equation 1 proposed by Frischknecht et al.[4], and by equating the median value with the geometric mean [5], the lognormal probability distribution (PDF) can be directly constructed.

$$GSD = \exp\sqrt{(\ln U_R)^2 + (\ln U_C)^2 + (\ln U_{TiC})^2 + (\ln U_{GC})^2 + (\ln U_{TeC})^2} \quad (1)$$

2.3 Monte Carlo Simulation

In the context of LCA, Monte Carlo simulations handle the probabilistic nature of input data by creating multiple scenarios that reflect potential variations. This process helps provide a more realistic assessment of uncertainty across different stages of the production process [8]. The reliability of this approach increases as

Table 2. Default uncertainty factors used in combination with the pedigree matrix

Score	1	2	3	4	5
U_R	1.00	1.05	1.09	1.20	1.49
U_C	1.00	1.02	1.05	1.09	1.20
U_{TiC}	1.00	1.03	1.09	1.20	1.49
U_{GC}	1.00	1.01	1.02	1.05	1.09
U_{TeC}	1.00	1.05	1.20	1.49	2.00

more simulations are conducted, with the results gradually converging toward a stable outcome as the number of steps increases [6].

The Monte Carlo simulation generates a range of possible outcomes by running repeated random sampling. In this study, it allows us to estimate the uncertainty of total impact results, individual contributions to total impact and inventory data based on the PDFs. For each input parameter random values are drawn from these PDFs to simulate the variability of the data.

For this study, a 1,000-step Monte Carlo simulation was conducted, allowing us to derive the probability distributions for each environmental impact factor and identify key uncertainty hotspots in the production of hydrogen tanks. The simulation results provide a probabilistic range of environmental impacts, offering insights into how input variability influences the overall assessment. The analysis was performed using Brightway2.5, an open-source framework specifically designed for LCA calculations and uncertainty assessments [9].

3 Case Study: Hydrogen Tank Production

LCA has been standardized by ISO 14040 series [10], which includes goal and scope definition, inventory analysis, impact assessment, and interpretation.

3.1 Goal and Scope

The goal of this LCA is to evaluate the global warming potential of Type IV hydrogen tank production in Germany, which is used in fuel cell vehicles. Type IV vessels are commonly used in automotive industry because of its lightweight [11]. The tank is made of a polymeric liner fully wrapped with Carbon Fiber Reinforced Plastics (CFRP) and consists of six components: CFRP, liner, valve, metal boss, drop protection and fire protection. Therefore, the scope of the analysis covers the production of each component.

Climate change impacts are a primary concern due to their extensive effects on ecosystems, human health, and economic stability. By evaluating the global warming potential, this study aims to address one of the most pressing environmental issues and contribute to the development of more sustainable hydrogen tank solutions. Additionally, focusing on climate change allows for a more streamlined analysis. Therefore, this study assesses the endpoint global warming potential over 100 years (GWP100) using the IPCC 2021 method. The functional unit is defined as *"one hydrogen tank"*.

3.2 System Boundary

The aim of this study is to evaluate the uncertainty of process data to identify key issues during the manufacturing design phase. For this purpose, the system boundary is defined as the cradle-to-gate production of the hydrogen tank, including all components (Fig. 3). The processes such as assembly, testing, and transportation of components are ignored.

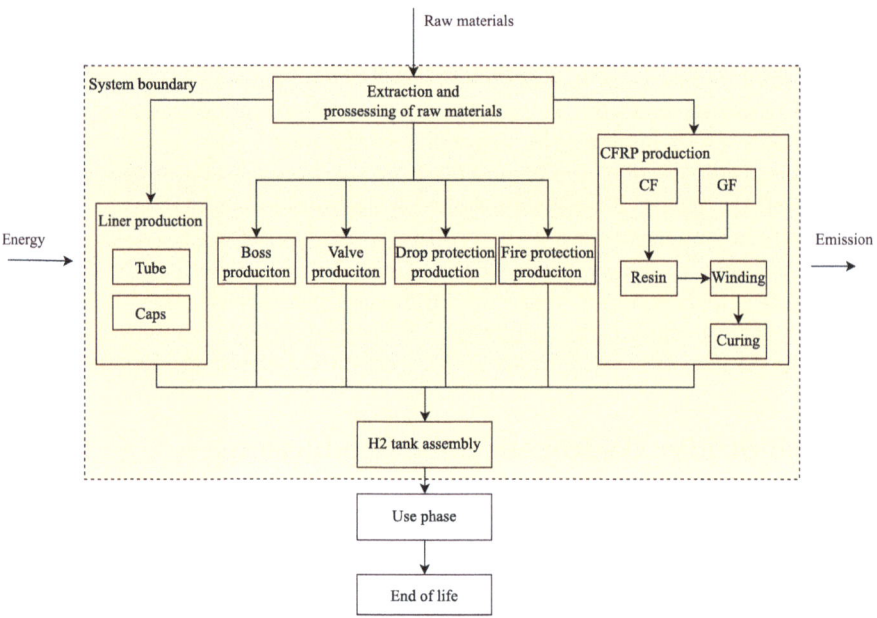

Fig. 3. System boundary for production of hydrogen tank

3.3 Life Cycle Inventory

The life cycle inventory (LCI) involves collecting data on the inputs and outputs of the entire model. Table 3 shows the material flow of the hydrogen tank. Notably, due to the limited availability of known data for carbon fiber in the Ecoinvent database, the input-output LCI data for carbon fiber was referred from the production process of petroleum-based polyacrylonitrile (PAN) fiber in relevant studies [12,13]. PAN fibers are processed into carbon fiber through stabilization, carbonization, and surface treatment. Due to the confidential nature of the company's information, the values in Table 3 were adjusted by coefficients ranging from 0% to 20%, and the processing methods for each component are not listed. The materials used for drop protection and fire protection are made from polymer. Due to confidentiality, the process used in the Ecoinvent database is not shown in the table. The data pertains to a test product by Hexagon, with the tank manufactured in Germany in 2024. Background data was obtained from the Ecoinvent v3.9.1 database.

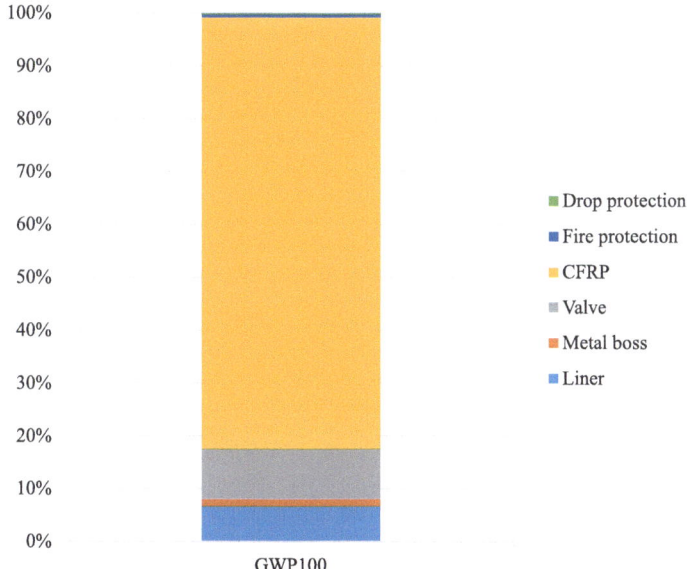

Fig. 4. Contribution analysis in the global warming potential category

3.4 Contribution Analysis

Figure 4 illustrates the contribution of each component for hydrogen tank in the global warming potential category. Notably, the production of CFRP contributes the highest CO_2-equivalent emissions, accounting for over 80% of the total impact. The valve is the second-largest contributor, representing approximately 10% of the impact. Other components have relatively minor contributions. The contribution analysis highlights the primary sources of environmental impact in the production of the hydrogen tank.

3.5 Uncertainty Analysis

The first step is to evaluate DQI scores for the given LCI. Taking liner production as an example, the material for it is PA6, and the corresponding activity in

Table 3. Material inputs of the hydrogen tank production

Components	Input Material	Quantity
Liner	Polyamid 6 (PA6)	2.3kg
CFRP	Glass fibers	1.1kg
	Carbon Fibers	6.9kg
	Resin components	3.3kg
Metal boss	Aluminium alloy	0.28kg
Valve	Aluminium alloy	1.7kg
Drop protection	Polymer	1.5kg
Fire protection	Polymer	0.56kg

Ecoinvent database is *market for nylon 6 (RER)*. This market activity represents the average supply of nylon 6 in Europe. Based on the definition of pedigree matrix in Table 1, the DQI scores are evaluated as listed in Table 4. Reliability and completeness are determined from the suppliers' information on PA6, and the other three DQIs are determined following the rules in pedigree matrix.

Then, by looking up the corresponding uncertainty factors in Table 2, GSD can be calculated using Eq. (1), in this case equal to 1.06. In this way, the PDF for PA6 in liner production can be obtained, which is a lognormal distribution with a geometric mean of 2.3 and a GSD of 1.06. Likewise, the statistical information of other processes in the LCI can be obtained. Based on the PDFs, 1,000 values for climate change category are generated using the Monte Carlo simulation, and the distribution of GWP100 for the hydrogen tank can be generated as shown in Fig. 5. The median value of the Monte Carlo simulation result is 325.43 kg CO2-equivalent, with a standard deviation of 47.53 kg CO2-equivalent, and a coefficient of variation of 14.41%.

Table 4. DQI score for liner production

DQI	Score	Comment
R	1	Data based on measurement
C	2	Representative data from markets
TiC	1	>3years
GC	2	Average value from Europe
TeC	2	Data of same technology

A further breakdown of the hydrogen tank reveals insights into the GWP100 outcomes for each component, as shown in Fig. 6. This figure presents the results of Monte Carlo simulations for individual components. The Y-axis of the graph represents the GWP associated with each component, while the X-axis indicates

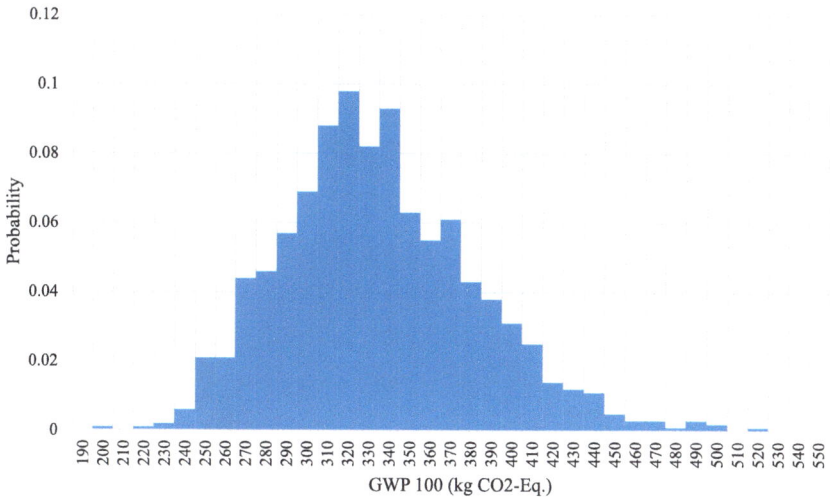

Fig. 5. Results of the uncertainty analysis for climate change category for hydrogen tank

their respective contributions to the final product. This visualization enables a quantitative assessment of the variability in results for each component and their contributions. Notably, the outcomes for CFRP significantly affect the overall GWP of the hydrogen tank, with data showing considerable dispersion. The GWP values for CFRP are consistently above 150 kg CO2 Eq., primarily concentrated in the range of 250-300 kg CO2 Eq. Furthermore, during each iteration, the contribution of CFRP to climate change associated with the hydrogen tank consistently exceeds 70%, typically falling between 75% and 85%. On the other hand, components such as drop protection have a relatively minor impact on the results, exhibiting lower dispersion. Additionally, both the valve and metal boss show a high degree of variability in the figure. One potential underlying cause for this is the use of the aluminum alloy process from the Ecoinvent database titled market for aluminum, wrought alloy (GLO), which may lead to a lack of representativeness for specific aluminum alloy materials.

Figure 6 provides a qualitative analysis of the environmental impact and uncertainties associated with each component from a visualization perspective. To identify the uncertainty hotspots within the hydrogen tank model, the combined results of uncertainty and contribution for each component are illustrated in Fig. 7. A four-quadrant matrix has been established to facilitate analysis, with quadrant boundaries set at 50% contribution and a coefficient of variation of 0.1. The components of the hydrogen tank are represented in the diagram based on the results of the calculations. The matrix enables a meaningful evaluation by combining both the uncertainty and contribution results. Components situated in Quadrant 1 indicate processes with high uncertainty and significant influence on the final result, thus identified as uncertainty hotspots requiring later modification. Conversely, processes in Quadrant 4 contribute minimally to the

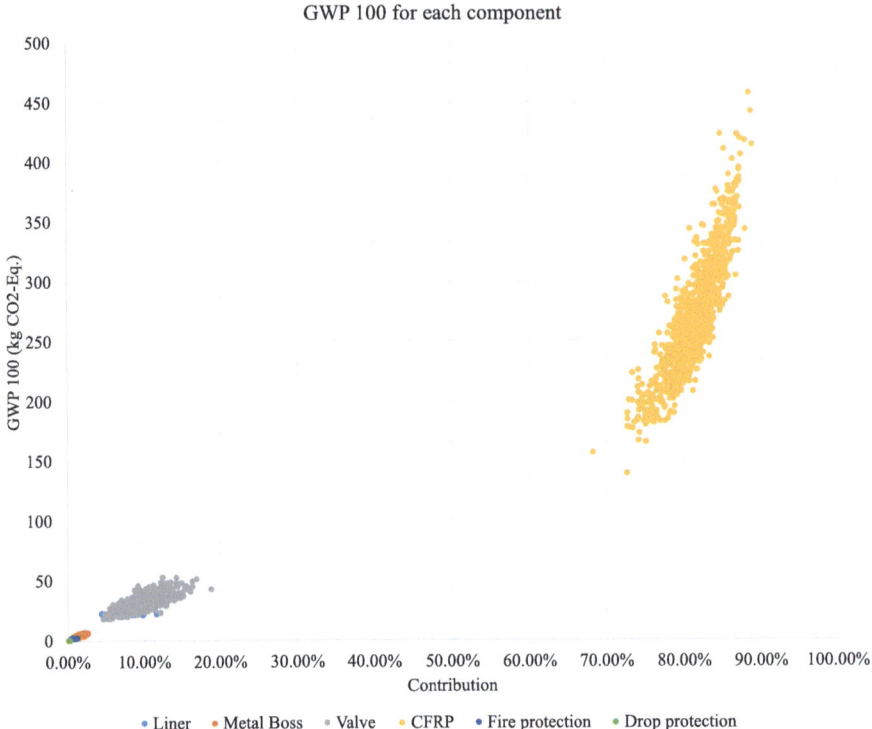

Fig. 6. Results for climate change category for each component of hydrogen tank

final result's uncertainty. Furthermore, improving data quality for components in Quadrants 2 and 3 enhances the overall robustness of the results.

Using the quadrant matrix, the criteria for data quality during the design phase are defined, helping to identify areas for data improvement. In this case, the processes in Quadrant 1 should be prioritized for updates, either by refining the process with more detailed information or by obtaining higher-quality data. Refining the process involves implementing a set of sub-processes to help identify the primary causes of uncertainty and the contribution of each component. This means that the process in Quadrant 1 can be disaggregated into several sub-processes, which may fall into other quadrants, allowing for the pinpointing of the primary reasons for uncertainty in the main process. On the other hand, collecting superior data directly diminishes process uncertainty, effectively moving the process from Quadrant 1 to Quadrant 3.

In the case study, the results indicate that CFPR production should be considered as the primary uncertainty hotspot for the hydrogen tank. When breaking down CFRP production into sub-processes, it becomes evident that the carbon footprint is primarily influenced by the production of carbon fibers, whereas the impact of glass fibers and resin components on the final result is minimal. On the other hand, while the sub-processes show lower contribution and uncertainty,

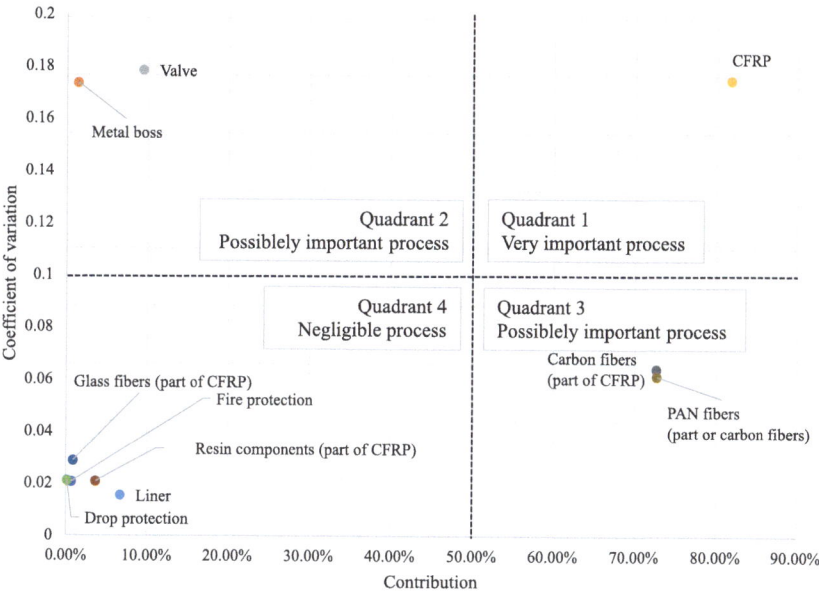

Fig. 7. Combining uncertainty of and contribution toward process to identify uncertainty hotspots

the resulting composite process has significantly high value. This indicates that a portion of the uncertainty in CFRP production stems from the uncertainty of the sub-processes, while a larger portion arises from the compatibility of each sub-process with the specific production process. For example, the degree to which the carbon fiber production data extracted from the literature aligns with the carbon fiber data used in CFRP production for the hydrogen tank. Therefore, to reduce the uncertainty of the final result most efficiently, acquiring more accurate data related to the carbon fiber production process is crucial. Simultaneously, enhancing the quality of data concerning the aluminum alloy used in the production of the metal boss and valve components is also meaningful, which can reduce the uncertainty of the final result and increase the robustness of the whole model. Nevertheless, any endeavors to enhance the data quality for the components in Quadrant 1 (such as drop protection) may not significantly impact the final result.

4 Discussion

The aim of this study is to identify uncertainty hotspots in the hydrogen tank production process by analyzing data uncertainty and integrating it with the contribution analysis of LCA results. In the case study, carbon emission results of products are highly uncertain due to a lack of data, particularly for CFRP. By combining contribution and uncertainty analysis, the key factors influencing confidence in predictions regarding the environmental impacts of the hydrogen tank are identified. This approach avoids iterative updates of the LCA model with each design change and instead focuses on updating the model only when hotspot data is updated (uncertainty reduced), significantly enhancing the efficiency of conducting LCA during the design process. Additionally, the methodology is designed to assist non-experts in better understanding and utilizing LCA tools during the design phase.

The first step involves collecting data and assessing its quality. To achieve a semi-quantitative evaluation of data uncertainty, the pedigree matrix is utilized. However, the pedigree matrix heavily relies on expert judgement. Although many scholars argue that data quality scores should be independent of uncertainty analysis [14], using this approach during the early design phase remains a reasonable method for evaluating data uncertainty. At this early stage, a rough indication is often sufficient, as the focus is more on orders of magnitude rather than precise values. However, the use of the pedigree matrix for data quality assessment is subject to limitations such as the experts' background knowledge and the accuracy of data description. Additionally, this study did not consider the fundamental uncertainty of foreground data. Therefore, achieving an accurate assessment of data quality remains a persistent challenge.

The contribution analysis results presented in Sect. 3.4 illustrate the climate change impact of components in the hydrogen tank (Fig. 4), which is beneficial for the designers to identify the environmental impact early in the phase of design, enabling them to reduce the impact during the design phase instead of making major changes in the end.

Combining the contribution analysis results with the uncertainty analysis results in Sect. 3.5, a four-quadrant matrix has been established for analysis, revealing that CFRP production is identified as the hotspot for the hydrogen tank model (Fig. 7). While some components, like the metal boss and valve, exhibit relatively high uncertainty, they are not anticipated to significantly impact the overall results. Nevertheless, improving the data for these components could enhance the robustness of the findings, making them potentially significant. As a result, the LCA model should be updated whenever more reliable data for these key components becomes available. In cases where data for other components is revised, the iterative process of the LCA model can be skipped.

Additionally, the results provide designers with a clearer understanding of the LCA results, guiding them to effectively address data gaps by enhancing the quality of the identified data.

5 Conclusion

This study presents a novel approach for identifying product LCA hotspots by integrating uncertainty analysis and contribution analysis in the early stages of design. This method elucidates the interplay between data quality and environmental impact through Monte Carlo simulation during product development, thereby precisely pinpointing the crucial factors of LCA results. It also pinpoints processes that require optimization during development to reduce the number of iterations of the LCA model throughout the development process.

In a case study examining carbon emissions in the production process of hydrogen tanks, we found that CFRP production accounted for the majority of CO2 equivalent emissions. Moreover, CFRP production exhibited the highest level of uncertainty, emphasizing the critical need for more accurate data collection during the design phase.

Although this study only considered climate change as an environmental impact category, the proposed methodology can be expanded to encompass other impact categories such as water scarcity and land use. Expanding the scope of analysis allows for a deeper understanding of the environmental consequences associated with product development processes.

Acknowledgments. This work has been financed in the framework of the Digitalization for Sustainability (DigiTain) project, funded by the German government and the European Union as part of the economic stimulus package, point 35c in module b "New, innovative products as the key to vehicles and mobility of the future."

References

1. Roberts, M., Allen, S., Coley, D.: Life cycle assessment in the building design process - a systematic literature review. Build. Environ. **185**, 107274 (2020). https://doi.org/10.1016/j.buildenv.2020.107274
2. Weidema, B.P., Wesnæs, M.S.: Data quality management for life cycle inventories-an example of using data quality indicators. J. Clean. Prod. **4**(3–4), 167–174 (1996). https://doi.org/10.1016/S0959-6526(96)00043-1
3. Funtowicz, S.O., Ravetz, J.R.: Uncertainty and Quality in Science for Policy. Springer, Dordrecht, The Netherlands (1990). https://doi.org/10.1007/978-94-009-0621-1
4. Frischknecht, R., et al.: The ecoinvent database: overview and methodological framework (7 pp). Int. J. Life Cycle Assess. **10**(1), 3–9 (2005). https://doi.org/10.1065/lca2004.10.181.1
5. Weidema, B.P., et al.: Overview and methodology: data quality guideline for the ecoinvent database version 3, Swiss Centre for Life Cycle Inventories, Ecoinvent Report, vol. 3, no. 1 (2013)
6. Muller, S., Lesage, P., Ciroth, A., Mutel, C., Weidema, B.P., Samson, R.: The application of the pedigree approach to the distributions foreseen in ecoinvent v3. Int. J. Life Cycle Assess. **21**(9), 1327–1337 (2016). https://doi.org/10.1007/s11367-014-0759-5
7. Rosenbaum, R.K., Georgiadis, S., Fantke, P.: Uncertainty management and sensitivity analysis. In: Hauschild, M.Z., Rosenbaum, R.K., Olsen, S.I. (eds.) Life Cycle Assessment, pp. 271–321. Springer, Cham (2018). https://doi.org/10.1007/978-3-319-56475-3_11
8. Raynolds, M., Checkel, M.D., Fraser, R.A.: Application of Monte Carlo analysis to life cycle assessment. In: International Congress & Exposition (1999)
9. Mutel, C.: Brightway: an open source framework for life cycle assessment. J. Open Source Softw. **2**(12), 236 (2017). https://doi.org/10.21105/joss.00236

10. ISO 14040:2006: Environmental management – Life cycle assessment – Principles and framework. International Organization for Standardization (2006)
11. Barthélémy, H.: Hydrogen storage - industrial prospectives. Int. J. Hydrogen Energy **37**(22), 17364–17372 (2012). https://doi.org/10.1016/j.ijhydene.2012.04.121
12. Khalil, Y.F.: Eco-efficient lightweight carbon-fiber reinforced polymer for environmentally greener commercial aviation industry. Sustain. Prod. Consumption **12**, 16–26 (2017). https://doi.org/10.1016/j.spc.2017.05.004
13. Duflou, J.R., De Moor, J., Verpoest, I., Dewulf, W.: Environmental impact analysis of composite use in car manufacturing. CIRP Ann. **58**(1), 9–12 (2009). https://doi.org/10.1016/j.cirp.2009.03.077
14. Edelen, A., Ingwersen, W.W.: The creation, management, and use of data quality information for life cycle assessment. Int. J. Life Cycle Assess. **23**(4), 759–772 (2018). https://doi.org/10.1007/s11367-017-1348-1

Open Access This chapter is licensed under the terms of the Creative Commons Attribution 4.0 International License (http://creativecommons.org/licenses/by/4.0/), which permits use, sharing, adaptation, distribution and reproduction in any medium or format, as long as you give appropriate credit to the original author(s) and the source, provide a link to the Creative Commons license and indicate if changes were made.

The images or other third party material in this chapter are included in the chapter's Creative Commons license, unless indicated otherwise in a credit line to the material. If material is not included in the chapter's Creative Commons license and your intended use is not permitted by statutory regulation or exceeds the permitted use, you will need to obtain permission directly from the copyright holder.

The Role of Metal Additive Manufacturing in a Circular Economy

Matthias Duve[✉], David Petasch, Bernd Lüdemann-Ravit, and Frieder Heieck

Institute of Production and Informatics (IPI), Mittagstraße 28a,
87527 Sonthofen, Germany
matthias.duve@hs-kempten.de

Abstract. This paper provides an overview and examples of the potential, but also the challenges of Metal Additive Manufacturing in the future Circular Economy paradigm. The Circular Economy, aimed at reducing resource consumption, waste generation, and the overall carbon footprint of our industries, presents significant challenges to traditional, linear production systems. In this context, Additive Manufacturing emerges as a highly flexible technology with the potential to overcome many of these hurdles: The ability to reduce material usage through optimized lightweight and functional designs, the use of recycled materials, Additive Manufacturing-based repair and remanufacturing of tools and products or the optimization of production through hybrid manufacturing approaches (e.g. combinations of subtractive and additive manufacturing) are just a few examples.

However, the lack of automated and digitalized production processes, complex process chains with potential health and safety concerns, and high demands on material and component qualification still pose significant hurdles to the widespread adoption of Metal Additive Manufacturing in the Circular Economy framework. In addition, the Circular Economy alone does not necessarily mean a lower carbon footprint over the entire product life cycle. On the contrary, depending on the specific process used, the carbon footprint of metal additive part production is often higher than in conventional manufacturing scenarios due to the high energy requirements of the additive process and long production times. Therefore, in this study, a comprehensive literature review is performed to outline the potential of Additive Manufacturing for Circular Economy principles, providing several beneficial examples of the so-called "R-strategies" covered with the help of Additive Manufacturing. In addition, an outlook is given on the development of new technologies to increase the potential of Additive Manufacturing to make an important contribution to circular production scenarios.

Keywords: Circular Economy · Metal Additive Manufacturing · R-Strategies · Sustainability

1   Introduction: Circular Economy and R-Strategies

The concept of a Circular Economy (CE) is gaining traction as a means to create a more sustainable future by minimizing waste and maximizing the reuse of resources. Metal Additive Manufacturing (MAM) is increasingly recognized as a technology that could play a critical role in this transition. This paper explores the synergies between Additive Manufacturing (AM) and CE, discussing the most beneficial R-Strategies for AM-applications, assessing the potential for Carbon Footprint (CF)-reductions and highlighting the existing challenges for a broader industrial acceptance.

The German Institute for Standardization (DIN) recommends the implementation of CE approaches using the so-called R-strategies as a method to make products more sustainable [1]. The application of R-strategies within a product lifecycle is visualized in Fig. 1. As part of the Corporate Sustainability Reporting Directive (CSRD), which itself is part of the European Green Deal, CE is a so-called 'topic that must be analyzed' [2]. EU law requires all large companies and all listed companies to provide information on what they see as the risks and opportunities arising from social and environmental issues, and on the impact of their activities on people and the environment [2].

According to the review by "Kirchherr et al.", CE was defined with great clarity by N. van Buren et al. (2016, p.3) [3]:

"Unlike the current economy, which is largely based on the principle "take-make-waste" (linear economy), the focus point in a CE is to not unnecessarily destroy resources." [4] This definition extends the focus of CE beyond the pure reduction of waste through recycling and emphasizes the following key points: Reducing the consumption of raw materials, applying eco-design principles to products so that they can be easily dismantled and reused after use, extending the use phase of products through maintenance and repair, using recyclable materials in products, and recovering raw materials from waste streams [4]. Van Buren et al. add economic and social dimensions to the definition of a CE, making it a very holistic approach to manufacturing principles: "A CE aims for the creation of economic value (the economic value of materials or products increases), the creation of social value (minimization of social value destruction throughout the entire system, such as the prevention of unhealthy working conditions in the extraction of raw materials and reuse) as well as value creation in terms of the environment (resilience of natural resources)" [4]. On a critical note, it must be mentioned that the definition uses the R9-framework and therefore lacks the R-Strategy "Rethink" (see Fig. 2), does not include references to future generations, and fails to include business models as an enabler of CE [3].

The practical implementation of circular products is based on and guided by the so called R-strategies, which have evolved from R3 (reduce, reuse, recycle) to R9 or R10 frameworks, depending on the source [3]. Ocicka et al. [5] claim that the most common conceptions are R10 frameworks, structured in the R9 notation, starting with R0. Potting et al. [6] clustered the 10 strategies in 3 fields of application, to "strategies for smarter product use and manufacture" (R0-R2), "extend lifespan of product and its parts" (R3-R7) and "useful application of materials" (R8, R9). They also define an order of priority for the strategies with a rising level of circularity as a rule of thumb, displayed in Fig. 2.

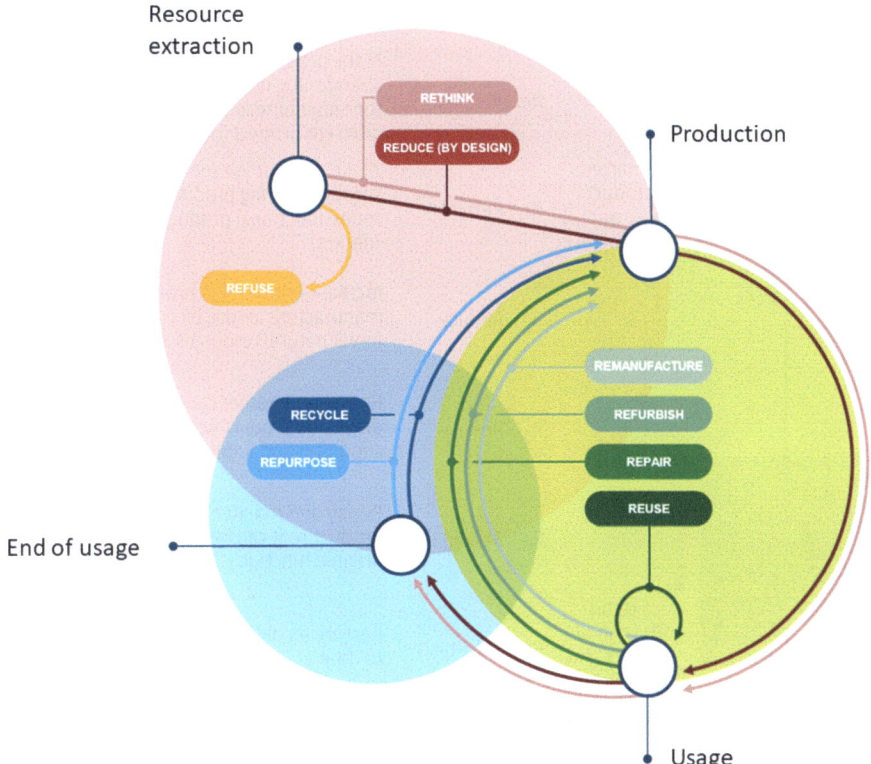

Fig. 1. R-strategies as a framework for the circular economy. Image provided by the German Institute for Standardization (DIN), partially translated [1].

The combination and implementation of R-Strategies in a products life-cycle is visualized in Fig. 1. This representation underlines the clustering approach by Potting et al. [6] and shows the relevance along the comprehensive manufacturing stages, starting from resource extraction and convoying manifold steps and repeated life-cycles until the final end of usage. The contribution of AM along the R-strategies is discussed in the following sections.

In additive manufacturing, a digital model is manufactured by adding layers of material, creating the possibility to decentralize the economic production of very complex parts in low to medium sized batch numbers from various materials. AM has the potential to disrupt traditional, linear manufacturing life cycles by offering a more flexible and less wasteful production method. AM has, thus, been considered a driving force in the fourth industrial revolution, also know as the transition to industry 4.0 [7].

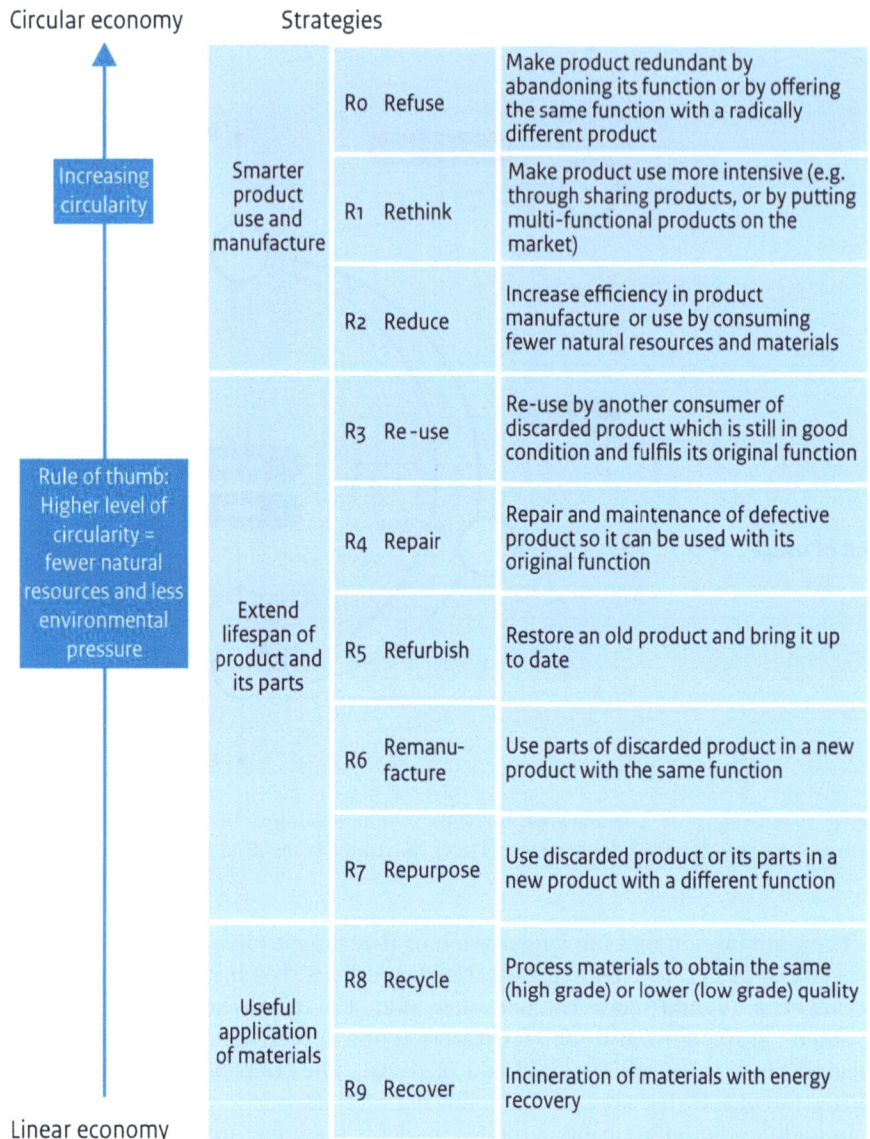

Fig. 2. Circular or R-strategies within the production chain in order of priority. Image from Potting et al. [6]

The most commonly used production methods for the additive manufacturing of metallic components, MAM, are Laser Powder Bed Fusion (LPBF) and Direct Energy Deposition (DED). LPBF is a Powder Bed Fusion (PBF) process in which pulverized raw materials are melted in hatches using a laser as heat source, as displayed in Fig. 3 picture (A), thereby creating components with very good structural-mechanical characteristics, and a comparably good surface quality. This additive process enables the generation of very complex and precise parts, which would be impossible or very difficult to produce through Conventional Manufacturing (CM) methods. DED processes supply and melt powder or wire materials, applying material locally only where needed. This task is realized, utilizing the degrees of freedom of a robotic arm, as displayed in Fig. 3 picture (B) and Fig. 5. Due to high productivity rates, DED processes can produce parts with bigger volumes and are very suitable for repair and remanufacturing processes.

Any material thicknesses and forms can be created on existing workpieces [8]. The advantages lie in direct application of material on the original component and reduced material and energy input compared to new products, as well as the possibility of new business cases [8]. Disadvantages lie in possibly reduced structural strength, high demands on technical data and documentation [8]. However, repair processes and hybrid parts production using AM combined with CM base parts show good structural strength [9]. The relatively coarse surface texture of components, as depicted in Fig. 5, generally necessitates extensive machining during post-processing in the majority of applications.

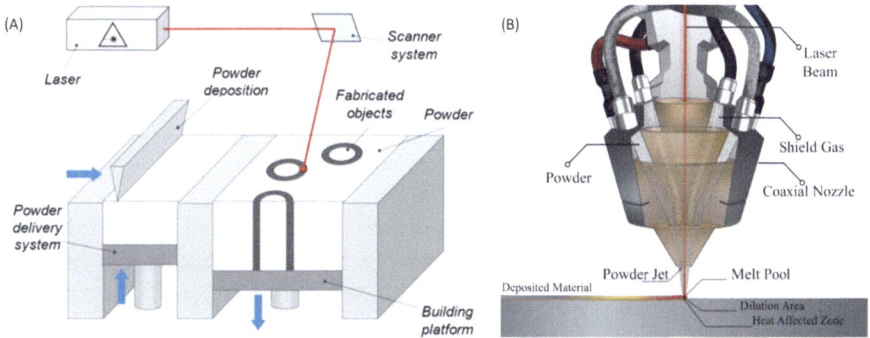

Fig. 3. Powder based fusion (A) and Direct Energy Deposition (B) both based on powder metal raw materials. Image from Selema et al. [10]

While manufacturing via DED inhabits the possibility to deposit material on complex existing geometries, LPBF processes require a horizontal surface and typically a subtractive preparation. Figure 4 illustrates the complex production setup, including a horizontal working plane and the necessary, product-specific clamping devices on the build platform. MAM parts usually need at least limited amounts of post processing procedures like blasting, machining or thermal treatments.

Fig. 4. Repair process of an angle lever via Powder based fusion. Images from Wurst et al. [11]

Both processes can be used to add AM-specific structures and geometry to pre-fabricated, incomplete parts. High volume areas of components can be pre-produced using manufacturing methods with a more beneficial CF per kg processed material, while additively completing the part with more complex geometries.

To maximize the benefits of MAM processes, it is often essential to modify or entirely redesign a component using Design for Additive Manufacturing (DfAM) principles [12]. The need for in-depth knowledge of the necessary design modifications and the resulting environmental impacts, as well as the challenges of comparing AM and CM parts, necessitate a discussion on the ecological evaluation of additive processes and possibilities for the correct comparison to conventional manufacturing.

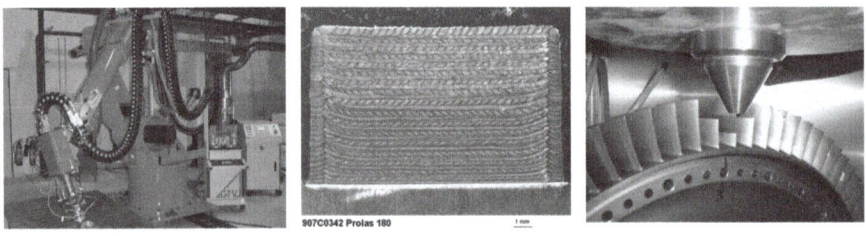

Fig. 5. Repair process of a turbine via Direct Energy Deposition. Images from Nowotny et al. [13]

2 Ecological Evaluation

When discussing the ecological evaluation of Additive Manufacturing (AM), several metrics for sustainability assessment come across, covering different dimensions or scopes. The Circular-Life Cycle Sustainability Assessment (C-LCSA) was introduced lately as an integrated method for the Circular Economy (CE)-specific context, considering both the technical circularity and the complexity of the three dimensions of sustainability (environment, economy, and social) [14]. To achieve this, Circularity Assessment (CA) was added as an additional dimension to the established Life Cycle Sustainability Assessment (LCSA) framework, alongside e Life Cycle Assessment (LCA), Life Cycle Costing (LCC), and Social Life Cycle Assessment (S-LCA) [14].

The LCA framework is an essential and standardized method (set out in ISO 14040/44), with the help of which additively manufactured products can be made more sustainable [15]. An LCA allows for the compilation and evaluation of the inputs, outputs, and the potential environmental impacts of a product system throughout its life cycle. It is based on four main phases including 1. Goal and scope definition, 2. Inventory analysis 3. Impact assessment and 4. Interpretation.

An LCC is performed alongside LCAs to evaluate the economic performance of manufacturing technologies. The LCC summarizes all costs over the life cycle of a product in which different actors are involved [16]. In the CE context, the LCC must take into account the material supplier, the manufacturer, the consumer, and the End of Life actor [17]. In the context of a CE, Social Life Cycle Assessment S-LCA assesses social impacts using United Nations Environment Programme (UNEP) guidelines and several subcategories like employment, training, and social acceptance to address the challenges and opportunities of circular practices. This approach highlights both positive impacts, such as job creation, and negative impacts, such as job losses in traditional sectors, ensuring a comprehensive evaluation of CE initiatives [18]. Circularity Assessment (CA) is a holistic evaluation approach designed to measure the circularity performance of products, businesses, or cities by assessing various aspects such as material flows, energy use, waste management, and stakeholder engagement. It employs a framework that includes specific metrics and standards, such as the ISO 59020, to identify opportunities for improvement and support the transition from a linear to a CE [19]. According to Luthin et al.(2024) [14] the C-LCSA framework can be described by adapting the traditional LCSA formula as shown in formula 1:

$$C - LCSA = LCA + LCC + S - LCA + CA \tag{1}$$

The C-LCSA is intended to be used by LC(S)A experts and will serve as a reference for future studies to achieve consensus and comparability in assessing CE concepts [14]. This shall enhance the research on CE assessment. The framework is intended to help LCSA assessors to consider the implications of their decisions and to identify chances for improvement as the framework allows the identification of trade-offs between an enhanced circularity and resulting impacts on sustainability performance [14]. However, the included LCC is not sufficient for a complete economic evaluation. To achieve a realistic assessment of R-strategies the respective business case must also be considered. For example, possible revenues must be estimated. Only a positive business case, where the difference between revenue and costs is above a desired threshold, allows for the considered R-strategies to realistically be used in an industrial environment.

While AM has the potential to be more sustainable, especially for small-to-medium series production, it has not been conclusively proven that AM processes are generally more environmentally friendly than CM processes. It is therefore only logical to aim for comparability between AM and Conventional Manufacturing (CM) processes. This goal comprises several challenges that need to be addressed, including the typically necessary geometrical differences for functionally identical components. Similar system boarders for additive and subtractive manufacturing methods are necessary, as well as a consideration of the impact the design differences have on the use phase. Studies have shown that the global warming potential is the most investigated environmental indicator when comparing AM to CM, with a focus on carbon footprint and emissions [20].

80% of the product-related environmental costs of AM components are already determined in the product planning and design phase [21] and can lead to ecological disadvantages compared to CM processes if not sufficiently taken into account in the component design [22]. This emphasizes the need for ecodesign approaches and specific "design for R-strategies"-frameworks to be considered in early product conception phases [21]. The complexity of these requirements needs to be reduced to make them tangible for product development engineers. The reduction to easily accessible key figures and early, integrated feedback systems in product development software are necessary. For the same reason, in the context of CE designs, focusing on the most promising R-strategies in term of global warming reduction potential for AM application designs is crucial. Consequently, in this paper, the assessed AM application studies are clustered by potential (CF)-reduction to rate the applied R-Strategy.

3 Assessing Applications of R-Strategies with Additive Manufacturing and Their Environmental Impact

AM with metal materials is considered a potentially sustainable technology, which under certain conditions can significantly reduce CO_2 equivalent emissions along the entire product life cycle of industrially manufactured components [23]. Comparative studies, however, attribute higher ecological costs per kilogram of processed metal to additive manufacturing, than to comparable CM methods such as milling, casting, and forging [23]. Ecological advantages of AM components arise mainly from a reduced use of primary materials, energy, shortened supply chains, and shifted waste streams [24].

In order to systematically implement AM as CE applications and to ensure a positive impact on the CF of products, it is essential to consistently employ R-Strategies and assess the associated ecological costs. This approach maximizes the potential benefits of AM, thereby enhancing the likelihood of achieving a positive outcome. The following classification aims to improve the selection of the most suitable R-Strategy for a given use case, considering the available Metal Additive Manufacturing (MAM) technology. To evaluate the implementation of CE within R-strategies in the context of MAM, it is necessary to categorize and specify the various strategies and their applications according to their capacity to diminish multicriterial impacts. This categorization serves as a guideline for scientists and product developers to identify the most effective R-strategies when enhancing circularity in components, assemblies, products, and studies. The environmental impact of AM processes is assessed within the CE framework, utilizing life cycle analysis and carbon footprint assessments. R-strategies can mitigate environmental impacts and promote sustainable practices in AM by optimizing resource use, recycling, and sustainable production methods.

The results of this evaluation are summarized in Table 1.

(0) Refuse: Making a product redundant by abandoning its function or by offering the same function with a completely different product

For this paper, applications of the refuse strategy with MAM take into account the consolidation of assemblies. AM offers unparalleled design flexibility, enabling the production of complex geometries in an economically attractive manner. This allows for the integration of multiple parts into a single component, reducing the need for assembly processes and the number of components overall.

Industrial and scientific examples:

- In 2024, Airbus replaced a 7 part assembly with a single series-qualified AM part [27].
- In the generic process of methodically reducing assemblies, several parts are refused, which, under certain preferences, results in environmental advantages [25,26].
- In an analogue way CM can be utilized to refuse metal springs, connecting elements, and joints [34].
- Casted metal parts in specific volumes can be substituted with additively manufactured components, replacing molds.
- AM reduces the need for inventory and transportation, as spare parts can be printed on demand and decentralized.

(1) Rethink: Intensifying product use (e.g. by sharing products or marketing multi-functional products)

In the context of new business models using product sharing concepts, MAM can support the value proposition by prolonging the use phases utilizing repair processes. Consequently these concepts become more financially attractive for prospective vendors.

Table 1. Assessment of applications of R-Strategies via AM in comparison with Conventional Manufacturing CM

R-Strategy	Application	Literature	CF-reduction compared to CM
(0) Refuse	Assembly reduction by functionality integration and part consolidation	[25]	Equal
		[26]	89,1%
		[27]	not defined
	AM spare part production	[28]	75%
(1) Rethink	Bike sharing: extending physical life and value of EBikes	[29]	not defined
(2) Reduce	Aircraft "cradle to gate" reductions	[30]	230–900t CO_{2eq}
	Material via topology optimization	[31]/[32]	12%/65% weight
	Material via generative design	[33]	38% weight
	Material via compliant mechanisms	[34]	not defined
	Mold Inserts printed on preform	[35]	not defined
(3) Reuse	no specific source found		
(4) Repair*	Repair of hot forging Tools	[36]	not defined
	Repair of Turbine Blades	[37]	45%
(5) Refurbish*	Improving product durability	[38]	not defined
	Multiple case study on AM repair and refurbishment with PBF	[11]	LCIA** results given
(6) Remanufacture*	Remanufacturing of a stamping tool with DMD Restoring a complex duct AM-built on a high-volume base part	[39] [35]	50% Energy Reduction not defined
(7) Repurpose	Repurposing an obsolete mold for a different component	[40]	97,5%
(8) Recycle	Recycling of electric circuit components	[29]	not defined
	Recycling of powder metal in the LPBF-process	[41]	not defined
(9) Recover energy	no specific source found		

*Since the terms repair, refurbish, and remanufacture are frequently utilized in ways that do not align with the definitions established in the R9 Framework (e.g., by Potting et al. [6]), the sources have been categorized according to the author's judgment.
***Life Cycle Impact Assessment.

Industrial and scientific examples:

- Bike-sharing, considered in combination with the remanufacturing of electric bicycles, is a result oriented product-service system that integrates a network of cooperation with organizational, logistic, and technological innovations [29]

(2) Reduce: Increase efficiency in product manufacture or use through reduced consumption of natural resources and materials

AM is known for its ability to significantly reduce waste. Compared to conventional methods, AM requires less material and generates less waste because it adds material only where needed. Furthermore, the high freedom in product design enables the use of innovative lightweight design methods.

Industrial and scientific examples:

- Reviewing the potentials of MAM for aircrafts, Huang et al. identified a potential of 1650–2840 kg mass reduction. This results in 4–7% less weight and 230–900 metric tons less equivalent cradle to gate emitted CO2, based on an average empty aircraft operating mass of 40,622 kg.
- The use of innovative lightweight construction methods, such as topology optimization [33], generative design [12], compliant mechanisms [34], hybrid designs using CM preforms to print on [35] and reprocessing strategies, as well as their combinations, can contribute significantly to the reduction of emissions bound in the component and arising during the usage phase [23].
- Reduction of support structures using heat dissipation and warping simulation as well as novel support generation methods.

(3) Re-Use: Re-use of a discarded product that is still in good condition and fulfills its original function by another consumer

For re-use, no specific studies were found that address a different topic than the re-usability of powder materials and the respective impact on the environment. These studies were classified as "Recycling", as they focus on the material level rather than the products, as defined for "Re-Use" in CE.

(4) Repair: Repair and maintenance of a defective product so that it can be used with its original function

MAM processes such as Laser Powder Bed Fusion (LPBF) and Direct Energy Deposition (DED) can be used for component repair, by locally applying material using a laser.

Industrial and scientific examples:

- The ecological and economic advantages of additive repair processes in the LPBF process have been demonstrated using the example of high-priced turbine blades [42].
- In the aerospace sector, GE Aviation has utilized metal AM to repair parts such as turbine blades and fuel nozzles. This practice extends the life of components made from expensive metals, reduces waste, and cuts material costs by up to 50%, demonstrating both economic and environmental benefits [43].

(5) Refurbish: Restore an old product, bringing it up to date

The restoration to a good-as-new state or better is possible for products and raw materials, for example in powder regrading.

Industrial and scientific examples:

- Powder rejuvenation to preserve and restore powder material properties could reduce the quantity of powder being sent to landfill, while also reducing the amount of virgin powder that needs to be produced [41].
- The additive restoration process not only extends the life of components, which are often made from expensive and high-grade metals, but also significantly reduces waste and the need for virgin materials. Using AM to refurbish aircraft parts can reduce material costs by up to 50%, providing a direct economic benefit in addition to environmental sustainability [43].

(6) Remanufacture: Use parts of a discarded product in a new product with the same function

Due to the strict distinction between Refurbish and Remanufacturing in the R9 framework, several reviewed studies titling "Remanufacturing" via additive manufacturing were assigned to the "Repair" and "Refurbish" strategies. Generic use cases and business cases include the following:

- Special channels, regulating the temperature, can be created for molds in which only the areas containing these channels are printed. Meanwhile the main part of the mold is produced using conventional methods, saving time and resources and thus enabling the production of geometries that could not be achieved in any other way [35].
- Remanufactured products usually cost 40 to 80% less then equivalent new products [29].
- Remanufacturing enhances the interception of bottlenecks or interruptions, leading to an increased supply chain resilience in the short term [29].

(7) Repurpose: Use of a discarded product or its parts in a new product for a different purpose

A notable use case of MAM in a CE is the repurposing of products for different usage scenarios. This involves taking an existing product and modifying it to serve a new function, thereby extending its lifecycle and reducing waste.

Industrial and scientific examples:

- Repurposing an obsolete mold, using it for a completely different component while retaining a considerable percentage of the mold in the circle. This leads to a reduction of 98% in environmental damage and 97,5% in specific global warming potential when the mold piece is refurbished with the DED technology, instead of being replaced with a new mold piece [40].

(8) Recycle: Recycle Process materials to obtain the same (high grade) or lower (low grade) quality

Additive remanufacturing enables the recycling of materials and components, resulting in a positive impact on the environmental and social production costs. In addition, production costs can be reduced by maintaining added value and using less material. This leads to cost savings or higher profit margins, which can be passed on to the customer via lower sales prices [29].

Industrial and scientific examples:

- Additive remanufacturing enables the recycling of electric bike components and has a positive impact on environmental and social compatibility [29].
- Recycling powders in powder bed MAM processes may result in a slight increase in powder particle size distribution [29]
- Recycling and Re-using powders leads to a total loss of only 12.5% waste powder from virgin powders after several use cycles [41].

(9) Recover energy: Incineration of residual flows

The recovery of energy from metal products through incineration is typical for non-recyclable powder waste. This waste occurs, for example, as leftover in the sieving process and in the printing process. For example by filling the filters of the printers, the peripheral devices and the obligatory air conditioning system. But there are more possible applications for energy recovery along the AM process chain:

- Harvest residual heat from chillers and heat emitting processes like low-stress annealing.

While many sources claim to examine CE use cases, most of them do not stick to the definitions provided by the presented R-strategy frameworks. Many of the studies mix up, for example, R4 - Repair, R5- Refurbish and R6- Remanufacturing, using repaired parts in different contexts without providing exact data about the status of the final product. This makes the separation and assignment to specific R-strategies complex and results in low numbers of studies in some of the R-strategies.

With the help of this classification of AM applications in a CE-context and assessed by CF-reduction potential the possible advantages are easier to assess, supporting the selection of the most promising R-Strategy for a present use case early in the innovation phase. As the advantages of AM and CE were known previously on a more general level, it is necessary to discuss what hinders the wider industrial roll out and how to address these hurdles.

4 Discussion: Prerequisites for Wider Industrial Rollout of R-Strategies with Additive Manufacturing

MAM holds the potential to make holistic contributions to the sustainability and circularity of products, components and systems with all the complementary benefits. However, it is remarkable, that the acceptance and application in the industrial context remain relatively slow. Grzesik and Ruszaj's 2021 study [44] highlights the ecological uncertainty associated with the use of AM in the automotive sector, while Prochatzki et al. in 2023 [45] report that the implementation of the CE in the automotive sector continues to be generally inadequate. One notable exception is low-value recycling. Thus, the authors recommend the development of new product development methods to strengthen higher-value strategies [45]. This delay is due to several significant hurdles that need to be overcome when starting with metal AM in general and with the introduction of CE methods, such as the R-strategies. These hurdles include knowledge transfer and transparency in several areas, a lack of digitilization and automation along the entire digital and rather complex analogue process chain, the availability and distribution of AM-infrastructure, potential health and safety issues as well as high demands on material and component qualification.

When discussing the current state of the distribution of additive applications in a CE-context, many topics can be broken down to a lack of knowledge transfer and transparency. To identify suitable use cases for circular MAM and to assess possible resulting benefits, expert knowledge is necessary. If the infrastructure for MAM is not already existent, necessary information about health and safety measures as well as all required process steps and essential periphery for production readiness need to be researched. In a next step, product developers and industrial designers need to be firm with the capabilities of the technology and Design for Additive Manufacturing (DfAM) guidelines for the specific MAM process. This is necessary to utilize the great freedom in component design through optimized lightweight and functional designs to its best potential.

These issues can be addressed by improved training offers, early in the educational paths of designers, and by a higher degree of transparency along the holistic branch of industry, but especially on side of manufacturers of MAM raw materials and printers. Better availability of data about the manufacturing steps, environmental impacts of raw materials and the consumption of process media is crucial for the swift assessment of life cycles.

The lack of digitilization and automation along the digital and manufacturing process chain is a significant hurdle to reduce the reliance on expert knowledge on all levels of AM-implementations and can limit the scalability and efficiency of AM. Therefore, improvements in this area could reduce the entry hurdles for entrepreneurs into additive applications and enhance the possibility to embed new and existing AM-infrastructure in existing digital landscapes of companies.

An improved availability of information for engineers and managers can contribute to the wider distribution of MAM printing devices. Functional, economical, and ecological beneficial applications and demonstrations make it more likely for the field of MAM to experience growth.

Potential health and safety concerns need to be addressed on manifold levels. Knowledge and transparency as well as standardization and regulation are key factors to address the uncertainty that is part of this discourse. A higher degree of automation can additionally reduce the need for direct contact with hazardous materials and media in the manufacturing processes.

The stringent requirements for material and component qualification necessitate rigorous evaluation tailored to specific application areas. National and international standardization and regulatory bodies play a critical role in addressing these challenges. Concurrently, the development of industry-specific benchmarks and best practices is essential to establish robust industrial standards.

If these hurdles are overcome, a wider industrial adoption of additive manufacturing is likely to be facilitated, enabling the full potential of metal AM to contribute to sustainability and circularity in various industries.

5 Conclusions

The CE and R-strategies provide a framework for sustainable resource management and production processes in an industrial context. Additive manufacturing emerges as a powerful tool within this framework, enabling efficient use of energy, resources, waste reduction, and the creation of sustainable business models. However, fully exploiting the potential of AM within the CE requires overcoming challenges that hinder broader acceptance, including issues related to knowledge transfer and transparency, digitalization and automation along the holistic AM process chain, optimizing availability and accessibility of AM infrastructure, addressing potential health and safety concerns, and meeting the demands of material and component qualification. Approaches to addressing these hurdles were discussed in this paper.

(0) The assessment of AM in the context of CE shows the greatest potential for CF-reduction, especially through the following strategies and examples:
- Repurpose: Using obsolete molds and repurposing them via MAM for new products, rather than producing new molds from scratch, shows high ecological potential. A reduction in environmental impacts of up to 97–98% [40], depending on the scope, is possible. Therefore, product designers should consider the necessary preparation early in the product layout.
- Refuse and Reduce: Great potential for CF-reduction during the product innovation phase lies in reducing components or even eliminating them by utilizing the design freedom of MAM. The utilization of innovative product design methods to fully exploit this design freedom, making components significantly lighter, reducing the need for support structures, and using conventionally manufactured preforms for less complex parts can greatly reduce the CF of products.
- Repair / Refurbish / Remanufacture: Extending the lifespan of products by restoring them to a usable or even up-to-date condition, or using old, restored parts for new products, can save up to 45% of the carbon footprint compared to conventional manufacturing, according to the studies referenced.

(1) There is a lack of data and/or methods for the following strategies, highlighting the need for more research, transparency, and transfer to industrial partners:
- Rethink & Reuse: Leveraging the advantages of MAM to explore novel business models and intensify product use. This can be achieved by making products more versatile for different user groups or by extending the physical life of products, making rent and share models more attractive. These business models are particularly applicable to high-value products.
- Recover Energy: Recovering energy from degraded metal powder materials, removed support structures, and products that can no longer remain in the life cycle requires innovative business models. Tracking and collecting high-value materials in pure forms, emphasizing the advantages of monolithic structures and mono-material printing, are crucial for this strategy. The potential economic benefits need to be evaluated.

(2) For broader application, there are technological challenges and areas requiring improvement:
- Knowledge Transfer: Enhancing accessibility and simplifying MAM, starting with the safe implementation of infrastructure and the design phase.
- Process Chain: Increased automation and digitalization of the holistic AM process chain to lower entry barriers, reduce safety concerns, and minimize the need for expert knowledge.
- Developing comprehensive, easy-to-use, predictive assessment tools for evaluating the functional, economic, and environmental impacts of designs, simplifying DfAM.
- Reduction of disposables from the manufacturing process, such as personal protective equipment, cleaning products, and filter elements, as well as process media, such as energy, inert gas, compressed air, and blasting media.
- Achieving higher degrees of standardization, regulation, and transparency to increase scalability and efficiency in industrial applications.
- The Qualification of materials and components, particularly in repair, refurbishment, and remanufacturing processes, to meet the same technical criteria as newly manufactured parts.

The use of additive manufacturing in the context of R-strategies presents a promising approach to waste reduction and resource efficiency, aligning well with the principles of the CE. While it offers many advantages over conventional manufacturing, including design flexibility and the potential for on-demand production, the environmental and economic impacts must be carefully assessed. This study focused on the CF of the examined case studies for reasons of comparability and due to the limited number of holistically assessing studies. In particular, there is insufficient data for comparison to CM processes. It is strongly recommended that future research assess the holistic impact of additive part production using extended LCA frameworks, such as the C-LCSA.

References

1. Engelt, A.: Circular thinking in standards, Technical report, Version 03, DIN (2024). https://www.din.de/de/forschung-und-innovation/themen/circular-economy/normenrecherche/modell-der-r-strategien
2. European Commission: Corporate sustainability reporting (2024). https://finance.ec.europa.eu/capital-markets-union-and-financial-markets/company-reporting-and-auditing/company-reporting/corporate-sustainability-reporting_en
3. Kirchherr, J., Reike, D., Hekkert, M.: Conceptualizing the circular economy: an analysis of 114 definitions. SSRN Electron. J. (2017)
4. Van Buren, N., Demmers, M., Van der Heijden, R., Witlox, F.: Towards a circular economy: the role of Dutch logistics industries and governments. Sustainability **8**, 647 (2016)
5. Ocicka, B., Wieteska, G., Wieteska-Rosiak, B.: Toward circular product lifecycle management through industry 4.0 technologies. In: Handbook of Sustainable Development Through Green Engineering and Technology. CRC Press (2022)
6. Potting, J., Hekkert, M., Worrell, E., Hanemaaijer, A.: Circular economy: measuring innovation in the product chain, Report 2544. PBL - Netherlands Environmental Assessment Agency, The Hague (2017)
7. Vaidya, S., Ambad, P., Bhosle, S.: Industry 4.0 - a glimpse. Procedia Manuf. **20**, 233–238 (2018)
8. Bergs, T., Brimmers, J.: Manufacturing for a circular economy. In: Empower Green Production Conference Proceedings. Fraunhofer-Gesellschaft (2023)
9. Zhang, X., Li, W., Chen, X., Cui, W., Liou, F.: Evaluation of component repair using direct metal deposition from scanned data. Int. J. Adv. Manuf. Technol. **95**, 3335–3348 (2018)
10. Selema, A., Ibrahim, M.N., Sergeant, P.: Metal additive manufacturing for electrical machines: technology review and latest advancements. Energies **15**(3), 1076 (2022)
11. Wurst, J., Ganter, N.V., Ehlers, T., Schneider, J.A., Lachmayer, R.: Assessment of the ecological impact of metal additive repair and refurbishment using powder bed fusion by laser beam based on a multiple case study. J. Clean. Prod. **423**, 138630 (2023)
12. Junk, S., Klerch, B., Hochberg, U.: Structural optimization in lightweight design for additive manufacturing. Procedia CIRP **84**, 277–282 (2019)
13. Nowotny, S., Scharek, S., Beyer, E., Richter, K.-H.: Laser beam build-up welding: precision in repair, surface cladding, and direct 3D metal deposition. J. Therm. Spray Technol. **16**, 344–348 (2007)
14. Luthin, A., Traverso, M., Crawford, R.H.: Circular life cycle sustainability assessment: an integrated framework. J. Ind. Ecol. **28**(1), 41–50 (2024). https://onlinelibrary.wiley.com/doi/pdf/10.1111/jiec.13446
15. Ribeiro, I., et al.: Framework for life cycle sustainability assessment of additive manufacturing. Sustainability **12**(3), 929 (2020)
16. Hunkeler, D., Lichtenvort, K., Rebitzer, G.: Environmental Life Cycle Costing. CRC Press, Boca Raton (2008)
17. Jansen, B.W., van Stijn, A., Gruis, V., van Bortel, G.: A circular economy life cycle costing model (CE-LCC) for building components. Resour. Conserv. Recycl. **161**, 104857 (2020)
18. Luthin, A., Traverso, M., Crawford, R.H.: Assessing the social life cycle impacts of circular economy. J. Clean. Prod. **386**, 135725 (2023)

19. ISO: ISO 59020:2024, Circular Economy - Measuring and Assessing Circularity Performance (2024)
20. Mecheter, A., Tarlochan, F., Kucukvar, M.: A review of conventional versus additive manufacturing for metals: life-cycle environmental and economic analysis. Sustainability **15**(16), 12299 (2023)
21. N.u.R. Deutschland Bundesministerium für Umwelt, Ökodesign von Produkten: Gestaltungsauftrag für mehr Umweltschutz und Innovation. Politische Ökologie München, Ökom-Verlag (2005)
22. Yi, L., et al.: An eco-design for additive manufacturing framework based on energy performance assessment. Addit. Manuf. **33**, 101120 (2020)
23. Faludi, J., Bayley, C., Bhogal, S., Iribarne, M.: Comparing environmental impacts of additive manufacturing vs traditional machining via life-cycle assessment. Rapid Prototyping J. **21**, 14–33 (2015)
24. Naser, A.Z., Defersha, F., Pei, E., Zhao, Y.F., Yang, S.: Toward automated life cycle assessment for additive manufacturing: a systematic review of influential parameters and framework design. Sustain. Prod. Consum. **41**, 253–274 (2023)
25. Yang, S., Min, W., Ghibaudo, J., Zhao, Y.F.: Understanding the sustainability potential of part consolidation design supported by additive manufacturing. J. Clean. Prod. **232**, 722–738 (2019)
26. Tang, Y., Yang, S., Zhao, Y.F.: Sustainable design for additive manufacturing through functionality integration and part consolidation. In: Muthu, S.S., Savalani, M.M. (eds.) Handbook of Sustainability in Additive Manufacturing. EFEPP, pp. 101–144. Springer, Singapore (2016). https://doi.org/10.1007/978-981-10-0549-7_6
27. Liebherr-Aerospace & Transportation SAS: Additive manufacturing: Liebherr reaches major milestone (2024). https://www.liebherr.com/de/deu/aktuelles/news-pressemitteilungen/detail/3d-druck-liebherr-erreicht-weiteren-meilenstein.html
28. Li, Y., Jia, G., Cheng, Y., Hu, Y.: Additive manufacturing technology in spare parts supply chain: a comparative study. Int. J. Prod. Res. **55**, 1498–1515 (2017)
29. Koop, C., Grosse Erdmann, J., Koller, J., Döpper, F.: Circular business models for remanufacturing in the electric bicycle industry. Front. Sustain. **2**, 785063 (2021)
30. Huang, R., Riddle, M., Graziano, D., Warren, J., Das, S., Nimbalkar, S., Cresko, J., Masanet, E.: Energy and emissions saving potential of additive manufacturing: the case of lightweight aircraft components. J. Clean. Prod. **135**, 1559–1570 (2016)
31. Bierdel, M., Pfaff, A., Kilchert, S., Köhler, A.R., Baron, Y., Bulach, W.: Ökologische und ökonomische Bewertung des Ressourcenaufwands - Additive Fertigungsverfahren in der industriellen Produktion, Technical report, VDI Zentrum Ressourceneffizienz (2019)
32. Gebisa, A.W., Lemu, H.G.: A case study on topology optimized design for additive manufacturing. IOP Conf. Ser. Mater. Sci. Eng. **276**, 012026 (2017)
33. Barbieri, L., Muzzupappa, M.: Performance-driven engineering design approaches based on generative design and topology optimization tools: a comparative study. Appl. Sci. **12**(4), 2106 (2022)
34. Howell, L.L., Magleby, S.P., Olsen, B.M. (eds.): Handbook of Compliant Mechanisms, 1 edn. Wiley (2013)
35. Oros Daraban, A.E., et al.: A deep look at metal additive manufacturing recycling and use tools for sustainability performance. Sustainability **11**(19), 5494 (2019)
36. Foster, J., Cullen, C., Fitzpatrick, S., Payne, G., Hall, L., Marashi, J.: Remanufacture of hot forging tools and dies using laser metal deposition with powder and a hard-facing alloy Stellite 21®. J. Remanuf. **9**, 189–203 (2019)

37. Wilson, J.M., Piya, C., Shin, Y.C., Zhao, F., Ramani, K.: Remanufacturing of turbine blades by laser direct deposition with its energy and environmental impact analysis. J. Clean. Prod. **80**, 170–178 (2014)
38. Buican, G., Oancea, G., Manolescu, A.: Remanufacturing of damaged parts using selective laser melting technology. Appl. Mech. Mater. **693**, 285–290 (2014)
39. Morrow, W.R., Qi, H., Kim, I., Mazumder, J., Skerlos, S.J.: Environmental aspects of laser-based and conventional tool and die manufacturing. J. Clean. Prod. **15**, 932–943 (2007)
40. Gouveia, J.R., et al.: Life cycle assessment and cost analysis of additive manufacturing repair processes in the mold industry. Sustainability **14**, 2105 (2022)
41. Powell, D., Rennie, A., Geekie, L., Burns, N.: Understanding powder degradation in metal additive manufacturing to allow the upcycling of recycled powders. J. Clean. Prod. **268**, 122077 (2020)
42. Walachowicz, F., et al.: Comparative energy, resource and recycling lifecycle analysis of the industrial repair process of gas turbine burners using conventional machining and additive manufacturing. J. Ind. Ecol. **21**(S1), S203–S215 (2017). https://onlinelibrary.wiley.com/doi/pdf/10.1111/jiec.12637
43. GE Additive: GE aviation singapore first offer metal additive engine component repair (2024). https://www.ge.com/additive/press-releases/ge-aviation-singapore-first-offer-metal-additive-engine-component-repair
44. Grzesik, W., Ruszaj, A.: Hybrid additive and subtractive processes. In: Hybrid Manufacturing Processes. Springer Series in Advanced Manufacturing, pp. 191–208. Springer, Cham (2021)
45. Prochatzki, G., et al.: A critical review of the current state of circular economy in the automotive sector. J. Clean. Prod. **425**, 138787 (2023)

Open Access This chapter is licensed under the terms of the Creative Commons Attribution 4.0 International License (http://creativecommons.org/licenses/by/4.0/), which permits use, sharing, adaptation, distribution and reproduction in any medium or format, as long as you give appropriate credit to the original author(s) and the source, provide a link to the Creative Commons license and indicate if changes were made.

The images or other third party material in this chapter are included in the chapter's Creative Commons license, unless indicated otherwise in a credit line to the material. If material is not included in the chapter's Creative Commons license and your intended use is not permitted by statutory regulation or exceeds the permitted use, you will need to obtain permission directly from the copyright holder.

Overview of the Challenges in High-Pressure Type V Hydrogen Tanks for Automotive Applications

Santwana Pati[✉], Akshay Deshmane, Maximillian Korff, and Tobias Dickhut

Chair of Composite Materials and Engineering Mechanics, Institute for Aeronautical Engineering, Bundeswehr University Munich, Werner-Heisenberg-Weg 39, 85577 Neubiberg, Germany
santwana.pati@unibw.de

Abstract. Hydrogen emerges as a pivotal element in achieving fossil-free transportation, offering a sustainable alternative to conventional fossil fuels and significantly reducing environmental impacts. This report outlines the development of a cutting-edge Type V high-pressure vessel employing carbon fiber composite material, aimed at tackling the challenges associated with liner less hydrogen pressure vessel. Current research focuses on addressing two major challenges, namely manufacturing complexity and leak tightness of the tank. Manufacturing complexity of the tank is thoroughly addressed, covering various removable mandrel technologies and further a novel integral mandrel technique is introduced, featuring a carbon fiber-reinforced polymer (CFRP) structure.

Another significant challenge is ensuring the leak tightness of the tank without a polymer liner. The current article explains this issue in detail and discusses various solutions for a liner less, leak-tight tank structure. While Type V pressure vessels offer advantages in weight and storage capacity, their integration into automotive applications requires addressing manufacturing complexity, leak tightness, material compatibility, and cost-effectiveness. By overcoming these obstacles, the full potential of Type V tanks in hydrogen vehicles for the transportation sector can be realized, fostering sustainable mobility solutions.

Keywords: Filament winding · Type V pressure vessels · Permeability First Section

1 Introduction

The automotive industry is integrating futuristic research and focusing on sustainability, autonomous driving, smart connectivity and electric and hydrogen powered vehicles. The development of hydrogen fuel cell vehicles can enhance safety, efficiency and sustainability of the transportation sector. According to the Global EV outlook 2023 report, the number of hydrogen based vehicles has increased by 40% in 2022 [1]. Although hydrogen offers a clean energy source and high energy efficiency, the safe and compact storage is a significant factor to consider in order to replace the fossil fuels completely. As part of the DigiTain research project, a high pressure hydrogen composite overwrapped pressure vessel is being developed in compliance with EU Regulations No. 134 [2, 3].

Lightweight and high strength pressure vessels are widely used in various sectors like aviation, space and transportation etc. The evolution of the various types of pressure vessels has been a focus of research throughout decades [4]. The initial versions (Type I-III) had a significant metal component in it. Type I completely made up of metal, Type-II is metallic with composite hoop-wrapped and Type-III is completely composite overwrapped on a metallic liner. In the case of Type IV, the fuel is stored in a plastic liner which also serves as a mandrel for the filament winding process. The latest Type V technology offers a tank that is made up of only CFRP. The main challenge with a metallic structure is the heavy weight. Additionally the hydrogen embrittlement issue is also prevailing in most metals that leads to degradation of the tank [5]. The composite tanks overcome these challenges and provide lightweight construction but face limitations of their own like matrix microcracks [6, 7]. The latest type of pressure vessel is a Type V composite pressure vessel that offers a complete carbon fiber-reinforced polymer (CFRP) based tank structure. The Fig. 1 illustrates the five different kinds of pressure vessels.

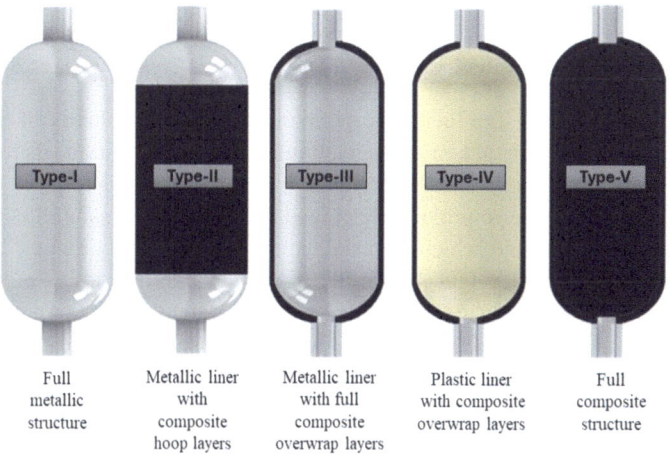

Fig. 1. Illustration of the various types of pressure vessels

Several researchers are working on the development of Type V tanks. Liner less tanks were fabricated in 1999 by Meyer et al. for cryogenic fluid storage in space applications [8]. Although there was no leakage or microcracks in the structure, the tank developed wrinkles due to collapse of the hollow sand mandrel. Similarly, Mallick et al. reported in 2004 about the development of a liner less tank for space application, focusing on crack resistant material development and failure analysis [9]. However, the application of Type V pressure vessels in automotive sector is quite limited in the available literature. Hassan et al. provide a detailed discussion on the various types of pressure vessels and elaborate on the research trends in the current scenario [10]. Air et al. report on the design and manufacturing of a Type V pressure vessel using the automated fiber placement technology [11, 12]. The tank had several leakage points and therefore held much lower hydrostatic pressure, but it laid foundation for the fabrication of a pressure vessel. This report highlights the significant challenges involved in developing a Type V pressure vessel for automotive applications.

2 Motivation

The DigiTain project aims to develop models for digital product development to demonstrate sustainable architecture [3]. One of the major components of this initiative is the development of a 700 bar hydrogen storage pressure vessel. The Type V pressure vessel is expected to have significant impact on the hydrogen economy by demonstrating lightweight design and high vehicle efficiency through very high strength to weight ratio. However, the development of a Type V pressure vessel poses several challenges [13]. Some of the significant challenges are (a) Hydrogen compatibility (b) Durability and fatigue resistance (c) Regulatory compliance (d) Cost effectiveness considering materials, manufacturing and system integration; especially for automotive applications (e) Gas permeation and (f) Manufacturing complexity. Gas permeation and manufacturing complexity are the major challenges and are discussed in details in this article.

2.1 Manufacturing Complexity

Introducing a liner less design increases the manufacturing complexity of the Type V tanks. Without the liner, the tank material itself must meet the requirements for strength, permeability and compatibility with hydrogen. In order to manufacture the tank structure, a mandrel is essential to provide shape and dimensions. A mandrel is a structure that serves as a core around which materials are wrapped to obtain composite structures at the end. The Type V tanks generally use removable mandrels to manufacture the tank [14–16]. These are made of sand or plaster or sometimes 3D printed materials. These mandrels are extensively developed for the purpose of removal post curing the tank. However, they face certain limitations. Firstly, it is difficult to remove the complete material post-curing especially with geometries that have smaller polar openings. Secondly, the sand based mandrels are manufactured based on the specific geometry of the tank. Therefore, they come with a high price, especially for research production. Even for later series production, the cost of the mandrel is a significant factor aiming at the automotive applications. Thirdly, the 3D printed resin material mostly has lower melting temperature than the curing temperature of the overwrap towpreg. Therefore, the mandrel disintegrates during the tank curing. Finally, another major challenge is the limitation to attach barrier film or coating for a removable mandrel. These limitations, listed above, pose research opportunities for the development of a new mandrel concept.

Therefore, a new integral mandrel technology has been introduced in this project that makes the mandrel out of CFRP and does not require removal from the tank. This mandrel is expected to act as a partial load bearing structure in the tank and reducing the total number of overwrap layers compared to a Type IV pressure vessel, demonstrating weight saving potential for the automotive sector. Additionally, it can be a cost-effective process since it is manufactured with a reusable metallic mold. Finally, as the mandrel eliminates the use of sand, plaster or removable materials, it can be a sustainable solution that encourages low wastage. Thereby, looking at these major advantages, the integral mandrel has been studied as an alternative technique for Type V tank development. This article discusses the initial investigations made in this direction. The detailed research and development of the mandrel is in the future scope of the research.

2.2 Gas Permeation

One of the primary concerns with a liner less Type V pressure vessels is the permeation of hydrogen outside of the tank owing to the lack of a polyamide-based liners as in the Type IV tanks [17–19]. This can lead to the loss of the stored amount of gas and also pose potential safety risks. Therefore, it is a crucial challenge for a Type V pressure vessel to ensure safe and efficient storage of hydrogen. Hydrogen is the smallest and lightest molecule and has the potential to permeate through various materials. Figure 2 illustrates the leak path formation and subsequent loss of fuel through the composite. The driving range of the vehicle depends largely on the amount and pressure of the hydrogen stored in the pressure vessels. Therefore, premature leaking and structural degradation need to be carefully avoided especially without a liner. The tank structure needs to make sure that hydrogen doesn't permeate out while sustaining high pressure and the environmental loads. Consequently, designing the pressure vessel with these criteria in mind is a crucial step in the development process.

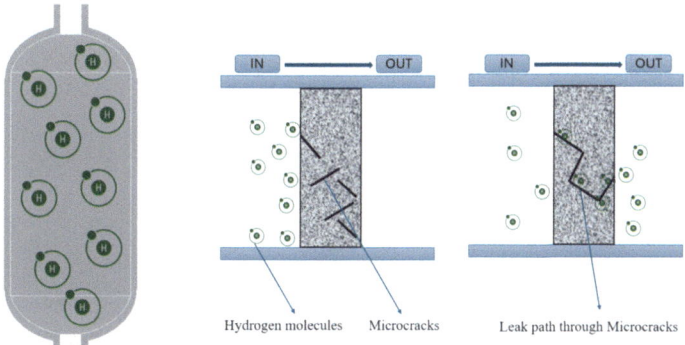

Fig. 2. Hydrogen molecules tend to leak out of the composite walls of the pressure vessels. The microcracks are developed owing to the cyclic loading and they tend to form leak paths through these microcracks that leads to gas leakage.

3 Preliminary Investigations and Results

3.1 Design of the Mandrel

The preliminary model was designed using μwind software by Mefex GmbH [16]. The model was developed considering the installation space and dimensions required within the framework of the DigiTain project. The laminate thickness was determined to be 8.55 mm. The inner diameter of the tank is 135 mm, and the cylinder length is 1050 mm as shown in Fig. 3. The modeling utilized towpreg material with T700 24K carbon fibers. The primary objective was to ensure the structure was free from fiber failure. Initial evaluations indicate that the design can withstand a burst pressure of 1400 bar without fiber failure, as depicted in Fig. 4. Further assessment will focus on Inter-fiber-failure as this of greater importance in Type V tank development considering the absence of the liner. Taking into account the current design setup the plan involves using the initial two hoop layers as the CFRP tube and attaching CFRP domes to it and beginning the overwrap from the third layer.

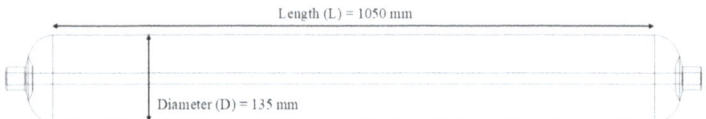

Fig. 3. Dimensions of the 1400 bar pressure vessel according to the design based on installation space.

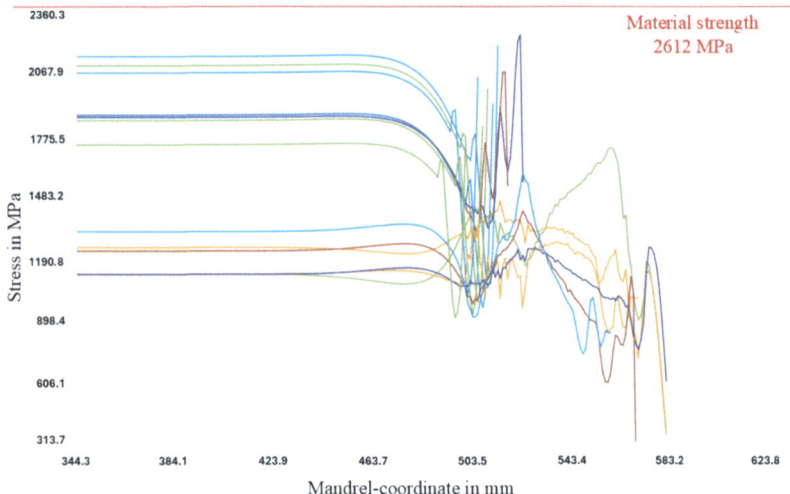

Fig. 4. Stress distribution along the tank geometry calculated by the analytical method using µwind software and the red line denotes the material strength from the datasheet.

3.2 Manufacturing Trials

The integral mandrel is divided into three different sections, one long tubular CFRP section and two dome caps. The metallic bosses are typically designed to match the external piping and valve connections. They are attached to the dome caps. The manufacturing steps for a mandrel development and subsequent tank manufacturing is shown in the Fig. 5.

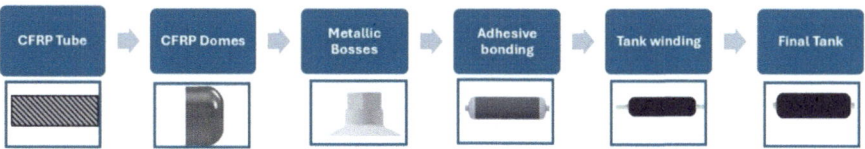

Fig. 5. The manufacturing flowchart for the development of a CFRP based mandrel. The dome caps and tube are adhesively bonded together with the boss and then filament winding is done on the structure which is then cured to obtain the complete tank structure.

CFRP Tube

The metallic tooling made of Aluminum alloy is used. This metallic tooling provides precision to the interior profile of the finished composite structure. The CFRP tube is obtained by winding towpreg on this metallic tooling. After curing, the thin composite tube is taken off the metallic tooling. The metallic tooling is made of AlMgSi0,5 with a roughness average (Ra) of 1 μm. The towpreg used here is made of 24k T700 towpreg by Kümpers [20]. The tube is manufactured using a robotic filament winding machine by Roth Composite Machinery GmbH [21].

CFRP Dome Caps

The dome caps are made up of thin composite parts using a 3D-printed tooling for preliminary investigations. The inner surface of the tooling is lined with silicone for easy removal of the composite part and to provide enhanced surface finish. The conventional and easily available twill weave (2 X 2) fibers and epoxy resin system are used for the initial investigations (Epikote RIMR 426 with hardener RIMH 435). They enable an ambient cure system and medium low viscosity. The twill weave fibers provide drapability, which is the ability to conform to different three-dimensional shapes, which helps to manufacture dome-shaped parts. The tooling is designed to provide full outer shape to the dome caps. Figure 6 shows the CAD design of the tooling and images of the actual tooling used for the manufacturing. The tooling is divided into two parts that are joined using screws and leads to easier removal of the composite part after curing.

A simple hand layup technique was applied for these initial investigations. Dry fibers were placed and the resin was applied using a brush; subsequently the hand roller was used to ensure uniform resin distribution and obtaining the required thickness. Figure 7 illustrates the process used for the dome fabrication. To avoid voids in the composite part, vacuum bagging was done above the mold during the curing process.

The initial investigations provide results that are essential for further validation. The CFRP tube is manufactured with precise geometry and surface finish owing to the metallic removable tooling. The process is sustainable due to the reusable metallic tooling. The technique is reproducible, as the geometry is repeatable. Figure 8 shows a sample CFRP tube manufactured by filament winding on metallic tooling.

Regarding the dome caps, initial results of the manufacturing trials suggest a few limitations. Firstly, the dome structure is technically complicated to manufacture using hand layup and therefore requires a significant labor-intensive approach. The use of a more advanced and simple technique is essential, such as RTM. Secondly, the slots for boss attachment required holes in the dome caps and it is difficult to manufacture a

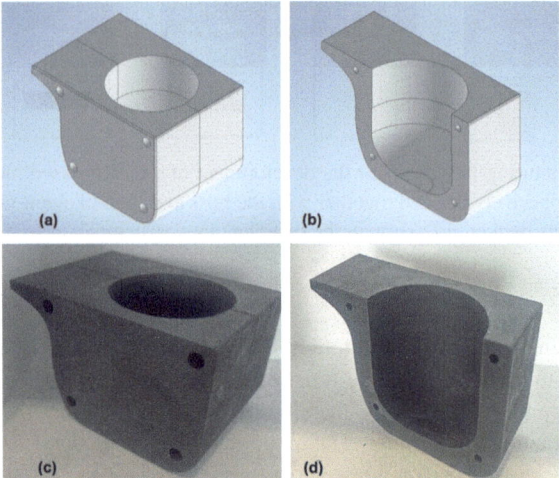

Fig. 6. CAD details of the tooling designed for the manufacturing of the dome caps. (a) shows the two portions of the tooling attached and (b) shows the inner surface of the tooling with slot for boss attachment and slit for tube attachment. (c) and (d) show the actual images of the tooling

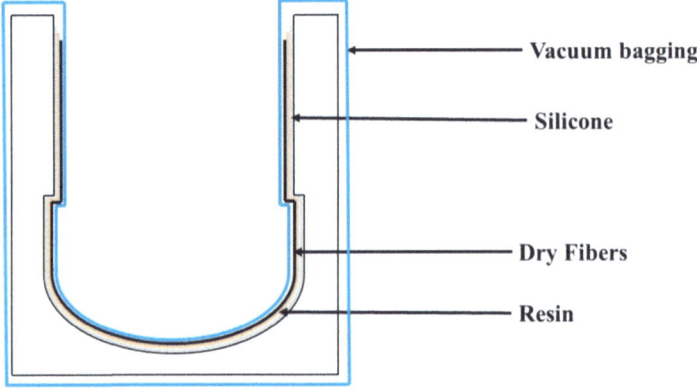

Fig. 7. Hand layup process for the dome caps manufacturing using the 3D printed tooling.

dome cap with hole using the currently utilized technique. Thirdly, the reproducibility of the composite part seems challenging since the hand layup technique offers lower precision in geometry. Figure 9(a) shows the images of the dome caps as manufactured. Figure 9(b) shows the assembly of CFRP dome caps and tube.

The preliminary trials for the integral CFRP mandrel have been completed. The tubular section of the mandrel is manufactured by winding and is therefore a promising solution. However, the dome caps are complicated composite parts and hence show limitations for reproducible and precise fabrication. The next step could involve using a different technique such as resin transfer molding (RTM) to produce the dome caps.

Fig. 8. CFRP Tube manufactured by winding towpreg on a metallic tooling.

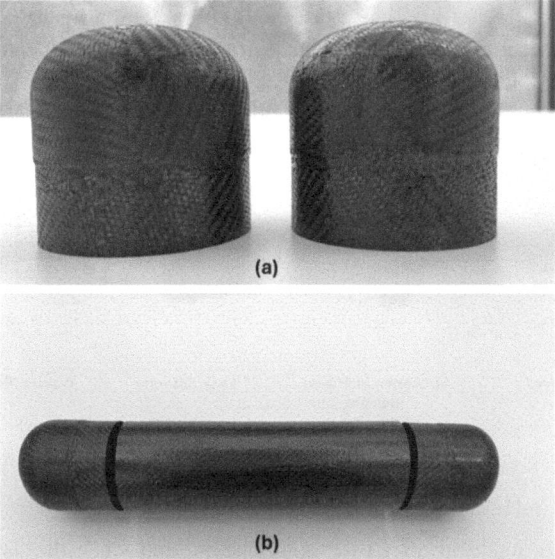

Fig. 9. Images of the fabricated CFRP integral mandrel components in the initial investigations (a) The image of the dome caps manufactured using hand layup technique. (b) CFRP tube fabricated using prepregs wound around a metallic tooling and the dome caps are attached on both ends to illustrate the CFRP based integral mandrel concept.

Furthermore, additive manufacturing also offers numerous advantages in this regard, especially for production of complicated designs like the dome caps. An improved step could be a 3D- printed dome cap structure integrated with a CFRP tube to produce the final mandrel. The 3D- printed part however needs to be high temperature resistant so that it doesn't disintegrate during the curing of the final tank.

3.3 Gas Permeation Through the Tank Walls

The challenge of gas permeation is a significant factor for Type V pressure vessels in the absence of a liner. The integral mandrel is expected to provide structural stability for tank winding and also mitigate the hydrogen permeation through the tank walls. From the structural perspective, the fibers define the stress and strain limits of the full composite. Whereas the leakage component is defined by the matrix part of the composite. As explained in the Sect. 2.2, the composite is subjected to the operating pressure of the tank and tends to form microcracks. Additionally, the resin system needs to hold the laminates together to avoid delamination of the composite structure in the tank. Hence, it is essential that the resin system is toughened to be crack-resistant until the operating pressure of 700 bar [22, 23].

The gas permeation is analyzed at sample level using the in-house diffusion test set up as shown in the Fig. 10. This set up required samples to be within 1mm thickness and less than 50mm in dimensions and need to be fully dried for 48 h prior to the testing. The specimens are only subjected to a low pressure of 4 bar since the leakage through a sample is pressure dependent, but the material specific permeation property is not significantly affected by it. The test set up uses helium instead of hydrogen for the initial test on the samples. They both have similar molecular dimensions and helium is non-flammable, inert and non-reactive. Additionally, helium is not common in the environment allowing high detection accuracy.

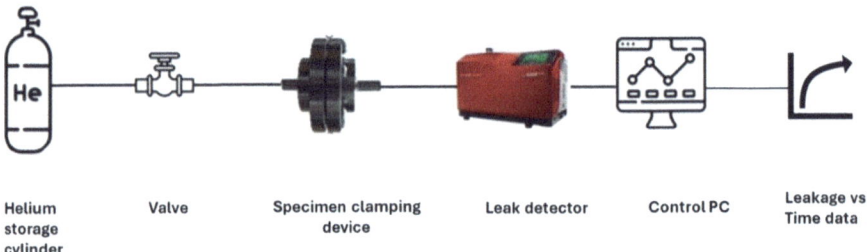

Fig. 10. Schematic diagram showing the inhouse permeation test set up.

The mandrel is made up of CFRP components, and since the Type V is a liner less design, the mandrel itself should ensure that the tank is leak-tight. Therefore, it is essential to evaluate its permeation properties. The flat laminate sample of 0.8 mm thickness was fabricated using vacuum-assisted curing of four layers of twill (2 x 2) prepreg. The measurement was done for 24 h and against a pressure of 4 bar of helium on a specimen prepared using a sample twill CFRP prepreg as used for the dome cap manufacturing trials. The leak rate with respect to time in hours is plotted in the Fig. 11.

The permeation is a property of the material and it is evaluated from the experimentally obtained leak rate by using the Eq. 1.

$$P = \frac{Q \times d}{\Delta p \times A} \quad (1)$$

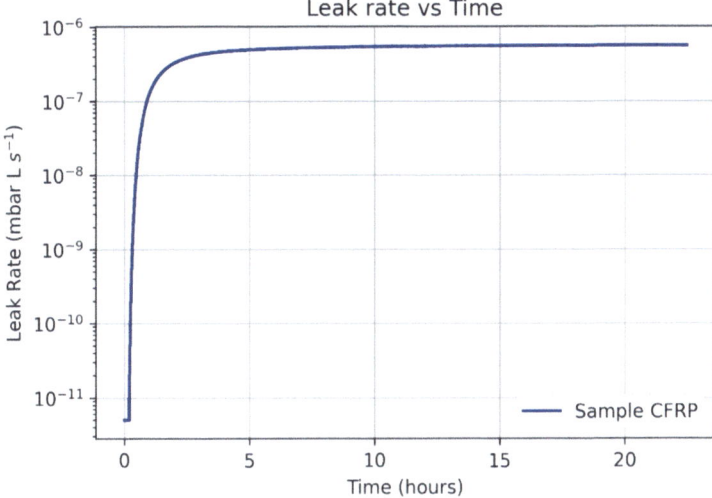

Fig. 11. The leak rate of the CFRP sample measured against time for a 24h duration.

In Eq. (1), P is the permeability, Q is the leak rate obtained through the experiments, d is the thickness of the sample, Δp is the pressure difference of the gas on both sides of the sample and A is the surface area of the sample that comes in contact with the gas [24]. The maximum permeability value obtained is 4.14×10^{-14} mm^2/s which is lower than that of liner samples due to the presence of fibers but is expected to worsen as microcracks develop in the resin [19]. Gas leakage through composites has been an extensively analyzed topic, and various reasons can be identified. Firstly, diffusion is when the gas molecules seep through the resin system. Secondly, sorption can be another factor in which the gas molecules are attached to the surface and change the mechanical properties over time. This can lead to further gas permeation through the composite material. Thirdly, and the most vital cause can be the formation of microcracks. They are induced due to mechanical, thermal and environmental stresses and lead to formation of leak path [25]. Therefore, liner less tank structure requires significant reinforcement for a barrier layer. This barrier can be integrated into the composite either through modified resin or through layer of braided plies as shown in the patent by Cronin et al. [26]. In another aspect it can also be a thin coating material that can decrease the permeation of gas. Further steps for a leak tight tank using the integral mandrel can be integrating the barrier layer as foil in the CFRP structure at mandrel level. The overall summary is that extensive research is still needed for the development of a Type V tank with a leak tight structure.

4 Conclusion and Critical Discussion

Type V pressure vessels are a groundbreaking advancement in hydrogen storage technology. They differ from Type IV vessels by eliminating the internal polymer gas barrier. Instead, Type V vessels use carbon fiber laminate to provide the necessary structural properties and prevent gas leakage. Notably, Type V vessels have found utility in launch vehicles and spacecraft applications and are now emerging as a promising solution for achieving lightweight and efficient fuel storage in automotive applications owing to its high strength to weight ratio and increased capacity resulting due to absence of a liner. This article explains the various hurdles in the development of a Type V pressure vessel for automotive application. The discussion delves into two primary limitations: manufacturing complexity and gas permeation, offering detailed insights into each aspect.

This article proposes an integral mandrel concept in which a CFRP-based structure is used as a mandrel upon which filament winding can be done to obtain the full tank structure. This mandrel provides additional load-bearing capacity and hence the number of overwrap layers are expected to be reduced. Secondly, a CFRP based structure is expected to provide cost effectiveness as compared to sand or plaster based removable mandrels paving the way for low cost mass production. Finally, the integration of the mandrel structure into the tank design holds the potential for sustainability and reduced waste, as it streamlines manufacturing processes and material usage. The integral mandrel is anticipated to offer the aforementioned functional advantages and, as such, holds potential for integration into the automotive industry. The initial investigations have been conducted to assess the manufacturing process of the integral mandrel, with subsequent analysis aimed at predicting the permeation behavior of the tank. Permeability assessments revealed detectable leakage in the sample within 24 h, highlighting the necessity of incorporating a permeation barrier layer. In conclusion, addressing the challenges of high-pressure Type V hydrogen tanks is critical for advancing automotive applications, requiring innovations in materials, permeation control, and safety standards to support hydrogen's role in fossil-free transportation.

5 Future Scope

Anticipating future advancements, there is ample room for further development of the integral mandrel. While the tube structure has been successfully obtained, the dome cap structure presents a few manufacturing limitations. A potential next step could involve utilizing different techniques such as resin transfer molding for the dome caps manufacturing. Additionally, additive manufacturing could also be employed to obtain a 3D printed dome cap. Permeation studies on the composite mandrel indicate a need for enhanced barrier properties. One approach could involve modifying the resin to imbue it with inherent barrier characteristics. In that direction, nanomaterials that can block the leakage like Graphene can be added to the epoxy resin, followed by the permeation analysis at the sample level. Alternatively, integrating a thin barrier film into the mandrel's structure could ensure complete leak tightness. Metallic foils can offer full barrier against fuel leakage but their integration into a tubular tank wall structure will require extensive research and development efforts. In summary, advancing the integral mandrel through innovative manufacturing techniques and enhanced barrier properties is essential for overcoming current challenges and improving the performance of Type V hydrogen tanks.

Acknowledgement. The research is done within the framework of the project DigiTain. DigiTain is funded by the Federal Ministry for Economic Affairs and Climate Protection on the basis of a decision by the German Federal Assembly under Grant FKZ 19S22006M.

References

1. HydrogenInsight: The number of hydrogen fuel-cell vehicles on the world's roads grew by 40% in 2022, says IEA report
2. Directive, E.U.: 28/EC. Retrieved **3**, 2013 (2009)
3. ARENA 2036: DIGITALIZATION FOR SUSTAINABILITY (DIGITAIN)
4. Olsson, R., Cameron, C., Moreau, F., Marklund, E., Merzkirch, M., Pettersson, J.: Design, manufacture, and cryogenic testing of a linerless composite tank for liquid hydrogen. Applied Composite Materials (2024). https://doi.org/10.1007/s10443-024-10219-y
5. Okonkwo, P.C., et al.: A focused review of the hydrogen storage tank embrittlement mechanism process. Int. J. Hydrogen Energy **48**, 12935–12948 (2023)
6. Ma, Q., Rejab, M.R., Azeem, M., Hassan, S.A., Yang, B., Kumar, A.P.: Opportunities and challenges on composite pressure vessels (CPVs) from advanced filament winding machinery: a short communication. Int. J. Hydrogen Energy **57**, 1364–1372 (2024)
7. Yan, Y., Zhang, J., Li, G., Zhou, W., Ni, Z.: Review on linerless type V cryo-compressed hydrogen storage vessels: resin toughening and hydrogen-barrier properties control. Renewable and Sustainable Energy Reviews (2024). https://doi.org/10.1016/j.rser.2023.114009
8. Linerless composite cryogenic technology. Space Technology Conference and Exposition
9. Ultralight linerless composite tanks for in-space applications. Space 2004 Conference and Exhibit (2004)
10. Hassan, I.A., Ramadan, H.S., Saleh, M.A., Hissel, D.: Hydrogen storage technologies for stationary and mobile applications: review, analysis and perspectives. Renewable and Sustainable Energy Reviews (2021). https://doi.org/10.1016/j.rser.2021.111311
11. Air, A., Shamsuddoha, M., Prusty, B.G.: A review of Type V composite pressure vessels and automated fibre placement based manufacturing. Compos. B Eng. **253**, 110573 (2023)
12. Air, A., Shamsuddoha, M., Oromiehie, E., Prusty, B.G.: Development of an automated fibre placement-based hybrid composite wheel for a solar-powered car. The Int. J. Advanced Manufacturing Technol. **125**, 4083–4097 (2023)
13. Azeem, M., et al.: Application of filament winding technology in composite pressure vessels and challenges: a review. J. Energy Storage **49**, 103468 (2022)
14. Reynolds, H.: Pressure vessel design, fabrication, analysis, and testing. In: Composite filament winding, pp. 115–148. ASM International Ohio (2011)
15. Shen, F.C.: A filament-wound structure technology overview. Mater. Chem. Phys. **42**, 96–100 (1995)
16. Munro, M.: Review of manufacturing of fiber composite components by filament winding. Polymer Composites **9**, 352–359 (1988)
17. Su, Y., Lv, H., Zhou, W., Zhang, C.: Review of the hydrogen permeability of the liner material of type iv on-board hydrogen storage tank. World Electric Vehicle Journal **12**, 130 (2021)
18. Li, X., Huang, Q., Liu, Y., Zhao, B., Li, J.: Review of the hydrogen permeation test of the polymer liner material of type iv on-board hydrogen storage cylinders. Materials **16**, 5366 (2023)
19. Condé-Wolter, J., et al.: Hydrogen permeability of thermoplastic composites and liner systems for future mobility applications. Compos. A Appl. Sci. Manuf. **167**, 107446 (2023)
20. Kümpers GmbH: Kümpers GmbH. https://kuempers-composites.com/

21. Roth Composite Machinery: Machines Filament Winding. https://www.roth-composite-machinery.com/en/filament-winding-prepreg/machines
22. Nair, A., Roy, S.: Modeling of permeation and damage in graphite/epoxy laminates for cryogenic tanks in the presence of delaminations and stitch cracks. Compos. Sci. Technol. **67**, 2592–2605 (2007)
23. Ebermann, M., Bogenfeld, R., Kreikemeier, J., Glüge, R.: Analytical and numerical approach to determine effective diffusion coefficients for composite pressure vessels. Compos. Struct. **291**, 115616 (2022)
24. Humpenöder, J.: Gas permeation of fibre reinforced plastics. Cryogenics **38**, 143–147 (1998)
25. Liner-less tanks for space application-design and manufacturing considerations. 5th Conference on Aerospace Materials, Processes, and Environmental Technology (2003)
26. John Cronin, Denver, CO (US): Damage and leakage barrier in All composite pressure vessels and storage tanks Patent US8074826B2

Open Access This chapter is licensed under the terms of the Creative Commons Attribution 4.0 International License (http://creativecommons.org/licenses/by/4.0/), which permits use, sharing, adaptation, distribution and reproduction in any medium or format, as long as you give appropriate credit to the original author(s) and the source, provide a link to the Creative Commons license and indicate if changes were made.

The images or other third party material in this chapter are included in the chapter's Creative Commons license, unless indicated otherwise in a credit line to the material. If material is not included in the chapter's Creative Commons license and your intended use is not permitted by statutory regulation or exceeds the permitted use, you will need to obtain permission directly from the copyright holder.

Sustainable and Affordable Strategies to Reduce Traffic Emissions in Urban Areas

Ali Khan Muhammad[1] and Ali Khan Majid[2]()

[1] Nürtingen-Geislingen University (HfWU), Nürtingen, Germany
[2] Department of Computer Science, Technische Universität Chemnitz, Chemnitz, Germany
majidalikhn@gmail.com

Abstract. Due to rising global temperature, mobility sector is witnessing a transition phase. It is essential to cut tailpipe emissions significantly to reduce rising global temperature and avoid changing climate patterns. In response to climate change, innovations in automotive sector, sustainable mobility solutions and increased public awareness for global warming is needed. While significant progress has been made in reducing greenhouse gas emissions, the transportation industry remained largely unchanged. Europe aims to achieve carbon neutrality by 2050, necessitating a major shift in traditional transportation practices. Conventional means of transport burning fossil fuel in urban areas are major contributors to emissions, posing environmental and health risks. Emerging technological advancements in automotive sector offer a promising way to improve urban environments, in line with Europe's Green Deal and Paris Agreement. This study aims to suggest techniques to create more sustainable and liveable cities through affordable mobility innovations. The research explores the potential of integrating modern vehicle technologies with existing road infrastructure to reduce transport emissions.

Keywords: Sustainable Mobility · Liveable cities · Traffic flow · Tailpipe emissions · Urban traffic congestion · Minimizing urban emissions

1 Introduction

According to the Intergovernmental Panel on Climate Change (IPCC) reports and various scientific researches, substantive efforts are needed to mitigate global warming. The landmark Paris Agreement provides a structured framework for the international community to curtail Green House Gases (GHGs), targeting a 2°C limit, and a more ambitious goal to minimize these GHGs to 1.5°C above pre-industrial levels. The ongoing debate on carbon neutrality and pollutant emissions reduction by 2050 though, looks challenging considering the Paris Agreement, as previously set objectives are not being met, narrowing down the window to tackle the factors, as Sven unfolded [1].

Research shows that about one-third of the global energy consumption is attributed to the world's transport sector, contributing to the production of approximately 95% of direct tailpipe matter particles through the consumption of fossil fuels [2]. Moreover, it also contributes as the primary sector responsible for the emission of major greenhouse gases, significantly worsening the existing situation of global warming [3,4].

Europe is committed to reduce the harmful gases about 50% till 2030 compare to 1990 levels and become the first carbon neutral continent till 2050. This ambitious goal is only possible by its political will and following the objectives of European Green Deal and Paris Agreement, as experts say. [5,6]. However, automotive sector with many technological inventions for example, vehicle auto start-stop feature, tyre pressure monitoring sensor, cylinder deactivation technology and transition to electrify the mobility sector to make the transport sustainable, look unsuccessful comparing with other categories responsible for the GHG emissions. Despite 22.5% reduction in Europe's overall greenhouse gas (GHG) emissions between 1990 and 2018, there was an alarming surge of over 23% in total emissions from the transportation sector. The predominant contributor to this rise was transport emissions specifically the road-dominant sector, which increased roughly 27% during the mentioned period, constituting approximately 95% of the entire spectrum of transport emissions by the year 2018. The proportion of emissions attributed to road transport witnessed a rise, climbing from just under 13% in 1990 to nearly 21% by 2018, as researchers unfolded [7,8] (Fig. 1).

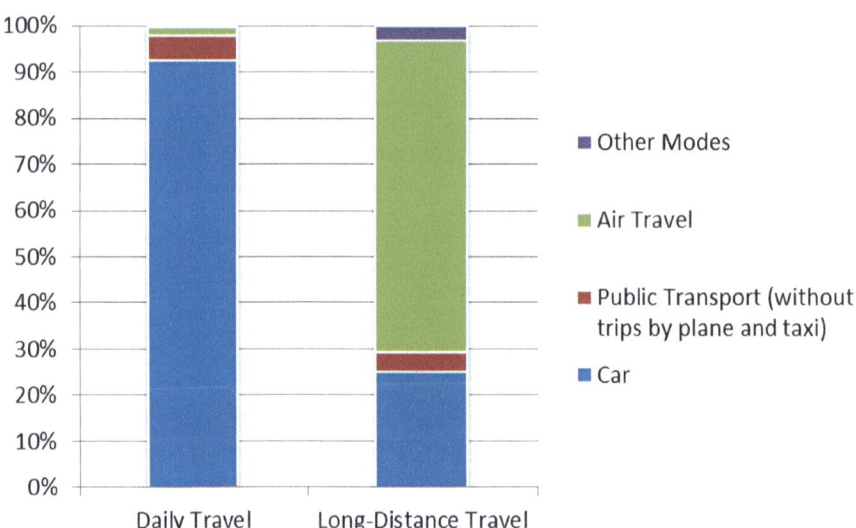

Fig. 1. CO2 emissions proportion by transport mode. (a) Reference: Proportions of daily travel and long-distance travel (Reichert, A. et al. 2016)

Furthermore, mega cities play a substantial role in economic development and contribute up to 60% of global GDP, according to the UN's 11th sustainable development goal "Sustainable Cities and Communities". Notably, with the economic progress, the mega cities are also responsible for 70% global GHG emissions, and the fundamental reason is rapid urbanization. [9,10].

According to UN's estimates, over 80% of European citizens will reside in the cities around the year 2050. The growing urban population of Germany is already evident, where between the years 2008 to 2016, mega cities of Frankfurt and Munich experienced a notable growth of 7%, as Mark describes [11]. The IPCC in its 2018 report declared that mobility is the major sector liable for 28% of the globe's final energy demand and 23% of its energy-related GHG emissions. The least diverse energy end-use sector, which accounted for 65% of the world's oil final energy demand and 92% of which was made up of oil products [12].

These statistics point to significant obstacles to de-carbonization. The major challenge lies in transitioning to sustainable and smart cities, a shift deemed very difficult without sustainable transportation, as noted by experts. This wider move towards "environmentally friendly cities" or "green urbanization" is currently influencing the global stage [13]. Sven describes that the effects of rising temperatures, traffic congestion, pollution, health inequalities, and limited access to these facilities are major concerns for authorities. He recognized that the current transportation status quo is not just unsustainable but also a major contributor to greenhouse gas emissions, a barrier to transform cities and make them more liveable [14].

Rapid urbanization is the reason for the increase in motorized vehicles; studies claim that by the year 2030, the number of vehicles will reach about two billion. Compared to suburbs, cities contribute higher greenhouse gases per kilometre. The total number of journeys taken and the distance travelled are aspects in determining an individual's carbon footprint. The more frequent the travel in propelled modes, the greater the tailpipe emissions produced. [15].

The discussion shows road transport a major contributor to GHGs at the local level and to urban air pollution, leading to various illnesses among neighbouring residents. Implementing new and advanced vehicle technologies is crucial to improve air quality, bridge them with road infrastructure and participate actively to minimize the carbon emissions from internal combustion engines (ICEs). This study focuses on the role of vehicle's engine auto start-stop feature and its potential to improve air quality. The paper also examines the benefits of combining different approaches, vehicle technology, road infrastructure and drivers active participation, to reduce urban traffic pollution.

2 Literature Review

2.1 Vehicle Auto Start-Stop Technology

Huff, S, et al. states that by letting the auto start-stop feature engaged, the technology reduces CO2 emissions, it has financial benefits too, according to the start-stop intervals because of increased fuel efficiency in busy cities where the idling is commonly observed [16]. In their research Thitipatanapong, S., et and Gaines, L., et al. say that this system is an economical way to minimize energy losses when a car is idling, which is important in cities with increasing traffic and it is proved by previous studies that traffic congestion and small stops consume 50 percent extra fuel, while instantly producing toxic emissions [17,18].

Due to the potential benefits, some manufacturers have put start-stop systems in their vehicles from, BOSCH, Valero, and Denso. Comparing the feature to cars without the start-stop function, the installation of this system saves about 12% on fuel. Zhu, T, et al. and Asekar, A.K found that in traffic with significant congestion, the start-stop technology can save about 8% on fuel, according to test results and figures released recently by BOSCH, a well-known manufacturer of auto start-stop systems. Zhu, T, et al. and Asekar, A.K. in their findings observed that significant congestion results in even larger fuel efficiency and financial discounts [19,20].

Karrouchi, M, et al. and Santos, N.D.S.A, et al. in their study argued that total fuel consumption is up to 10% by vehicle when its engine is at idle and an effective start-stop system has the potential to reduce fuel consumption by 7% to 9%, respectively [21,22]. Huff, S, et al. and Kropiwnicki, J. and Kneba explain the Start-Stop system's operation that it depends on the motor being turned off when the car is moving or motionless and not obtaining power from its engine. When the driver plans to move forward or accelerates, the vehicle's engine is started. Reduced tailpipe pollutants are due to improved fuel efficiency [16,23].

When gasoline is burned, it produces toxic gasses and CO_2, and for every gallon of gasoline burned, it produces 8,887 g of CO_2. To make it easier to understand, let us convert grams into metric tons. An easier way to do this is by multiplying by a very small number, which is 10 to the power of minus 3, written as 10^{-3}. So, 8,887 grams is the same as 8.887×10^{-3} metric tons. So, in a simpler form, it means that for every gallon of gasoline burned, it produces about 8.887 kg of CO_2 emissions [24]. These calculations highlight the significant contribution of CO_2 gasses from the combustion of gasoline from ICEs. Emphasizing initiatives towards sustainability, technology has the potential to play an important role in achieving the aims outlined in the European Green Deal (EGD) and the Sustainable Development Goals (SDGs) set by the United Nations, paving the way for a better future.

2.2 Traffic Signals with Countdown Timer

One of the many significant solutions to reduce GHGs in local areas is to turn off the vehicle engines during waiting in busy traffic or at signalized crossroads according to the duration, remaining. Jou, R.C.et al. in their research expressed that to lower the emissions from idling of vehicles, some of the states have prohibited the idling of the engines, and motorists have to turn the vehicles off during stops at intersections and traffic congestion, for example, this duration in different European states lays from 10 s to 3 min [25].

Kim, M and Kim, H.K in their findings observed that in account to keep the drivers well informed about the remaining red signal time, countdown timers can play significant role to reduce the CO_2 emissions at busy urban areas [26].

Krukowicz, T, et al. explained that the countdown timers are some kind of digital clocks, showing the duration of changing the red, green, or yellow signal at urban intersections and crossroads. The purpose of countdown signals is to keep the drivers well informed in their decisions, whether to brake or pass the junction, according to the time duration shown on the countdown timer [27]. Yan, W. et al. Islam, M. R. et al. and Islam, M. R. highlighted in their study that previous research has predominantly concentrated on analysing traffic flow and reducing red signal violations. They emphasized the significance of incorporating countdown timers at intersections for safety purposes [28–30]. Goel, A. and Kumar, P. referenced that frequent stop-and-go at intersections and busy traffic situations create extra delays, acceleration fluctuations, increased fuel consumption, exceeded emissions, and environmental degradation in local urban areas [31]. Goel, A., and Kumar unfolded in their research that because of acceleration and slow downing practices, particularly at heavy traffic locations, CO_2 emissions are increased because of higher load of engine and road frictions [32].

Studies have demonstrated that traffic signals equipped with countdown timer possess the potential to positively influence driver's behavior in busy traffic scenarios, alleviating anxiety and facilitating decision-making for future journeys. Another significant contributor to air pollution resulting from traffic is the occurrence of sudden stops and rapid accelerations at intersections. When drivers approach a junction and encounter changing the signal to red, it might be challenging to react promptly, often leading to abrupt stops. These sudden braking actions increase tire friction, consequently releasing pollutants into the environment and posing significant health risks, due to a lack of information about changing the duration of traffic signals. Additionally, drivers may attempt to rush through intersections before the signal changes to red, accelerating their vehicles to pass promptly and further contributing to air pollution. This behavior encourages the emission of pollutants, compounding the environmental impact and again opposing the state policies to minimize the tailpipe emissions [27].

2.3 The Proposed Approach

The approach suggested in this paper not only strengthens the results of existing studies, but also emphasizes the importance of active driver participation and their behavior towards climate change in the traffic congestion. As the strategies to reduce overall carbon footprint, by curtailing the tailpipe emissions needed to improve urban environment without studying the behavior of drivers will be difficult.

3 Methodology

3.1 Online Survey

This study is comprised of an online survey based on carefully developed questionnaire. Online survey made it fast and inexpensive to reach out to participants and collect primary data directly, it reduced the overall response time (from distribution to submission) and made it possible to avoid inconsistencies caused by an indirect data collection or paper based survey. The target group consisted of drivers between 18 to 55 years of age, with diverse driving experience. The survey avoided asking questions that could potentially lead to any social profiling, stereotyping, racial, ethnic, or gender bias [33–36].

The survey consisted of 24 multiple-choice questions to assess the use of auto start-stop system by drivers and their response based on the type of traffic signals they come across. The drivers were asked if they would like to have traffic signals equipped with countdown timers and whether these help them reduce emissions by switching engines off using auto start-stop system. Ethical practices were followed by obtaining consent from all participants, and making sure participants were aware of the purpose of the survey and their rights with all data collected adhering to General Data Protection Regulation (GDPR) guidelines.

3.2 Results

1. The survey data in Fig. 2 revealed that around 40% of participants owned vehicles manufactured before 2015, which lacked the auto start-stop feature. This means these cars continuously pollute the urban air in busy traffic situations or at traffic signals. By implementing countdown traffic signals at crossroads, drivers can be informed about the remaining duration before the traffic signal changes, allowing them to decide whether to pass, stop, keep the engine running, or turn it off according to the countdown timer.

   Additionally, it might be possible to invent prototype start-stop kits for retrofitting these vehicles. These kits would enable the start-stop feature to operate automatically based on preset measures, minimizing tailpipe emissions in busy urban traffic without driver intervention. What is the model year/manufacturing year of the vehicle?

Fig. 2. Manufacturing year of the vehicles driven by the survey participants. No. of Participants (N) = 74.

2. Figure 3 reveals the preferences of participants regarding the types of vehicles they own. It is evident that most vehicle owners prefer driving gasoline/benzene vehicles.

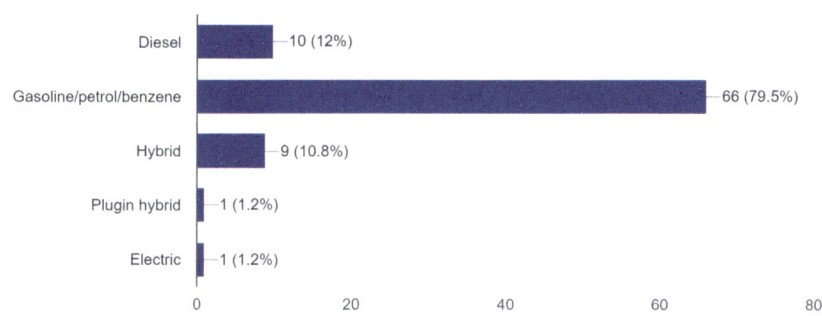

Fig. 3. Vehicle type the participants own and drive. No. of Participants (N) = 90

Out of 90 participants, only 3 own electric vehicles (EVs), indicating a significantly lower adoption rate compared to internal combustion engine (ICE) vehicles. This suggests that various barriers still hinder people from purchasing EVs over ICEs. To close this gap and make EVs more common on the roads, improvements in charging infrastructure, reduced charging times, and lower costs to buy EVs, are needed. These changes could increase EV acceptance among citizens and significantly contribute to reducing transport-related CO_2 emissions over time.

3. Figure 4 reveals the willingness of citizens to respond to traffic countdown timers. The data indicates that most people are aware of urban pollution and its causes.

Around 60 out of 83 participants are willing to turn off their vehicle engines at countdown timer traffic signals, even for duration of less than 10 s. This demonstrates their understanding of climate change and their commitment to reducing tailpipe CO_2 emissions. Such active participation is crucial for quickly reducing emissions, achieving the goal of sustainable cities, and aligning with state efforts to combat global warming.

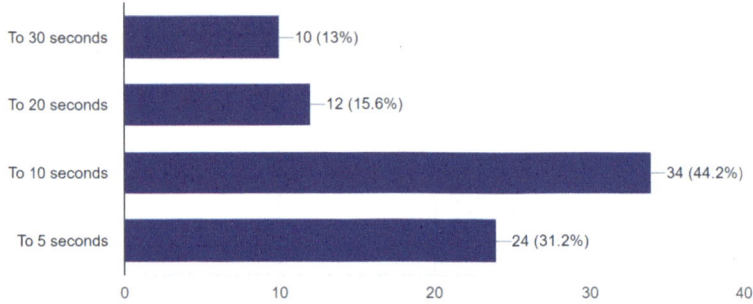

Fig. 4. Driver's preferences at the count-down timer traffic signals. No. of Participants (N) = 80

4 Methodological Limitations

On-line surveys even though are inexpensive, fast and efficient for data collection, they also have some limitations. A major issue is the difficulty to affectively engage the target group that receives the survey, thus limiting the control over responses. Another concern is that people don't actively participate and engage in the online surveys as traditional data collections methods, such as paper-based, in person or on a telephone call.

5 Analysis

The study reveals several critical factors regarding urban CO2 pollution and public perception. The data indicates that implementing both the start-stop feature and traffic countdown timers can significantly reduce CO2 emissions in urban areas, improve traffic flow, and minimize time wastage, thereby enhancing the overall transportation system. People are aware of urban pollution and related diseases, and they are willing to actively participate in efforts to reduce pollution and combat rising global temperatures. According to the collected data, there are many older vehicles on the roads lacking innovative features like start-stop. Citizens show a willingness to turn off the engines at busy traffic signals or during traffic jams if they know the waiting time is longer than usual. This awareness suggests the potential for behavioral change if the duration of stops is made clear through countdown timers.

The preference for internal combustion engine (ICE) vehicles over electric vehicles (EVs) or hybrids is evident due to reasons such as easier maintenance, greater flexibility, time-saving, stress-free operation, better infrastructure, and other benefits. While EVs promise lower emissions, their acceptance is hindered by factors such as limited charging infrastructure, high costs, time-consuming charging, and quick battery power drain. The data also shows that people are willing to turn off their vehicles if they are aware of the waiting duration, even if it is just 5 or 10 s remaining on the countdown timers. This motivation, combined with the auto start-stop feature, can significantly improve urban air quality, reduce pollutants, and lower disease rates. In the long term, it has economic benefits, too.

Overall, the study suggests positive attitude of participants to reduce CO2 pollution by implementing auto start-stop feature in vehicles and countdown timers on traffic signals. People are eager to adopt these measures for the sake of addressing climate change. Continued efforts could help achieve the goals of the European Green Deal (EGD) and the United Nations Sustainable Development Goals (UNSDGs).

6 Conclusion

This paper explores cost-effective and sophisticated solutions to combat urban air pollution, particularly due to the increasing number of vehicles on busy urban roads. Despite the continuous growth of traffic in urban areas, state policies and regulations aimed at reducing carbon emissions to improve urban air quality have not been very fruitful. However, by embracing technological advancements that could impact drivers' behaviour on busy intersections in urban areas and promoting environment friendly practices, greenhouse gas emissions can be substantially reduced. These initiatives not only offer viable alternatives to traditional practices but also pave the way for achieving essential urban development goals.

The paper also highlights the potential benefits to install countdown timer on traffic signals to reduce carbon emissions from urban transportation, reflecting public interest in combating air pollution through voluntary actions. It emphasizes the importance of aligning technological advancements, cost-effective solutions, and citizens' willingness to tackle urban air pollution, particularly from the increasing volume of cars in metropolitan areas due to growing urbanization. Public awareness plays a crucial role in fostering greater participation in minimizing urban air pollution efficiently. For policy-makers, transport authorities and experts, this research offers valuable insights into the implementation of countdown timer based traffic signals and vehicle auto start-stop feature's role to mitigate the GHGs, enhance traditional transport management methods, and promote sustainable practices in urban development.

The study also suggests avenues for further research, emphasizing the need for integrated approaches leveraging various technological innovations to effectively reduce greenhouse gas emissions for present and future generations.

References

1. Kesselring, S., Simon-Philipp, C., Bansen, J., Hefner, B., Minnich, L., Schreiber, J.: Sustainable mobilities in the neighborhood: methodological innovation for social change. Sustainability **15**(4), 3583 (2023)

2. Leach, F., Kalghatgi, G., Stone, R., Miles, P.: The scope for improving the efficiency and environmental impact of internal combustion engines. Transp. Eng. **1**, 100005100005 (2020)
3. Kenworthy, J., Kesselring, S., Lanzendorf, M.: Perspectives on mobility cultures in megacities (2013)
4. Wulfhorst, G., Kenworthy, J., Kesselring, S., Lanzendorf, M.: Perspectives on mobility cultures in megacities. In: Megacity Mobility Culture: How Cities Move on in a Diverse World, pp. 243–258 (2013)
5. Ringel, M., Bruch, N., Knodt, M.: Is clean energy contested? Exploring which issues matter to stakeholders in the European green deal. Energy Res. Soc. Sci. **77**, 102083 (2021)
6. Buberger, J., Kersten, A., Kuder, M., Eckerle, R., Weyh, T., Thiringer, T.: Total co2-equivalent life-cycle emissions from commercially available passenger cars. Renew. Sustain. Energy Rev. **159**, 112158 (2022)
7. Kilian, L., Owen, A., Newing, A., Ivanova, D.: Exploring transport consumption-based emissions: spatial patterns, social factors, well-being, and policy implications. Sustainability **14**(19), 11844 (2022)
8. Enzmann, J., Ringel, M.: Reducing road transport emissions in Europe: investigating a demand side driven approach. Sustainability **12**(18), 7594 (2020)
9. Crippa, M., et al.: Global anthropogenic emissions in urban areas: patterns, trends, and challenges. Environ. Res. Lett. **16**(7), 074033 (2021)
10. Kennedy, C., et al.: Greenhouse gas emissions from global cities (2009)
11. Ringel, M.: Smart city design differences: insights from decision-makers in Germany and the Middle East/North-Africa region. Sustainability **13**(4), 2143 (2021)
12. Paul, A., Haricharan, J., Mitra, S.: An intelligent traffic signal management strategy to reduce vehicles co2 emissions in fog oriented Vanet. Wirel. Pers. Commun. **122**(1), 543–576 (2022)
13. Lytras, M.D., Visvizi, A.: Information management as a dual-purpose process in the smart city: collecting, managing and utilizing information (2021)
14. Kesselring, S.: Networks, flows and the city of automobilities. In: Handbook of Urban Mobilities, pp. 31–40. Routledge London, UK (2020)
15. Reichert, A., Holz-Rau, C., Scheiner, J.: GHG emissions in daily travel and long-distance travel in Germany-social and spatial correlates. Transp. Res. Part D Transp. Environ. **49**, 25–43 (2016)
16. Technical report
17. Thitipatanapong, S., Noomwongs, N., Thitipatanapong, R., Chantranuwathana, S.: A comparison study on saving fuel by idle-stop system in Bangkok traffic condition, Technical report, SAE Technical Paper (2013)
18. Gaines, L., Rask, E., Keller, G.: Which is greener: idle, or stop and restart, Argonne National Laboratory, US Department of Energy (2012)
19. Zhu, T., Yang, W., Li, B., Zong, C., Li, J.: Simulation research on the start-stop system of hybrid electric vehicle. J. Adv. Veh. Eng. **3**(2), 55–64 (2017)
20. Anand Kishor Asekar: Stop-start system using microhybrid technology for increasing fuel efficiency. Int. J. Mech. Prod. Eng. **1**(6), 20–26 (2013)
21. Karrouchi, M., et al.: Practical investigation and evaluation of the start/stop system's impact on the engine's fuel use, noise output, and pollutant emissions. e-Prime-Adv. Electr. Eng. Electron. Energy **6**, 100310 (2023)
22. Santos, N.D.S.A., Roso, V.R., Faria, M.T.C.: Review of engine journal bearing tribology in start-stop applications. Eng. Failure Anal. **108**, 104344 (2020)
23. Kropiwnicki, J., Kneba, Z.: Carbon dioxide potential reduction using start-stop system in a car. Key Eng. Mater. **597**, 185–192 (2014)

24. Hoekstra, A.: Producing gasoline and diesel emits more co2 than we thought (2020). https://ioplus.nl/en
25. Jou, R.-C., Pai, C.-W., Yuan-Chan, W.: Idling stop fines for exceeding legal idling times-from the driver's perspective. Transp. Res. Part A Policy Pract. **66**, 88–99 (2014)
26. Kim, M., Kim, H.K.: Investigation of environmental benefits of traffic signal countdown timers. Transp. Res. Part D Transp. Environ. **85**, 102464 (2020)
27. Krukowicz, T., Firląg, K., Suda, J., Czerliński, M.: Analysis of the impact of countdown signal timers on driving behavior and road safety. Energies **14**(21), 7081 (2021)
28. Yan, W., et al.: An assessment of the effect of green signal countdown timers on drivers' behavior and on road safety at intersections, based on driving simulator experiments and naturalistic observation studies. J. Safety Res. **82**, 1–12 (2022)
29. Islam, M.R., Wyman, A.A., Hurwitz, D.S.: Safer driver responses at intersections with green signal countdown timers. Transp. Res. Part F Traffic Psychol. Behav. **51**, 1–13 (2017)
30. Islam, M.R.: Safety and efficiency benefits of traffic signal countdown timers: a driving simulator study (2014)
31. Goel, A., Kumar, P.: A review of fundamental drivers governing the emissions, dispersion and exposure to vehicle-emitted nanoparticles at signalised traffic intersections. Atmos. Environ. **97**, 316–331 (2014)
32. Goel, A., Kumar, P.: Zone of influence for particle number concentrations at signalised traffic intersections. Atmos. Environ. **123**, 25–38 (2015)
33. Andrade, C.: The limitations of online surveys. Indian J. Psychol. Med. **42**(6), 575–576 (2020)
34. Nayak, M.S.D.P., Narayan, K.A.: Strengths and weaknesses of online surveys. Technology **6**(7), 0837-2405053138 (2019)
35. Wardropper, C.B., Dayer, A.A., Goebel, M.S., Martin, V.Y.: Conducting conservation social science surveys online. Conserv. Biol. **35**(5), 1650–1658 (2021)
36. Hooley, T., Wellens, J., Marriott, J.: What is online research? Using the internet for social science research. Bloomsbury Academic (2012)

Open Access This chapter is licensed under the terms of the Creative Commons Attribution 4.0 International License (http://creativecommons.org/licenses/by/4.0/), which permits use, sharing, adaptation, distribution and reproduction in any medium or format, as long as you give appropriate credit to the original author(s) and the source, provide a link to the Creative Commons license and indicate if changes were made.

The images or other third party material in this chapter are included in the chapter's Creative Commons license, unless indicated otherwise in a credit line to the material. If material is not included in the chapter's Creative Commons license and your intended use is not permitted by statutory regulation or exceeds the permitted use, you will need to obtain permission directly from the copyright holder.

Software-Defined Value Networks: Industrial Requirements and Research Gap

David Dietrich[1()], Manuel Zürn[1], Ann-Kathrin Briem[2], David Koch[3], Werner Lober[4], Jannik Lind[5], Armin Lechler[1], and Alexander Verl[1]

[1] ISW, University of Stuttgart, Seidenstr. 36, 70174 Stuttgart, Germany
`david.dietrich@isw.uni-stuttgart.de`
[2] Fraunhofer Institute for Building Physics IBP, Nobelstraße 12, 70569 Stuttgart, Germany
[3] Fraunhofer Institute for Manufacturing Engineering and Automation IPA, Nobelstraße 12, 70569 Stuttgart, Germany
[4] DXC Technology, Schickardstraße 32, 71034 Böblingen, Germany
[5] ARENA2036 e.V, Pfaffenwaldring 19, 70569 Stuttgart, Germany

Abstract. Software-Defined Value Networks (SDVN) aim to increase flexibility, value-chain resilience through redundancy, scalability through standardized linkages and data formats as well as data consistency for sustainability requirements such as the carbon footprint, as opposed to current Global Value Chains. For a potential success of implementing SDVNs, the extent to which current research technologies are being utilized in companies and challenges of the companies regarding SDVNs need to be elaborated. This article presents a survey to identify industrial requirements, roadblockers and successfactors for implementing SDVN. The survey was conducted among industrial partners of ARENA2036, a research campus for future mobility and automotive production in Stuttgart, with two representatives, development and management, from each participating company The survey covers the interviewee's experience with automation systems, the use of emerging technologies throughout the company's value chains, perceived risks, sustainability factors, the potential of existing value chains, and requirements for further automation in SDVNs. The results of the survey are summarized and interpreted in the context of current technological developments, such as the Eclipse Dataspace Components, the Asset Administration Shell and solutions of Software-defined Manufacturing. They show the need of value-chain orchestration across companies including strategies for solving goal conflicts with e.g. sustainability requirements, as well as applicability of scientific developments and standardization processes to a broad industrial implementation.

Keywords: Software-defined Value Networks · Global Value Chains · sustainability · Industrial requirements · Asset Administration Shell · Eclipse Dataspace Components

© The Author(s) 2025
D. Holder et al. (Eds.): SCAP 2024, ARENA 2036, pp. 426–441, 2025.
https://doi.org/10.1007/978-3-031-88831-1_33

1 Introduction

Software-defined approaches, such as Software-defined Radio, have led to greater adaptability to changes of established technologies, by substitution of hardwired structures with programmable devices [1]. Transferred to industrial manufacturing, software-defined manufacturing can already be used to make production systems within a factory more adaptable [2], for example, by establishing further control loops [3]. An extension of this approach is the consideration of value chains. With the involvement of all stakeholders, the concept of Software-defined Value Networks (SDVN) enables a dynamic, diversified and adaptable ecosystem, which responds quickly to changes. Its approach is introduced in [4]. The consideration and orchestration of a value network enables, for example, the use of alternative suppliers or manufacturers in the value chain of a product. Challenges such as the evaluation of conflicting objectives, the integration of new production technologies, or the adaptation of the supply chain in the event of disruptions can also be addressed. For this purpose, as part of ARENA2036, a research campus for the future of mobility and automotive production in Stuttgart, research is to be carried out on SDVNs and an exemplary implementation is to be provided for its industrial and scientific partners.

Standardized interfaces, data models and formats are needed in a SDVN to make cross-company communication possible, but also to meet political requirements such as the transparency of supply chains, the digital product passport, the carbon footprint [5], or the european green deal [6]. Efforts are already being made by international research projects such as Gaia-X to establish a european data space [7], which enables sovereign data exchange among multiple partners and is realized by connectors for each participants based on their usage policies and implemented in the Eclipse Dataspace Components (EDC). In addition, efforts are being made by standardization organizations such as IDTA with the Asset Administration Shell (AAS). The AAS is a digital representation of an asset enabling standardized communication and interoperability with all relevant data [8]. Nevertheless, the extent to which current research technologies are being utilized in companies and their challenges regarding sustainability, disruptions and interoperable cross-company communication to establish value networks remains uncertain. This leads to the research questions: (i) What are the key industrial requirements, (ii) the roadblockers and (iii) the potential success factors for implementing SDVN that need to be addressed in future research? In order to answer this question, a survey was conducted among the ARENA2036 partners on the requirements for SDVN, whose results are presented in this paper. With its partners covering a wide range of the Southern German automotive and supplier industry the ARENA2036[1] represents a unique ecosystem between research and industry. The plain questions and results of the survey along with a Web-based evaluation tool are freely accessible on GitHub[2].

---

[1] Website and Members of ARENA2036: https://arena2036.de/en/members

[2] Dietrich, D. (2024). Industrial Requirements for Software-defined Value Networks: Survey Results, https://github.com/ARENA2036-Well-defined/survey-results_requirements_of_sdvn/

2　Design of the Study

The study was designed to collect requirements regarding SDVN of partners of the ARENA2036 research campus. In order to ensure a basic level of knowledge about the project and the SDVN approach among the survey participants, the survey was conducted at a bilateral meeting and the objective was first presented in a 5-minute presentation. The survey itself was then conducted anonymously as an online survey with a duration of about 15 min. It mainly consists of single- or multiple-choice questions with text supplement and a few open-text questions and was structured as follows:

- In a first question block regarding *information about the company*, such as the sector and size, the position in the supply pyramid, and the function of the participant and his or her experience with automation systems was collected.
- To meet the rising relevance of *sustainability* in value chains, the role of environmental aspects, drivers in sustainability efforts and challenges in their implementation are thematised in a second question block.
- In a third block, the *use of technology* is queried to determine currently used automation systems, the relevance of possible technological enablers for SDVN such as the AAS, or the Connector of the EDC, the current Supply-Chain Management (SCM) process and obstacles in the implementation of automation technologies.
- In order to enable an evaluation of past *disruptions* to the company's value chains, the fourth block of questions covers their resilience and dependencies as well as measures to improve them.
- The last question block covers the *SDVN* themself and is designed to extract a ranking for the most significant properties for the development of SDVN. It raises expectations and engagement in cross-company communication, the potential, expectations and conditions of SDVN, as well as missing key-factors for their successful implementation.

Further comments could be added at the end of the survey or were recorded in the bilateral meeting so that they could be taken into account in the interpretation of the results.

The survey went out to all ARENA2036 partners by posters on the shop-floor as well as bilateral invitations to contacts of current projects. Two people per company were requested, one with a management perspective and one with a development perspective. Out of 27 recipients of different companies, a total of 23 surveys from 15 different partners were conducted with mostly 2 participants per company. The survey can therefore be regarded as representative, at least for the ARENA2036 environment. Furthermore, it can also point to trend-setting indicators beyond this. All graphs are therefore based on 23 answers given. The sector of the participating companies can be seen in Fig. 1, and shows the broad industry participation with focus on machinery and plant engineering, software and information technology (IT) as well as automotive. The small share of the science sector are research partners from industrial research projects of

ARENA2036. All levels of the supply pyramid are present in the current structure of global value chains, but especially Original Equipment Manufacturer (OEM) and Tier 1 (OEM 26%, Tier 1: 30%, Tier 2: 17%, Tier 3: 4%, unsure 17%). Regarding their size about 65% of the companies have more than 250 employees, 26% have 50 to 249 employees and 9% have 10 to 49 employees. The distribution of respondents to development (56%) and management (44%) is almost the same. The personal work with automation systems of the automation pyramid are: input/output (field level) 30%, programmable logic controllers (PLC) 56%, supervisory control and data acquisition (SCADA) 13%, manufacturing execution system (MES) 48%, enterprise resource planning (ERP) 57%, supply chain management (SCM) 30%. Still 26% of the participants are unsure about their personal work with automation systems.

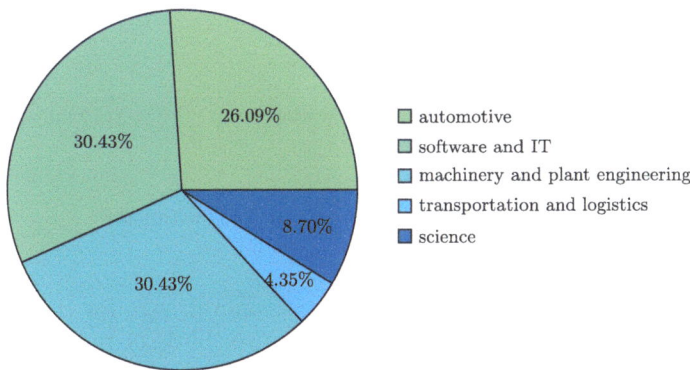

Fig. 1. Which sector would you assign your company to?

3 Survey Results: Industrial Requirements

This section highlights the results of the survey and concludes the industrial requirements. It is structured according to the design of the study as introduced in Sect. 2 and starts with the question block regarding sustainability. The proportions are rounded to whole percentages for presentation in the text.

3.1 Sustainability

The majority of participants confirmed the importance of environmental aspects and sustainability in all business areas surveyed (see Fig. 2). More than half stated that these aspects played either a central or major role in the development of their business strategy and technology investments. When asked about product development, the share of "central" answers was noticeably lower with

13%, but 43% of respondents assigned a major role. Similarly, the role of environmental and sustainability aspects in production was estimated to be central or major by more than half of the participants. In all the business areas surveyed the share of medium importance was similar and around 30%. The share of minor importance or "none" was between 8.6% and 13%.

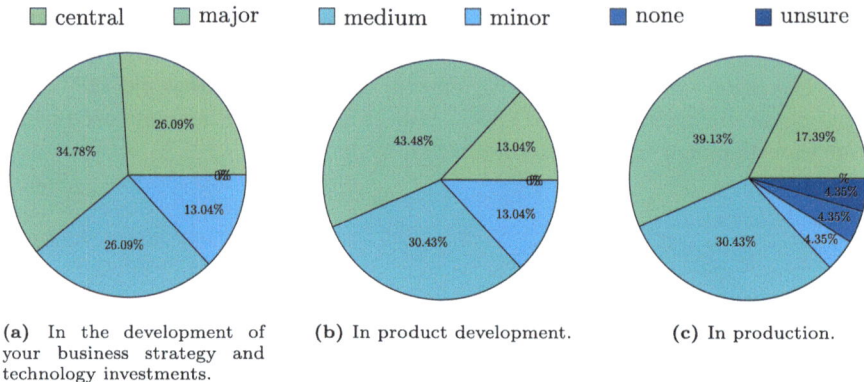

Fig. 2. Role of environmental aspects and sustainability.

The next survey question asked about the main drivers for the company's sustainability efforts. From the given options, two-thirds of respondents chose "customer requests", followed by "regulation" (56%) and "marketing" as well as "corporate values" (each 48%). Finally, "cost savings" were chosen the least (39%).

The three dimensions of sustainability have varying importance according to the survey participants, as shown in Fig. 3. Economical aspects were being the most important, environmental aspects ranking second and the social pillar being the least important regarding decision making, as indicated by the selection of the options "high" and "very high" regarding the importance of the three pillars of sustainability.

When asked about having encountered conflicts of interest between different sustainability aspects when implementing measures in their company, most participants confirmed that "yes" (43%) such conflicts have been encountered in the past. Some were unsure or didn't answer (39%). 17% of the participants said they had not yet encountered the mentioned conflicts.

Followed up by the question if the company has a clear strategy for achieving an overall optimum in such cases, on the one hand, 26% of the participants stated that, "yes", there is a clear strategy. On the other hand, 39% of the participants answered that they were unsure and 35% of the participants stated that there was no strategy, at least none that they knew of.

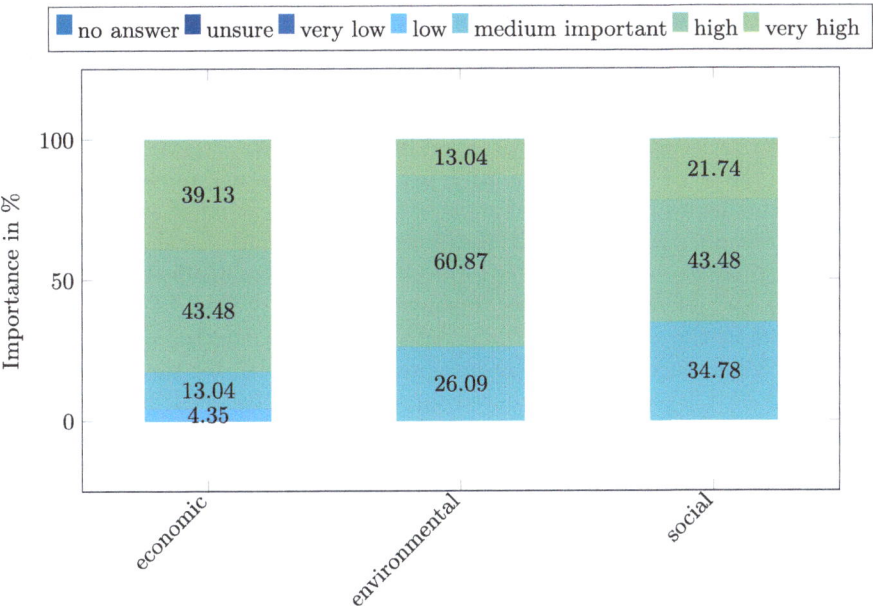

Fig. 3. How highly do you rate the relevance of the three dimensions of sustainability four your company?

The median year by which the participating companies that want to achieve climate neutrality want to have reached that goal is 2040. Overall, about 83% of the companies confirmed that their company has defined the goal to reach climate neutrality, 17% of the companies provided no answer.

The last question of the sustainability section asked about the challenges that the participants face when implementing sustainability efforts in their company. From the given options, the top answer chosen by a significant margin was "not enough internal resources" with 48%. All other given options were chosen between 21% and 30%, including regulatory "jungle", missing connection between company and product level, unclear responsibilities within the company, contradictory targets, data gaps or data management, conflicting goals between sustainability dimensions. Furthermore, some participants were unsure about these options (22%). Additionally, challenges pertaining to purchasing, IT and digitization, cost and Return on Invest, and conflicts between sustainability measures and product quality were mentioned.

3.2 Use of Technology

Continuing with the use of technologies, the first question asked about used automation systems of the company in contrast to their personal work with them in Sect. 2. Regarding the results: field-level (56%), PLC (74%), SCADA (35%), MES (61%), ERP (70%), SCM (48%) and unsure (13%), the distribution is similar to that of the personal experience in Sect. 2. However, it is more pronounced for the individual systems and shows less uncertainty. In a more detailed analysis, it was also determined that the personal experiences with automation systems largely coincide with the systems used by the company, which means that experience with the systems is given in each case.

Besides the conventional automation pyramid, the use and implementation of current technologies for intra- and cross-company data exchange as well as their relevance was queried in the next questions, whose results are shown in Fig. 4. In Fig. 4a, it shows that Open Platform Communications Unified Architecture (OPC-UA) and Message Queuing Telemetry Transport (MQTT) are already widely used in the companies, while AAS and the Connector of the Eclipse Dataspace Components (EDC) are less present. The plan for the implementation in the next 2 years is distributed for all of them between 21% and 31%, in a single free-text appendix artificial intelligence was mentioned. The great uncertainty regarding current (30%) and future implementation (35%) is striking. Considering the expected relevance of the AAS in Fig. 4b, high or very high is at more than 60%. Opposite to this in sum about 43% have it in use or plan to do so in the next 2 years.

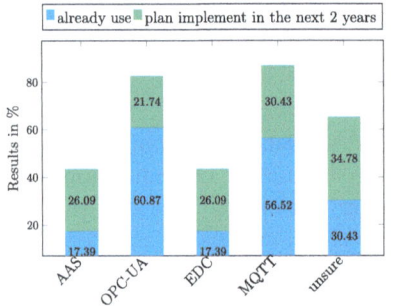
(a) Use of technology in the companies

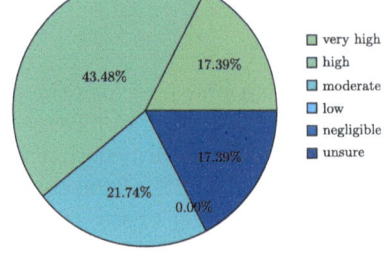
(b) What future relevance do you see for the asset administration shell?

Fig. 4. Technologies for intra- and cross-company data exchange

Concerning the state of the current SCM in the companies, the involvement of departments, the use of software for optimization and the integration of suppliers and customers was queried. The departments of purchase, quality, production, and development are currently engaged to a similar extent in SCM, with involvement levels ranging from 39% to 53%. 13% of the participants state that it "does not exist" and 17% are unsure about the involvement. Regarding the use of software to "optimize and manage your SCM processes" more than half of the participants state that software is used (56%), with one mention of "SAP, Jaggaer". 26% are unsure about the use and about 17% do not use any. For the integration of suppliers and customers into the value creation process, the main focus lies in "manual contract negotiations" (70%) followed by "automated ordering processes" (52%) and "supplier/customer portals" (43%). 26% are unsure about their integration.

Results about obstacles in the implementation, use of automation technologies, capacities for digital transformation and innovation are shown in Fig. 5. Obstacles in implementation are higher than in use, while the main obstacles are the "integration into existing systems", the "complexity of the technology" and "data protection and security". While, on closer inspection, the complexity of the technology in development plays a central role for developers alongside integration, for managers this is only in third place with the same rating of costs. However, there is no such difference in the obstacles regarding the use. "High costs" and "skilled labor shortage" play a minor role, "legal and regulatory hurdles and certification" were chosen the least. Nevertheless, capacities for digital transformation and innovation are assessed as high or very high with over 61%.

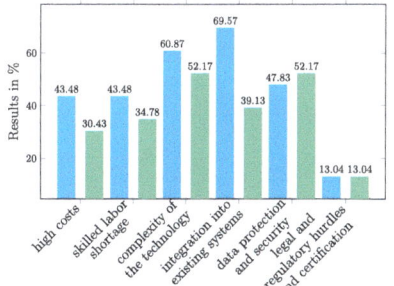
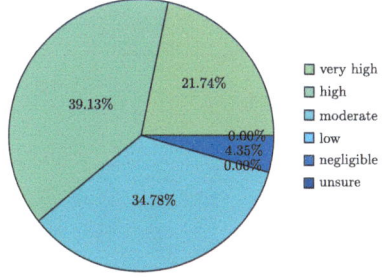

(a) What obstacles do you see in the implementation (blue) and use (green) of automation technologies in your company?

(b) How do you assess your company's current capacities in terms of digital transformation and innovation?

Fig. 5. Obstacles of and capacity for digital transformation

The aforementioned technologies, such as AAS or EDC, are being developed in open source projects. Therefore, the role and contribution to these were surveyed. The role of open source technologies in the corporate strategy for innovation is 13% central, 43% major, 26% medium and each 9% minor or none. Contrariwise, 22% of the participants contribute to open source technologies, 39% contribute as a company and another 39% generally do not.

3.3 Disruptions

Following up with the question block regarding disruptions and its results in Fig. 6, more than half of the supply chains have been impacted very high (26%) or high (35%) by the coronavirus pandemic, while the criticality of raw material for the company is for 17% central and for 43% major. Despite having invested heavily in preparation against uncertainty and volatility in supply and value chains over the past five years, the organization is still considered to be moderately resilient at 48%, 30% unsure and 21% highly resilient.

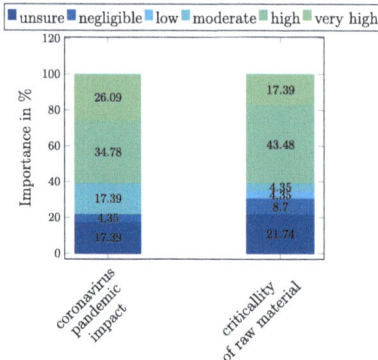

(a) Impact of the coronavirus pandemic and criticality of raw material

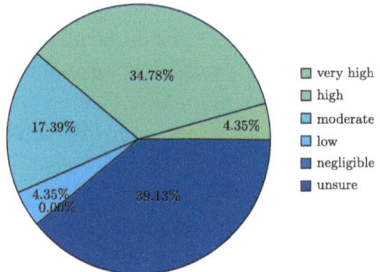

(b) To what extent have you taken measures in the last 5 years to prepare your supply and value chains for uncertainty and volatility?

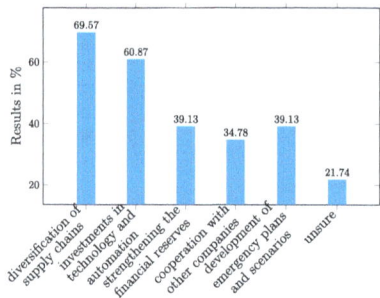

(c) What strategies are you pursuing to increase your company's resilience to future crises?

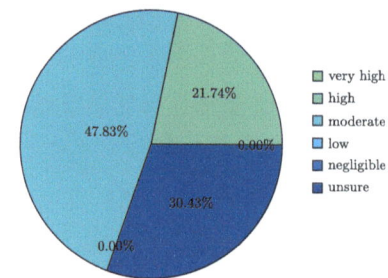

(d) How resilient do they generally consider your supply chains to be?

Fig. 6. Resilience of value chains

3.4 Software-Defined Value Networks

Lastly, the results of the question block regarding SDVN are shown. As cross-company communication plays a major role in SDVN's concept, expected factors regarding cross-company communication from customers and suppliers, as well as their engagement in these factors, are shown in Fig. 7.

Across all factors, the expectations of suppliers and customers largely coincide with their perceived engagement. Factors to be highlighted are the security and compliance of cross-company communication, while interoperability and technological compatibility have a slightly lower significance.

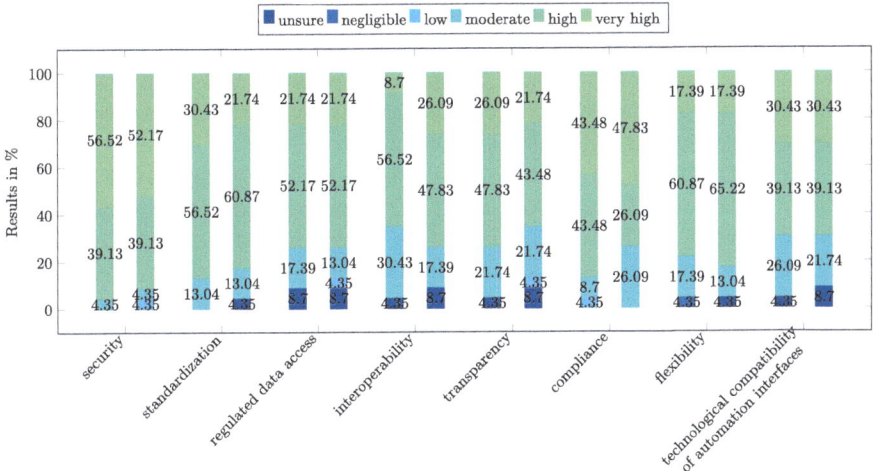

Fig. 7. With regard to cross-company communication, which of the following factors do you expect from your suppliers or customers (left bar) and how important is it to them to engage in the points (right bar)

9% rate the potential of value networks compared to conventional global value chains as very high and a further 52% as high, which leads together to more than 60% with high expectations. A further 22% rate this as moderate and 17% are unsure. More than half of the participants would let a higher-level system orchestrate their value creation processes and make data and services available to network partners, as long as it contributes to a comprehensive improvement in value creation with clarified responsibility (22%) or if it contributes to the company's value creation (35%). 17% would enable the orchestration with restricted configuration options and 4% not at all, with another 22% being unsure about it.

Regarding the existing automation systems the participants would build SDVN on, 17% would concern the field level, 17% the PLC, 9% SCADA, 30% MES, 35% ERP and 30% SCM Systems, whilst a high uncertainty with 44% "unsure" exists. A more detailed analysis in consideration of the personal experience shows on the one hand, that about 50% of the participants would each recommend their personally used systems to build SDVN on. On the other hand, each system is recommenced by participants with experience in them except SCM, which shows that which shows that there is no generalized delegation of authority. Also all participants with unsure personal experience in the automation systems choose the unsure option regarding systems to build SDVN on. These correlations reinforce the aforementioned uncertainty regarding technological implementation based on current automation systems.

The expectations regarding SDVN are overall clear (unsure with 0%) and concern greater flexibility (74%) as main objective followed by cost savings and better data consistency with 65%, improved reliability through redundancy (61%) and faster response times (56%). Scalability plays a comparatively minor role with 44%, while transparency and accurate decision-making were added in the free-texts. Regarding the concept of data rooms for cross-company information management about one-third are in detail (4%) or moderately (26%) familiar with them. The high proportion of little to non-existent knowledge ("a bit" 35%, "heard of" 13% and "not at all" 22%) may be due, for example, to the current complexity of the topic and must be made understandable to cope with SDVN. Missing key factors for the successful implementation of SDVN are shown in Fig. 8 and further challenges for the integration of new suppliers or the introduction of new products in existing supply chains are mentioned as: "quality, security, availability", "high integration time", "reliability, automotive certifications necessary", "setting up clear communication channels", "complexity" and "missing standards".

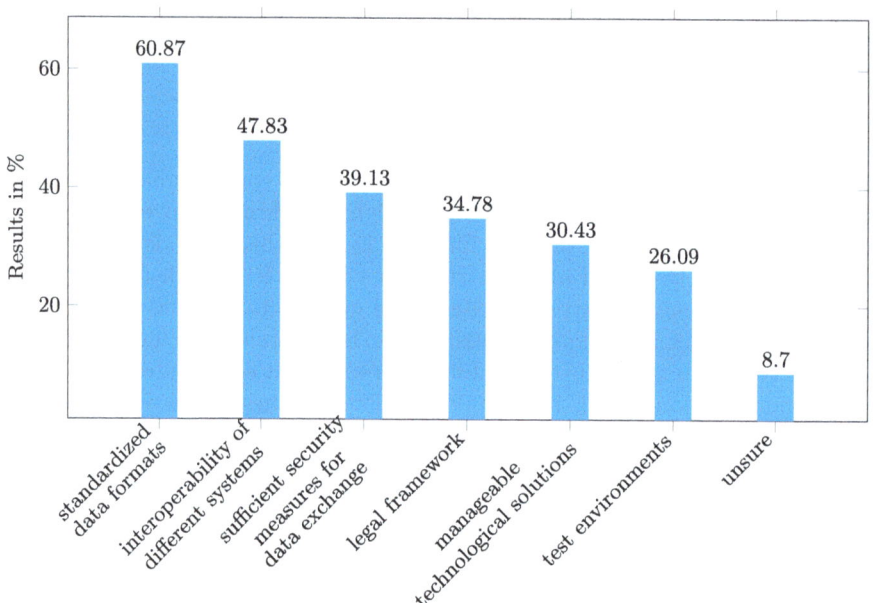

Fig. 8. In your opinion, what key factors are still missing for the successful implementation of automated value creation networks?

The figure shows that standardized data formats with interoperability of different systems as well as their secure exchange are regarded as main factors, while manageable technological solutions and test environments play a minor role. This is noteworthy, as mentioned obstacles such as the complexity of the technology, but also mentioned challenges of integrating new suppliers into existing value chains like, such as integration time, complexity, or missing standards, would lead to the assumption of required test environments. However, this contradicts in part the factors of cross-company communication as well as the implementation of current technologies from Subsect. 3.2, in which interoperability and the integration of cross company standards for data exchange like AAS and EDC has a lower priority. Additional free-text entries regarding missing factors regard the "trust, clearness of benefits", the "power to enforce a western standard" and the "needed mindset" and thus all refer to the human factor in addition to the technological ones which were primarily recorded in this survey.

4 Research Gap

This chapter evaluates the presented results of the survey and further identifies current deficits.

4.1 Survey Evaluation

Sustainability plays an ever more important role in business strategy, product development and production. Aside from the potential impact on cost, multiple other challenges exist such as in the regulatory framework as well as the implementation of sustainability aspects in the company strategy and organizational setup.

As can be seen from the disruptions, value chains are still not considered resilient, despite heavy investment in previous years. This paradox stems from the increasing interconnectedness of global markets, which continues to introduce new and unforeseen challenges. The still missing resilience corresponds with a high rating of SDVN, and therefore a demand on their development to increase value chain resilience.

Surprisingly, most companies could imagine having their value chains orchestrated by a higher-level system. The clarity of the requirements for cross-company data exchange and missing key factors should be emphasized. These points in combination with the highly estimated capacity for digital transformation and the minor obstacle of high costs leads us to expect a moment of upheaval. Still noticeably, further understanding and detailing of SDVN needs to be established regarding their ratings of moderate and unsure.

Nevertheless, high uncertainty highlights that the implementation of these requirements with cross-company standards, existing automation systems and open source technologies is still unclear. The survey shows the role of current standardization processes for secure and interoperable data exchange, especially regarding the AAS and EDC. Even though their role is not emphasized in the factors for cross-company communication they are seen as most relevant for a successful implementation of SDVN. The reasons for this may be both, in the complexity of the technologies and their applicability to industrial problems. At the moment their integration effort might not be worth due to inner company standardization, but gains future importance as soon as heavy connected value networks will develop, which we forecast due to, for example, regularization, current disruptions and resilience benefits. Although this differs from the participants' assessments, these complex standards can be made tangible in sample applications or test environments.

Regarding the interpretation of the study on the one hand a bias in its participants needs to be assumed, which are close to the ARENA2036 research campus. On the other hand a bias regarding the midpoint (moderate) of the scale in combination with the unsure option can be reflected [9]. This can especially be noticed in the questions regarding the impact of the coronavirus pandemic, the considered resilience of supply chains and the potential of SDVN, when unsure and moderate form together a significant part of the result. According to the above-mentioned study, this may indicate additional uncertainty, which is remedied by the moderate assessment. On the one hand, this uncertainty may be due to the background of the interviewee, who may not have an overview of the value chains, especially in large companies. On the other hand, this can demonstrate the novelty of the topic combined with the paradigm shift of software-definedness, which needs to be clarified in future research.

4.2 Identification of Current Deficits

This subsection identifies the resulting research gaps for a successful implementation of SDVN based on industrial requirements. These are complementary to research questions already elaborated on a literature basis in [4] and thus broaden the perspective on the industrial feasibility of SDVN research.

From a sustainability standpoint, based on the replies to the survey, there appears to be a research gap regarding the resolution of conflicts of goals between the different pillars of sustainability. In particular, there is a need for a systematic approach to identify and cope with such conflicts. This need can also be confirmed by other literature [10].

The expected moment of upheaval should be used by research and standardization organizations to clarify strategies and develop tools and technologies for the need of cross-company resilience in global value chains with the help of SDVN. Regarding for example the perceived relevance of the AAS in this context, which fulfills the required key factors, research needs to show their applicability to industrial problems but also industrial requirements need to be included into the standardization processes. A clear lack in the transfer of scientific developments and standardization processes to a broad industrial implementation is visible. Even the possible biases of the survey reinforces the uncertainty shown with regard to the current state of value chains and possible technological realizations of the clear requirements and expectations.

However, to make this a success, the following gaps need to be considered in research:

1. There is a need in developing an advanced orchestration system for value chains that seamlessly integrate with a company's existing infrastructure without replacing current factory management systems.

2. There is a need in creating a streamlined approach to mapping and simplifying the orchestration of supply chain diversification strategies within the broader landscape of global industrial production, ensuring they are comprehensible and easily implementable.
3. There is a need in embedding manual contract negotiation processes into the architecture and orchestration of SDVNs, ensuring the active participation of all relevant stakeholders across company departments. This necessitates the development of effective mechanisms for transitioning from manual to automated systems, facilitating seamless integration and collaboration among involved parties.
4. Last, there is a need for application-specific solutions to render complex cross-company specifications tangible and actionable. For this the collaboration between academia and industry needs to be fostered to master the complex system of global production.

5 Conclusion and Outlook

Political requirements, disruptions in supply chains and technological enablers lead to a rethink towards SDVN. To answer the posed research question, a survey among ARENA2036 partners was conducted to identify key industrial requirements, roadblockers and potential success factors for implementing SDVN. Based on the motivation of cross-company software-definedness of value chains and the need to identify industrial requirements, the design and background of the conducted survey was presented. Due to the given participation of the ARENA2036 partners, the survey results are of significant importance for the ARENA2036 environment and thus represent the southern German automotive industry. The shown results confirm the need for cross-company orchestration of value creation taking sustainability requirements into account, while at the same time the technological implementation remains uncertain. The conducted survey thus underscores in its elaborated research gap the posed research questions in [4] regarding required appropriate standards for interoperability, technological integration into existing systems and the application-oriented use of data spaces with according tools. Even if the number of surveys rather reveals trends, as discussed in Sect. 2, the potential bias mentioned in Sect. 3 leads to a higher assessment of uncertainty or may overestimate a broad industrial implementation, and thus reinforces the need for further investigation. Addressing these issues will pave the way for SDVN, ensuring a successful, sustainable and resilient production in uncertain environments.

Acknowledgments. This research and development project is funded by the German Federal Ministry of Education and Research (BMBF) within the "Research Campus - Public-Private Partnership for Innovation" funding initiative (02P23Q820 ff) and managed by the Project Management Agency Karlsruhe (PTKA). The authors are responsible for the content of this publication.

References

1. Sadiku, M.N.O., Akujuobi, C.M.: IEEE Potentials. Software-defined radio: a brief overview. **23**(4), 14–15 (2004). https://doi.org/10.1109/MP.2004.1343223
2. Neubauer, M., Frick, F., Ellwein, C., Lechler, A., Verl, A.: Ein Paradigmenwechsel für die industrielle Produktion/a paradigm shift for industrial production – software-defined manufacturing. **112**(06), 383–389 (2022). https://doi.org/10.37544/1436-4980-2022-06-33
3. Dietrich, D., Neubauer, M., Lechler, A., Verl, A.: Iterative planning as a holistic framework for production system-wide optimization control loops. In: Silva, F.J.G., Pereira, A.B., Raul, D., Campilho, S.G. (eds.) Flexible Automation and Intelligent Manufacturing: Establishing Bridges for More Sustainable Manufacturing Systems. LNME, pp. 611–621. Springer, Switzerland (2024). https://doi.org/10.1007/978-3-031-38241-3_69
4. Dietrich, D., Zürn, M., Reiff, C., Neubauer, M., Lechler, A., Verl, A.: Software-defined value networks: motivation, approaches, and research activities. In: Bauernhansl, T., Verl, A., Liewald, M., Möhring, H.-C. (eds.) Production at the Leading Edge of Technology. LNPE, pp. 514–524. Springer, Switzerland (2024). https://doi.org/10.1007/978-3-031-47394-4_50
5. Poulle, J.-B., Kannan, A., Spitz, N., Kahn, S., Sotiropoulou, A.: Corporate sustainability reporting directive. In: Poulle, J.-B., Kannan, A., Spitz, N., Kahn, S., Sotiropoulou, A. (eds.) EU Banking and Financial Regulation, pp. 648–653. Edward Elgar Publishing (2024). https://doi.org/10.4337/9781035301959.00096
6. Fetting, C.: The European green deal. ESDN Rep. **2**(9), 53 (2020)
7. Braud, A., Fromentoux, G., Radier, B., Le Grand, O.: The road to European digital sovereignty with Gaia-X and IDSA. IEEE Netw. **35**(2), 4–5 (2021). https://doi.org/10.1109/MNET.2021.9387709
8. Bader, S., Barnstedt, E., Bedenbender, H., Berres, B., Billmann, M., Ristin, M.: Details of the asset administration shell - part 1 : the exchange of information between partners in the value chain of Industrie 4.0 (version 3.0RC02) (2022). https://www.plattform-i40.de/IP/Redaktion/EN/Downloads/Publikation/Details/_of/_the/_Asset/_Administration/_Shell/_Part1/_V3.html
9. Kulas, J.T., Stachowski, A.A., Haynes, B.A.: Middle response functioning in Likert-responses to personality items. J. Bus. Psychol. **22**(3), 251–259 (2008). https://doi.org/10.1007/s10869-008-9064-2
10. Koch, D., Sauer, A.: Identifying and dealing with interdependencies and conflicts between goals in manufacturing companies' sustainability measures. Sustainability **16**(9), 3817 (2024). https://doi.org/10.3390/su16093817.PII:su16093817

Open Access This chapter is licensed under the terms of the Creative Commons Attribution 4.0 International License (http://creativecommons.org/licenses/by/4.0/), which permits use, sharing, adaptation, distribution and reproduction in any medium or format, as long as you give appropriate credit to the original author(s) and the source, provide a link to the Creative Commons license and indicate if changes were made.

The images or other third party material in this chapter are included in the chapter's Creative Commons license, unless indicated otherwise in a credit line to the material. If material is not included in the chapter's Creative Commons license and your intended use is not permitted by statutory regulation or exceeds the permitted use, you will need to obtain permission directly from the copyright holder.

Author Index

A
Ajdinović, Samed 137
Araya-Martinez, J. Moises 283
Araya-Martinez, Jose M. 112

B
Bähr, Philipp 13
Bartels, Jonathan 160, 175
Baur, Johannes 344
Beise, Hans-Peter 303
Berger, Ulrich 191
Bischoff, Manfred 1
Bragmann, Daniel 319
Braiger, Jonas 344
Braun, Axel 206, 223
Briem, Ann-Kathrin 426

C
Carosella, Stefan 257, 344
Christiansen, Alfred Thoft 58
Constantinescu, Carmen 257

D
Dam, Elias Thomassen 58
Deng, Ruiyang 368
Deshmane, Akshay 402
Deubert, Darius 88
Dickhut, Tobias 402
Dietrich, David 426
Duve, Matthias 383

E
Eisenbart, Boris 31, 335

F
Faisal, Sheharyar 31
Falter, Jan 335
Fischer, Marc 150
Fleischer, Jürgen 353

G
Genikomsidis, Dimitrios 183
Gerlicher, Ansgar 206, 223
Grieser, Pascal 183
Gugliuzza, Jakob 344

H
Haas, Alexander 183
Hanzelmann, Andreas 183
Hartmann, Stefan 13
Heieck, Frieder 383
Heinemann, Daniel 206
Heizmann, Michael 319
Hengel, Katharina 160
Hölzle, Markus 121
Huber, Marco 319
Huse, Timo 39

J
Janik, Simon 206
Jazdi, Nasser 257

K
Kamble, Pradnil 39
Kernbach, Andreas 319
Kessler, Daniel 303
Khelil, Abdelmajid 243
Kilchert, Sebastian 368
Kläb, Mara I. 319
Klare, Stefan 68
Klingel, Lars 68, 76
Knüpper, Jonas 183
Koch, David 426
Koessler, Florian 353
Komesker, Simon 160, 175
Korff, Maximillian 402
Kraus, Werner 319
Kreimeyer, Matthias 31, 335

© The Editor(s) (if applicable) and The Author(s) 2025
D. Holder et al. (Eds.): SCAP 2024, ARENA 2036, pp. 443–445, 2025.
https://doi.org/10.1007/978-3-031-88831-1

Krüger, Jörg 283
Kuhn, Korbinian 223

L
Lambert, Mats 283
Lambrecht, Jens 283
Lang, Bastian 175
Lechler, Armin 137, 150, 319, 426
Leidinger, Lukas 13
Liebgott, Florian 303
Lind, Jannik 426
Lober, Werner 426
Lück, Matthias 257
Lüdemann-Ravit, Bernd 68, 183, 268, 383

M
Madsen, Ole 58
Maisch, Nicolai 137
Majid, Ali Khan 415
Mantz, Daniel 76
McMahon, Troy 58
Merzhäuser, Elias 206
Middendorf, Peter 257, 344
Motsch, William 160
Muhammad, Ali Khan 415
Muhammad, Hassaan 335

N
Nadeem, Eiman 31
Neher, Philipp A. 150
Neubauer, Michael 137, 150
Neumann, Rebekka 137
Nistler, Maximilian 76

P
Pati, Santwana 402
Petasch, David 383
Pinnow, Lea 206

R
Riedel, Oliver 137
Rosenthal, Philipp Niklas 243
Rudolph, Stephan 344
Ruskowski, Martin 160, 175

S
Sabah, Narmeen 335
Saeed, Muhammad 31, 335
Salem, Makki Ben 243

Sardari, Sarvenaz 283
Schilling, Maximilian 1
Schleipen, Miriam 183
Schmerbeck, Carsten 319
Schmitt, Fabian 183
Schnierle, Maximilian 257
Scholta, Joachim 121
Scholz, Johannes 353
Schön, Alexander 257
Schou, Casper 58
Seitz, Andreas 303
Selig, Andreas 88
Shramenko, Volodymyr 68
Sommer, Silke 13
Staudenrausch, Tim 268

T
Töper, Florian 283
Track, Dominik 303

U
Uhlemayr, Theresa 121
Unger, Nicolas 39
Usta, Tolga 1

V
Verl, Alexander 68, 76, 88, 150, 319, 426
Voekler, Sascha 191
von Scheven, Malte 1

W
Wagner, Achim 160, 175
Wagner, Markus 31, 335
Walker, Moritz 137, 150
Walker, Zack 206, 223
Weiß, Matthias 257
Weißenfels, Christian 104, 112
Werz, Martin 76
Weyrich, Michael 257
Willmann, Tobias 1

Y
Yaman, Alper 319

Z
Zak, J. Alexander 283
Zak, Jan A. 104, 112
Zürn, Manuel 319, 426

The manufacturer's authorised representative in the EU is Springer Nature Customer Service Centre GmbH, Europaplatz 3, 69115 Heidelberg, Germany. If you have any concerns regarding our products, please contact ProductSafety@springernature.com

Printed and bound by CPI Group (UK) Ltd, Croydon, CR0 4YY

26/03/2026

02078942-0008